CARBYNE AND CARBYNOID STRUCTURES

Physics and Chemistry of Materials with Low-Dimensional Structures

VOLUME 21

The titles published in this series are listed at the end of this volume.

Carbyne and Carbynoid Structures

Edited by

Robert B. Heimann

Department of Mineralogy,
Freiberg University of Mining and Technology,
Freiberg, Germany

Sergey E. Evsyukov

A.N. Nesmeyanov Institute of Organo-Element Compounds,
The Russian Academy of Sciences,
Moscow, Russia

and

Ladislav Kavan

J. Heyrovsky Institute of Physical Chemistry,
Prague, Czech Republic

SPRINGER-SCIENCE+BUSINESS MEDIA, B.V.

A C.I.P. Catalogue record for this book is available from the Library of Congress.

ISBN 978-94-010-5993-0 ISBN 978-94-011-4742-2 (eBook)
DOI 10.1007/978-94-011-4742-2

Printed on acid-free paper

HEIMANN - EVSYUKOV - KAVAN

CARBYNE AND CARBYNOID STRUCTURES

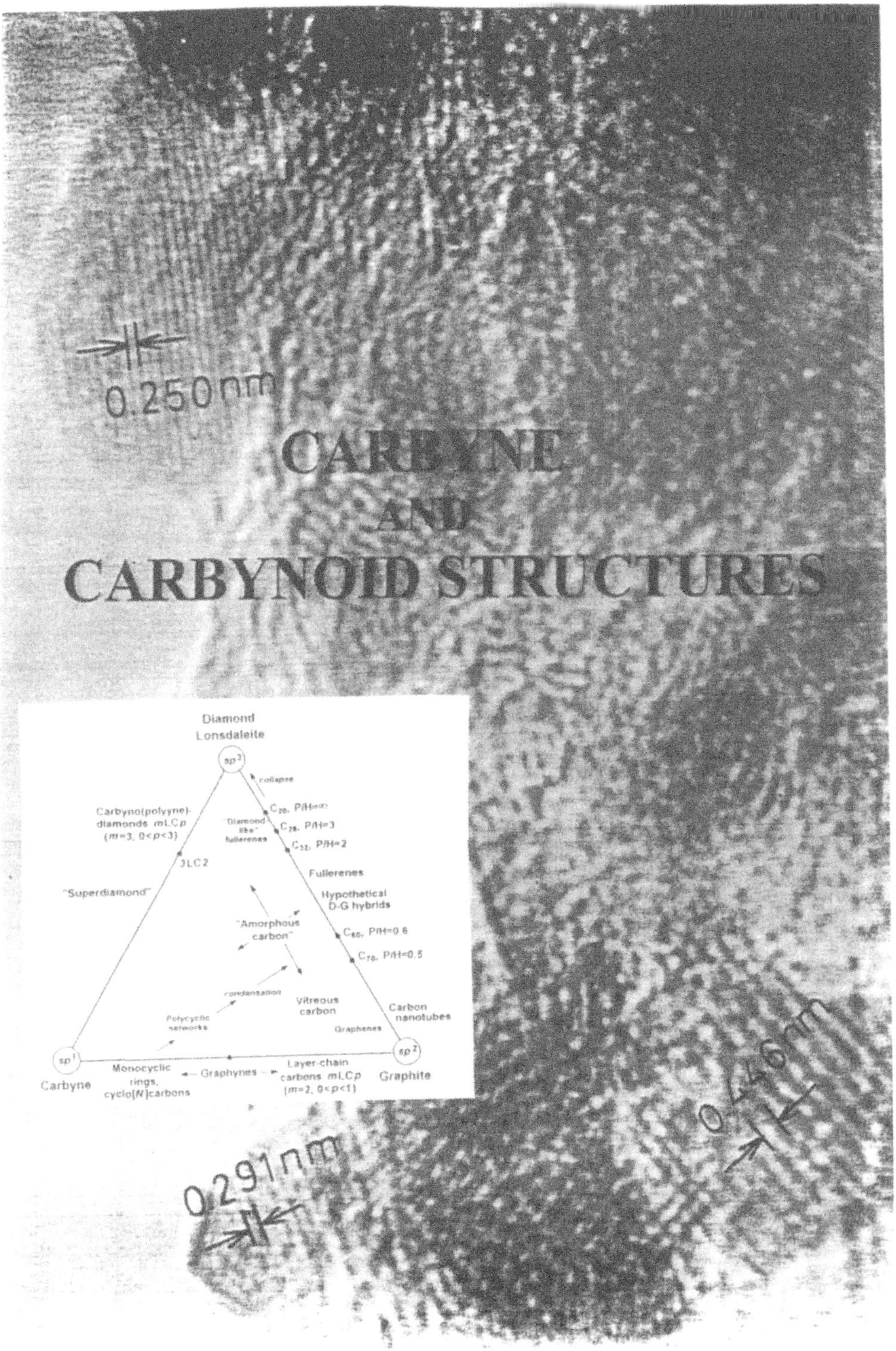

Table of Contents

LIST OF CONTRIBUTORS

Babaev, V.G. Department of Physics, Moscow State University, 119899 Moscow, Russia
e-mail: bab@pelc60A.phys.msu.su

Baitinger, E.M. Department of Physics, Chelyabinsk State Pedagogical University,454080 Chelyabinsk, Russia
e-mail: cspi@cspi.urc.ac.ru

Ertchak, D.P. Department of Physics, Byelorussian State University, 220080 Minsk, Byelorussian Republic
e-mail: dov@inp.minsk.by

Evsyukov, S.E. (co-editor) A.N.Nesmeyanov Institute of Organometallic Compounds, The Russian Academy of Sciences 117813 Moscow, Russia (on leave)

Present Address: BASF Aktiengesellschaft, Polymer Laboratory ZKS/A - B 1, 67056 Ludwigshafen, Germany
e-mail: serguei.evsioukov@msm.basf-ag.de

Fitzgerald, A.G. Department of Applied Physics and Electronic & Mech. Eng., University of Dundee, Dundee DD1 4HN, Scotland, UK
e-mail: a.g.fitzgerald@dundee.ac.uk

Gilkes, K. Department of Earth Sciences, The Open University, Milton Hall Milton Keynes MK7 6AA, UK

Guseva, M.B. Department of Physics, Moscow State University, 119899 Moscow, Russia
e-mail: via V.G. Babaev

Heimann, R.B. (Editor-in-Chief)
Department of Mineralogy,
Chair of Technical Mineralogy,
Freiberg University of Mining and Technology,
D09599 Freiberg, Germany
e-mail: heimann@mineral.tu-freiberg.de

Kastner, J.
PROFACTOR Produktionsforschung
GmbH, Wehrgrabengasse 1, A-54400
Steyr, Austria
e-mail: jkastn@novell.pr-steyr.ac.at

Kavan, L. (co-editor)
J.Heyrovsky Institute of Physical Chemistry
The Czech Academy of Sciences,
182 23 Prague 8, Czech Republic.
e-mail: kavan@jh-inst.cas.cz

Kijima, M.
Institute of Materials Science, University of
Tsukuba, Tsukuba, Ibaraki 305, Japan
e-mail: kijima@ims.tsukuba.ac.jp

Kirpatovsky, V.I.
Laboratory for Experimental Modelling of
Urological Diseases, Institute of Urology, 3rd
Parkovaya str. 51, 105425 Moscow, Russia
e-mail: kirp@chat.ru

Kleiman, J.I.
Integrity Testing Laboratory Inc.,
c/o Institute for Aerospace Studies (UTIAS),
University of Toronto, 4925 Dufferin Street, North
York, Ontario, M3H 5T6 Canada
e-mail: jkleiman@utias.utoronto.ca

Kudryavtsev, Yu.P.
A.N.Nesmeyanov Institute of Organoelement
Compounds, The Russian Academy of Sciences,
117813 Moscow, Russia
e-mail: bystrova@ineos.ac.ru

Lebedev, B.V. Scientific Research Institute of Chemistry, Nizhnii Novgorod State University,
603600 Nizhnii Novgorod, Russia
e-mail: lebedevb@ichem.unn.runnet.ru

Olszyna, A. Warsaw University of Technology,
Department of Materials Science and Engineering,
02-524 Warsaw, Poland
e-mail: AO@inmat.pw.edu.pl

Pesin, L.A. Department of Physics, Chelyabinsk State Pedagogical University, 454080 Chelyabinsk, Russia
e-mail: pesin@cspi.urc.ac.ru

Pillinger, C.T. Department of Earth Sciences
The Open University, Milton Hall
Milton Keynes MK7 6AA, UK
e-mail: C.T.Pillinger@open.ac.uk

Sawaoka, A.B. Materials and Structures Laboratory
Tokyo Institute of Technology, Midori,
Yokohama 226, Japan
e-mail: sawaoka@rlem.titech.ac.jp

Shirakawa, H. Institute of Materials Science, University of Tsukuba, Tsukuba, Ibaraki 305, Japan
e-mail: hideki@ims.tsukuba.ac.jp

Sokolowska, A. Warsaw University of Technology, Department of Materials Science and Engineering,
02-524 Warsaw, Poland
e-mail: ASOK@inmat.pw.edu.pl

Springborg, M. Fakultät für Chemie, Universität Konstanz,
D7750 Konstanz, Germany
e-mail: mcs@chclu.chemie.uni-konstanz.de

Tanuma, S. College of Science and Engineering, Iwaki Meisei University, Iwaki 970, Japan

Udod, I.A. Department of Chemistry, Moscow State University, Moscow, Russia (on leave)

Present Address: Allied Signal Inc., 101 Columbia Rd., Morristown, NJ 07962, USA
e-mail: Iudod@aol.com

Yabe, A. National Institute of Materials and Chemical Research (NIMC), Tsukuba, Ibaraki 305, Japan
e-mail: yabe@nimc.go.jp

Yamada, K. Department of Chemistry, The National Defence Academy, Hasirimizu, Yokosuka 239, Japan

„There are more things in heaven and Earth, Horatio, than are dreamt of in your philosophy"
(Hamlet, 1st act, 5th scene)

Preface

CARBYNE AND CARBYNOID STRUCTURES

The chemistry and materials science of carbon is a fascinating field of endeavour with a plethora of technological applications ever since Tennant in 1797 experimentally proved that diamond is an allotrope of carbon. The history of the synthesis of diamond, the application of carbon fibers for mechanically superior composite materials, the research into intercalated graphite structures, and the recent discovery and exploration of fullerenes and fullerenoids are but a few highlights of this development.

Much less known is the fact that besides the well-known and copiously researched carbon allotropes graphite with sp^2- and diamond with sp^{3-} carbon bond hybridization an allotrope seems to exists with linear sp-carbon bond hybridization that was discovered only in 1960 (Chapter 1.1.). Linear carbon chains with sp-hybridization are present in carbon vapour above 5000 K as well as in molecular clouds and cool carbon stars. Curiously, the search for such molecules triggered the discovery of fullerenes. However, it is an irony of science that fullerenes were never directly observed in outer space to date (Chapter 2.1.). On the other hand, carbyne was discovered in terrestrial impact craters, and related organic structure can be found ubiquituously in biological species (Chapter 2.2.).

The classification scheme of carbon allotropes based on the concept of the hybridization type of the valence orbitals of the carbon atoms was later confirmed by the discovery of mixed bonding types, and graphite and graphene intercalation compounds. The relationship of differently hybridized carbon allotropes is shown in the figure below.

Since there was, and still is a strong resistance, chiefly among organic chemists, to accept carbyne as a stable allotrope of carbon because of the notorious reactivity of carbon double and triple bonds, much work was expended to confirm its existence. Consequently there exists a rich but still highly controversial literature that also includes technological applications of carbyne and carbyne-like materials. Today some researchers consider carbyne the „third allotrope of carbon" even though no unambiguous evidence for its existence as a bulk material is available yet (Chapter 1.2.).

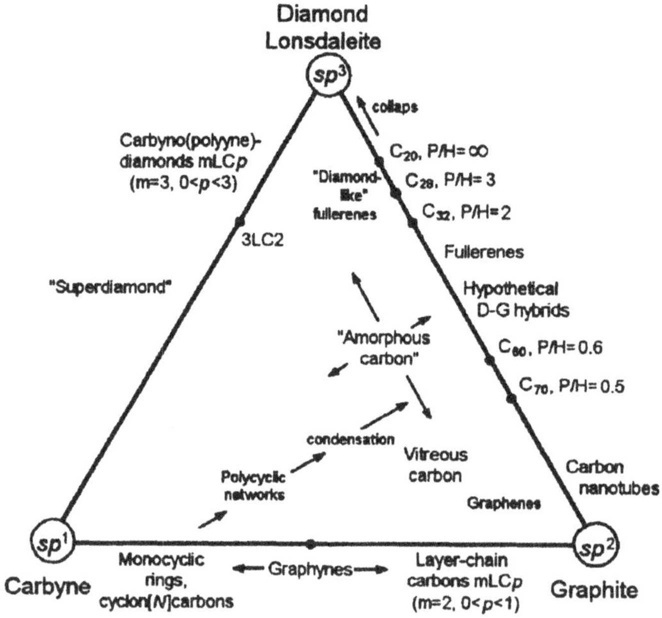

Nevertheless, a sizeable portion of the papers dealing with carbyne confirmed the presence of linear sp-hybridized carbon chains in the solid state, in which the carbon atoms are connected either by conjugated triple bonds (polyyne-type isomer) or cumulated double bonds (polycumulene-type isomer). Many ways to synthesize carbyne were explored by polymer scientists and solid state chemists during the last thirty five years (Chapter 3).

It was found that carbyne *sensu strictu* can be obtained today only by condensation of carbon vapour (Chapters 3.4.1., 3.4.2.), whereas chemical degradation reactions of polymers with a sizeable portion of conjugated multiple carbon bonds generally lead to *carbynoid structures* (Chapters 3.2., 3.3.2., 3.5.). These structures can be defined as carbon-rich chain-like poly(or oligo)mers with both polyyne- and cumulene-type moieties, extended interchain cross-linking as well as end-capping and/or pendant side groups, and intercalated larger atoms or ions to stablized the conformation. Hence these structures would hardly qualify for a carbon allotrope.

On the other hand, products obtained by condensation of carbon vapour (Chapters 3.4.1., 3.4.2.) under closely controlled conditions in an vacuum or inert gas environment approach the proposed ideal structural conformation of carbyne to a high degree. Long carbon chains terminated by stabilizing bulky organic groups or larger atoms and assembled in a van der Waals-bonded parallel array with occasional crosslinking could be considered a 'pure' carbon allotrope whose surface is terminated by heteroatoms the same way the surfaces of graphite and diamond are terminated by oxygen atoms and hydroxyl group to satisfy free, *i.e.* dangling bonds (Chapter 4).

Recently the carbyne research was pointed in a new direction. A series of papers were published dealing with potentially novel technological applications of carbyne intercalation compounds (CIC's) in the field of electrical and electronic devices (Chapter 4.3.). The intercalation of alkali metal atoms appears to split longer carbon chains of the precursor into shorter segments that form a 2D-lattice with allegedly tetragonal symmetry. Treatment with mineral acids leads to residual intercalation compounds in which metal atoms at the ends of the short chains prevent lateral cross-linking thus stabilizing the sp-hybridized carbyne structure. According to these studies carbyne can be considered a quasicrystalline structure with long range order in <hk.0> direction but short range order in <00.1> direction parallel to the chains.

Very recent work showed that it is possible to synthesize, by a modified Krätschmer-Huffmann experiment, carbyne with a chain length $n > 100$ using a „capping" reaction of reactive conjugated acetylene units with trifluoromethyl and cyano radicals. Since similar experiments were performed to synthesize fullerenes, interesting parallels abound. The explanation of the reaction involves the easy low activation energy transformation of polycyclic polyyne ring isomers into C_{50}-C_{76} fullerenes by a „spiral-and-zipper" mechanism.

In the course of research on shock-induced transformation of graphite to diamond (Chapter 3.4.3.), several carbyne polytypes with various carbon chain conjugation lengths and bonding type were discovered in the shock-compressed products and classified by a novel model (Chapter 7.1.). This initially phenomenological model of the carbyne conformation explained not only the lattice metric of crystalline carbyne by the existence of carbon chains with differing numbers of carbon atoms but also postulated stiff linear carbon chains arranged in parallel fashion along the c-axis of the carbyne structure and held together rather loosely by van der Waals forces and lateral crosslinks, respectively.

Because the carbon chains are thought to contain kinks, carbynes may form sheet structures in which the carbon chains of adjacent sheets are offset by „kinks". This kink model explains rather well many properties of carbyne, and recently several authors were able to confirm this heuristic model theoretically and experimentally (Chapter 4.2.).

Since then the heuristic-speculative kink hypothesis is being improved and expanded. Based on theoretical calculations it was assumed, and later tentatively confirmed by experimental means, that spin density waves and solitons, respectively exist in the polyyne chains that can travel along the one-dimensional carbon chain. Such waves modulate the electron density and shift carbon atoms to new equilibrium positions. Strong electron-phonon coupling is presumably the driving force for a commensurate-incommensurate transition whose elastic term leads to lattice kinks. These lattice kinks are the domain walls of the transition, and hence the soliton separates two energetically equivalent but structurally different segments of the polyyne chain, *i.e.* the postulated „kinks“. This explanation was confirmed in part by a Patterson synthesis of electron density measurements in carbyne and by ESR studies of oriented carbyne films, that revealed a spin-Peierls soliton lattice (Chapter 6.2.).

As an interesting new aspect of a structural model of carbyne appears the notion that carbyne may be related to chiral poly(diacetylene). Indeed, the comparison of the lattice constants of different carbyne polytypic forms reported in the literature with the lattice constants of chiral poly(diacetylene) shows close similarities. This conjecture is also confirmed by the frequently observed fact that the structural building unit of carbyne may be a linear string of just four carbon atoms (diacetylene or butadiynyl units) (Chapter 4.2.).

The unambiguous chemical and physical characterization of such materials with low-dimensional structures is difficult and loaded with contradictions. The extremely small carbyne crystals formed during the syntheses approaches described in Chapter 3 were predominantly identified and classified by electron diffraction studies, and microscopic and microprobe techniques (Chapter 4.4.). Because carbyne crystals with well-defined structure and sufficient size are not available to date, complete x-ray diffraction analyses are still lacking. This is indeed the major obstacle to general acceptance of carbyne as a true carbon allotrope (Chapter 1.2.). Instead, molecular and electron spectroscopy is widely used to characterize carbynoid materials, *i.e.*, amorphous and/or mesoscopic solids with short-range order containing carbon chains of varying lengths and bond multiplicity (Chapter 6). Not only the smallness of the crystals but also their impurity content and inhomogeneities in the carbon skeleton make the study of all chemical and physical properties complex and the interpretation of the results subject to debate (Chapters 5 and 6).

Notwithstanding these limitations carbyne may have exciting practical applications as precursor materials for diamond (Chapter 7.1.) and biocompatible material for use in reconstructive surgery (Chapter 7.2.).

Challenging tasks are still ahead, and despite a host of papers dealing with different aspects of the physics and chemistry of carbyne, many questions remain to be solved. We believe that a critical assessment of these papers is required but we also emphasize that continuing efforts will eventually result in carbyne single crystals large enough to permit in-depth structural analysis. Hence it is not the intention of this text to reveal completely new and stunning research results on carbyne but to provide a rather complete, ordered and concise summary of what is known with certainty today and what is still ambiguous or downright wrong. Hence this book should be considered an important stepping stone towards an understanding of carbyne, an aspect many researchers around the world have attempted hitherto. It endeavours to show the research community what has been confirmed, what is still doubtful, what is or was wrong, and most importantly, where do we go from here. As the reader will easily notice, there are some overlaps among the individual contributions to this book. While in a multi-authored text dealing with the same class of materials overlaps can never be completely avoided, we consider most of these overlaps fortuitous in that they allow to view many aspects of the subject matter from different angles. Thus we are left with a multifaceted approach to carbyne that will hopefully assist in charting a course towards future endeavours in the field of one of the most fascinating and controversial carbon allotropes in existence.

Since the opinions expressed in the individual chapters are those of the authors, they do not necessarily reflect those of the editors. However, we are proud to have succeeded in assembling a distinguished board of contributors. The text is a joint international endeavour of scientists from many countries: nine authors come from Russia, six from Japan, three from the United Kingdom, two each from Poland and Germany, and one each from Austria, Belorussia, Canada and the Czech Republic.

Many people assisted the editors in many stages of the preparation of this text. In particular, we are very much indebted to Ms. Jutta Hofmann of the Department of Mineralogy, Freiberg University of Mining and Technology for her very dedicated and able help in editing many chapters.

June 1998

Robert B.Heimann, Freiberg
Sergey E.Evsyukov, Ludwigshafen
Ladislav Kavan, Prague

Chapter 1: Introduction

1.1. THE DISCOVERY OF CARBYNE

YU.P. KUDRYAVTSEV
A.N. Nesmeyanov Institute of Organoelement Compounds, Russian Academy of Sciences, 117813 Moscow, Russia

Abstract – The history of the discovery of carbyne is briefly recalled. The existence of carbyne was first disclosed by Russian researchers in 1960. It was obtained for the first time *via* oxidative dehydropolycondensation of acetylene based on the Glaser coupling of ethynyl compounds.

1. Introduction

The polymeric nature of carbon was first pointed out by Mendeleev. He wrote: "The molecules of coal, graphite, and diamond are very complicated, and carbon atoms exhibit the capability of binding one to another to form complex molecules in all compounds of carbon. None of the elements possesses an ability of complicating in such an extent as does carbon. There is still no basis to define the polymerization degree of the coal, graphite, or diamond molecules. One should believe, however that they contain C_n species, where '*n*' is a large value" [1].

Until the 1960s only two allotropic forms of carbon were known, *viz.*, graphite and diamond, including their polymorphous modifications. For a long time 'amorphous carbon' was also included among the simple forms. Presently, however, the structure of amorphous and quasi-amorphous carbons (such as carbon blacks, soot, cokes, glassy carbon, *etc.*) is known to approach that of graphite to various degrees [2].

Elemental carbon is known to exist in three bonding states corresponding to sp^3, sp^2, and *sp* hybridization of the atomic orbitals. According to the classification scheme based on the types of chemical bonds in carbon [3,4], each valence state should correspond to a certain form of a simple substance. When considering carbon as a polymeric system in terms of this classification, one can distinguish three types of polymers. The three-dimensional (spatial) polymer of carbon, diamond, corresponds to the first state (sp^3); the two-dimensional (planar) polymer of carbon, graphite, corresponds to the second one (sp^2). Hence, it did make sense to assume the existence of a third state, *i.e.*, a chain-like (one-dimensional) polymer, whose linear molecules consist of *sp*-hybridized carbon atoms. For a long time this problem was the subject of great interest among scientists – both theoreticians and experimentalists.

R.B. Heimann et al. (eds.), Carbyne and Carbynoid Structures, 1–6.

2. Oxidative Coupling

The first serious attempt to synthesize a one-dimensional polymer of carbon was made by Baeyer back in 1885 [5,6]. As a synthesis approach he used an oxidative coupling reaction of ethynyl compounds disclosed earlier by Glaser [7] in 1870. Acetylenecarboxylic acid was used as the starting monomer. Bayer, the pioneer in the field of synthesis of polyynes, hoped to obtain the elusive chain-like polymer through a stepwise sequence of transformations: acetylenecarboxylic acid → diacetylenedicarboxylic acid → (decarboxylation) → diacetylene → tetraacetylenedicarboxylic acid → (decarboxylation) → tetraacetylene, and so forth. However, he did not succeed in isolating even tetraacetylene, which is extremely unstable. Hence the instability of the lower polyynes led Baeyer to develop a strain theory, in which he postulated the impossibility of preparing the chain-like polymer of carbon [6]. It is noteworthy that Baeyer referred to the hypothetical linear carbon as "explosive diamond".

Baeyer's reputation severely dampened the interest among scientists in the synthesis of polyynes for a long time. An important stimulus for renewed work in this field was the discovery of natural compounds of the polyyne series isolated from some plants and fungi [8–12][1]. The efforts by Bohlmann [13] and Jones *et al.* [14] resulted in the preparation of polyyne chains containing up to 10 conjugated triple bonds. Bulky *tert*-butyl or trialkylsilyl substituents were also used as protective terminal groups preventing the polyyne chains from intermolecular interaction, *i.e.*, chain collapse [15]. Later, using a similar approach, Walton *et al.* [16–18] synthesized polyyne with 16 triple bonds [16]. Organometallic bulky substituents have also been reportedly used as protective groups in studies on polyynes and cumulenes [19–22]. Up to now, however, the stepwise coupling methods have neither afforded carbyne nor longer polyynes or cumulenes.

In 1959–1960 systematic investigations of the oxidative (Glaser [7]) coupling of bis(ethynyl)arylenes were carried out at the A.N.Nesmeyanov Institute of Organoelement Compounds of the Russian (at that time USSR) Academy of Sciences. It was established that the Glaser coupling – a key reaction in the chemistry of polyynes – can be conducted with any bis(acetylene) compounds to form polymers whose repeating unit retains the structure of the starting bis(acetylene) without the terminal hydrogen atoms. This version of the oxidative coupling reaction was named 'oxidative dehydropolycondensation' [23], and a series of linear poly(arylenediynylene)s was synthesized using this approach (*Scheme 1*) [24].

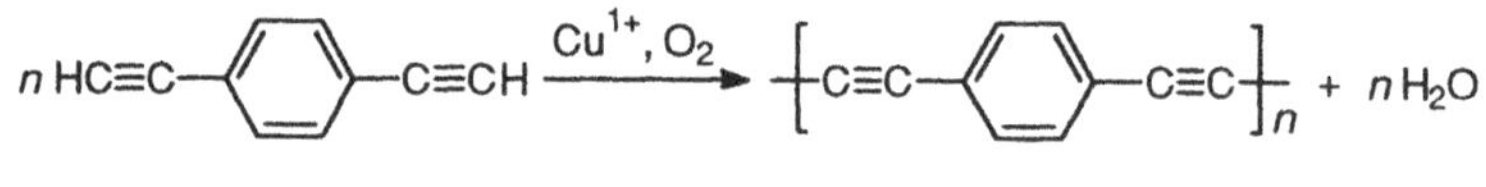

Scheme 1

Since acetylene may be regarded a bifunctional compound with two ≡C—H bonds, it was logical to assume that it can be used as a monomer for oxidative dehydro-

[1] *cf.* Chapter 2.2

polycondensation. Further research into this field[2] culminated in the discovery of the elusive third (linear) allotropic form of carbon [24,25]. The new form was labeled 'carbyne' [26], from the Latin word *carboneum* (carbon) and the suffix 'yne', customarily used in organic chemistry to designate an acetylenic bond C≡C. Admittedly, this name was an unfortunate choice because of the existence of its namesake – a term adopted by IUPAC [27] to designate organic carbon-centered triradicals[3]. Nevertheless, the name 'carbyne' has taken root and is now widely used by researchers working in the field of chemistry and physics of carbon. In 1971, the discovery was registered (patented) with priority dated back to November 4, 1960 (Fig. 1) [30]. Carbyne was supposed to occur in two isomeric forms (provided that the end groups are neglected), *viz.* a conjugated triply bonded form (polyyne, polyethynylene) and a cumulated doubly bonded form (polycumulene, polyethenediylidene). In 1968, carbyne was found for the first time as a natural mineral [31] and named 'chaoite' in honour of the respected American scientist E.C.T. Chao [32].

3. Other Approaches

It should be noted that, presumably, an analogous polycondensation reaction was performed already as far back as 1897 by Söderbaum [33], who passed acetylene through an ammoniacal solution of a Cu(II) salt, and obtained a black, insoluble, explosive precipitate. Söderbaum believed he had found a previously unknown Cu(II) acetylide. The decomposition of this product on treatment with hydrochloric acid resulted in the formation of a black insoluble substance with the empirical formula $C_{12}H_4O_3$, which the author assigned to a class of humic acids.

Fifty years later, this reaction was replicated with some modification by Klement and Köddermann-Gros [34]. The remainder formed upon decomposing the black explosive product with hydrochloric acid was interpreted as 'amorphous graphite'. It is remarkable that the authors were not surprised by the fact that a graphite-like structure was formed under so mild a condition.

Intriguing results were obtained by Schlubach and Wolf [35] in their study on the synthesis of diacetylene. In the course of the oxidation of sodium acetylide with potassium permanganate they observed the formation of a black polymer as a byproduct along with the target diacetylene. The isolated polymer, as the authors put it, resembled anthracite and did not burn in the flame of a Bunsen burner. Since the conditions employed to perform an oxidative coupling reaction appeared to be conducive to the formation of a polyyne chain, it only remains to regret that the authors did not study the polymer they obtained more closely.

Finally and amazingly, the linear carbon allotrope was 'rediscovered' by Lagow *et al.* [36] in 1995. The authors employed laser-induced vaporization of graphite to produce carbyne chains consisting of ~150 triple bonds. It should be noted, however,

---

[2] *cf.* Chapter 3.1.1 of this book.

[3] Perhaps, this is one of the reasons why carbyne was not included in a list of terms recently recommended by IUPAC for the description of carbon as a solid [28], another reason being, of course, the lack of rigorous structural evidence. On the other hand, one might argue in gist that lack of evidence was not an obstacle for the Nobel Committee to award a 1997 prize in Physiology and Medicine for the discovery of prions [29].

СССР

ДИПЛОМ
НА ОТКРЫТИЕ

В соответствии с Положением об открытиях, изобретениях и рационализаторских предложениях Комитет по делам изобретений и открытий при Совете Министров СССР установил, что граждане СССР

КУДРЯВЦЕВ Юрий Павлович
СЛАДКОВ Алексей Михайлович
КАСАТОЧКИН Владимир Иванович
КОРШАК Василий Владимирович

сделали открытие, определяемое следующей формулой:

„Экспериментально установлено неизвестное ранее явление существования новой кристаллической формы углерода-карбина, характеризующейся в отличие от алмаза и графита, цепочечным (линейным) строением углеродных макромолекул“.

* *
*

Настоящее открытие зарегистрировано в Государственном реестре СССР 7 декабря 1971 г. № 107 с приоритетом 4 ноября 1960 г.

Председатель Комитета *Ю. Максарев*

1973 г

Figure 1. Inventor's certificate (diploma) No. 107 issued by the USSR Committee of Inventions and Discoveries (in Russian). The disclosure is claimed as follows: "Experimentally established previously unknown phenomenon of the existence of a new crystalline form of carbon – carbyne that is characterized by a chain-like (linear) structure of the carbon macromolecules as distinct from diamond and graphite." Each co-author received the same certificate but with different order of co-authors starting with himself.

that laser-assisted synthesis of carbyne was employed by Kasatochkin *et al.* [37,38] and Whittaker *et al.* [39,40] in the early 1970s.

4. References

1. Mendeleev, D.I.: *Principles of Chemistry*, 10th edn., Vol. 1, Moscow-Leningrad, 1931, (in Russian).
2. Kasatochkin, V.I.: Transitional forms of carbon, in: *Structural Chemistry of Carbon and Coals*, Kasatochkin, V.I., Ed., Nauka, Moscow, 1969, pp. 7–16 (in Russian).
3. Terms concerned with carbon as chemical element and its reactivity, *Proc. 5th London Int. Carbon and Graphite Conf.*, 18–22 Sept., 1978, Vol. 3, Publ. 1979, pp. 103–108.
4. Heimann, R.B., Evsyukov, S.E., and Koga, Y.: Carbon allotropes: A suggested classification scheme based on valence orbital hybridization, *Carbon* **35** (1997), 1654–1658.
5. Baeyer, A.: Ueber Polyacetylenverbindungen, *Ber. Deuts. Chem. Ges.* **18** (1885), 674–681.
6. Baeyer, A.: Ueber Polyacetylenverbindungen, *Ber. Deuts. Chem. Ges.* **18** (1885), 2269–2281.
7. Glaser, C.: Untersuchungen über einige Derivate der Zimmtsäure, *Ann. Chem. Pharm.* **154** (1870), 137–171.
8. Vil'yams, V.V., Smirnov, V.S., and Gol'mov, V.P.: Concerning the nature of a crystalline matter from essential oil of *Lachnophyllum Gossypinum*, *Zh.Obsch. Khim.* **5** (1935), 1195-1204 (in Russian).
9. Anchel, M.: Identification of an antibiotic polyacetylene from *Clitocybe Diatreta* as a suberamic acid enediyne, *J. Am. Chem. Soc.* **75** (1953), 4621–4622.
10. Bu'Lock, J.D.: Acetylenic compounds as natural products, *Quart. Rev.* **10** (1956), 371–394.
11. Nakaminami, G.: The structures of natural acetylenic compounds and their biogenesis, *J. Soc. Org. Synth. Chem., Japan*, **21** (1963), 751–765 (in Japanese).
12. Bohlmann, F., Burkhardt, T., and Zdero, C.: *Naturally Occurring Acetylenes*, Academic Press, London, 1973.
13. Bohlmann, F.: Konstitution und Lichtabsorption, VI. Mitteil.: Zur Deutung von Polyacetylen-Spektren, sowie Darstellung von Bis-*tert.*-butyl-decapentain-(1.3.5.7.9), *Ber. Deuts. Chem.Ges.* **86** (1953), 63–69.
14. Jones, E.R.H., Lee, H.H., and Whiting, M.C.: Researches on acetylenic compounds. Part LXIV. The preparation of conjugated octa- and deca-acetylenic compounds, *J. Chem. Soc.* (1960), 3483–3489.
15. Sladkov, A.M. and Kudryavtsev, Yu.P.: Polyynes, *Usp. Khim.* **32** (1963), 509–538 (in Russian) [*Russ. Chem. Rev.* **32** (1963), 229–243 (Engl. Transl.)].
16. Eastmond, R., Johnson, T.R., and Walton, D.R.M.: Silylation as a protective method for terminal alkynes in oxidative couplings. A general synthesis of the parent polyynes $H(C{\equiv}C)_nH$ (n = 4–10,12), *Tetrahedron* **28** (1972), 4601–4616.
17. Johnson, T.R. and Walton, D.R.M.: Silylation as a protective method in acetylene chemistry. Polyyne chain extensions using the reagents, $Et_3Si(C{\equiv}C)_mH$ (m = 1,2,4) in mixed oxidative couplings, *Tetrahedron*, **28** (1972), 5221–5236.
18. Eastmond, R., Johnson, T.R., and Walton, D.R.M.: Base-catalysed cleavage of silyl-substituted polyynes. Attenuation of hydrocarbon acidity and transmission of substituent electrical effects in long-chain conjugated polyacetylenes, *J. Organomet. Chem.* **50** (1973), 87–92.
19. Bartik, T., Bartik, B., Brady, M., Dembinski,R., and Gladysz, J.A.: A step-growth approach to metal-capped one-dimensional carbon allotropes: Synthesis of C_{12}, C_{16}, and C_{20} μ-polyynediyl complexes, *Angew. Chem. Int. Ed. Engl.* **35** (1996), 414–417.
20. Le Narvor, N., Toupet, L., and Lapinte, C.: Elemental carbon chain bridging two iron centers: synthesis and spectroscopic properties of $[Cp^*(dppe)Fe{-}C4{-}FeCp^*(dppe)]^{n+}\cdot n[PF_6]^-$. X-ray crystal structure of the mixed valence complex (n=1), *J. Am. Chem. Soc.* **117** (1995), 7129–7138.
21. Touchard, D., Haquette, P., Daridor, A., Toupet, L., and Dixneuf, P.H.: First isolable pentatetraenylydene metal complex containing the $Ru{=}C{=}C{=}C{=}C{=}CPh_2$ assembly. A key intermediate to provide functional allenylidene complexes, *J. Am. Chem. Soc.* **116** (1994), 11157–11158.
22. Zhou, Y., Seyler, J.W., Weng, W., Arif, A.M., and Gladysz, J.A.: New families of coordinated carbon: oxidative coupling of an ethynyl complex to isolable and crystallographically characterized $MC{\equiv}CC{\equiv}CM$ and $^+M{=}C{=}C{=}C{=}C{=}M^+$ assemblies, *J. Am. Chem. Soc.* **115** (1993), 8509–8510.
23. Kudryavtsev, Yu.P.: Oxidative dehydropolycondensation – a new method to synthesize polymers with triple bonds, in: *Progress of Polymer Chemistry*, Korshak, V.V., Ed., Nauka, Moscow, 1969, pp. 87–112 (in Russian).

24. Kudryavtsev, Yu.P.: Synthesis and properties of polymers containing conjugated triple bonds in the chain, *Ph.D. Thesis*, Institute of Organoelement Compounds, USSR Academy of Science, Moscow, 1964 (in Russian).
25. Kudryavtsev, Yu.P.: Synthesis and properties of carbyne, *D.Sc. Thesis*, Institute of Organoelement Compounds, USSR Academy of Science, Moscow, 1988 (in Russian).
26. Sladkov, A.M. and Kudryavtsev, Yu.P.: Diamond, graphite, carbyne – allotropic forms of carbon, *Priroda* (5) (1969), 37–44 (in Russian).
27. Glossary of Terms Used in Physical Organic Chemistry, *Pure Appl. Chem.* **55** (1983), 1281–1371.
28. Recommended Terminology for the Description of Carbon as a Solid, *Pure Appl. Chem.* **67** (1995), 473–506.
29. Coles, H.: Nobel panel rewards prion theory after years of heated debate, *Nature* **389** (1997), 529.
30. Kudryavtsev, Yu.P., Sladkov, A.M., Kasatochkin, V.I., and Korshak, V.V., Inventors certificate No. 107 of Dec. 7, 1971, priority date Nov. 4, 1960 (Fig. 1, in Russian) [*Bull. Izobr.* (6) (1972), 3 (in Russian)].
31. El Goresy, A. and Donnay, G.: A new allotropic form of carbon from the Ries crater, *Science* **161** (1968), 363–364.
32. El Goresy, A.: Eine neue Kohlenstoff-Modifikation aus dem Nördlinger Ries, *Naturwiss.* **56** (1969), 493–494.
33. Söderbaum, H.G.: Ueber die Einwirkung des Acetylens auf Cuprisalze, *Ber. Deuts. Chem. Ges.* **30** (1897), 760–765.
34. Klement, R. and Köddermann-Gros, E.: Die Oxydationsprodukte des Kupfer(I)-acetylides, *Z. Anorg. Chem.* **254** (1947), 201–216.
35. Schlubach, H.H. and Wolf, V.: Über Polyacetylene, *Justus Liebigs Ann. Chem.* **568** (1950), 141–159.
36. Lagow, R.J., Kampa, J.J., Wei, H.-Ch., Battle, S.L., Genge, J.W., Laude, D.A., Harper, C.J., Bau, R., Stevens, R.C., Haw, J.F., and Munson, E.: Synthesis of linear acetylenic carbon: the "sp" carbon allotrope, *Science* **267** (1995), 362–367.
37. Kasatochkin, V.I., Kazakov, M.E., Savranskii, V.V., Nabatnikov, A.P., and Radimov, N.P.: Synthesis of a new allotropic form of carbon from graphite, *Dokl. Akad. Nauk SSSR* **201** (1971) 1104–1105 (in Russian).
38. Kasatochkin, V.I., Savranskii, V.V., Smirnov, V.N., and Mel'nichenko, V.M.: A study of carbyne condensed from carbon vapour, *Dokl. Akad. Nauk SSSR* **217** (1974), 796–799 (in Russian).
39. Whittaker, A.G. and Wolten, G.M., Carbon: a suggested new hexagonal crystal form, *Science* **178** (1972), 54–56.
40. Nelson, L.S., Whittaker, A.G., and Tooper, B.: The formation of new polymorphs of carbon and fluid flow patterns by irradiating solid carbons with a CO_2 laser, *High Temp. Sci.* **4** (1972), 445–477.

1.2 THE NATURE OF CARBYNE- PROS AND CONS

R.B. HEIMANN
Chair of Technical Mineralogy, Department of Mineralogy, Freiberg University of Mining and Technology, D09596 Freiberg, Germany

Abstract - There is still controversy whether carbyne is an all-carbon sp-hybridized allotrope of carbon, i.e. the 'third form of carbon'. However, more and more papers appear that strongly suggest carbyne may be reality rather than fiction.

Until the early sixties of this century it has been generally held that the only carbon allotropes in existences were the naturally occuring two- and three-dimensional carbon polymers, i.e. hexagonal graphite and cubic diamond with sp^2- and sp^3-hybridization, respectively. However, it was frequently noted and considered odd by many researchers in the field of carbon chemistry that the element with the richest chemistry of all should be limited to just two allotropes. Even considering polytypic variants such as hexagonal diamond (lonsdaleite) and rhombohedral graphite would not account for the fact that the one-dimensional sp-hybridized allotrope was conspicuously missing. Hence this problem was tackled by theoreticians and experimentalists alike for over a hundred years (see Chapter 1.1). The situation changed after the alleged discovery by Russian scientists of a linear chain-like all-carbon material coined 'karbin (carbyne)', with the suffix '-in' poiting to the observed preponderance of carbon triple bonds [1]. This was a rather unfortunate choice of a name since it conflicts with the generally accepted IUPAC terminology for the free radical species $H\text{-}C^{3\bullet}$ and $H\text{-}C^{\bullet}$: [2]. In lieu of a better name the term 'carbyne' will be retained for the time being. Notwithstanding the fact that today the zero-dimensional closed-cage carbon molecules, i.e. fullerenes are universially called the third carbon allotrope [3], for historical and logical reason carbyne should be given this distinction [4].

From a wealth of synthesis methods and analyses of the sp-hydridized material the following definition can be given. An ideal 'carbyne' is a systems consisting of straight [1] or kinked [5] carbon chains held together by van der Waals forces to form a crystal [6]. Theoretically, linear carbon chains with conjugated triple (polyyne-type) and/or cumulated double (polycumulene-type) intrachain bonds are assembled in a hexagonal array without any strong chemical bonds (σ-bonds) between the chains ('pencil-in-box' model). Weak interchain bonding is achieved between sufficiently close chains by overlapping of molecular π-orbitals of one chain with empty antibonding π*-orbitals of the neighbouring chain. This is similar to the bonding suggested for solid C_{60} molecules [7].

R.B. Heimann et al. (eds.), Carbyne and Carbynoid Structures, 7–15.

In 'real' carbyne, however, σ-type cross-linking with various periodicities as well as kinks will frequently occur resulting in a layered lattice arrangement (see Chapter 4.2). Hence the definition above can be extended by stating that carbyne is a "partially cross-linked three-dimensional polymer having a more or less regular (layered) crystalline structure and (is) predominantly composed of extensive fragments of linearly polymerised carbon" [6]. Since in these materials hybrid polyyne, and polycumulene or polyene moieties exist simultaneously, we will call these 'real' carbynes *'carbynoid structures'* as indicated in the title of this treatise. In the spirit of this, graphite and diamond, respectively are limiting conformations of network and chain structures that are regularly periodically cross-linked by three (graphite)- and four (diamond)-function branching points. Indeed, the graphite plane can be viewed as composed of 2D-cross linked polyyne [8] or polycumulene [9] chains whereas the diamond structure consists of a 3D-array of an equal number of screw axes of opposite chirality, *i.e.* of helically twisted carbon chains [10].

Acceptance of the existence and significance of carbyne is still a slow process. Organic chemists tend to look at the pile-up of double and triple bonds connecting the carbon atoms in the chains and take cover. To avoid an explosive collapse of the conjugated chains, stabilization is required by 'spacers' that keep the neighbouring chains apart. Such spacers were found to be metal atoms such as potassium, iron [11], copper [9] or rhenium [12], bulky alkyl or aryl groups [13], end-cap CO molecules [14] or alkali metal fluorides [15]. It might be argued that incorporation of hetero-atoms or foreign compounds invalidate the notion of carbyne being an all-carbon allotrope. However, for sufficiently long carbon chains the electronic contribution of foreign atoms at kink sites within the chains as well as bulky organic groups terminating the chains would be small. Hence, despite the presence of stabilizing end-capping groups carbyne ought to be considered a carbon allotrope in the same sense graphite and diamond are considered „pure" carbon despite the fact that their two- and three-dimensional surfaces, respectively are terminated by hydrogen, OH or oxygen groups [16]. Apart from this, carbyne should not be regarded within the narrow framework of classical organic chemistry just mechanistically as 'wisps' of non-interacting chain-like carbon molecules. This would be as meaningless as considering graphite and diamond as aggregates of, benzene and adamantane molecules respectively and attempting to look for direct analogies in properties [17].

Carbyne has semiconducting properties with a band gap width of 1-2 eV as determined by EELS investigations [18]. Thus its conductive properties are intermediate between those of graphite with a specific electrical conductance of $\sigma = 8.3$ ohm^{-1} cm^{-1} parallel <00.1> [19] and the almost perfect electrical insulator diamond. It has been described as a silvery white, light-reflecting opaque material interspersed with graphite in impact-shocked graphitic gneisses of the Ries crater, Nördlingen, Bavaria, Germany and coined 'chaoite' [20]. Chaoite was accepted by the IMA Commission for New Minerals as a new carbon modification. Subsequent experiments on sublimation of carbon vapour obtained by resistive heating of carbon yielded identical material that was called „white carbon" [21,22] and another carbyne polytypic form, Carbon VI, was obtained from laser-ablated graphite [23] (cp. Chapter 3.4.1.1). The latter material was found to be optically transparent with some evidence of a faintly orange-red colouration in macroscopic clusters that, according to the authors, could be due to impurities. Recent experiments by Tanuma and Pal'nichenko [24] yielded a carbyne-analoguous material called 'carbolite' that was described as slightly orange-coloured hexagonal crystals (see Chapter 4.2). The white colour of carbyne most likely indicates that the forbidden band gap width in the electronic spectrum should exceed the maximum energy of a photon in the optical range (3.3 eV) [25]. However, most carbyne samples obtained experimentally to date were black or gray. This attests to a defect structure of the material being due to a substantial contribution of sp^2-hybridized strings within the carbon chains, for example at kink sites.

Counterevidence for the existence of carbyne as a distinct carbon allotrope was given by Nagornyi *et al.* already in 1975 [26]. These researchers considered the appearance of the electron diffraction pattern of some carbyne microcrystals as the product of double scattering of electrons in twinned graphite with specific twinning angles. Assuming that the electron beam diffracted by the first crystal undergoes additional scattering by a second crystal rotated by an angle γ against the c-axis, the geometrical conditions required for scattering are

$$2\pi(\mathbf{K}' - \mathbf{K}_0) = \mathbf{H}'; \; 2\pi(\mathbf{K}'' - \mathbf{K}') = \mathbf{H}'' , \tag{1}$$

where $\mathbf{K}_0$, $\mathbf{K}'$ and $\mathbf{K}''$ are the wave vectors of the incoming beam, the diffracted beam after leaving the first crystal, and the doubly reflected beam after passing both crystals, respectively, and $\mathbf{H}'$ and $\mathbf{H}''$ are the reciprocal lattice vectors of the first and second crystal:

$$H' = h_1 \cdot \mathbf{a}^{*\prime}_1 + k_1 \cdot \mathbf{a}^{*\prime}_2 + l_1 \cdot \mathbf{a}^{*\prime}_3 \tag{2a}$$
$$H'' = h_2 \cdot \mathbf{a}^{*\prime\prime}_1 + k_2 \cdot \mathbf{a}^{*\prime\prime}_2 + l_2 \cdot \mathbf{a}^{*\prime\prime}_3 \tag{2b}$$

with $a_i^{*\prime,\prime\prime}$ = basic vectors of the reciprocal lattice. The combined diffraction process can be described by the equation

$$2\pi(\mathbf{K}'' - \mathbf{K}_0) = \mathbf{H}' + \mathbf{H}''. \tag{3}$$

To recast the lattice metric in terms of the basic vectors it is useful to eliminate the vectors $\mathbf{a}^*_1$ and $\mathbf{a}^*_2$.

Then

$$\mathbf{H} = H \cdot \mathbf{a}_1 + K \cdot \mathbf{a}_2 + L \cdot \mathbf{a}^*_3, \tag{4}$$

where
$$H = (h_1 + h_2) \cos \gamma/2 + (h_1 + 2k_1 - h_2 - 2k_2)(\sin \gamma/2)/\sqrt{3},$$
$$K = (k_1 + k_2) \cos \gamma/2 + (-2h_1 - k_1 + 2h_2 - k_2)(\sin \gamma/2)/\sqrt{3},$$
$$L = l_1 + l_2.$$

The diffraction pattern will look very complex because of the presence of two multiperiods, proportional to cos γ/2 and (sin γ/2)/√3. The correct hexagonal lattice yields, assuming the ratio of these periods being rational,

$$\sin\gamma/2\,[\sqrt{3}\cos\gamma/2]^{-1} = m/n \quad \text{or} \quad \mathrm{tg}\,\gamma/2 = (\sqrt{3}\,m)/n, \tag{5}$$

where m,n are small integers.

From this assumption one obtains a set of rational twinning angles. The parameter a is defined as

$$a = a_{graph}\,(n^2 + 3m^2)^{1/2} \tag{6}$$

In particular, a twinning angle $\gamma = 27^o48'$ yields $a = a_{graph} \cdot \sqrt{13} = 887$ pm, close to the a_0-lattice parameter of α-carbyne (892 pm). Figure 1 shows the calculated electron diffraction pattern of a graphite twin projected onto (hk0). The circles denote the diffraction pattern of a graphite single crystal, the solid dots show the superposed pattern of two graphite single crystals rotated against their c-axes by an angle of 27°48', and the crosses are indicative of the observed diffraction maxima of carbyne. On first sight, the pattern are conspicuously similar. However, the diffraction pattern of carbyne shows only a small portion of spots close to the positions calculated for the twinned graphite. Furthermore, strong graphite reflections at d_{100} = 212 pm and d_{200} = 106 pm are absent in the electron diffraction pattern of carbyne [27]. In addition, it would be difficult to explain, by the concept of twinning, the existence of a polytypic series of carbyne with a_0-parameters ranging from 824 to 940 pm [5,9].

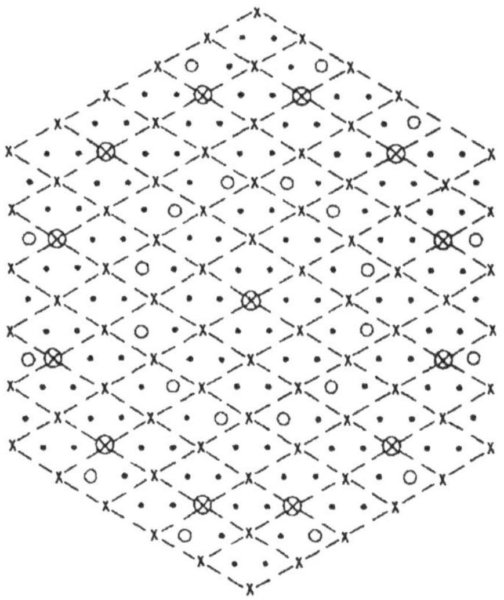

Figure 1 Twofold electron diffraction of graphite (hk0 plane) [27].
o single crystal graphite, • superposition of two graphite crystals in twin position, × carbyne

It was not surprising that the idea of the existence of an sp-hybridized linear carbon allotrope with highly reactive conjugated triple and double bonds in close proximity was greeted with great scepticism by organic chemists. Ironically, however, a wave of rejection was triggered by a paper of two geologists, Smith and Buseck [28] in which the authors sweepingly denied the existence of any carbyne, based on their failure to detect anything but airborne sheet silicate (nontronite, muscovite, talk, kaolinite) and quartz contamination in their TEM mounts of samples [29] provided by El Goresy [20], Setaka and Sekikawa [30], Whittaker [31] and others. In their zealous pursuit to discredit carbyne the authors did not even spare their own earlier papers in which they at least tentatively identified α-carbyne and/or ĉhaoite [32]. Similar criticism was raised by Lumpkin [33] who reanalyzed samples obtained by acid treatment and physical separation of a carbonaceous chondrite (Allende). Apart from the rather unlikely fact that the electron microscopy laboratories of all researchers detecting carbyne in their samples should be awash in airborne sheet silicates (if so, antigorite asbestos would be the most likely contender but was never reported), there were other concerns with S & B's critique. On investigating the original chaoite sample they reported values directly opposite to El Goresy's findings as far as optical reflectivity, hardness (polish) and impurity content are concerned. Hence the latter feels that S&B likely confused a graphite lamella with an adjacent chaoite lamella [34]. It is also difficult to understand why grains that reportedly contain less than 0.5% silicon and chlorine [20] ought to be a mere mixture of quartz and nontronite as claimed by Smith and Buseck.

However, work done by other authors seem to indicate that carbyne is indeed fact rather than fiction. Carbyne obtained by heating very carefully purified pyrolytic graphite with an impurity content of only 6 ppm [35] as well as shock compression of > 99% pure amorphous acetylene black [36] and pure diamond powder [37] yielded little indication of silicate contamination but resulted instead in all-carbon material consistent with the carbyne structure. A thorough study of hundreds of grains in shock-compressed graphite/copper powders by TEM [38] revealed dozens of hexagonal platelets with electron diffraction pattern closely resembling those reported in the literature as being indicative of carbyne polytypes. Many of those particles were also investigated with a scanning transmission electron microscope (STEM) to record their electron diffraction pattern as well as their energy-dispersive X-ray (EDX) fingerprints [9]. The only non-carbon element found was copper which can be ascribed to the mounting grid material. Hence, following Smith and Buseck's advise to always check the elemental composition of the particles before making an assignment based on electron diffraction results it could be shown that in this experiment all-carbon particles with a lattice structure of carbyne were present in shock compressed graphite samples.

There is another argument to refute the contamination hypothesis. Gilkes *et al.* [39] acquired HRTEM images, SEAD patterns, and EDX and EELS spectra of very small particles of the acid residue prepared from the Murchison carbonaceous chondrite. The particles were typically only 10 nm in size thus making it, in the authors' opinion, unlikely to be airborne contaminants.

An interesting paradox in the story of carbyne criticisms was recently pointed out by Kavan and Kastner [40]. The arguments used against the existence of carbyne by Smith and Buseck [28] also included a statement by Jansta and Dousek [15] questioning the thermodynamic stability against collapse of the proposed linear chain carbon structure. As it turned out later, Jansta and Dousek described in their paper a preparative method to obtain 'electrochemical carbon' in which neighbouring carbon chains are sterically stabilized by interspersed LiF crystals of 3.3 nm edge length [40].

The most recent attempt to reinterprete the carbyne structure yielded the interesting notion that carbyne polytypes may be crystalline C-(H-O-N) carbons [41]. The author objected to the fact of 'carbyne proliferation'. Therefore an internally consistent reinterpretation of carbyne diffraction data available in the literature was proposed as a way out of this predicament. His approach was based on an earlier observation by Whittaker [42] showing that carbyne occurs interstratified with graphite in natural samples on a worldwide scale. This alleged similarity of graphite and carbyne tempted the author to assume a graphite unit cell[1] for each carbyne form (carbon-VI, carbon-IX, α-carbyne and β-carbyne) and hence to deviate from the hitherto used convention to assign an (hkl) of (11.0) to the maximum d-value obtained. Instead, assuming the maximum values are the interplanar spacings of the (20.0) plane a different lattice metric is obtained. Unfortunately, this leads for carbon-IX to even more complexity in that this phase is now considered a mixture of three different 'carbynes' with $d_{200} = 472$, 441 and 419 pm, respectively. In conclusion, the author states that in his reinterpretation of interplanar spacings of different carbyne forms taken from the literature carbyne is not a carbon allotrope but impure, randomly interstratified, mixed layered crystalline C-(H-O-N) carbon with bulk C/(C+H+O+N) ratios ranging from 0.45 and 0.60. This ratio was determined from an empirical relationship between lattice metric and impurity content of pregraphitic carbon, based on the position of the d_{002} interplanar spacing found by the author previously [43]. It is difficult to follow Rietmeijer's arguments and to accept his somewhat numerological approach. Nevertheless, it sheds some new light on the carbyne riddle and thus deserves more detailed investigations.

These few critical papers quoted are more than balanced by over 200 papers confirming the existence of carbyne by many analytical methods (for example [4,17]). Of course, the number of papers on a subject is not proof of its real existence, and thus Smith and Buseck hastened to point out in a 1985 rebuttal [44] of a rebuttal of their 1982 paper [28] by Whittaker [45] that 'the study of something by many investigators, Russian or otherwise, provides no more compelling evidence .... than the hundreds of papers on polywater proved its existence'. Or on cold fusion, one may add. We consider the question of the existence of a thermodynamically stable carbyne structure in the solid state still open with mounting evidence for the presence of long-chained all-carbon material sterically stabilized by either intrachain hetero-atoms at kink sites or end-capping bulky 'spacers' [13]. Whether such structures deserve to be called carbon 'allotropes' is still a matter of conjecture since 'the temptation to fit all the multitude of non-diamond polymeric forms of carbon into the Procrustean bed of graphite is still strong' [6].

---

[1] Table 1 of Rietmeijer's paper shows a_0, b_0 and c_0 parameters suggesting an orthorhombic unit cell. Rietmeijer (personal written communication) apologizes for the fact that a wrong table was published in his original paper [41].

In this context is appears fitting to quote the final sentences of an older paper by this author that still have not lost their actuality: „ The present paper heavily relies on the reasonable, but by direct crystallographic method yet unproven structural element of an sp-hybridized carbon chain, which in its pure state is thermodynamically unstable. Whether impurities are vital to stabilize the carbyne structure at all, is only a matter of speculation at present. It is to be hoped that experimental methods will be developed soon to obtain single crystals of carbyne of suitable size and purity to permit structural determination“ [9]. It appears that carbyne stubbornly resists forming those larger crystals. A reason for this may be sought in its short range order in <00.1> direction, parallel to the carbon chains. This conjecture is consistent with the fact that so far only thin films, produced by either chemical dehydrohalogenation of halogenated polymers (see, for example [11) or condensation of carbon vapour obtained from laser ablated or sputtered [46] graphite surfaces, led to a well-ordered (micro)crystalline carbyne structure.

References

1. Sladkov, A.M., Kasatochkin, V.I., Korshak, V.V. and Kudryavtsev, Yu.P. (1971) Inventor's Certificate, No.107 (7 December 1971), (Priority date 4 November 1960).
   Sladkov, A.M. and Kudryavtsev, Yu.P. (1969) Diamond, graphite, carbyne - allotropic forms of carbon, *Priroda* **5** ,37-44 (in Russian);
   Sladkov, A.M. (1981) Carbyne-a new allotropic form of carbon, *Sov.Sci.Rev.B* **3** ,75-110.
2. Fitzer, E., Köchling, K.-H., Boehm, H.P. and Marsh, H. (1995) Recommended terminology for the description of carbon as a solid, *Pure Appl.Chem.* **67**, 473-506.
3. Session titled 'The Third Form of Carbon: Fullerenes - Properties and Prospects, in ASM/TSM Materials Week '92, Chicago, IL, Nov 2-5, 1992, ASM International, Materials Park, OH, 1992.
4. Kudryavtsev, Yu.P., Evsyukov, S.E., Guseva, M.B., Babaev, V.G. and Khvostov, V.V. (1993) Carbyne-the third allotropic form of carbon, *Russ.Chem.Bull.* **42**, 399-413.
   Heimann, R.B. (1994) Linear finite carbon chains (carbyne): their role during dynamic transformation of graphite to diamond, and their geometric and electronic structure, *Diamond Rel.Mater.* **3**, 1151-1157.
   Heimann, R.B., Evsyukov, S.E. and Koga, Y., (1997) Carbon allotropes: a suggested classification scheme based on valence orbital hybridization, *Carbon* **35** (10/11), 1654-1658.
5. Heimann, R.B., Kleiman, J. and Salansky, N.M. (1983) A unified structural approach to linear carbon polytypes, *Nature* **306**, 164-167.
6. Mel'nichenko, V.M., Sladkov, A.M. and Nikulin,Yu.N. (1982) Structure of polymeric carbon, *Russ. Chem. Rev.* **51 (5)**, 421-438.
7. Aoki, H., Yamawaki, Y., Kakudate,Y., Yoshida, M., Usuba, S., Yokoi, H., Fujiwara, S., Bae, Y., Malhotra, R. and Lorents, D. (1991) High-pressure FT-IR study of solid C_{60}, *J.Phys.Chem.* **95** (1991) 9037-9039.
8. Whittaker, A.G. (1978) Carbon: a new view of its high-temperature behavior, *Science* **200**, 763-764..
9. Heimann, R.B., Kleiman, J. and Salansky, N.M. (1984) Structural aspects and conformation of linear carbon polytypes, *Carbon* **22(2)**, 147-156.
10. Zdanov, G.S., Ilyushin, A.S. and Nikitina, S.V. (1980) Symposium "Difraktionnyi i Rezonansnyi Strukturnyi Analiz" (Diffraction and Resonance Methods of Structural Analysis), (Nauka, Moscow, 1980) p.45. Quoted after [6].
11. Kudryavtsev, Yu.P., Evsyukov, S.E., Babaev, V.G., Guseva, M.B., Khvostov, V.V. and Krechko, L.M. (1992) Oriented carbyne layers, *Carbon* **30**, 213-221.
    Udod, I.A., Shchurik, V.I., Bulychev, B.M., Sirotinkin, S.P., Guseva, M.B., Babaev, V.G., Kudryavtsev, Yu.P. and Evsyukov, S.E. (1993) Formation of carbyne in the interaction of polyacetylene with potassium under high quasi-hydrostatic pressure, *J.Mater.Chem.* **3**(4), 413-416.

12. Bartik, T., Bartik, B.,Brady, M., Dembinski, R. and Gladysz, J.A. (1996) Durch schrittweises Kettenwachstum zu eindimensionalen Kohlenstoffallotropen mit terminalen Metallkomplexfragmenten; Synthese von C12-, C16- und C20-μ-Polyindiylkomplexen, *Angew.Chem.* **108** (4), 467-469.
13. Lagow, R.J., Kampa, J.J., Wei, H-C., Battle, S.L., Genge, J.W., Laude, D.A., Harper, C.J., Bau, R., Stevens, R.C., Haw, J.F. and Munson, E. (1995) Synthesis of linear acetylenic carbon: the sp carbon allotrope, *Science* **267**, 362-367.
    Eastmond, R., Johnston, T.R. and Walton, D.R.M. (1972) Silylation as a protective method for terminal alkynes in oxidative couplings, *Tetrahedron* **28**, 4601-4616.
    Rohlfing, E.A., Cox, D.M. and Kaldor, A. (1984) Production and characterization of supersonic carbon cluster beams, *J.Chem.Phys.* **81**, 3322-3330.
14. van Thiel, M. and Ree, F.H. (1987) Properties of carbon clusters in TNT detonation products: graphite-diamond transition, *J.Appl.Phys.* **62**, 1761-1767.
15. Jansta, J. and Dousek, F.P. (1980) Some aspects of existence of elementary carbon with sp-hybridized bonds, *Carbon* **18**, 433-437.
16. Wilkie, C.A. and Haworth, D.T. (1978) The carbon -13 NMR solid state spectroscopy of various classes of coal, *J.Inorg.Nucl.Chem.* **40**, 1989-1991.
17. Kudryavtsev, Yu.P., Heimann, R.B. and Evsyukov, S.E. (1996) Carbynes: advances in the field of linear carbon-chain compounds, *J.Mater.Sci.* **31**, 5557-5571.
18. Korshak, V.V., Kudryavtsev, Yu.P., Khvostov, V.V., Guseva, M.B., Babaev, V.G. and Rylova, O.Yu. (1987) Electronic structure of carbynes studied by Auger and electron energy loss spectroscopy, *Carbon*, **25**(6), 735-738.
19. Dresselhaus, M.S. and Dresselhaus, G. (1981) Intercalation compounds of graphite, *Adv. Phys.* **30**, 139-146.
20. El Goresy, A. and Donnay, G. (1968) A new allotropic form of carbon from the Ries crater, *Science* **161**, 363-364.
21. Whittaker, A.G. and Kintner, P.L. (1969) Carbon: observations on the new allotropic form, *Science* **165**, 589-591.
22. Nelson, L.S., Whittaker, A.G. and Tooper, B. (1972) The formation of new polymorphs of carbon and fluid flow patterns by irradiating solid carbon with a CO_2 laser, *High Temp.Sci.* **4**, 445-477.
23. Whittaker, A.G. and Wolten, G.M. (1972) Carbon: a suggested new hexagonal crystal form, *Science* **178**, 54-56.
24. Tanuma, S. and Pal'nichenko, A. (1995) Synthesis of low density carbon crystal (carbolite) by quenching of carbon gas, *J.Mater.Res.* **10**(5), 1120-1125.
25. Bochvar, D.A., Nikerov, M.V, and Stankevich, I.V. (1987) Problems in the structural chemistry of crystalline carbon, *Sov.Sci.Rev. B. Chem.* **9**, 483-527.
26. Nagornyi, V.G., Nabatnikov, A.P., Frolov, V.I., Deyev, A.N. and .Sosedov, V.P. (1975) Tezisy Vsesoyuznoi konferentsii po elektronnoi mikroskopii (Abstr.All-Union Conf.Electron Microscopy, Tbilisi, 1973, *Zhurn. Fiz. Khim.* **49**, 840 (in Russian).
27. Kasatochkin, V.I., Mel'nichenko, V.M. and Elisen, V.M. (1975) Electron diffraction by single crystals of carbyne, *Vysokomolek.Soedin.* **17**(9), 1883-1888 (in Russian).
28. Smith, P.P.K. and Buseck, P.R. (1982) Carbyne forms of carbon: do they exist?, *Science* **216**, 984-986.
29. Smith, P.P.K. and Buseck, P.R. (1981) Carbon in the Allende meteorite: evidence for poorly graphitized carbon rather than carbyne, *Proc.Lunar Planet.Sci.* **128**, 1167-1175.
30. Setaka, N. and Sekikawa, Y. (1980) Chaoite: a new allotropic form of carbon produced by shock compression, *J.Am.Ceram.Soc.* **63**, 238-239.
31. Herr, W., Englert, P., Herpers, U., Watts, E.J. and Whittaker, A.G.(1980) A contribution to the riddle about the origin of certain glassy spherules, *Meteoritics* **15**, 300-305.
32. Smith, P.P.K. and Buseck, P.R. (1980) High resolution transmission electron microscopy of an Allende acid residue, *Meteoritics* **15**, 368-369.
    Smith, P.P.K. and Buseck, P.R., (1981) Graphitic carbon in the Allende meteorite: a microstructural study, *Science* **212**, 322-324.
33. Lumpkin, G.R. (1981) Electron microscopy of carbonaceous matter in Allende acid residues, *Proc.Lunar Planet.Sci.* **12B**, 1153-1166.
34. El Goresy, A., written communication 1982 (quoted in [9]).

35. Whittaker, A.G., Neudorffer, M.E. and Watts, E.J. (1983) Carbon: a rhombohedral carbyne form, *Carbon* **21**(6), 597-599.
36. Yamada, K., Kunishige, H. and Sawaoka, A.B., (1991) Formation process of carbyne produced by shock compression, *Naturwiss.* **78**, 450-452.
37. Yamada, K., Burkhard, G., Dan, K., Tanabe, Y. and Sawaoka, A.B. (1994) Microstructures of carbon polymorphs formed in shock compressed diamond powder utilizing an interaction of oblique shock waves, *Carbon* **32** (7), 1197-1213.
38. Kleiman, J., Heimann, R.B., Hawken, D. and Salansky, N.M. (1984) Shock compression and flash heating of graphite/metal mixtures at temperatures up to 3200 K and pressures up to 25 GPa, *J.Appl.Phys.* **56** (5), 1440-1454.
39. Gilkes, K.W.R., Gaskell, P.H., Russell, S.S., Arden, J.W. and Pillinger, C.T. (1992) Do carbynes exist as interstellar material after all?, *Meteoritics* **27**, 224.
40. Kavan, L. and Kastner, J. (1994) Carbyne forms of carbon: continuation of the story, *Carbon* **32**(8), 1533-1536.
41. Rietmeijer, F.J.M. (1993) Are crystalline C-(H-O-N) carbons the elusive meteoritic carbyne?, *Meteoritics* **28**, 242-245.
42. Whittaker, A.G. (1979) Carbon: occurence of carbyne forms of carbon in natural graphite, *Carbon* **17**, 21-24.
43. Rietmeijer, F.J.M. (1991) Mixed layering in disordered Sri Lanka graphite, *Carbon* **29**, 669-675.
44. Smith, P.P.K. and Buseck, P.R. (1985) Carbyne forms of carbon: evidence for their existence (reply), *Science* **229**, 486-487.
45. Whittaker, A.G. (1985) Carbyne forms of carbon: evidence for their existence, *Science* **229**, 485-486.
46. Onuma, Y., Kawai, Y., Nakao, M. and Kamimura, K. (1996) Highly oriented crystalline carbon thin films obtained by dc magnetron sputtering with a hot filament, *Jpn.J.Appl.Phys.* **35**, L1354-L1356.

Chapter 2: Carbyne and carbynoid structures in nature

2.1. CARBON - HOW MANY ALLOTROPES ASSOCIATED WITH METEORITES AND IMPACT PHENOMENA?

K.W.R. GILKES
School of Mathematics and Physics, University of East Anglia
NORWICH, NR4 7TJ, UK

C.T. PILLINGER
Planetary Sciences Research Institute, The Open University
Walton Hall, MILTON KEYNES, MK7 6AA, UK

Abstract. Virtually all the allotropic forms of carbon known naturally have been at some time or another recognised within meteorites or as a part of phenomena associated with meteorite impacts. Thus, three types of graphite the well known cubic and rhombohedral, plus a unique cubic form (cliftonite) have been encountered. There are several reported occurrences of buckminster-fullerene and its siblings. Diamonds are amongst the most interesting form of carbon in meteorites, their incidence having been noted for as long as the mineral was known to be found on Earth in kimberlites. Both the cubic and hexagonal (Lonsdaleite) forms are seen, the latter almost always denoting the advent of shock conditions. A particularly unusual type of crystal, which has been found to be ubiquitous in primitive unmetamorphosed chondrites, consists of nanometre sized grains. Originally the acid resistant residues which have transpired to be diamond were first thought to be the sp carbon allotrope generically known as carbyne. This identification proved to be highly controversial, but chaoite, one of the initial carbyne candidates, was definitely located in the impact breccia suevite found at the Ries Crater, S. Germany. Tiny crystals of chaoite seem to be elusive grains coexisting with meteorite nanodiamonds where they give EELS data which demonstrate they are different from other types of carbon. Molecules with multiple triple bonds as the main structural unit are common in the interstellar medium.

1. Introduction

Behind the seemingly innocent question posed by the title of this chapter, which could be asked of any school pupil, lurks a minefield for the unsuspecting. The majority of chemists, when introduced to the subject of allotropy in their teens, were probably taught using the example of carbon.

It therefore seems surprising that Ferguson [1] could ask in a letter to *Chemistry in Britain* in 1993 "Is there an undiscovered allotrope of carbon based on sp

R.B. Heimann et al. (eds.), Carbyne and Carbynoid Structures, 17–30.

hybridisation?" It is perhaps even more astonishing that the answer the Editor received [2] was a resounding maybe; it was pointed out that if there is (are) sp allotrope(s), then much credit for recognition in Nature should go to El Goresy and Donnay [3] who in 1968, whilst studying ejecta from the Ries meteorite crater, found a white carbon phase which they characterised and christened chaoite. Since both the discovery and the controversy over whether another form of carbon had been identified partly arose through meteoritics, then it seems appropriate to review the state-of-the-art, particularly with reference to the current activities involving meteorite research, and to catalog just what forms of elemental carbon have been found associated with meteorite phenomena.

2. What is an allotrope?

If Aristotle, who coined the term allotropy, had known about molecular orbital theory, had access to an electron microscope and studied crystallography, he might have given us a more precise definition, thus making the question in the title somewhat easier to respond to. Consultation of two text books on Inorganic Chemistry widely in use today does not help. Cotton and Wilkinson [4] suggest that diamond and graphite, because of their difference in physical and chemical properties, are two allotropes of carbon, with chaoite (carbyne) named as carbon-III. Greenwood and Earnshaw [5] employ a crystallographic definition which makes chaoite the fifth allotrope. Both textbooks are pre-1990, so neither mentions buckminsterfullerene.

The C_{60} community, e.g. Kroto and Walton [6], ignore chaoite and claim the third allotrope is theirs, even though some of the people involved had a background of polyyne research. The problem of what constitutes an allotrope is really beyond our remit, so for the purposes of the present discussion we are going to consider graphite, diamond, fullerenes and chaoite in that order. Chaoite in nature has never been seen isolated from other forms of carbon. We should say that some other plausible predictions concerning elemental carbon (metallic carbon, graphyne honeycomb graphite, glass-like carbon, layer chain carbons) have been made in the literature but none of these have been mentioned in connection with meteorites.

3. Graphite

What is loosely described as graphite exists in two crystalline forms: the most stable and abundant α or hexagonal graphite and β or rhomohedral graphite; the latter makes up 10% of what is found naturally. These differ according to the way planar repeating layers of sp^2 hybridised "chicken wire" are stacked, with replication every second layer for the former or every third layer for the latter. There is no known situation in which layers lie vertically above each other on a one for one basis. Graphites with a random mixing of the layers, a situation akin to a solid solution of one form in another, are common in meteorites [7]. Layer-chain carbons or intercalation of carbon in carbon (C atoms between graphite chains) are postulated to explain extended d-spacings sometimes seen; such a situation would undoubtedly partially pucker the ring system towards diamond. A completely different kind of graphite only seen in meteorites is called cliftonite [8,9]. As long ago as 1846, crystals with a cubic morphology, some possibly hollow, were found in iron meteorites; these must be pseudomorphs, *ie.* they crystallised as something else (diamond, iron carbide?) which subsequently changed to graphite through an unknown event.

We should point out that many amorphous carbons (carbon black, lamp black *etc.*) which used to be considered as separate forms of carbon, thanks to electron microscopy, are now recognised as graphite with a very low degree of crystallinity.

4. Diamond

Diamonds were found in meteorites by Ierofeiff and Latchinoff [10] way back in 1888 almost the same time as they were traced to kimberlite pipes (see Pillinger [11] for a review). When people talk about diamond (shown to be a form of carbon in 1797 by Tennant), it is usually sloppy terminology for what should correctly be referred to as cubic diamond, the more common form of the mineral. A by far lesser abundant form, hexagonal diamond or lonsdaleite, (named for Dame Kathleen Lonsdale), also occurs [12], but never in isolation. Again cubic and hexagonal diamond differ according to the stacking patterns, ABAB *vs.* ABCABC, respectively of their sp^3 network of six carbon "chairs".

Except for experiments involving high static or shock pressures (e.g. explosions) [13], lonsdaleite is restricted to natural situations where meteorites are implicated: It is found in microcrystalline aggregates, sometimes called carbonados, in iron or ureilite meteorites, possibly as a result of impacts on Earth or between planetessimals in space (See *inter alia* Vdovykin [14,15]; Clark *et al* [16]; Smith and Dawson [17]. The impact scenario is extended *via* a specific form of carbonado called yakutite (which may also include chaoite) which is widespread as a placer in Siberia where it can be traced to ejecta from the 100 kilometre plus diameter Popigai crater [18].

Enormous carbonados, with distinct melt rinds, have been found in chunks up to 500g in size in Brazil and the Central African Republic since early in the 18th century; they could have derived from a cataclysmic impact 2by ago, before Africa and South America separated by plate tectonics [19]. However calculations by De Carli [20] suggest there is insufficient time for such large aggregates of diamond to grow. Diamond does not have a systematically mixed crystal but it can exist as amorphous, mainly sp^3 bonded, material called diamond-like carbon made by vapour deposition in the laboratory by a huge variety of processes.

Unlike other forms of carbon, it is almost impossible to get diamond free of nitrogen existing as substitutional impurities; nitrogen contamination, upto several thousand atomic ppm has an enormous effect on the properties of diamond, particularly spectroscopic considerations. Carbonados have a relatively low nitrogen content and a different carbon isotopic composition which suggests that they were made on Earth from the transformation of crustal carbon. The impact origin is neither proved nor disproved and it might be that the conversion of organic matter to diamond occurs as a result of deep subduction [22].

The most abundant form of diamond found in meteorites is that which occurs in primitive (unmetamorphosed) chondrites; ordinary, carbonaceous, enstatite and anomalous chondrites all being carriers of up to 0.1wt% [23] (a good kimberlite might have 0.1ppm!). This material is exceedingly fine grained, average size 2-3nm; it has a distinctive carbon isotopic composition and is contaminated by very high abundances of nitrogen and noble gases all of which have non-terrestrial isotopic ratios. As a consequence this form of diamond is believed to constitute interstellar grains older than the solar system and probably synthesised in type II supernovae which exploded before our star was born [25]. The search for interstellar material has led to a variety of other diamond occurrences in meteorites being recognised [11].

Nanometric sized diamonds have also been isolated from the boundary layers marking the end of the Cretaceous geological epoch and the beginning of the Tertiary. These were at first wrongly identified [25] as residual from the meteorite which impacted at Chicxulub to produce a 200km diameter crater. The diamonds were probably synthesized in a chemical vapour deposition type mechanism using carbon from the target rocks [26].

5. Fullerenes

Rarely in science does a subject expand with such rapidity as the one which has developed following the large scale isolation of C_{60} and C_{70} [6]. There have been a trickle of very tentative natural occurrences, including in a meteorite and the Sudbury impact crater (see Becker *et al* [27] and references cited therein). In addition to C_{60} and C_{70}, there are potentially many higher carbon number closures based entirely on sp^2 hybridized bonding. It seems unrealistic to refer to each as a separate allotrope; the term polymorph as applied to compounds eg. SiC might be more appropriate. The ultimate fullerene, concentric onion-like structures [28] have been seen in the acid residues of primitive meteorites by the present authors (Fig. 1).

6. Chaoite

As already said, in 1968, El Goresy and Donnay [3] whilst looking for diamond in graphite-rich gneisses from the Nördlinger Ries impact crater, a 20 km diameter feature created 14.7 Myrs ago in Southern Bavaria, came across a strongly reflecting white material which they analysed with an electron probe and showed to be carbon intergrown with hexagonal graphite. After obtaining an X-ray patterns, quite distinct from any other types of carbon in respect of larger d spacings, they calculated dimensions for a primitive hexagonal cell containing 168 atoms. Their proposed name, chaoite (in honour of E.C.T. Chao, a geologist responsible for much of the mapping of the Ries), was accepted by the Mineralogical Society in 1969.

The work by El Goresy and Donnay seems to have been the stimulus for several other groups to announce their results. Within a couple of years, a group at the Aerospace Corporation, El Segundo, California, [29,30] published a series of papers to suggest the existence of chaoite related materials named as carbon-VI, VII, XI *etc.* formed in experiments involving attempts to vapourise graphite at 2550K and low pressure (cf. chapter 3.4.1.1. of this book). Chaoite was also reported by Vdovykin [14,15] intergrown with diamond/lonsdaleite in two ureilite meteorites Novo Urei and Havero.

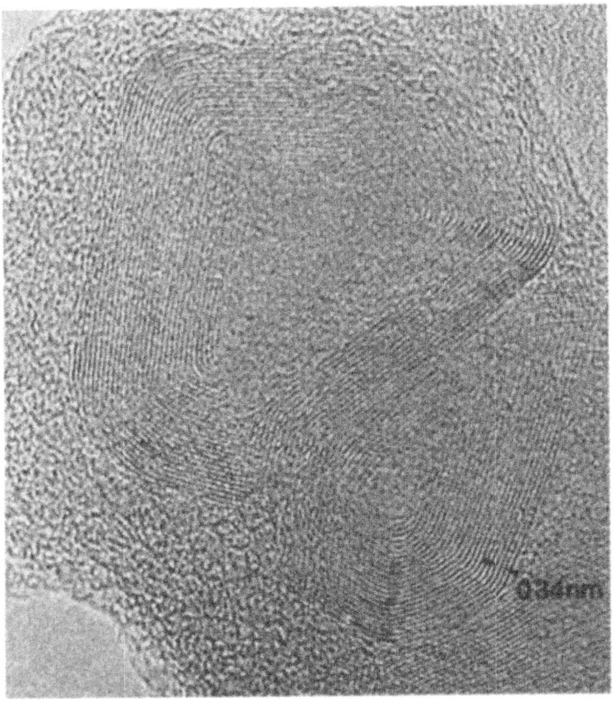

Figure 1. An onion shell structure seen by TEM in a Murchison meteorite acid resistant residue.

The latter work carried out in Russia, cited El Goresy but made no reference to the activities of the A.N. Nesmeyanov Institute of Organoelement Compounds, Russian Academy of Sciences, who had been working since 1961 attempting to chemically polymerise acetylene (see chapter 3.1.1 of this book). In 1968, they proposed the name "carbyne" and made the suggestion that it might be a sp-hybridised allotrope of carbon. In 1971, a silvery white material was isolated and the link through electron diffraction patterns made to the separate efforts of El Goresy, Whittaker and Vdovykin.

During the next few years the number of proposed carbynes (the name had now attained a generic significance)[1] proliferated to nearly ten without any obvious dissent that a new family of carbon allotropes was creeping into the scientific literature. It was pointed out that carbynes - including chaoite - could be very hard (harder than BN), resistant to chemical attack and a new version of the carbon phase diagram at high

[1] (Note by the Editor): The generic name "carbyne" is, however, not without principal terminological objections, see Chapter 1.2 for further discussion.

temperatures was proposed; it was even suggested carbynes could be found in association with natural graphites [31]. Alarm bells only started to ring when the possibility of a truly extraterrestrial significance was raised by the proposition that a family of carbynes might be the carriers of various noble gases having anomalous isotopic composition in carbonaceous chondrites [32]. The idea was seized upon to forge a link between presolar grains, primitive meteorites and observations of the interstellar medium [33].

However a damning series of reports, to pour cold water on the whole concept of carbynes, appeared over the period 1980-1985 from Smith and Buseck [7,34,35,36,37]. These authors disputed the interpretation of the crystallographic data after considering four situations where carbynes had been reported. In their opinion, and they included samples provided by El Goresy and Whittaker in their study, that hexagonal X-ray and electron diffraction patterns, characteristic of carbynes, may be obtained from graphite together with flakes of sheet silicates (nontronite and talc) or specks of quartz, all common laboratory contaminants. Another more recent view was that carbynes might be randomly stratified (or mixed-layered) carbons contaminated with variable amounts of C+H+O+N rather than true allotropes [38].

7. Interstellar grains in meteorites and chaoite

The idea that carbynes were prolific in primitive meteorites was promptly abandoned, quite rightly so as it turns out, because care isolation studies [39] led to the realisation [40] that nanometre sized diamond crystals, rich in nitrogen [41], were implicated in the noble gas puzzle. Such grains were concentrated from carbonaceous chondrites by dissolution of non-carbonaceous phases in HF/HCl, followed by destruction of organics and graphite in chromic and perchloric acids. Finally, by a quirk of fate, it was found that the tiny (2000 atom) diamond crystals could be separated from surviving oxide minerals and silicon carbide because they were acidic (surficial CO_2^- groups were invoked) and thus extractable in ammonia solution; in fact the grains carry so much oxygen they are only 70 wt% carbon [42].

Such diamond residues are ideal for study by various techniques associated with high resolution transmission electron microscopy (HRTEM). We [43] studied several isolated from the meteorites Allende, Cold Bokkeveld and in particular Murchison, a CM2 chondrite which fell in Australia in 1969. To load material for microscopy, the final diamond-rich product was resuspended as a colloid in 0.5N NH_4OH and pipetted on to a perforated carbon support film. The samples were baked at 160°C for several hours to remove trace ammonia salts. HRTEM was performed at 200 kV using a JEOL 200CX transmission electron microscope and a JEOL 2010 microscope fitted with an energy dispersive X-ray (EDX) analysis system allowing detection of elements $Z > 11$ (Na). Both microscopes have a point-to-point image resolution of less than 0.2 nm. Electron energy loss spectroscopy (EELS) has also been performed on selected specimens using an HB 501 scanning TEM (STEM) with a spectrometer capable of 0.4 eV resolution as defined by the FWHM of the zero-loss peak in the spectrum.

Axial bright field (BF) HRTEM images of the Murchison sample MIL were taken by selecting suitably thin regions of the sample which protruded over the edges of holes in the carbon support. A typical BF image is shown in Fig. 2 consisting of groups of lattice fringes with 0.2 nm spacings corresponding to the {111} planes of diamond. These diamond crystallites, average size 3-4 nm, are separated by structureless regions which might be crystallites not properly oriented for lattice imaging, or a disordered interfacial material, or a combination of the two. A selected area electron diffraction pattern (SAED) of regions like those shown is reproduced. Up to ten diffraction rings have been observed which agree well with the known d-spacings of cubic diamond.

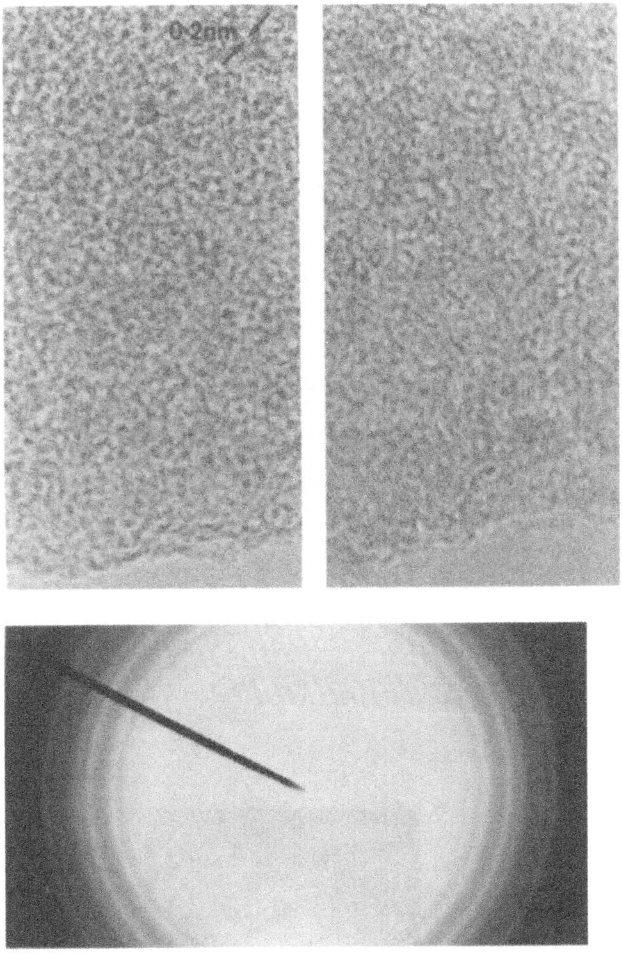

Figure 2. Lattice images of nanometre sized diamond crystals from the Murchison meteorite showing the spacing between successive carbon planes; a electron diffraction pattern taken on one of the ordered areas is also shown.

In addition to the diamonds, we have observed crystalline particles with interlayer planar spacings much larger than 0.34 nm (the lattice spacing for graphite). The size of these grains rarely exceeds 20 nm and they are extremely thin making it difficult to distinguish their detailed morphology. In general, the grains are identified in diffraction mode and HRTEM images taken from the areas selected by the diffraction aperture. The grains invariably suffer some degree of electron beam damage, even with the electron beam reduced to the lowest practical intensity; this poses the greatest difficulty in performing a full characterisation of each grain. However, SAED patterns have been successfully acquired for 26 grains.

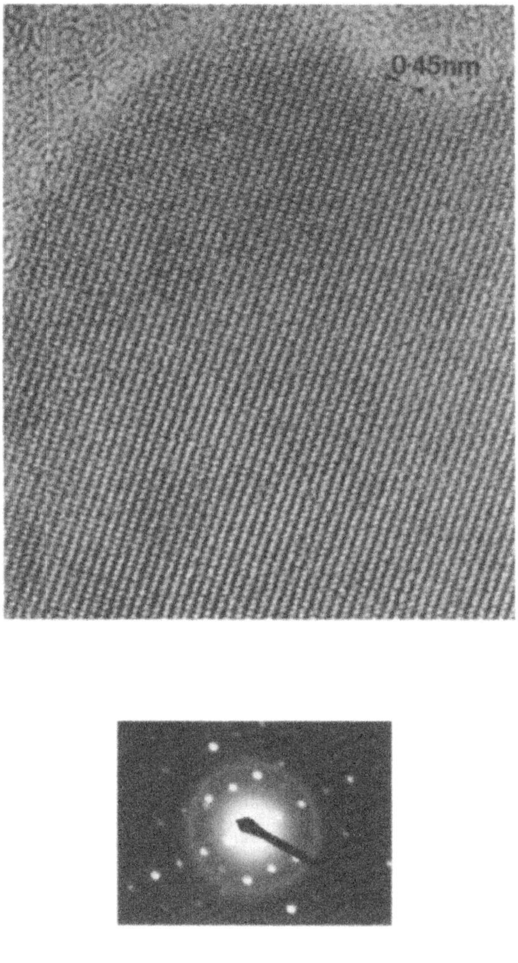

Figure 3. A lattice image of one of the grains from the Murchison meteorite giving a "chaoite" electron diffraction pattern (also shown).

Typical lattice images are shown in Fig. 3. The 0.45 nm lattice spacings in the single crystal of Fig. 3 agree with the (110) interplanar spacings of chaoite, whilst the smaller particle in Fig. 3 exhibits 0.3 nm fringes, matching the (203) chaoite spacings. A full list of d-spacings calculated from SAED (using the nanometre sized diamond as an internal standard) is given in Table 1 with errors approximately ± 5%. It is extremely unlikely that the microscopic grains encountered in these residues could be sheet silicates, as the samples were extensively treated with HF.

TABLE 1. Comparison of d-spacings (in nm) measured for MIL carbynes with published values for chaoite.

hkl	MIL	Chaoite
110	0.450	0.447
203	0.298	0.303
301	0.251	0.255
401	0.189	0.191
227	0.149	0.150
336	0.126	0.126
337, 523	0.120	0.120

Atmospheric contaminants do not generally exist as nanometre-sized fragments, but coagulate and precipitate out as aggregates. However, in order to check the purity of the specimens, a freshly prepared diamond-rich residue from Murchison was kept in solution in a sealed tube until immediately before HRTEM/EDX analysis, and a "blank" carbon support was subjected to the same exposure to the atmosphere as the microscope specimen. Although seven chaoite-like grains were identified in the Murchison sample, repeated examination of the blank support was unsuccessful in locating grains with similar SAED patterns. EDX spectra provide further evidence that the nanometre-sized crystallites are not extraneous silicates. Although a small peak at the energy of the Si Kα and Kβ emission is present in the spectra, this is not due to the grains under examination, as spectra taken from blank regions of the carbon support show the same feature, probably due to secondary electron bombardment of the SiLi detector.

To learn more about the nature of the bonding in chaoite associated with interstellar diamond, attempts have been made to obtain EELS data. These unfortunately were unsuccessful with individual crystals because of their small size and fragility. This, in itself, may be circumstantial evidence that the grains are not one of the more common forms of elemental carbon. EELS results (Fig. 4a), revealing a characteristic carbon K edge, distinct from amorphous carbon, graphite and diamond (Fig. 4b) have been obtained from delocalised and hence unspecified areas on the sample. The absence of any Si edge rules out the possibility of silicate contamination.

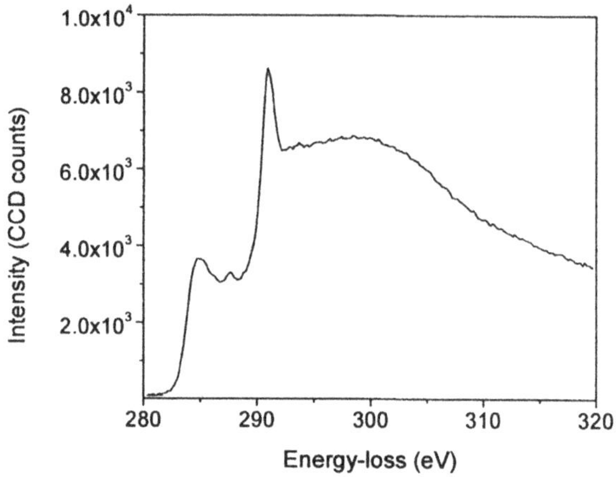

Figure 4a. An electron energy loss spectrum of a region having particles(s) of chaoite electron diffraction.

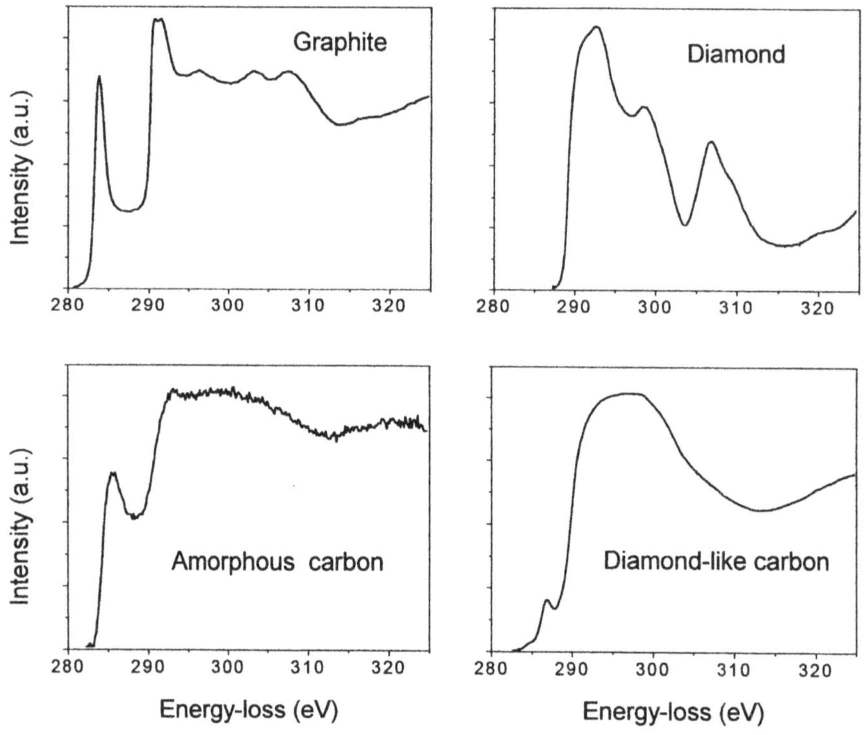

Figure 4b. Electron energy loss spectra from ordinary types of carbon for comparison.

Our results demonstrate that the characteristic hexagonal SAED patterns reported in the literature for chaoite and carbynes may be equated with a carbonaceous component whose EELS spectrum and hence bonding is different from diamond and graphite[2]. We therefore believe that chaoite not only exists, but co-exists, in primitive meteorites with nanometre sized diamond grains accepted to be of interstellar origin on the basis of isotopic measurements of noble gas and nitrogen. Our claims are intermediate between those of previous workers who have advocated all or nothing. We thought of trying to isolate chaoite from Ries crater ejecta for a similar study using our dissolution techniques; in a rather surprising reversal of fate, we found cubic and hexagonal diamond and SiC but no chaoite [44].

Structures having alternating single and triple bonds or cumulene bonding would make excellent candidate material for stellar condensates [45]. Linear hydrocarbons of the form:

C_nH (where n = 2 to 6)

and the cumulene isomers propadienylidene and diacetylene:

(H_2CCC)

are ubiquitous in space, for example in the region of the star CW Leonis (IRC + 10216).

CW Leonis is a type of star called an asymptotic branch Red Giant, the nearest of this sort to the solar system and, in the infrared, the brightest star in the sky after the sun. Optically, it is obscured by a carbon-rich dust envelope a light year across, shed as a part of the AGB evolutionary process. Given the structure of some of the molecules mentioned above, the dust cloud could include carbynes of immense interest for us.

A group of molecules associated with CW Leonis, and presumably AGB stars in general, are the cyanoacetylenes:

HC_nN where n = 3,5,7,9 and 11.

Not only do these have the triple bond structure required for carbynes but they have the cyano group as a means of terminating acetylenic chains (see below) thereby affording stability to the otherwise extremely reactive structure. It is assumed that the major nitrogen in nanometre sized diamonds in primitive meteorites is substitutional. The isotopic systematics show that several nitrogen containing components exist [46] that coexisting carbynes terminated by cyano groups are a more plausible ultimate carrier. It should be noted that Hayatsu *et al.* [47] recognised fragments with a $-(C{\equiv}C)_n-CN$ structure during pyrolysis of an HF/HCl resistant residue of the Allende meteorite.

Acetylenic chains of about 10 carbon atoms with CN terminators could easily be the cause accounting for only a few percent of the nitrogen carriers. A carrier within a carrier would greatly simplify problems concerning the apparent variability of diamonds from primitive meteorites and the fact that nitrogen and a distinct Xe component can be decoupled by heating experiments [42]. When diamonds were isolated [40] as the end product of extensive meteorite processing in the quest for

[2] For further discussion of EELS spectra of carbyne-like materials see Chapters 6.3 and 6.4 of this book.

interstellar grains, it was tacitly assumed that this mineral marked the end of the trail. Nanometre diamonds are undoubtedly more complex than at first thought [47] and still may contain minor mineral phases including the elusive and enigmatic sp bonded allotropic form of carbon.

Others have looked at meteorite acid insoluble residues using HRTEM; a study [48] (which includes Buseck, one of the original doubters concerning chaoite [7,34,35,36,37] (amongst its authors) has recognised the existence of lonsdaleite, even though this is supposed to be a shock generated species and interstellar diamonds are usually accepted as produced by low pressure processes. A private communication from Daulton [49], the first author of the above study, states that he also saw features which he believed were similar to chaoite in a Murchison acid residue and that at some time he hoped to carryout comparative studies on authentic chaoite from the Ries Crater.

8. Postscript

Virtually everyone involved in the carbyne story has at one stage or another stated that if only more material or better crystals were available, (for a full characterisation by a panoply of methods, other than just the diffraction techniques associated with electron microscopes) progress and general acceptance might follow. Without mass production, C_{60} would almost certainly have remained as just a nice idea. It would be somewhat ironic if as soon as the techniques for mass production of carbynes were recognised they demonstrated that chaoite was truly the third, or was that the fifth, allotrope of carbon.

9. REFERENCES

1. Ferguson, I.F. (1993) Undiscovered carbon allotrope, *Chemistry in Britain* **28**, 297.
2. Urben, P., Bruce, M.I., and McEnamey, B. (1993) Carbon allotropes, *Chemistry in Britain* **28**, 574.
3. Goresy, A. El. and Donnay, G.T. (1968) A new form of carbon from the Ries Crater, *Science* **161**, 363-364.
4. Cotton, F.A. and Wilkinson, G. (1988) *Advanced Inorganic Chemistry*, John Wiley and Sons, New York, fifth edition.
5. Greenwood, N.N. and Earnshaw, A. (1989) *Chemistry of the Elements*, Pergamon Oxford.
6. Kroto, H.W. and Walton, D.R.M. (1993) Polyynes and the formation of fullerene, *Phil. Trans. R. Soc.* Lond A **343**, 103-112.
7. Smith, P.P.K and Buseck, P.R. (1981) Carbon in the Allende meteorite: Evidence for poorly graphitized carbon rather than carbyne, *Proc Lunar Planet Science* **12B**, 1167-1175.
8. Grenville-Wells, H.J. (1952) The graphitization of diamond and the nature of cliftonite, *Mineralogical Magzine* **216**, 803-816.
9. Brett, R. and Higgins, G.T. (1969) Cliftonite: A proposed origin, and its bearing on the origin of diamonds in meteorites, *Geochim. Cosmochim. Acta* **33**, 1473-1484.
10. Ierofeieff, M.V. and Latchinoff, P.A. (1888) Météorite diamantére tombé le 10/22 Septembre 1886 en Russie, à Nowo Urei, Government de Penze. *C.r.Hebd. Seanc. Acad. Sci. Paris* **106**, 1679-1682.
11. Pillinger, C.T. (1993) Elemental carbon as interstellar dust, *Phil. Trans. R.Soc.* Lond A **343**, 73-86.
12. Bundy, F.P, and Kasper, J.S. (1967) Hexagonal diamond - a new form of carbon, *J. Chem. Phys.* **46**, 3437-3446.

13. De Carli, P.S. and Jamieson, J.C. (1961) Formation of diamond by explosive shock, *Science* **133**, 1821-1822.
14. Vdovykin, G.P. (1969) New hexagonal modification of carbon in meteorites, *Geochem Int.* **6**, 915-918.
15. Vdovykin, G.P. (1972) Forms of carbon in the new Haverö ureilite of Finland, *Meteoritics* 7, 547-552.
16. Clarke, R.S., Appleman, D.E., and Ross, D.R. (1981) An antarctic iron meteorite contains preterrestrial impact produced diamond and insdalite, *Nature* **291**, 396-398.
17. Smith, J.V. and Dawson, J.B. (1985) Carbonados: Diamond aggregates from early impacts of crustal rocks, *Geology* **13**, 342-343.
18. Kaminsky, R.V. (1995) Carbonado and Yakutite: Properties and possible genesis, preprint.
19. Shelkov, D., Verchovsky, A.B., Milledge, H.J., and Pillinger, C.T. (1997a) Carbonado origin: A comparison with other forms of microcrystalline diamond based in C, N, He data and inner morphology, Submitted.
20. De Carli, P.S. (1997) Pers. Comm., Research Seminar at the Open University.
21. McKenzie, D.R. (1996) Tetrahedral bonding in amorphous carbon. *Rep. Prog. Phys.* **59**, 1611-1664.
22. Shelkov, D., Verchovsky, A.B., Milledge, H.J., and Pillinger, C.T. (1997b) Carbonado a comparison between the Brazilian and Ubangui sources based on carbon and nitrogen isotopes, *Proc. 6th International Kimberlite Conf.* in press.
23. Russell, S.S., Arden, J.W. and Pillinger, C.T. (1991) Evidence for multiple sources of diamonds, *Science* **254**, 1188-1191.
24. Clayton, D.D., Meyer, B.S., Sanderson, C., Russell, S.S., and Pillinger, C.T. (1995) *Astrophysical J.* **447**, 894-905.
25. Carlisle, D.B. and Braman, D.R. (1991) Nanometre-sized diamonds in the Cretaceous/Tertiary boundary clay of Alberta, *Nature* **352**, 708-709.
26. Gilmour, I., Russell, S.S., and Pillinger, C.T. (1992) Terrestrial carbon and nitrogen isotopic ratios from cretaceous tertiary boundary nano-diamonds, *Science* **258**, 1624-1626.
27. Becker, L., Bada, J.L., Winans, R.E., and Bunch, T.E. (1994) Fullerenes in Allende meteorite, *Nature* **372**, 507.
28. Ugarte, D. (1992) Curling and closure of graphitic networks under electron-beam irradiation, *Nature* **359**, 707-707.
29. Whittaker, G.A., and Kintner, P.L. (1969) Carbon: Observations on the new allotropic form, *Science* **165**, 589-591.
30. Whittaker, G.A., and Wolten, G.M. (1972) Carbon: A suggested new hexagonal crystal form, *Science* **178**, 54-54-56.
31. Whittaker, G.A. (1979) Carbon: Occurrence of carbyne forms of carbon in natural graphite, *Carbon* **17**, 21-24.
32. Whittaker, G.A., Watts, E.J., Lewis, R.S., and Anders, E. (1980) Carbynes: Carriers of primordial noble gases in meteorites, *Science* **209**, 1512-1514.
33. Webster, A. (1980) Carbynes in meteorites: Detection, low-temperature origin, and implications for interstellar molecules, *Mon. Nat. Roy. Astron. Soc.* **192**, 7-9.
34. Smith, P.P.K. and Buseck, P.R. (1980) High resolution transmission electron microscopy of an allende acid residue, *Meteoritics* **15**, 368-369.
35. Smith, P.P.K and Buseck, P.R. (1981) Graphite carbon in the Allende meteorite: a microstructural study, *Science* **212**, 322-324.
36. Smith, P.P.K and Buseck, P.R. (1982) Carbyne forms of carbon: Do they exist?, *Science* **216**, 984-986.
37. Smith, P.P.K and Buseck, P.R. (1985) Carbyne forms of carbon: Evidence of their existence, *Science* **229**, 485-487.
38. Rietmeijer, F.J.M. (1993) Are crystalline C-(H-O-N) carbons the elusive meteoritic carbynes? *Meteoritics* **28**, 242-245.
39. Tang, M., Lewis, R.S., Ander, E , Grady, M.M., Wright, I.P. and Pillinger, C.T. (1988) Isotopic anomalies of Ne, Xe and C in meteorites I. Separation of carriers by density and chemical resistance, *Geochim. Cosmochim. Acta* **52**, 1221-1234.
40. Lewis, R.S., Tang, M., Wacker, J.F., Anders, E., and Steel, E. (1987) Interstellar diamonds in meteorites, *Nature* **326**, 160-162.
41. Lewis, R.S., Anders, E., Wright, I.P., Norris, S.J., and Pillinger, C.T. (1983) Isotopically anomalous nitrogen in primitive meteorites, *Nature* **305**, 767-771.
42. Verchovsky, A.B., Fisenko, A.V., Semjenova, L.F., and Pillinger, C.T. (1997) Heterogeneous distribtution of Xenon-HL within presolar diamonds, *Meteoritics* **32**, A131-A132.
43. Gilkes, K.W.R., Gaskell, P.H., Russell, S.S., Arden, J.W., and Pillinger, C.T. (1992) Do carbynes exist as interstellar material after all?, *Meteoritics* **27**, 224

44. Hough, R., Gilmour, I., Pillinger, C.T., Arden, J.W., Gilkes, K.W.R, Yuan, J., and Milledge, H.J. (1995) Diamond and silicon carbide in suevite from the Nördlinger Ries impact crater, *Nature* **378**, 41-44.
45. Millar, G.T. (1992) Up the starry chimney, *New Scientist* **1845**, 32-36.
46. Hayatsu, R., Scott, R.G., and Studier, M.H. (1980) Carbynes in meteorites: Detection, low-temperature origin, and implications for interstellar molecules, *Science* **209**, 1515-1518.
47. Verchovsky, A.B., Fisenko, A.V., Semjonova, L.F., Wright, I.P. and Pillinger, C.T. Farraday Society Discussions in press.
48. Daulton, T.L., Eisenhour, D.D., Buseck, P.R., Lewis, R.S., and Bernatowicz, T.J. (1994) High-resolution transmission electron microscopy of meteoritic and terrestrial nano-diamonds, *Lunar Planetary Science* **XXV**, 313-314.
49. Daulton, T.L. (1994) Personal letter to the authors.

2.2. OTHER NATURAL CARBYNOID STRUCTURES

L. KAVAN
J. Heyrovský Institute of Physical Chemistry,
Dolejškova 3, CZ-182 23 Prague 8, Czech Republic

R.B. HEIMANN
Chair of Technical Mineralogy, Department of Mineralogy,
Freiberg University of Mining and Technology, D-09596 Freiberg,
Germany

Abstract. Molecules containing several conjugated (-C≡C-) links occur naturally in space and in certain biological species. Natural carbyne crystals have been occasionally reported to occur in interstellar dust and in some terrestrial locations. After the pioneering work on natural carbynes in the late 60s, little progress has been achieved on their further identification.

1. Introduction

Although carbyne was first announced as an "artificial" allotropic carbon phase in 1961, seven years later, it was also found in nature [1]. The most important issues related to the discovery and subsequent studies of natural carbynes refer to phenomena associated with meteorites and impact phenomena (see Chapter 2.1 of this book). The aim of this chapter is to review previous work on natural carbynes, and to demonstrate the existence of various natural structures with conjugated sp-carbon chains mimicking that of carbyne. However, it is beyond the scope of this book to discuss all naturally occurring molecules and solids containing these structural moieties. Hence, we will show below only several typical examples. For a fundamental review on natural carbyne-like species see, e.g., Ref. [2].

2. Natural carbyne-like molecules

Species approaching carbyne are linear all-carbon molecules [3]:

$$\cdot C - (C{\equiv}C)_n - C\cdot \quad \text{or} \quad :C = (C = C)_n = C:$$

(I) (II)

occurring in gaseous state in carbon stars, the Sun, interstellar clouds and comets (the detection of C_3 in comets dates back to 1882) [3,4]. Molecule (II) is a dicarbene and is

R.B. Heimann et al. (eds.), Carbyne and Carbynoid Structures, 31–38.

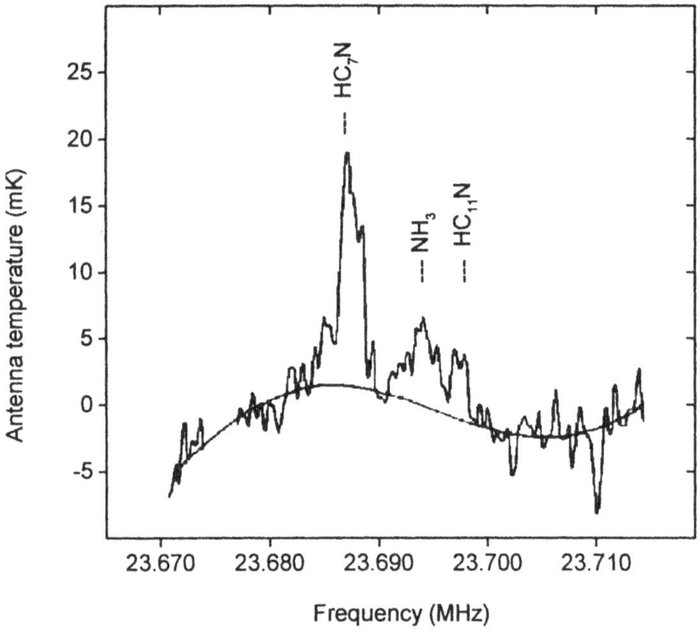

Figure 1. Microwave spectrum of the cool carbon star, IRC+10°216 (= CW Leo) in the region of the microwave inversion of ammonia, NH_3 (1,1). This spectral region also displays the rotational transitions of cyanohexatriyne, HC_7N (J = 21→20) and cyanodecapentayne, $HC_{11}N$ (J = 70→69). Adapted from Ref. [7].

likely a precursor of the cyclopropenylidene ring that has been reported to be ubiquitous in the Galaxy [5]. Also, simple oligoynes (III) and especially the corresponding radicals (IV) [6]:

$H\text{-}(C{\equiv}C)_n\text{-}H$, n = 1-3 $H\text{-}(C{\equiv}C)_n\cdot$, n = 1-3

(III) (IV)

are considered as probable interstellar molecules. Larger end-capped sp-carbon chains occur in polyynylcyanides (V):

$H\text{-}(C{\equiv}C)_n\text{-}C{\equiv}N$, n = 2 - 5

(V)

They were observed in interstellar clouds and cool carbon stars (e.g. IRC+10°216) [7,8] (Fig. 1). The detection of interstellar carbonaceous molecules is based on the analysis of interstellar optical extinction curves, and their deconvolution with regard to the theoretical or experimental spectra of the expected species [9]. Interstellar carbonaceous molecules have been reviewed by Leach [6].

There have been many attempts to reproduce the space-borne molecules (I-V) by a laboratory synthesis, either in the gas phase [3,8,10] or in inert matrices [11]. (Among the most important outputs of these astrophysically motivated efforts was the discovery of fullerenes [8]). Pure all-carbon molecules (I, II) can also be end-capped *in-situ*, e.g. by potassium (VI) [10] or by some radicals (VII) [12]:

$K - (C{\equiv}C)_n - K; \quad 3 \leq n \leq 12$ (VI)

$R - (C{\equiv}C)_n - R; \quad R = CF_3$ or $-CN$, $n \approx 150$ (VII)

Although it is generally expected that long-chained oligoynes will decompose spontaneously (even explosively) the polyyne (VII) is surprisingly stable and soluble in organic solvents [12]. The structure of shorter end-capped oligoynes is known in more detail. For instance, a complete X-ray diffraction analysis of phenyl-capped pentayne revealed the lengths of carbon single and triple bonds to be 136.8 - 142.3 pm and 119.2 - 120.8 pm, respectively. It is interesting that the carbon chain is not perfectly straight, the bond angles being 178.08 - 178.67 ° [13]. A similar pattern follows from the X-ray structure of the lower homologues, e.g., end-capped tetrayne [12]. The synthesis of such molecules has been reviewed by Diederich and Rubin [14,15].

Under terrestrial conditions, carbyne-like molecules have been found in various biological species [2,16-19]. For instance, oligoynes occur frequently in common garden and wild plants of the *Compositae* family. More than 150 years ago, crystals of the dehydromatricaria ester (VIII) were isolated, while the matricaria ester (IX) was found in the scentless mayweed *Matricaria inodora* L. [2].

$H_3C - (C{\equiv}C)_3 - CH{=}CH{-}COOCH_3$ (VIII)

$H_3C - (CH_2)_2 - (C{\equiv}C)_2 - CH{=}CH{-}COOCH_3$ (IX)

The methyltriyne endgroup (cf. VIII) has been detected in a large variety of other naturally occurring derivatives.

Early work on naturally occurring oligoynes relied mostly on the isolation of essential oils made from plants by steam distillation. However, these procedures often failed to sequester the unstable oligoynes, which easily decompose thermally and photochemically. Sometimes even violent explosions have been reported to occur during distillation of essential oils [2]. Nevertheless, the low-temperature solvent extraction, chromatography and the common structural characterization methods have shown these species to be, by no means, rare in many higher plants, algae and microorganisms.

It is certainly exciting to explore the links between natural oligoynes and the carbonaceous molecules occurring in space (I-V). For instance, a derivative (X) of the "cosmic" hexatriyne (III, IV, n = 3) occurs in tropical weeds *(Bidens pilosa L., Coreopsis lanceolate*, etc.) [19]:

$H - (C{\equiv}C)_3 - Ph$; Ph = phenyl

(X)

Also, the "cosmic" polyynylcyanides (V) are reproduced in some terrestrial plants. Substituted hexatriynylcyanide (XI) has been found in large concentrations in the fungus *Lepista glaucocana* [16]:

$CH_3OOC\text{-}CH{=}CH\text{-}(C{\equiv}C)_3\text{-}C{\equiv}N$

(XI)

Even octatetrayne moieties occur naturally; e.g., in the plant antibiotic *caryoynecin* extracted from *Pseudomonas caryophylli* [17] (VIII):

$H - (C{\equiv}C)_4\text{-}(CH = CH)_2\text{-}CH(OH)\text{-}(CH_2)_4 - COOH$

(XII)

Most of the naturally occurring oligoynes were also synthesized in the laboratory [16-19], and their reactions studied to understand their biological activity, e.g. the phototoxicity of (X) towards micro-organisms [19], and antibacterial or antifungal activity of (XI, XII) and other oligoynes [16-18].

In 1984, Heimann *et al.* [20] expressed a provocative idea that carbyne-related compounds may well be among the primordial carbon-based stuff from which life on the Earth originated. This speculation is supported by the fact that I-V are the most abundant carbonaceous molecules in space, and that they can, under suitable conditions, easily transform into more complex derivatives owing to their high reactivity. On the other hand, we have to admit that the "cosmic" oligoynes are also produced in plants by biogenetic routes from fatty acids, e.g. from oleic acid, which yields upon reduction a variety of important methyltriynes including the classical dehydromatricaria ester (VIII) [2]. Although this does not strictly exclude the above speculation, there are little novel arguments to continue a discussion of the role of carbynoid molecules in the evolution of life. For instance, a significant feature of naturally occurring oligoynes (VIII-XII, ...etc.) is that they nearly always exhibit an unbranched carbon chain. Although this can be smoothly interpreted in terms of their biogenesis from fatty acids, the "cosmos-borne" synthesis from (I-V) would be supported by the same argument as well.

3. Natural carbynoid solids

Whereas the existence of carbynoid molecules, both pure and end-capped, in space appears to be supported by spectroscopic methods (cf. previous section), the occurrence of solid carbyne in space has never been demonstrated unambiguously. Nanocrystalline carbon grains as possible candidates for cosmic dust are mostly identified as "amorphous" carbon [21,22], diamond [23] or lonsdaleite [24]. Raman spectra of interplanetary dust particles show evidence in favor of amorphous or graphite-like carbon [21]. For a review of various forms of elemental carbon (including carbyne) in interstellar dust see Ref. [25].

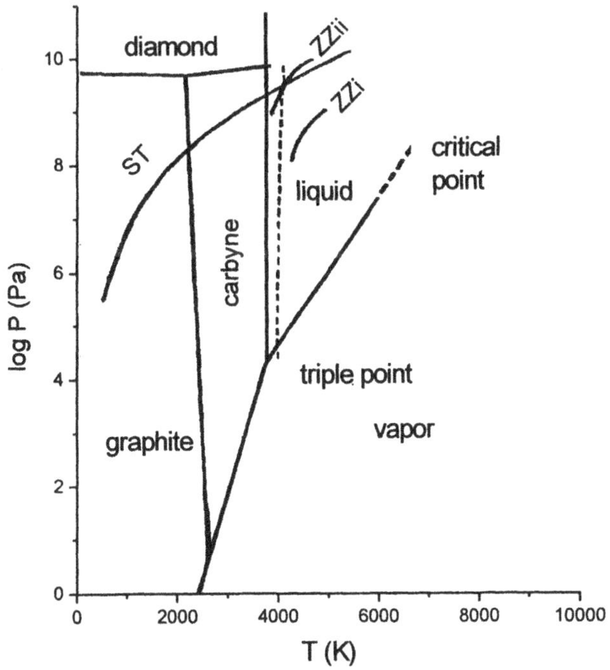

Figure 2. Intersections of the EOS of Shapiro and Teukolsky's (ST) and Zhilyaev and Zubko (ZZ) [31] white dwarf model atmospheres with Whittaker's carbon p-T diagram [29] modified by [31] (dashed line). According to Ref. [28].

A first suggestion, that carbyne might occur ubiquitously in grains of interstellar dust, was presented in the early 80s by Webster [26] and Forrest *et al.* [27]. However, not much work has been done on that problem in the subsequent twenty years [25]. This apparently reflects a declining interest in carbynes after their very existence was questioned nearly at the same time (cf. Chapter 1.2). The carbyne grains were reportedly identified in the 16-30 μm spectra of several carbon stars and planetary nebulae [27]. Albinson and Evans [28] have suggested that carbyne might form in the atmosphere of white dwarf in a cataclysmic variable system.

According to this suggestion the tentative equation of state (EOS) of the stellar atmosphere intersects the field of stability of carbyne in the Whittaker's p-T diagram [29] (Fig. 2). Carbyne whiskers, either pre-existing a classical nova outburst or formed by collisional processes in the accretion disk, may provide nucleation centers for growth of layer carbyne grains. Finally, the combined radiation pressures of the accretion disk and the white dwarf can effectively blow grains out of the system. As the authors note [28], carbyne needles are expected to have IR absorption features in the 2 – 6 μm wavelength range. Such strong features are frequently found in the 3.3 μm IR band of planetary nebula and galaxies [30].

Under terrestrial conditions, solid carbyne (chaoite) has been discovered by ElGoresy and Donnay [1,32] in 1968 in the Ries crater. The geological environment in this locality indicates that the carbynes found are at least 60 million years old [33] (cf. Chapter 2.1 for further details). Natural carbyne has been identified also in marble from Santa Rosa mountains in Southern California [34] and Ticonderoga (New York) [33]. A more systematic study of natural carbynes has been presented by Whittaker [33], who found carbynes also in natural graphites from Sri Lanka and several locations in California (Fossil Canyon, Coyote Mountains, Imperial County, Pacoima Canyon, San Gabriel Mountains, Los Angeles County, Coon Creek Canyon, San Bernardino Mountains). All natural carbynes were typically intergrown in graphites and gneiss as white-gray lamellae.

Based on the discovery of ElGoresy and Donnay [1,32], it was initially thought that natural carbyne was formed as a result of a meteoritic impact on graphitic gneiss (cf. Chapter 2.1). However, Whittaker had suggested that carbynes might have existed in the gneiss also before the impact; hence, the impact would have had nothing to do with carbyne formation [33]. This speculation is supported by the fact that carbynes have been identified worldwide distributed in a variety of geographical locations, and that it is not very probable that all these sites were meteorite targets. On the other hand, if carbyne at the Ries crater site preceded the impact, it is not clear why and how these pre-existing carbyne grains remained intact and did not convert to diamond in the high-pressure, high-temperature shock wave.

Sladkov et al. [35] have reported on layer-chain carbons (carbyno-diamonds) found in Kuzbas coal and Donbas anthracite. These carbons contain, reportedly, both sp and sp^3 moieties, and their structure can be derived from graphite via intercalation of short polyyne chains perpendicular to the puckered ex-graphene layers. These materials have been tentatively characterized by X-ray diffraction, but their deeper investigation is still missing. They have also been suggested to be intermediary structures of carbyne, subjected to shock wave, in a transformation sequence towards diamond [36].

4. References

1. El Goresy, A. and Donnay, G.: A new allotropic form of carbon from the Ries crater, *Science* **161** (1968), 363-4.
2. Bohlmann, F., Burkhardt, G. and Zdero, C.: *Naturally occurring acetylenes*, Academic Press Inc., London, 1973.
3. Weltner, W. and Van Zee, R.J.: Carbon molecules, ions, and clusters, *Chem.Rev.* **89** (1989), 1713-47.
4. Bernath, P.F.: Laboratory astrophysics and molecular astronomy of pure carbon molecules, *Adv.Space Res.* **15** (1995), 15-23.
5. Bohme, D.K.: Ionic origin of carbene in space, *Nature* **319** (1986), 473-4.
6. Leach, S.: Study of fullerenes and linear carbon chains in space, *Planet.Space Sci.* **43** (1995), 1153-64.
7. Bell, M.B., Feldman, P.A., Kwok, S. and Matthews, H.E.: Detection of $HC_{11}N$ in IRC+10°216, *Nature* **295** (1982), 389-91.
8. Kroto, H.W.: C60: Buckminsterfullerene, The celestial sphere that fell to Earth, *Angew.Chem.* **31** (1992), 111-29.
9. Thaddeus, P.: Carbon chains and the diffuse interstellar bands, *Astrophys.Space Sci.Lib.* **202** (1998), 369-78.

10. Rohlfing, E.A., Cox, D.M. and Kaldor, A.: Production and characterization of supersonic carbon cluster beams, *J.Chem.Phys.* **81** (1984), 3322-30.
11. Weltner, W. and Van Zee, R.J.: Matrix-isolated polycarbon molecules, *J.Mol.Struct.* **222** (1990), 201-7.
12. Lagow, R.J., Kampa, J.J., Wei, H.C., Battle, S.C., Genge, J.W., Laude, D.A., Harper, C.J., Bau, R., Stevens, R.C., Haw, J.F. and Munson, E.: Synthesis of linear acetylenic carbon: The sp carbon allotrope, *Science* **267** (1995), 362-7.
13. Rubin, Y., Lin, S.S., Knobler, C.B., Anthony, J., Boldi, A.M. and Diederich, F.: Solution-spray flash vacuum pyrolysis: A new method for the synthesis of linear polyyne with odd numbers of C-C bonds from substitued 3,4-dialkynyl-3-cyclobutene-1,2-diones, *J.Am.Chem.Soc.* **113** (1991), 6943-9.
14. Diederich, F.: Carbon scaffolding: building acetylenic all-carbon and carbon-rich compounds, *Nature* **369** (1994), 199-207.
15. Diederich, F. and Rubin, Y.: Synthetic approaches toward molecular and polymeric carbon allotropes, *Angew.Chem.* **104** (1992), 1123-5.
16. Hearn, M.T.W. and Turner, J.L.: The carbon-13 NMR of the antibiotic polyacetylenic nitrile, diatretyne 2 (7-cyanohept-*trans*-2-ene-4,6-diynoic acid), and related compounds, *J.Chem.Soc.Perkin II* (1975), 1027-9.
17. Yamaguchi, M., Park, H.J., Hirama, M., Torisu, K., Nakamura, S., Minami, T., Nishihara, H. and Hiraoka, T.: Synthesis and reactions of monosilylated 1,3,5,7-octatetrayne. Total synthesis of caryoynencins, *Bull.Chem.Soc.Jpn.* **67** (1994), 1717-25.
18. Grandjean, D., Pale, P. and Chuche, J.: A general access to optically pure epoxypolyynes: Asymmetric synthesis of antifeedant natural products, *Tetrahedron* **49** (1993), 5225-36.
19. Shim, S.Ch. and Lee, T.S.: Photocycloaddition reaction of some conjugated hexatriynes with 2,3-dimethyl-2-butene, *J.Org.Chem.* **53** (1988), 2410-3.
20. Heimann, R.B., Kleiman, J. and Salansky, N.M.: Structural aspects and conformation of linear carbon polytypes, *Carbon* **22** (1984), 147-56.
21. Strazzulla, G., Baratta, G.A. and Spinella, F.: Production and evolution of carbonaceous material by ion irradiation in space, *Adv.Space Res.* **15** (1995), (3)385-(3)399.
22. Bussoletti, E., Colangeli, L. and Orofino, V.: Interstellar amorphous carbon, *Astrophys.J.* **321** (1987), L87-L90.
23. Lewis, R.S., Ming, T., Wacker, J.F., Anders, E. and Steel, E.: Interstellar diamonds in meteorites, *Nature* **326** (1987), 160-2.
24. Rietmeijer, F.J.M. and Mackinnon, I.D.R.: Metastable carbon in two chondritic porous interplanetary dust particles, *Nature* **326** (1987), 162-5.
25. Pillinger, C.T.: Elemental carbon as interstellar dust, *Phil.Trans.R.Soc.Lond.A* **343** (1993), 73-86.
26. Webster, A.S.: Meeting of the Royal Astronomical Society, *Mon.Not.R.Astron.Soc.* **192** (1980), 7-9.
27. Forrest, J.W., Houck, J.R. and McCarthy, J.F.: A far IR emission in carbon-rich stars and planetary nebulae, *Astrophys.J.* **248** (1981), 195-200.
28. Albinson, J.S. and Evans, S.: Possible role of the white dwarf in grain formation in cataclysmic variable systems, *Astrophys.Space Sci.* **131** (1987), 443-7.
29. Whittaker, A.G.: Carbon: a new view of its high-temperature behaviour, *Science* **200** (1978), 763-4.
30. Knacke, R.F.: Carbonaceous compounds in interstellar dust, *Nature* **269** (1977), 132-4; 763-4.
31. Zhilyaev, B.E. and Zubko, V.G.: The formation of condensed carbon particles in the atmospheres of white dwarfs, *Astrophys.Space Sci.* **105** (1984), 99-107.
32. El Goresy, A.: Eine neue Kohlenstoff-Modifikation aus dem Nordlinger Ries, *Naturwiss.* **33** (1969), 493-4.
33. Whittaker, A.G.: Carbon: occurrence of carbyne forms of carbon in natural graphite, *Carbon* **17** (1979), 21-4.
34. Whittaker, A.G. and Kintner, P.L.: Carbon: analysis of spherules and splats formed from the liquid state and of the forms produced by quenching gas and solid, *Carbon* **23** (1985), 255-62.
35. Melnitchenko, V.M., Nikulin, Y.N. and Sladkov, A.M.: Layer-chain carbons, *Carbon* **23** (1985), 3-7.

36. Heimann, R.B.: Linear finite carbon chains (carbyne): their role during dynamic transformation of graphite to diamond, and their geometric and electronic structure, *Diamond Rel.Mater.* **3** (1994), 1151-7.

Chapter 3: Syntheses of carbyne and carbynoid structures

3.1. Catalytic and electrochemical polycondensation reactions

3.1.1. DEHYDROPOLYCONDENSATION OF ACETYLENE

YU.P. KUDRYAVTSEV
A.N. Nesmeyanov Institute of Organoelement Compounds,
Russian Academy of Sciences,
117813 Moscow, Russia

Abstract – Syntheses of carbyne *via* oxidative dehydropolycondensation of acetylene and butadiyne will be described and the role of terminal copper atoms in the stabilization of lower polyynes will be discussed.

1. Introduction

The oxidative coupling reaction of ethynyl compounds catalyzed by copper(I) salts was discovered by Glaser back in 1870 [1]. This coupling along with its modifications proved to be one of the key reactions in the chemistry of polyynes and acetylene derivatives [2,3]. Moreover, this reaction turned out to be a versatile method in polymer chemistry as well. When conducted with aromatic bis(ethynyl) compounds, it afforded a series of linear conjugated poly(arylenebutadiynylenes) (*Scheme 1*) [4],

$$n\ \mathrm{HC{\equiv}C{-}Ar{-}C{\equiv}CH} \xrightarrow[\mathrm{Ox}]{\mathrm{Cu^{1+}}} \left[\mathrm{C{\equiv}C{-}Ar{-}C{\equiv}C}\right]_n + n\ \mathrm{H_2O}$$

Scheme 1

where Ar = arylene, and Ox = oxidizer. This version of the oxidative coupling was named 'oxidative dehydropolycondensation' [5]. Similarly to diethynylarylenes, the molecule of acetylene has two ≡C—H bonds. Therefore, it was thought to be capable of reacting on both sides in a polycondensation fashion [6].

2. Synthesis of Carbyne

During the investigation of the possibility of synthesizing chain-like carbon by oxidative dehydropolycondensation, it was found that the passage of acetylene through an aqueous ammoniacal solution of a Cu(II) salt results in the rapid formation of a black insoluble powder, which explodes on heating in the dry state and on detonation in the wet state [7,8]. Previous investigations into the oxidative

R.B. Heimann et al. (eds.), Carbyne and Carbynoid Structures, 39–45.

dehydropolycondensation of bis(acetylenes) in the presence of Cu(II) salts revealed that the polycondensation process produces polymeric acetylides of Cu(I) (but not of Cu(II)), which are then oxidized by an excess of Cu(II) to polyarylenepolyynes [5]. Therefore, in analogy to the polycondensation of bis(acetylenes) the process of the oxidative dehydropolycondensation of acetylene can be schematically represented as follows [8] (*Scheme 2*).

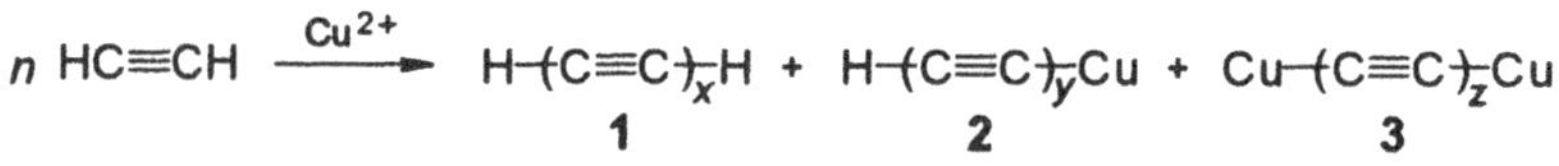

Scheme 2

where $x + y + z = n$. The main difficulty in conducting such a reaction was to retain acetylenic repeat units intact during the entire synthesis. Since experiments showed [6] that the rate of the oxidative coupling at pH=3 markedly decreases with increasing number of conjugated triple bonds, the reaction was carried out in an alkaline medium. In the presence of an excess of Cu ions and binding of the eliminated protons by ammonia, a mixture of polyynes and Cu(I) polyacetylides of various molecular weight is formed. The terminal copper atoms probably stabilize the lower polyynes as do the bulky trialkylsilyl substituents used by Walton *et al.* [9,10] as stabilizing protective groups in the stepwise synthesis of the lower polyynes R–(–C≡C–)$_n$–R, (n = 4–10, 12). Furthermore, the complexation of the copper atoms with triple bonds of the polymer chains may, probably, also stabilize the polyynes being formed (*Scheme 3*).

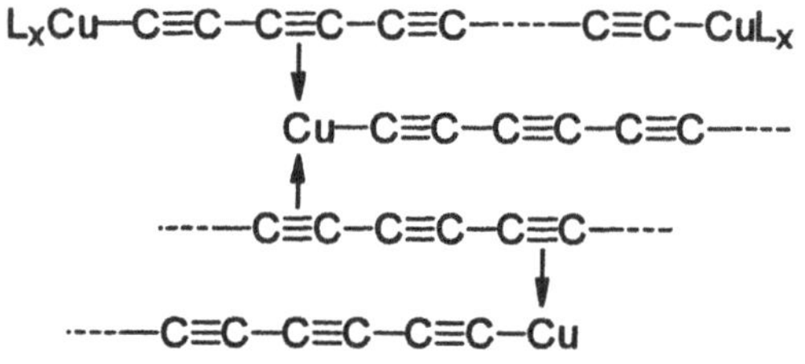

Scheme 3

where L is a ligand.

In order to obtain polyynes with increased molecular weights, an additional oxidation of the products obtained in the first step (**1**, **2**, and **3** in *Scheme 2*) was performed (*Scheme 4*).

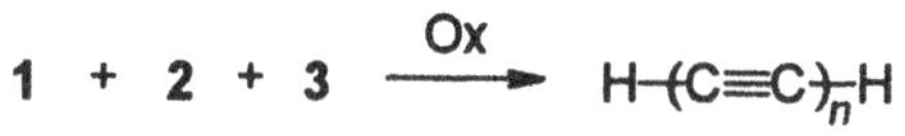

Scheme 4

where Ox is an oxidizer. Originally, the reaction was carried out using $K_3[Fe(CN)_6]$. However, the products obtained in such a way contained appreciable amounts of nitrogen, which was very difficult to remove. Since it was required to anneal the product *in vacuo* at 800 °C to eliminate nitrogen, in subsequent experiments $FeCl_3$ was used instead as the oxidizing agent. The oxidation gave a product which did not explode on heating or shock, but contained a large amount of copper. Copper was removed from the product by refluxing it alternately with dilute HCl and an aqueous solution of NH_3 for several days [11–15]. The data from elemental analysis of the product freed from copper and annealed *in vacuo* at 1000 °C for several hours showed a carbon content of 99.9%. The heating rate was found to play a crucial role: at a rate of 8 °C/min partial graphitization occurred[1], whereas at 2 °C/min the polyyne retained its linear structure.

Based on the stepwise polycondensation mechanism, one can assume that carbyne prepared by the procedure described above is a mixture of polymer homologs with different chain lengths. This conjecture was supported by measurements of the specific electrical resistance (ρ) as a function of temperature of polyyne samples containing 1.5% of copper [8]. At room temperature ρ was found to be 10^6 Ohm·cm. As the temperature increased, ρ decreased, suggesting the accumulation of free copper due to the decomposition of polymeric acetylides. However, ρ continued to decrease beyond 2000 °C, when copper was evaporated from the sample, but according to x-ray analyses no graphite was formed. In our view such a behaviour can be accounted for by the pyrogenetic synthesis of carbyne accompanied by the growth of the carbon chains. Furthemore, the annealing of the polyyne at 1000 °C and 10^{-4} Torr resulted in the formation of small amounts of a white deposit on the walls of a quartz tube. This substance gave rise to bright blue coloration when reacted with an aqueous ammoniacal solution of a Cu(I) salt, which is characteristic of copper acetylides.

Confirmation of the chain-like nature of the carbon obtained required a detailed study of its structure and properties using vibrational [12–16] and electron [17] spectroscopy, as well as x-ray and electron diffraction analysis [12–15,18–20]. When the x-ray diffraction patterns of a polymer prepared by the oxidation of copper polyacetylides were studied, two interference spacings corresponding to d = 390 and 262 pm were found. The former was assigned to intermolecular interference in a bundle of closely packed polyyne chains of carbon, and the latter is attributable to intramolecular interference in a carbon chain. The considerable background of independent scattering in the x-ray patterns attests to the existence of a poorly ordered portion of molecular carbon chains with a mean intermolecular distance of 400 pm. These data are attributable to the presence of a considerable amount of amorphous matter [8,12,13]. The x-ray pattern of samples subjected to prolonged annealing at 1000°C at reduced pressure exhibited numerous sharp diffraction lines, indicating the presence of a crystalline phase with a mean crystal size of ~ 10^{-5} cm. Fused aromatic structures of carbon were absent in both the amorphous and the crystalline phases of the samples obtained. When the annealed product was treated with strong oxidizing

[1] The complete transformation of carbyne into graphite was observed at 2300 °C.

agents, selective oxidation of the dark coloured amorphous impurities occurred. As a result, a light gray carbon material was obtained. A study of the crystalline phase using electron microscopy and microdiffraction revealed single crystals of two types possessing different parameters of a hexagonal unit cell: $a = 892$ pm; $c = 1536$ pm (α-modification) and $a = 824$ pm; $c = 768$ pm (β-modification) [18–20]. To explain the existence of two modifications, they were rather speculatively assigned to the polyyne and cumulene forms of carbyne, respectively. However, in view of the existence of more than ten distinct crystallographic modifications of carbyne reported to date [21,22], this assignment appears to be highly questionable, and the use of crystallographically derived terms, *viz.*, α- and β-carbyne, for describing the type of chemical bonds in carbynes may turn out to be incorrect.

The polyynic nature of bonds in the molecules of the chain-like carbon was substantiated by the results of a study of its oxidation reaction with ozone [15]. Ozonolysis for 30 days resulted in the complete oxidation of the product under formation of oxalic acid thus attesting to the presence of conjugated triple bonds in the chains.[2]

Finally, it was concluded rather indirectly[3] on the basis of a large body of experimental data that the product of the oxidative dehydropolycondensation of acetylene was a new, the third allotropic form of carbon with a chain-like structure, and coined *carbyne*[4] [14,15].

3. Further Synthesis Attempts

In 1969 Hay [23] reported on similar oxidative polycondensation of acetylene carried out in pyridine in the presence of copper(I) chloride and *N,N,N′,N′*-tetramethylethylenediamine with the use of oxygen as an oxidant. The reaction afforded a finely divided black product containing 62.3% of carbon, whose infrared spectrum indicated the presence of triple C≡C bonds. The unusually low carbon content was ascribed to the oxidation of the conjugated triple bond system in the presence of oxygen, but no analytical data on the contents of copper and nitrogen were reported. However, as discussed above, the products of such a reaction should be washed thoroughly to remove the copper impurities whose presence may be surmised from the low content of carbon (82.8%) even in a sample heated to 600°C [23].

Employing an essentially similar synthetic route, Nakanishi *et al.* [24] obtained a polymer with polyyne structure (**4**) using butadiyne, HC≡C—C≡CH, as a starting monomer. Butadiyne and oxygen were bubbled into the solution of cuprous chloride and *N,N,N′,N′*-tetramethylethylenediamine in dimethylacetamide (DMAc) at room temperature (*Scheme 5*).

---

[2] *cf.* Chapter 5.1 of this book.

[3] Rigorous x-ray structural data are still missing.

[4] *cf.* Chapter 1.1 of this book.

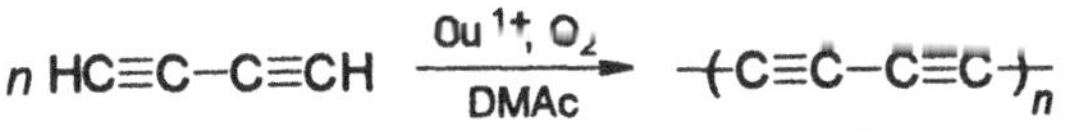

4

Scheme 5

In contrast, when butadiyne was bubbled into a solution of cuprous chloride in DMAc under argon atmosphere, a three-dimensionally coordinated polymer (5) of copper diacetylide was claimed to have been formed (*Scheme 6*).

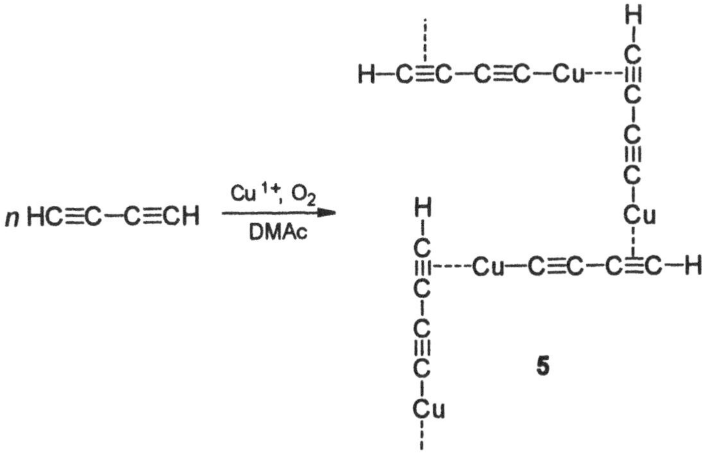

5

Scheme 6

Finally, the bubbling of butadiyne into the methanolic ammoniacal solution of cupric chloride was reported to yield linear poly(metal-ynes) (6) incorporating the copper atoms into the chain (*Scheme 7*).

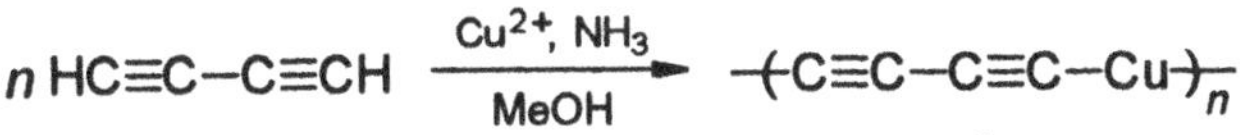

6

Scheme 7

Noteworthy was the explosive decomposition of polymers **5** and **6** when subjected to a heat or mechanical shock.

Infrared absorption of triple bonds in these polymers occurred in three different ways: a strong and broad peak at 2180 cm^{-1} in polymer **4**, very shallow diffused peaks between 1900 and 2200 cm^{-1} in polymer **5**, and distinct double peaks at 1950 and 2100 cm^{-1} in polymer **6**. This means that the electronic states of the triple bonds are different in these three polymers [24].

Surprisingly, the infrared spectrum of a polymer synthesized recently by Kijima *et al.* [25] *via* oxidative dehydropolycondensation of acetylene using the analogous approach did not exhibit any appreciable absorption in the region of stretching vibrations of triple C≡C bonds.

4. References

1. Glaser, C.: Untersuchungen über einige Derivate der Zimmtsäure, *Ann. Chem. Pharm.* **154** (1870), 137–171.
2. Sladkov, A.M. and Kudryavtsev, Yu.P.: Polyynes, *Usp. Khim.* **32** (1963), 509–538 (in Russian) [*Russ. Chem. Rev.* **32** (1963), 229–243 (Engl. Transl.)].
3. Diederich, F.: Carbon scaffolding: building acetylenic all-carbon and carbon-rich compounds, *Nature* **369** (1994), 199–207.
4. Kudryavtsev, Yu.P.: Synthesis and properties of polymers containing conjugated triple bonds in the chain, *Ph.D. Thesis*, Institute of Organoelement Compounds, USSR Academy of Science, Moscow, 1964 (in Russian).
5. Kudryavtsev, Yu.P.: Oxidative dehydropolycondensation – a new method to synthesize polymers with triple bonds, in: *Progress of Polymer Chemistry*, Korshak, V.V., Ed., Nauka, Moscow, 1969, pp. 87–112 (in Russian).
6. Kudryavtsev, Yu.P.: Synthesis and properties of carbyne, *D.Sc. Thesis*, Institute of Organoelement Compounds, USSR Academy of Science, Moscow, 1988 (in Russian).
7. Korshak, V.V., Sladkov, A.M., and Kudryavtsev, Yu.P.: The synthesis of polymeric acetylenides, *Vysokomolek. Soed.* **2** (1960), 1824–1827 (in Russian).
8. Korshak, V.V., Kasatochkin, V.I., Sladkov, A.M., Kudryavtsev, Yu.P., and Usenbaev, K.: Synthesis and properties of polyacetylene, *Dokl. Akad. Nauk SSSR* **136** (1961), 1342–1344 (in Russian).
9. Eastmond, R., Johnson, T.R., and Walton, D.R.M.: Silylation as a protective method for terminal alkynes in oxidative couplings. A general synthesis of the parent polyynes $H(C{\equiv}C)_nH$ (n = 4–10,12), *Tetrahedron* **28** (1972), 4601–4616.
10. Johnson, T.R. and Walton, D.R.M.: Silylation as a protective method in acetylene chemistry. Polyyne chain extensions using the reagents, $Et_3Si(C{\equiv}C)_mH$ (m = 1,2,4) in mixed oxidative couplings, *Tetrahedron* **28** (1972), 5221–5236.
11. Sladkov, A.M. and Kudryavtsev, Yu.P.: Is the third form of carbon feasible? *Priroda* (5) (1964), 88–94 (in Russian).
12. Sladkov, A.M., Kasatochkin, V.I., Kudryavtsev, Yu.P., and Korshak, V.V.: Synthesis and properties of chain-like polymers of carbon, *Izv. Akad. Nauk SSSR, Ser. Khim.* (12) (1968), 2697–2704 (in Russian).
13. Kasatochkin, V.I., Sladkov, A.M., Kudryavtsev, Yu.P., and Korshak, V.V.: Chain-like polymer of carbon – carbyne, in *Structural Chemistry of Carbon and Coals*, Kasatochkin, V.I., Ed., Nauka, Moscow, 1969, pp. 17–26 (in Russian).
14. Sladkov, A.M. and Kudryavtsev, Yu.P.: Diamond, graphite, carbyne – allotropic forms of carbon, *Priroda* (5) (1969), 37–44 (in Russian).
15. Korshak, V.V., Kudryavtsev, Yu.P., and Sladkov, A.M.: Carbyne – a new allotropic form of carbon, *Vestn Akad. Nauk SSSR* (1) (1978), 70–78 (in Russian).
16. Kasatochkin, V.I., Sladkov, A.M., Aseev, Yu.G., Kudryavtsev, Yu.P., Egorova, O.I., and Korshak, V.V.: Infra-red spectra of polyynes, *Dokl. Akad. Nauk SSSR* **153** (1963), 346–349 (in Russian).
17. Sergushin, I.P., Kudryavtsev, Yu.P., Élizen, V.M., Sadovsky, L.P., Sladkov, A.M., Nefedov, V.I., and Korshak, V.V.: X-ray photoelectron and x-ray emission study of carbyne, *Zh. Strukt. Khim.* **18** (1977), 698-700 (in Russian).
18. Kasatochkin, V.I., Sladkov, A.M., Kudryavtsev, Yu.P., Popov, N.M., and Korshak, V.V.: Crystalline forms of a linear modification of carbon, *Dokl. Akad. Nauk SSSR* **177** (1967), 358–360 (in Russian).
19. Kasatochkin, V.I., Korshak, V.V., Kudryavtsev, Yu.P., Sladkov, A.M., and Sterenberg, L.E.: On crystalline structure of carbyne, *Carbon* **11** (1973), 70–72.
20. Kasatochkin, V.I., Korshak, V.V., Kudryavtsev, Yu.P., Sladkov, A.M., and Élizen, V.M.: On polymorphism of carbyne, *Dokl. Akad. Nauk SSSR* **214** (1974), 587–589 (in Russian).
21. Heimann, R.B., Kleiman, J., and Salansky, N.M.: Structural aspects and conformation of linear carbon polytypes (carbynes), *Carbon* **22** (1984), 147–156.
22. Man, L.I., Malinovskii, Yu.A., and Semiletov, S.A.: Crystal phases of carbon, *Kristallografiya* **35** (1990), 1029–1039 (in Russian).
23. Hay, A.S.: Oxidative polymerization of diethynyl compounds, *J. Polym. Sci.: Part A-1* **7** (1969), 1625–1634.

24. Matsuda, H., Nakanishi, H., and Kato, M.: Conducting polyyne and poly(metal-yne) from butadiyne, *J. Polym. Sci., Polym. Lett. Ed.* **22** (1984), 107–111.
25. Kijima, M., Sakai, Y., and Shirakawa, H.: Electrochemical synthesis of carbyne catalyzed by nickel complex, Synth. Met. 71 (1995), 1837–1840.

3.1.2 POLYCONDENSATION REACTIONS OF HALIDES

M. KIJIMA and H. SHIRAKAWA
Institute of Material Science, University of Tsukuba,
Tsukuba, Ibaraki 305, Japan

Abstract - In this chapter approaches are summarized to electrochemical syntheses of carbyne by reductive polycondensation. As a typical example, the cathodic electrolysis of diiodo-acetylene and hexachloro-1,3-butadiene resulted in semiconducting carbynoid materials with mixed polyyne- and cumulene structures. These species were assumed to be stabilized by n-doping and/or incorporation of an electrolyte during the cathodic process.

1. Introduction

Cathodic reduction of a carbon-halogen σ–bond in halides frequently causes intermolecular coupling to form a carbon-carbon bond. A general formulation of the cathodic reduction process of halides is shown in scheme 1. The ease of reduction of carbon-halogen bonds generally follows the order I > Br > Cl > F and $CX_4 > CHX_3 > CH_2X_2 > CH_3X$. Synthetic uses of these reactions have been extensively investigated [1].

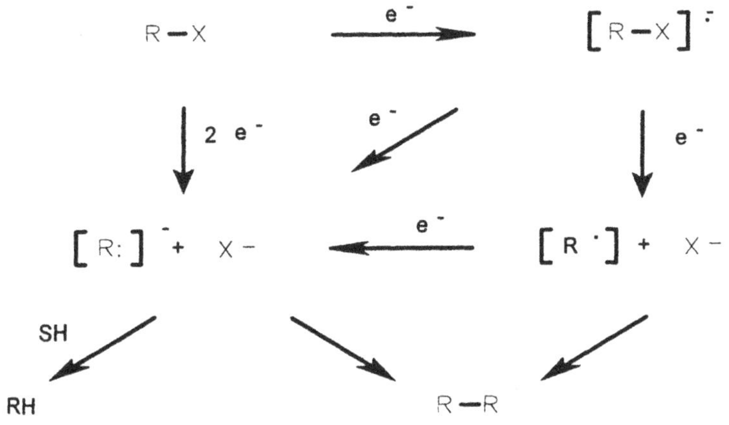

Scheme 1

R.B. Heimann et al. (eds.), Carbyne and Carbynoid Structures, 47–54.

Synthesis of carbyne and carbynoid materials may be achieved by utilization of the cathodic reduction processes. Cathodic polycondensations of some halides has been attempted, resulting in either π-conjugated or non-conjugated polymers. Controlled reduction at potentials around –1.0 V *vs.* SCE of several 1,4-dihalomethylbenzenes in dioxane-H_2O-HCl under a nitrogen atmosphere with respectively, X= Cl, Br and Y= H, Cl, F gave polymers (eq. 1) in good yields [2]. Electroreduction at –3.0 V (*vs.* Ag/Ag^+) of α,α,α',α'-tetrabromo-*p*-xylene in polar aprotic solvents containing a tetrabutylammonium salt gave a thick and uniform poly(*p*-phenylene vinylene) films (eq. 2) with a conductivity of 5 x 10^{-2} S cm^{-1} on SO_3 doping [3]. Electrolysis of α,α,α,α',α',α'-hexachloro-*p*-xylene at –1.7 V (*vs.* Ag/Ag^+) with a copper electrode in propylene carbonate containing *n*-Bu_4NBF_4 afforded a free-standing film of poly-(*p*-phenylene ethynylene) (eq. 3) with a conductivity of 70 S cm^{-1} on doping with SO_3 [4].

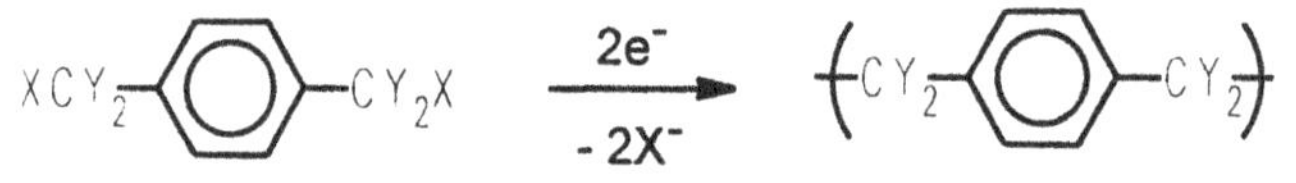

(1)

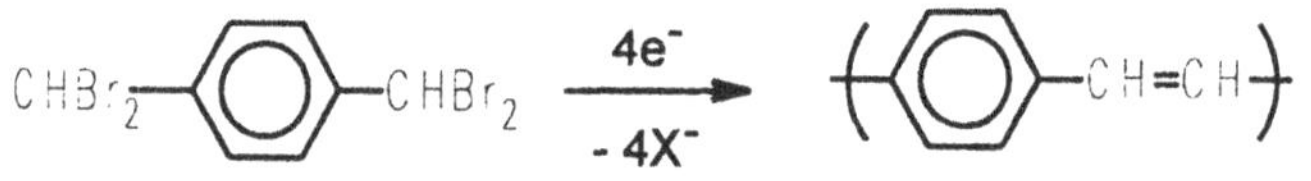

(2)

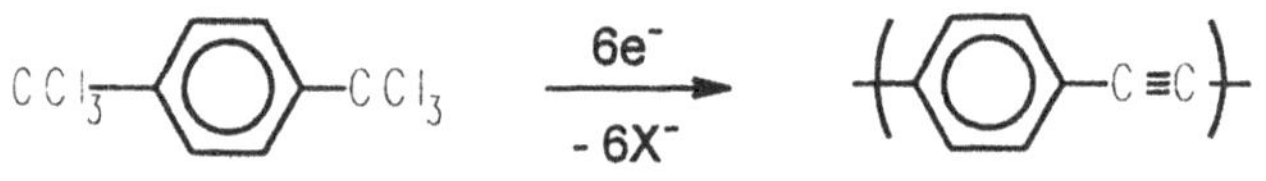

(3)

These results show that the cathodic reduction of halides can produce three types of intermolecular carbon-carbon bond linkages, although these products contain a phenylene unit in the polymer chain.

First attempts to carbonize halides by cathodic electrolysis was carried out by Nishihara *et al.* [5]. Electrochemical reduction of hexachloro-1,3-butadiene in aprotic solvents yielded an electroactive black polymer film consisting predominantly of carbon (> 62 %). Raman spectroscopy and XRD analysis of the film suggested that a considerable part of the film had a turbostratic graphite structure and that carbynoid structures were absent. However, hexachloro-1,3-butadiene is thought to be a good starting material for synthesis of carbyne and carbynoid materials because expected reductive intermediates of hexachlorobutadiene are close to a carbyne structure (eq. 4). If the reduction is carried out under mild conditions, a carbyne-like material may be produced at the cathode through these intermediates as well as during isolation of a reductive intermediate, α,α,α',α'-tetrachloro-*p*-xylylene, in the cathodic reaction of α,α,α,-α',α',α'-hexachloro-*p*-xylene at low temperature of -10°X (eq. 5) [2]. These conditions are milder than those of polycondensation.

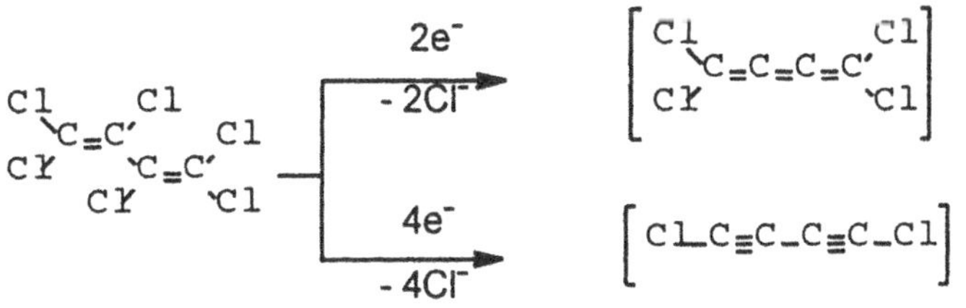

(4)

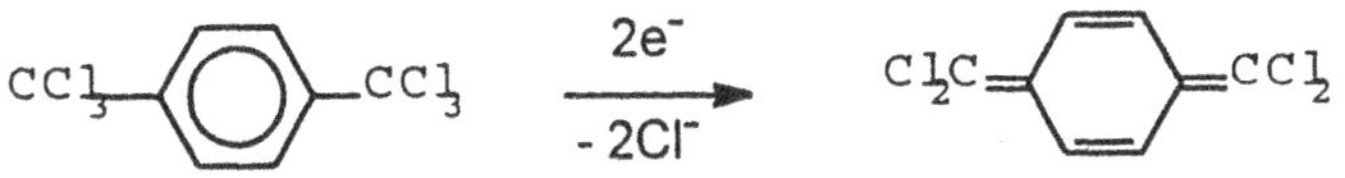

(5)

The first detection of a carbynoid species in a cathodic reaction product was during electroreductive polycondensation of diiodoacetylene catalyzed by a nickel complex (eq. 6) [6]. Secondly, a film consisting of carbynoid-like carbon was found to be formed by a cathodic electrolysis of hexachloro-1,3-butadiene under mild conditions (eq. 7) [7]. These reports suggest that polyhalides are potential compounds for synthesis of electrochemical carbons with carbynoid structures. There are readily available polyhalides other than hexachloro-butadiene such as tetrachloromethane, tetrabromomethane, hexachloroethane, tetrachloroethylene, hexachloropropene. Most of them have a reduction peak at around $-1.0 \sim -2.0$ V (*vs.* SCE) and some of them form a black material deposit on the cathode during cathodic electrolysis [8].

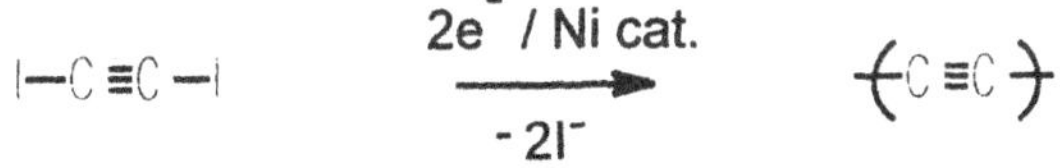

(6)

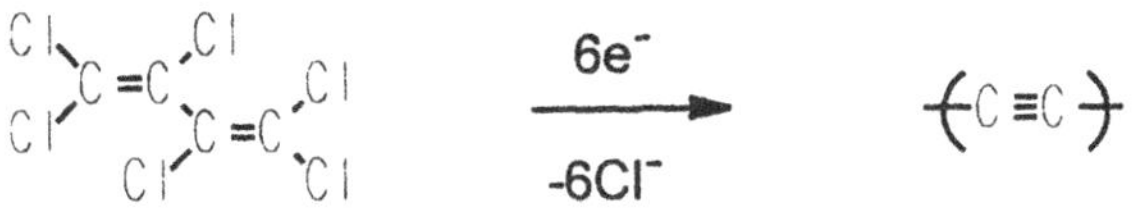

(7)

2. Electrolysis of Diiodoacetylene

Diiodoacetylene [6] was prepared by passing acetylene gas into a stirred 10 % KOH solution with simultaneous addition of an aqueous solution of I_2-KI mixture at room temperature [9]. Cyclic voltammograms of diiodoacetylene in DMF containing *n*-Bu_4NClO_4 (TBAP) showed a cathodic wave around -1.0 V. Direct cathodic electrolysis of diiodoacetylene in the DMF solution was very slow giving only a trace amount of a black deposit on a Pt cathode. On addition of a Ni complex to the electrolyte, the cathodic reaction smoothly proceeded giving a black powdery product in a good yield. The electrolytic conditions were a constant potential with a Pt cathode at –2.0 V (*vs.* SCE) in a DMF solution containing 0.1 mol dm^{-3} diiodoacetylene and 0.1 mol dm^{-3} of a TBA^{+} salt in a presence of 5 mmol dm^{-3} of the Ni(II) complex such as NiI_2(diphenylphosphinoethane) at room temperature under an argon atmosphere.

The nickel complex seems to work as an electron-transfer mediator for the reduction of the carbon-iodine linkages in diiodoacetylene rather than as a coupling catalyst since homo-coupling of iodophenylaetylene using equivalent Ni(cyclooctadiene)$_2$ is very slow.

The product consisted of 63 % carbon suggesting contamination with impurities such as the electrolyte and the Ni complex. The XRD pattern suggested that the product is an amorphous carbon. The existence of carbynoid species in the product were confirmed by infrared (IR) and Raman spectroscopy.

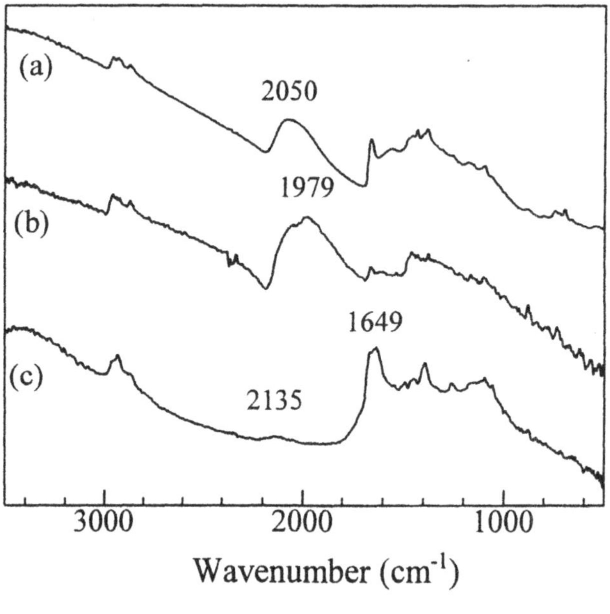

Figure 1. IR spectra of products obtained by cathodic electrolyses of (a) diiodoacetylene at a constant potential of –2.0 V with a Ni mediator, (b) hexachloro-1,3-butadiene at –2.0 V, and (c) hexachloro-1,3-butadiene at a constant current of 1 mA cm^{-2}. On the y-axis the absorption is given in artificial units.

Characteristic broad and strong IR absorption centered at 2050 cm^{-1} (Figure 1(a)) and Raman peak at 2150 cm^{-1} were observed and are assumed to be due to a stretching vibration of a carbon-carbon triple bond system ($\nu_{C\equiv C}$). A stable form of carbyne is considered to be a polyyne form. Conjugation numbers of π-conjugated carbon-carbon triple bonds in this product are estimated to be in excess of 14 obtained from the equation $\nu_{C\equiv C}(N) = 2104 + 698/(N+1)$, given by Akagi et al. [10], and about 5 estimated from another equation, $\nu_{C\equiv C}(N) = 1750 + 3980/N$, developed by Kastner et al. [11]. The IR absorption band typical for the carbynoid species decreased by heat or oxygen treatment. Thus the carbynoid species is thought to be unstable to heat and under oxygen.

The electrical conductivity of a pellet sample is 3 x 10^{-5} S cm^{-1} with little doping effect of I_2, which is comparable to that of I_2-doped materials containing carbynoid species prepared by other methods [12, 13, 14]. Since the sample shows a negative thermoelectric power, the carbynoid materials prepared by this manner is considered to be already *n*-doped by TBA^+ during the cathodic reaction. Actually, the presence of TBA^+ is confirmed by IR absorptions around 2900cm^{-1} due to $\nu_{C\text{-}H}$.

3. Electrolysis of Hexachloro-1,3-butadiene

Hexachloro-1,3-butadiene (50 mmol dm^{-3}) was electrochemically reduced at –2.0 V *vs.* SCE with a Pt cathode in DMF containing 0.1 mol dm^{-3} *n*-Bu_4NClO_4 (TBAP) at 0°X in a two-compartment cell under argon atmosphere [7]. A black film was deposited on the cathode.

Elemental analysis of the product shows 78% content of carbon. The solution-side of the film has a cauliflower morphology observed by SEM as common for electrochemically deposited materials. X-ray powder diffraction analysis shows that the material is an amorphous carbon. The IR spectrum shows a strong broad absorption in the range 2195-1705 cm^{-1} with the peak maximum centred at 1979 cm^{-1} (Figure 1(b)). The Raman spectrum shows a peak at 1950 cm^{-1}. These IR and Raman peaks are due to the stretching vibration of a carbon-carbon triple-bond system ($\nu_{C\equiv C}$).The extent of conjugation number of carbon-carbon triple bonds is estimated by the Kastner equation [11] to be approximately 10. The EPR spectrum shows a signal at g = 2.0036 suggesting a presence of radical species. The electrical conductivity of the pristine film is 5 x 10^{-8} S cm^{-1}, and it increased to 5 x 10^{-5} S cm^{-1} upon I_2 doping. The as-grown film sample shows a negative thermoelectric power, whereas the I_2-doped sample shows a positive value, suggesting that the as-grown product is electrochemically *n*-doped by TBA^+ during the cathodic process. The presence of TBA^+ is confirmed by IR and solid-state ^{13}C NMR.

A schematic structure of the product is suggested from these results and shown below. The product consists of a linear decaacetylenic chain with a carbon moiety *n*-doped by TBA^+ and incorporated TBAP. The doped moiety may cause cross-coupling of the linear chains or cumulene formation by combination.

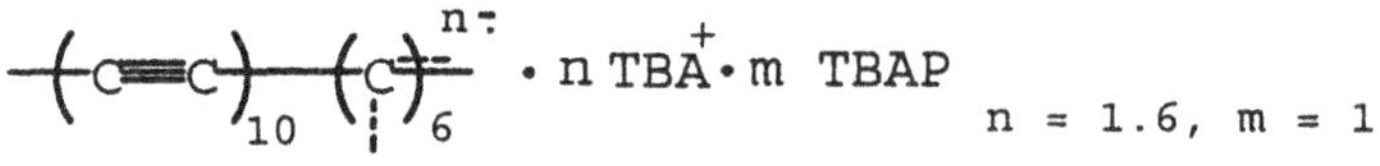

The constant potential electrolysis gave a polyyne-type carbynoid species (α-carbyne), while a constant current electrolysis resulted in another carbon film with a different chemical structure, maybe of cumulene-type (β-carbyne). The galvanostatic electrolysis was similarly carried out with a constant current of 1 mA cm^{-2} at 0°X under argon atmosphere [7]. In this case, the electrolytic potential changed from –1.5 to –3.0 V.

In the IR spectrum of the product obtained by the galvanostatic method (Figure 1(c)), the characteristic peak due to $\nu_{C\equiv C}$ is broad and very weak, and the peak maximum is at 2135 cm^{-1}, hence showing a higher wavenumber than that of products obtained by the potentiostatic methods. An intensive IR peak at 1649 cm^{-1} may be due to a conjugated carbon-carbon double bond system of cumulenic bonds as similarly assigned by Russian groups [15]. The electrical conductivity is 1 x 10^{-4} S cm^{-1} without the doping effect of I_2. From the elemental analysis, the carbon content of the product was determined to be 79 %, and the composition was estimated to be C_{24} · 2.1 TBA^+ · 0.8 TBAP.

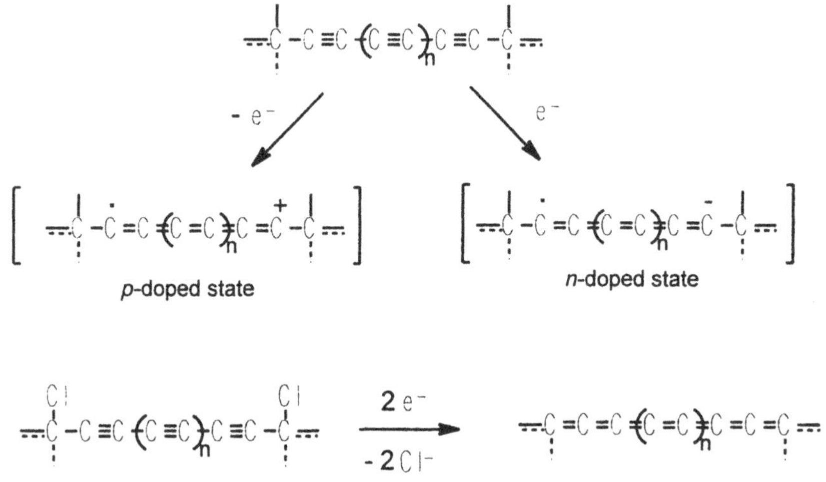

Scheme 2

The cumulene linkages in carbynoid materials arise either by doping processes or by reductive elimination reactions at an isolated polyyne segment as illustrated in scheme 2. These cumulenic states will contribute to an increase in conductivity. Practically, the product obtained by the potentiostatic method showed an increase of conductivity by *p*-type doping of I_2, while the highly reduced product obtained by the galvanostatic electrolysis exhibited a higher conductivity without an I_2-doping effect.

4. Conclusion

In this Chapter, attempts were described to synthesize carbyne by electro-reductive polycondensation of halides using diiodoacetylene and hexachloro-1,3-butadiene as precursor materials. In both cases, the cathodic products seem to be carbynoid materials. The products obtained by the potentiostatic electrolyses resemble each other in their IR spectra, Raman spectra, and some properties. Thus, the products obtained by controlled potential electrolysis basically posses a similar chemical structure, consisting of linear polyyne chains, and cross-linked and *n*-doped carbon moieties. On the other hand, the product obtained by constant current electrolysis of hexachlorobutadiene has a similar elemental composition but a different chemical structure, consisting of linear cumulenic carbon chains and cross-linked and *n*-doped carbon moieties. The characteristics displayed by the electrochemically deposited carbyne can be summarized as follows. They posses an electrochemically *n*-doped state, involving a radical species, have semiconducting properties showing electrical conductivities in the range of about 10^{-4} to 10^{-6} S cm^{-1} in the doped state, and are unstable to heat and under oxygen.

In order to synthesize quasi-one dimensional carbyne with long conjugated polyyne and cumulene structures, stabilization of carbon chains will be significant, since adjacent linear carbon chains in the solid state or in a solvent are reactive and hence transform to higher dimensional materials by inter-chain cross-linking. Success in detecting carbynoid species in the cathodic products is assumed to lie in a stabilization of growing linear carbon chains by the bulky *n*-dopant or the electrolyte incorporated during the cathodic process. However, the presence of the bulky TBA^+ or TBAP incorporated in the carbon chains causes low electrical conductivities of the materials. Modifications of the reaction will improve the electrical properties of the carbynoid material. Furthermore, the design of polymerization so as to allow an ordering of the growing linear carbon chains will result in a synthetic carbyne with long π-conjugated polyyne and cumulene segments. It is believed that the cathodic polycondensation of halides is one of several potential methods to synthesize carbyne with long conjugated carbon chains, since various modifications and strict control of the reaction are possible in a solvent.

5. References

1. Casanova, J. and Eberson, L. (1973) Electrochemistry of the carbon-halogen bond, in S. Patai (ed), The Chemistry of the Carbon-Halogen Bond,.Wiley, London, Part 2, chapter 15, pp. 979-1047 ; Becker, J. Y. 1983) Electrochemical oxidation, reduction and formation of the C-X bond direct and indirect processes, in S. Patai and Z. Rappoprt (eds), Supplement D, The Chemistry of Halides, Pseudo-Halides and Azides, Wiley, London, Part 1, Chapter 6, pp 203-286.

2. Gilch, H. G. (1966) Preparation of poly-p-xylylenes by electrolysis, J. Polym. Sci: Part A-1, 4, 1351-1357.
3. Nishihara, H., Tateishi, M., Aramaki, K., Ohsawa, T., and Kimura, O. (1987) Electrochemical synthesis of poly(p-phenylene vinylene) films, Chem. Lett., 539-542.
4. Tateishi, M., Nishihara, H., and Aramaki, K. (1987) Electrochemical Synthesis of poly(p-phenylene xylidene). A new electrically conducting polymer, Chem. Lett., 1727-1728.
5. Nishihara, H., Harada, H., Kaneko, S., Tateishi, M., and Aramaki, K. (1990) Synthesis of graphite films by electrochemical reduction of hexachlorobuta-1,3-diene, J. Chem. Soc., Chem. Commun., 26-27; Nishihara, H., Harada, H., Tateishi, M., Ohashi, K., and Aramaki, K. (1991) New approach to graphite synthesis: electrochemical carbonization of hexachlorobuta-1,3-diene, J. Chem. Soc. Faraday Trans., 87, 1187-1192.
6. Kijima, M., Sakai, Y., and Shirakawa, H. (1994) A novel approach for synthesis of carbyne by electroreductive polymerization of diiodoacetylene, Chem. Lett., 2011-2014; Kijima, M., Sakai, Y., and Shirakawa, H. (1995) Electrochemical synthesis of carbyne catalyzed by nickel complex, Synth. Met., 71, 1837-1840.
7. Kijima, M., Toyabe, T., and Shirakawa, H. (1996) Novel approach for synthesis of a carbyne film by electrochemical reduction of hexachlorobuta-1,3-diene, J. Chem. Soc., Chem. Commun., 2273-2274; Kijima, M., Toyabe, T., Shirakawa, H., Kawata, S., Kyotani, H., Kyotani, M., Nakamura, Y., and Endo, T. (1997) Synthesis of carbyne analogues by cathodic electrolysis of hexachlorobutadiene, Synth. Met., 86, 2279-2280.
8. Kijima, M., Ohmura, K., Toyabe, T., and Shirakawa, H., unpublished results.
9. Mahapatra, M., Rahman, M. S., Maiti, M. M., and Maiti, S. (1989) New conducting polymers. 4. Synthesis of polyethynyl sulfide, J. Polym. Mater., 6, 213-215.
10. Akagi, K., Nishiguchi, M., Shirakawa, H., Furukawa, Y., and Harada, I. (1987) One-dimensional conjugated carbyne-synthesis and properties, Synth. Met., 17, 557-562.
11. Kastner, J., Kuzmany, H., Kavan, L., Dousek, F. P., and Kurti, J. (1995) Reductive preparation of carbyne with high yield. An in situ Raman scattering study, Macromolecules, 28, 344-353.
12. Matsuda, H., Nakanishi, H., and Kato, M. (1984) Conducting poly-yne and poly(metal-yne) from butadiyne, J. Polym. Sci. Polym. Lett. Ed., 22, 107-111.
13. Dias, A. J. and MacCarthy, T. J. (1985) Dehydrofluorination of poly(vinylidene fluoride) in dimethylformamide solution: synthesis of an operationally soluble semiconducting polymer, J. Polym. Sci., Polym. Chem. Ed., 23, 1057-1061.
14. Kise, H. and Ogata, H. (1983) Phase transfer catalysis in dehydrofluorination of poly(vinylidene fluoride) by aqueous sodium hydroxide solutions, J. Polym. Sci., Polym. Chem. Ed., 21, 3443-3451.
15. Sladkov, A. M., Korshak, V. V., and Nepochtykh, V. P. (1968) Polymers with cumulated double bonds in the chain, Izv. Akad. Nauk SSSR, Ser. Khim., 196-197; Korshak, V. V., Kudryavtsev, Y. P., Korshak, Y. V., Evshukov, S. E., Khvostov, V. V., Babaev, V . G., and Guseva, M. B. (1988) Formation of β-carbyne by dehydrohalogenation, Makromol. Chem. Rapid. Commun., 9, 135-140; Vointseva, I. I., Gilman, L. M., Kudryavtsev, Y. P., Evsyukov, S. E., Pesin, L. A., Gribov, I. V., Moskvina, N. A., and Khvostov, V. V (1996) Chemical dehydrofluorination of polytrichlorobutadienes-A new route to carbynes, Eur. Polym. J., 32, 61-68.

3.2. Chemical, photo-, and electrochemical transformation of polymers

3.2.1. CHEMICAL DEHYDROHALOGENATION OF POLYMERS

S.E. EVSYUKOV
A.N. Nesmeyanov Institute of Organoelement Compounds, Russian Academy of Sciences, 117813 Moscow, Russia

Abstract – Chemical dehydrohalogenation of halogen-containing polymers has been considered as a method for synthesizing carbyne. General principles of the method, various dehydrohalogenating agents, the structure of original polymers and reaction products, as well as the nature of side reactions have been discussed. Based on vibrational, electron, and nuclear magnetic resonance spectroscopic data, it has been concluded that the method does not provide an approach to the synthesis of perfect (*i.e.*, defect-free) carbyne structures, and that dehydrohalogenation products should be regarded as carbynoid polymeric materials containing short carbyne fragments along with various defects.

1. Introduction

Amongst polymer-analogous transformations, one of the most convenient and accessible methods for synthesizing carbynoid structures is the chemical dehydrohalogenation of some halogen-containing polymers [1]. The main distinction between this approach and polycondensation processes is that the carbon backbone is formed beforehand by polymerizing relevant monomers, and the task for synthesizing carbyne is confined to the complete elimination of the hydrogen halide with retention of the linear structure.

2. Original Polymers

The formation of carbyne by means of dehydrohalogenation is possible only if equal numbers of the halogen and hydrogen atoms are attached to neighbouring carbon atoms in the original polymer chain. Theoretically, this allows complete elimination of side substituents to leave a 'naked' carbon chain, *i.e.*, chemical carbonization of the original polymer. This requirement is satisfied in poly(vinylidene halides) (PVDH) [2,3], poly(1,2-dihaloethylenes) [4,5], as well as in poly(1,1,2- and 1,2,3-trichloro-butadiene) [6,7]. All these types of polymers were used as original materials in numerous attempts of synthsizing carbyne by the dehydrohalogenation reaction.

R.B. Heimann et al. (eds.), Carbyne and Carbynoid Structures, 55–74.

2.1. POLY(VINYLIDENE HALIDES)

For the first time, the chemical dehydrochlorination of poly(vinylidene chloride) (PVDC) was proposed by Korshak *et al.* [8] as an alternative approach for synthesizing carbyne back in 1964. Indeed, the exhaustive dehydrohalogenation of PVDH should, seemingly, yield one-dimensional carbon chains provided that the linear structure of original polymers is retained during the reaction. The dehydrohalogenation of PVDH is supposed to proceed in two stages. At the first stage, the one molecule of hydrogen halide (HX) is eliminated from each repeating link of PVDH to afford a halogen-substituted polyene structure (*Scheme 1*, X = F, Cl, Br).

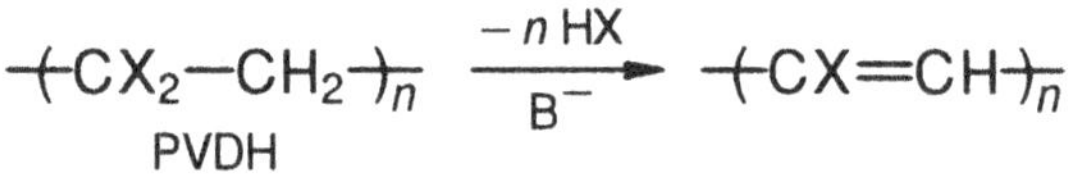

Scheme 1

The complete dehydrohalogenation at the second stage can be expected to yield carbon chains of both polyyne and cumulene type. The elimination of HX from neighbouring carbon atoms at (1)–(2), (1′)–(2′), *et seq.* positions should result in the formation of the polyyne structure (*Scheme 2*).

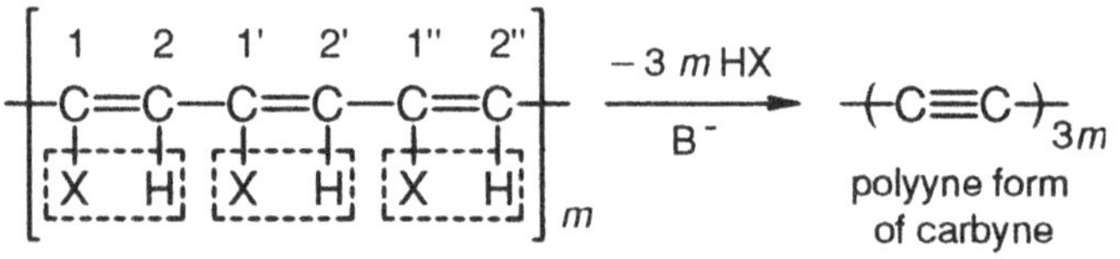

Scheme 2

On the other hand, elimination of HX from neighbouring positions (2)–(1′), (2′)–(1″) and further must give rise to the cumulene form of carbyne (*Scheme 3*).

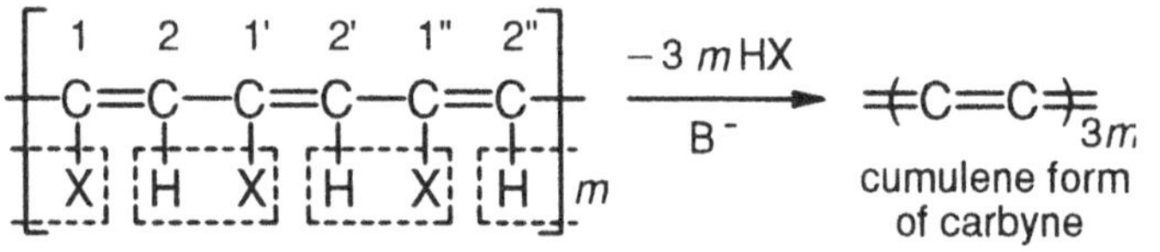

Scheme 3

The rate of dehydrohalogenation of PVDH depends on the nature of halogen. Generally, PVDH can be arranged in the following order with respect to the reaction rate: poly(vinylidene bromide), PVDB > PVDC > poly(vinylidene fluoride), PVDF [3]. This can be accounted for by increasing strength of the carbon–halogen bond in the sequence: C—Br < C—Cl < C—F [9].

2.2. POLY(1,2-DIHALOETHYLENES)

Poly(1,2-dihaloethylenes), $(\text{–CHX–CHX–})_n$, where X=Cl or Br, cannot be synthesized directly by radical polymerization of corresponding 1,2-dichloro-(or dibromo)ethylene because of steric hindrances [10]. Therefore, recourse to polymer-analogous transformations is necessary to obtain such a structure. Shirakawa *et al.* [4] have disclosed that the prolonged doping of polyacetylene with ferric chloride results in a stereospecific addition of chlorine to double bonds to afford poly(1,2-dichloroethylene) (PDCE) in a *trans*-gauche conformation (*Scheme 4*).

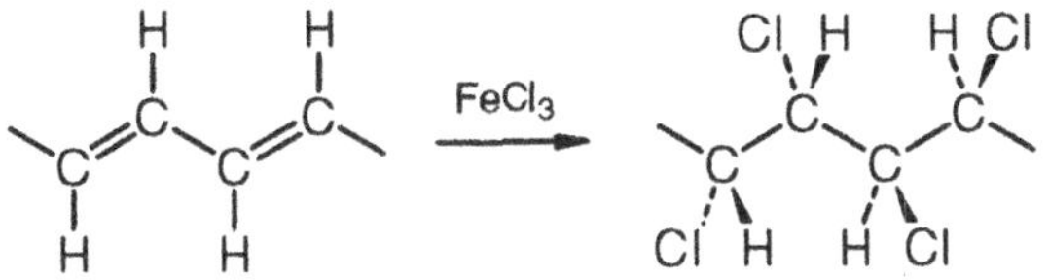

Scheme 4

The synthesis of poly(1,2-dibromoethylene) (PDBE) on the basis of poly(vinyl chloride) (PVC) have been reported by Evsyukov *et al.* [5]. For this purpose, the original PVC was dehydrochlorinated, and the polyene thus obtained was brominated to form a saturated polymer having predominantly the structure of PDBE (*Scheme 5*).

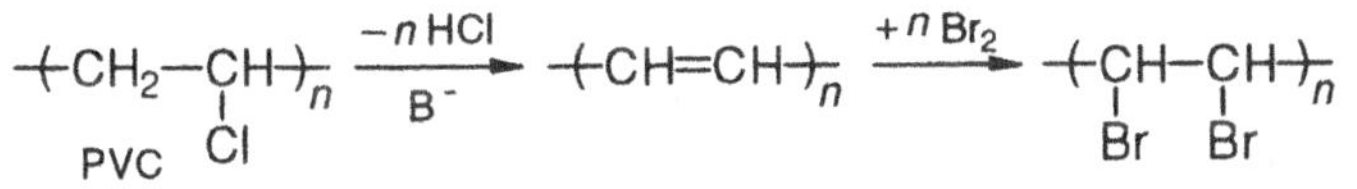

Scheme 5

Obviously, the dehydrohalogenation of poly(1,2-dihaloethylenes) at the first stage should bring about the formation of the halogen-substituted polyene structure (*Scheme 6*) being analogous to that resulted from the dehydrohalogenation of PVDH at the first stage (*cf. Scheme 1*).

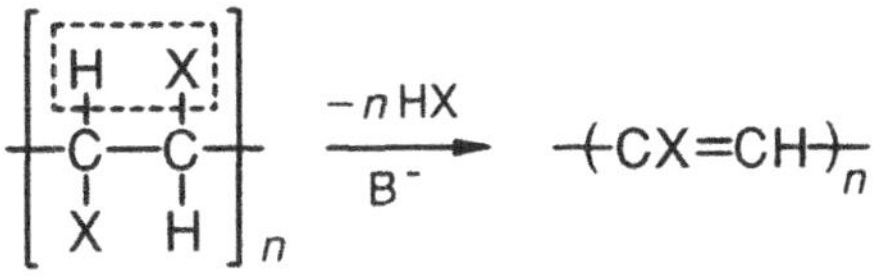

Scheme 6

Therefore, further dehydrohalogenation can similarly be expected to give rise to both polyyne and cumulene forms of carbyne (*cf. Scheme 2* and *3*).

2.3. POLYTRICHLOROBUTADIENES

Considering theoretically possible pathways of the dehydrohalogenation reaction of polytrichlorobutadienes (PTCB), one can conclude that poly(1,1,2-trichlorobutadiene-1,3) (P-1,1,2-TCB) and poly(1,2,3-trichlorobutadiene-1,3) (P-1,2,3-TCB) may be used as starting materials for synthesizing different polymorphous modifications of carbyne [6,7]. Indeed, complete dehydrochlorination of P-1,1,2-TCB should result in the formation of the polyyne form of carbyne. Owing to an allylic activation effect, the chlorine atoms at position (1) in each repeating unit of PTCB exhibit an enhanced lability[1], which causes their high reactivity on interacting PTCB with various compounds, including polymers [12]. Because of this, at the first stage of dehydrochlorination the hydrogen chloride elimination must occur from neighbouring positions (4′)–(1) and (4)–(1′) to form the halogen-substituted polyene structure of poly(1,2-dichloro-2-butene-1,4-diylidene) (*Scheme 7*).

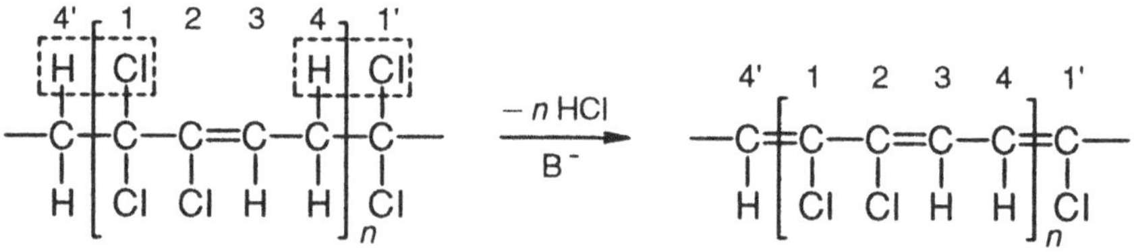

Scheme 7

It is evident that further elimination of HCl is possible only from neighbouring atoms at (4′)–(1), (2)–(3), and (4)–(1′) positions, *i.e.*, complete dehydrochlorination of this polymer must result in the formation of the polyyne structure (*Scheme 8*).

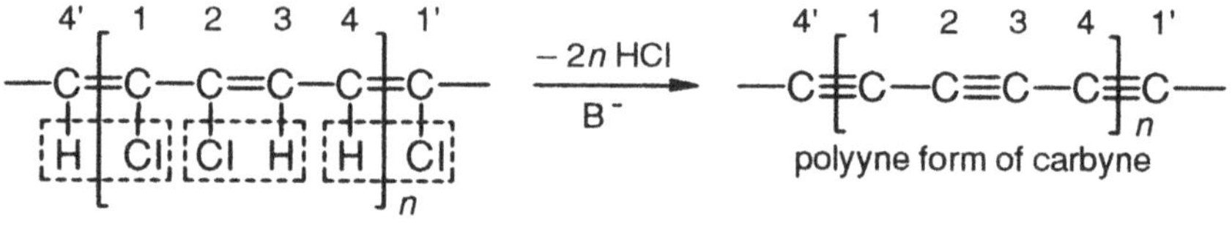

Scheme 8

On the other hand, dehydrochlorination P-1,2,3-TCB should bring about the formation of the cumulene form of carbyne. Analogously to the reaction with P-1,1,2-TCB, at the first stage of dehydrochlorination of P-1,2,3-TCB the allylic chlorine and hydrogen atoms must be splitted off from neighbouring positions (4′)–(1) and (4)–(1′) to form the halogen-substituted structure of poly(2,3-dichloro-2-butene-1,4-diylidene) (*Scheme 9*).

---

[1] The energy of the C—X bond in allylic compounds, $H_2C{=}CH{-}CH_2X$, is 50-105 kJ/mol less than that in saturated analogues [11].

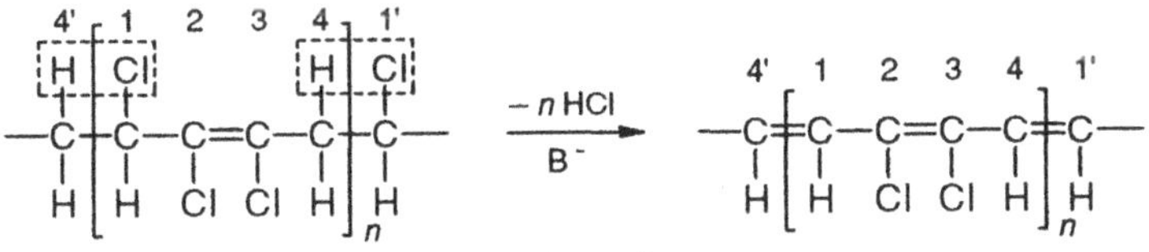

Scheme 9

It is easy to see that further elimination of HCl is possible only from neighbouring atoms at (1)–(2) and (3)–(4) positions, and complete dehydrochlorination must result in the formation of the cumulene form of carbyne (*Scheme 10*).

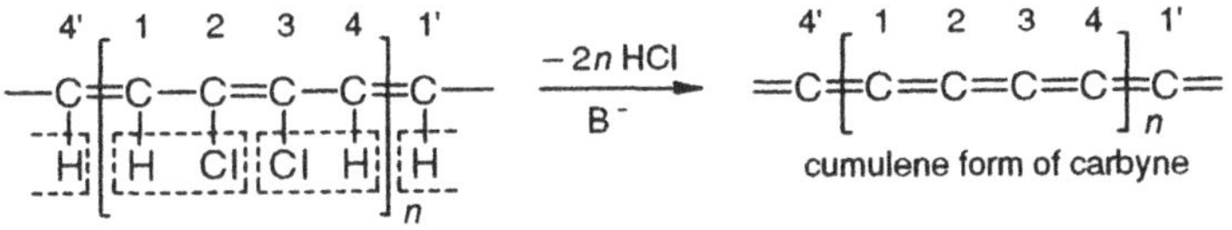

Scheme 10

3. Dehydrohalogenating Agents

Basic reagents being used for synthesizing carbynoid structures by the dehydrohalogenation reaction should meet two main requirements. Firstly, they should be strong enough to effect exhaustive elimination of hydrogen halides at the second stage of the reaction (*i.e.*, the dehydrohalogenation of halogen-substituted polyenes, *Schemes 2*, *3*, *8*, and *10*). Secondly, their nucleophilicity must be low to avoid side nucleophilic substitution reactions of halogen atoms. Furthermore, these reagents should be easily diffusible, *i.e.*, capable of penetrating in a polymer substrate (particle or film). Generally, the diffusibility of reagents is governed by the nature of organic solvents (or co-solvents) being used as a reaction medium. In each specific case the efficiency of dehydrohalogenating system (base + solvent) depends on the relation between solubility parameters (surface energies) of the system and the polymer being treated [1,13]. Adding various polar co-solvents (*e.g.*, THF [2,6], acetone [13]) or phase-transfer catalysts [14–18] allows one to adjust solubility parameters of the reacting medium and polymeric substrate, thereby making it possible to perform the reaction efficiently. It should be noted, however, that none of the basic reagents known to date fully comply with the requirements mentioned above. Dehydrohalogenating agents and systems that were reported to effect elimination of hydrogen halides yielding carbynoid structures (or their fragments) are listed in Table 1.

TABLE 1. Dehydrohalogenating agents and systems

Substrate	Base	Solvent	Co-Solvent or PTC [a]	Temperature	Reaction time, hr	Refs.
PVDB [b]	KOH	EtOH	THF	r.t. [c]	0.5	2,3
PVDC [b]	KOH	EtOH	THF	r.t.	1	2,3
	KOH	EtOH	–	reflux	168	19

TABLE 1. Dehydrohalogenating agents and systems (continued)

Substrate	Base	Solvent	Co-Solvent or PTC [a]	Temperature	Reaction time, hr	Refs.
PVDC [b]	$NaNH_2$	NH_3	–	-55 – 0 °C	10–100	8,20
	KNH_2	NH_3	–	-55 – 0 °C	10–30	20
	$NaNH_2$	THF	–	heating [d]	– [d]	8
	MeOK	MeOH	–	reflux	– [d]	8
	t-BuOK	DMSO	–	reflux	48	19
	Triton B [e]	Pyridine	–	reflux	48	19
PVDC [f]	KOH	EtOH	THF	r.t.	10	3,21
PVDC [f]	$NaNH_2$	NH_3	–	-33 °C	160–400	22
	KOH	EtOH	Pyridine	r.t./78 °C	112/2	22
	KOH	Benzene	18-C-6 [g,h]	80 °C	14	14
	KF/CaO	Benzene	18-C-6 [g,h]	80 °C	14	14
PVDC [i]	$NaNH_2$	NH_3	THF	50 °C [j]	20	23,24
	DBU [k]	DMA	–	70 °C	5	25
PVDF [b]	KOH	EtOH	THF	r.t.	10	2,3
	KOH	*i*-PrOH	–	70–100 °C	5	15,16
PVDF [b]	NaOH	H_2O	TBAB [l,h]	90 °C	24	15,16
	TBAH [m]	MeOH	–	65 °C	24	16
	t-BuOK	*t*-BuOH	–	80 °C	24	16
	DBU [k]	EtOH	–	80 °C	5	16
PVDF [f]	KOH	EtOH	Acetone	r.t.	1	13
	NaOH	H_2O	TBAB [l,h]	90 °C	– [d]	16
	TBAH [m]	MeOH	–	65 °C	– [d]	16
	KOH	*i*-PrOH	–	80 °C	– [d]	16
	DBU [k]	EtOH	–	80 °C	– [d]	16
PVDF [i]	KOH	DMA	QAS [n,h]	– [d]	– [d]	17
	KOH	DMF	*i*-PrOH [o]	r.t.	< 24	26
	LTMP [p]	DMF	Ether [o]	r.t.	< 24	26
	TMAH [q]	DMF	MeOH [o]	r.t.	< 24	26
	t-BuOK	DMF	–	r.t.	< 24	26
	t-BuO^- [r]	DMF	TBAP [s]	20 °C	– [d]	27
PDCE [f]	DBU [k]	DMSO [s]	–	70 °C	4	4
PDBE [f]	KOH	EtOH	THF	r.t.	3	5
PTCB [f,i]	KOH	EtOH	THF	r.t.	2–120	6,7
PTCB [i]	*t*-BuOK	THF	–	r.t.	2–120	7
Saran [t,b,f]	K(Na)OH	H_2O	TPAB [u,h]	70 °C	5–24	18
Saran [i]	K(Na)OH	H_2O/THF	TBAB [l,h]	r.t.	3–22	17,18
Saran [b,f]	KNH_2	NH_3	–	-55 – 0 °C	10–30	28
CPVC [v,b,i]	$NaNH_2$	NH_3	THF	50 °C [j]	20	23,24

[a] phase transfer catalyst; [b] reaction with a polymer powder; [c] room temperature; [d] the value has not been reported; [e] benzyltrimethylammonium hydroxide· [f] reaction with a polymer film (or fibre); [g] 18-crown-6; [h] used as a phase transfer catalyst; [i] reaction in (or with) solution; [j] the temperature indicated suggests the use of a high-pressure technique; the authors, however, did not report it [24]; [k] 1,8-diazabicyclo[5.4.0]-undec-7-ene; [l] tetrabutylammonium bromide; [m] tetrabutylammonium hydroxide; [n] quaternary ammonium salt; [o] solvent used to dissolve the base; [p] lithium 2,2,6,6-tetramethylpiperidide; [q] tetramethylammonium hydroxide; [r] *tert*.-butoxide species are supposed to be formed due to electrochemical reduction of *t*-BuOH during cathodic electrolysis in DMF; [s] anisole, DMF, THF, and toluene were also used, but the best results were obtained with DMSO; [t] Saran is a copolymer of vinylidene chloride and vinyl chloride; [u] tetrapropylammonium bromide; [v] chlorinated poly(vinyl chloride).

4. Characterization of the Dehydrohalogenation Products

4.1. VIBRATIONAL SPECTROSCOPY

4.1.1. *Infra-Red Spectroscopy*

As the dehydrohalogenation of halogen-containing polymers proceeds, the absorption bands, typical of C—Hal vibrations (mostly in the 400–700 cm^{-1} frequency region), disappear from the IR spectra of the reaction products. New absorption bands appear therewith near 1600 (strong), 1720 (mostly appears as a shoulder), and 2100–2200 (medium) cm^{-1}, being attributable to stretching vibrations of double C=C-bonds [1–8,14–29], carbonyl groups [5,14,15,19,25,29], and triple C≡C-bonds [1–8,14–29], respectively. Additional absorption bands (weak to medium) corresponding to C—H stretching vibrations in aliphatic groups appear in the 2850–3000 cm^{-1} frequency range

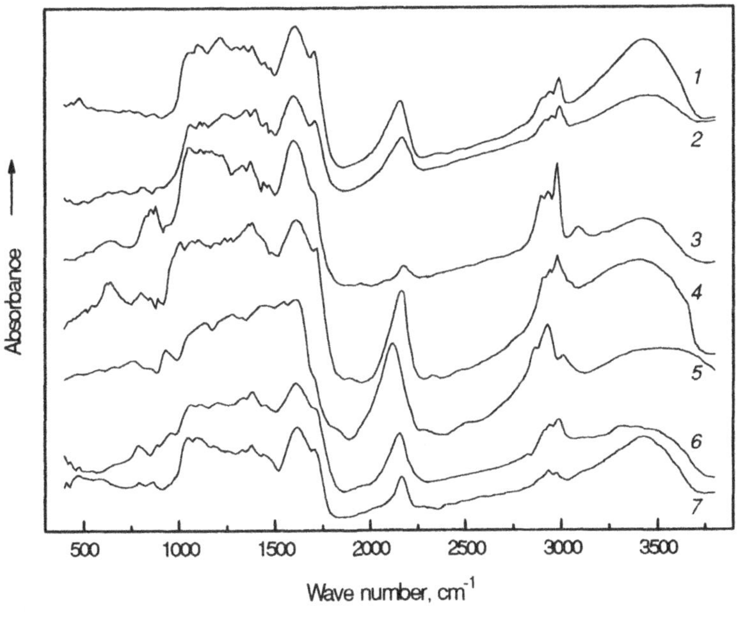

Figure 1. IR spectra of carbynoid materials produced by dehydrohalogenation of PVDB [2,3] (*1*); PVDC [2,3] (*2*); PVDF [2,3] (*3*); PDBE [5] (*4*); PDCE [4] (*5*); P-1,1,2-TCB [6,7] (*6*); and P-1,2,3-TCB [6,7] (*7*).

along with a strong broad band near 3400 cm^{-1} that can be assigned to both free (H_2O) and bonded hydroxy-groups [15,17–19,29]. The peaks near 870 and 750 cm^{-1} are attributable to out-of-plane bending of aromatic protons [14,18,29], and the band at 1100–1110 cm^{-1} [15,29] is likely to result from C—O—C stretching vibrations in ether groups. Aromatic structures can arise in the reaction products due to dehydrohalogenation of six-membered rings formed *via* the Diels-Alder type cyclization [14,18,29], whereas the ether linkages are produced as a consequence of the alkoxylation side reactions through the nucleophilic substitution of chlorine atoms by ethoxide species [15,16,29] or polymeric alkoxide ions [15,29].

It should be emphasized that the elucidation of chemical structure of the carbyne fragments on the basis of IR spectroscopic data is still troublesome. The absorption band in the region of 2100–2200 cm^{-1} was traditionally attributed in most studies to the stretching vibrations of triple C≡C-bonds [4,8,14–28] (including polyyne structures [4,8,21,27]), whereas the band near 1600 cm^{-1} was assigned to conjugated double C=C-bonds appeared in the reaction products owing to incomplete [14–18,27] or inter-chain [22] dehydrohalogenation, as well as side reactions of cross-linking [14,18], nucleophilic substitution [15,16] and addition [20]. However, in 1988 Evsyukov *et al.* suggested that vibrations of cumulated double bonds, =C=C=, formed by dehydro-halogenation can also contribute to the absorption band near 1600 cm^{-1} [2,3]. This assumption substantiated by previously reported data on IR spectra of some cumulenes was made in order to explain results of an Auger spectroscopic study that were interpreted within the framework of a cumulene-type model (see below). Subsequently, Babaev *et al.* [30] have calculated the phonon spectrum, *i.e.*, the fundamental vibrational frequencies for a model cumulene carbon chain having regular kinks[1] between linear fragments of four carbon atoms (a polybutatriene-like model) interacting with each other according to a harmonic law. The calculated data were shown to be in good agreement with the experimental spectrum of dehydrochlorinated PVDC, thus suggesting that the absorption of cumulated double bonds can occur both near 1600 cm^{-1} and in the 2000–2200 cm^{-1} region [30]. Recent *ab initio* calculations of the IR spectra of short cumulenes, *viz.*, $H_2(=C=C=)_nH_2$ (n = 1–5) performed by Kürti *et al.* [33] have confirmed this conclusion in principle, although the band at around 2100 cm^{-1} was reported to be the strongest one, what is quite opposite compared to experimental spectra of dehydrohalogenated polymers (Fig. 1). Analogous *ab initio* calculations for short polyynes, $H(—C≡C—)_nH$ (n = 1–5), revealed characteristic absorption bands in the 2000–2200 cm^{-1} region and no absorption near 1600 cm^{-1}. The most conspicuous absorption features of polyynes were shown to occur in the near infra-red region [33].

Finally, it is worthy of particular note that the IR spectra of the dehydrohalogena-tion products of various polymers at the later stages of the reaction are very similar (Fig. 1), even if the structure of these polymers allows one to expect the formation of different isomers of carbyne, *viz.*, polyyne and cumulene (*cf.* Fig. 1, *6* and *7*).

4.1.2. *Raman Spectroscopy*

Analogous contradictions exist in interpretation of Raman spectra of the dehydrohalogenation products. Spectral features reported to date by different groups

---

[1] According to a modem structural hypothesis proposed by Heimann *et al.* [31,32], the carbon chains in carbynes have regular defects (kinks) alternating with rectilinear fragments.

are very similar. The spectrum of dehydrochlorinated PVDC reported by Komatsu *et al.* [25] exhibits two pronounced lines at 1566 and 2159 cm^{-1} (514.5 nm excitation wavelength of an Ar^{+} laser) that were attributed to conjugated double and triple bonds, respectively. In addition to the peaks around 1550 and 2150 cm^{-1}, the spectrum of dehydrochlorinated PDCE reported by Shirakawa *et al.* [4] exhibits a line around 1140 cm^{-1}. The peak near 2150 cm^{-1} was assigned to C≡C stretching mode in polyyne fragments of carbyne, whereas the peaks around 1550 and 1140 cm^{-1} were attributed to C=C stretching vibrations in conjugated *trans*-polyenes being present in the reaction product due to imperfect chlorination of original polyacetylene [4]. On the other hand, the Raman spectrum of dehydrochlorinated PVDC showing virtually the same features, *viz.*, two marked peaks at 1550 and 2150 cm^{-1}, was interpreted by Berdyugin *et al.* [34] in terms of kinked polycumulene model. Recent first principles calculations by Kürti *et al.* [33] have shown Raman bands of both polyyne and cumulene carbon chains to occur at around 2000 cm^{-1}. The authors emphasized that distinguishing on the basis of Raman spectra between the polyyne and cumulene structures appears to be unlikely.

The frequency of the Raman bands in the spectra of dehydrohalogenated PVDC [34,35], PVDB [35], and PDCE [4] was reported to depend on the excitation wavelength. This suggests a wide length-distribution of the carbyne fragments formed by dehydrohalogenation. Based on empirical relation, Shirakawa *et al.* [4] have calculated the length of the carbyne (polyyne) fragments to be 12 (2157 cm^{-1}) to 65 (2113 cm^{-1}) conjugated triple C≡C-bonds. However, taking into account recent quantum chemical calculations of the dependence of Raman frequency on the conjugation length accomplished by Kuzmany *et al.* [36], this evaluation seems to be rather exaggerated.

4.2. ELECTRON SPECTROSCOPY

Electronic structure of carbonaceous materials produced by chemical dehydrohalogenation of some halogen-containing polymers was studied by various methods of electron spectroscopy including Auger electron spectroscopy (AES) [2,3,5–7,37–41], electron energy loss spectroscopy (EELS) [2,3,40,41], x-ray photoelectron spectroscopy (XPS) [6,7,42], and x-ray emission (C K_{α}) spectroscopy [42] in order to corroborate one-dimensional structure of these materials as well as determine the nature of carbon-carbon bonds therein.

4.2.1. *Auger electron spectroscopy*

Auger electron spectroscopy (AES) was employed in studies of dehydrohalogenated polymers by Khvostov *et al.* [37–41]. Experimental Auger spectra were normally transformed into curves of electronic density of states in the valence band (density of valence states, DOVS) that provide a most useful information on electronic structure of the valence band. The DOVS curve of a sample prepared by dehydrochlorination of the PVDC film (Fig. 2) exhibits four pronounced features at 12–13 (shoulder), 16–17,

19–20, and 25–26 eV [2,3,37–41]. The maxima at 16–17, 19–20, and 25–26 eV agree with the energy values corresponding to calculated bottom of the π-subband, top and bottom of the σ-subband, respectively, of carbon chains with an accuracy of 1 eV. The position of the σ-subband bottom at 25–26 eV attests to a stronger electron binding in the samples studied as compared to that in diamond and graphite. Such a stronger bond for carbon might be an *sp*-bond in the linear carbon chain [37,38]. It should be noted, however, that the position of the σ-subband within the 20–25 eV energy range is characteristic of all carbon, hydrocarbon, and halocarbon chains [43].

The π-subband of various chain-like molecules with up to 6 carbon atoms was reported to lie within the energy range of 7 to 11 eV for polyenes, 10 to 14 eV for polyynes, and 11 to 16.5 eV for cumulenes [40]. Therefore, the position of the π-subband bottom in carbynoid materials produced by dehydrohalogenation of PVDC [2,3,39,40] and PDBE [5] at 16–17 eV was interpreted as an evidence of the cumulene structure. Analogously, the peak at 14 eV in the DOVS curve of dehydrochlorinated P-1,1,2-TCB [6] was invoked to claim the presence of polyyne structures.

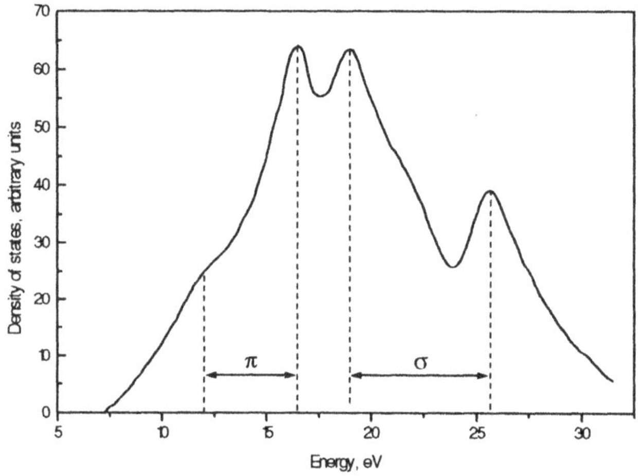

Figure 2. Density of states in the valence band of dehydrochlorinated PVDC film.

It should be noted that the DOVS curves of samples dehydrohalogenated at room temperature [2,6,39] as well as those prepared via multi-step polymer-analogous transformations [5] reveal electronic states in the 3–8 eV energy range (in contrast to samples prepared at –15°C) that was accounted for by the presence of non-carbyne (presumably graphite-like) structures.

4.2.2. *Electron Energy Loss Spectroscopy*

In order to study interband electronic transitions in dehydrohalogenated polymers, a method of the electron energy loss spectroscopy (EELS) was employed [40,41]. The

EELS data were found to be in conformity with the conclusion about one-dimensional nature of the material prepared by dehydrochlorination of PVDC. They revealed one-electron transitions and plasmons corresponding in energy to a one-dimensional chain. In addition, the EELS data substantiated the splitting of the π-subband and the metal-dielectric transition characteristic of one-dimensional systems with an energy spectrum of metallic type.

The electron energy loss is known to be described theoretically by a loss function, which is associated with real (ε_1) and imaginary (ε_2) parts of the dielectric constant of a solid. Plots of ε_2 and ε_1 calculated on the basis of experimental EELS data are shown in Fig. 3. The imaginary part ε_2 of the dielectric constant characterizes the energy spectrum of the one-particle electronic excitations, *i.e.*, it is proportional to the density of one-electron transitions from occupied states of the valence band to excited states. The $\varepsilon_2(E)$ curve (Fig. 3, solid line) exhibits a sharp maximum near the threshold (~1 eV) and two maxima at the energies of 7.8 and 9.9 eV. The first maximum is associated with a transition in the π-subband between states separated by a gap. The forbidden gap ΔE observed in dehydrohalogenated materials (1–2 eV) is formed due to either Peierls or Mott-Hubbard transitions predicted for a one-dimensional carbon chain [44]. The energy of the second maximum of $\varepsilon_2(E)$ at ~8 eV coincides with that of the transition from the bonding to the antibonding σ-subband. The third maximum at ~10 eV is caused by transitions from the bottom of the π-subband to the antibonding σ*-subband [40,41].

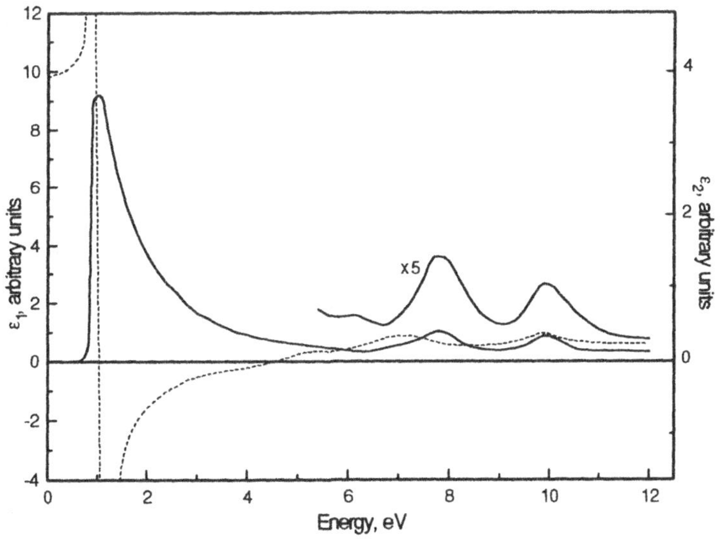

Figure 3. Real (ε_1) and imaginary (ε_2) parts of the dielectric function of dehydrochlorinated PVDC film.

The real part $\varepsilon_1(E)$ expresses collective nature of the electron oscillations in a solid. The condition for appearance of collective plasma oscillations (plasmons) is the equality of $\varepsilon_1(E)$ to zero, which is satisfied at the energy of *ca.* 4.5 eV in the samples studied (Fig. 3, dashed line) [40,41]. It is notable that the energy of the π-plasmons in dehydrohalogenated polymers is considerably larger than that in graphite and is close to that in a one-dimensional chain of polyacetylene [45].

Thus, the AES and EELS studies have corroborated the linear chain structure of the dehydrohalogenation products. Such features of the electronic structure as the continuity of the energy spectrum within the band, which attests to a large chain length, on the one hand, and a band width corresponding to short chains, on the other hand, suggest that the dehydrohalogenated chains consist of short, periodically repeated linear carbyne fragments [41].

4.2.3. *X-Ray Photoelectron Spectroscopy*

The results of XPS studies were shown to be consistent with those obtained by the AES method [42]. Experimental data were interpreted using semiempirical calculation of a band structure based on the strong binding method.

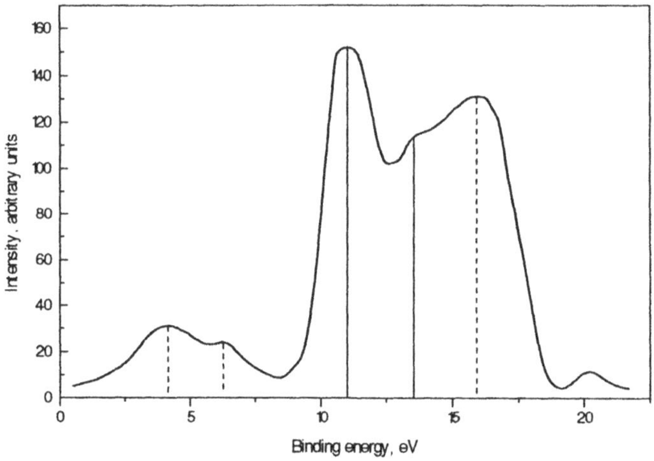

Figure 4. X-Ray photoelectron spectrum of dehydrofluorinated PVDF film.

The total XPS spectrum width (that can be identified as the valence band width) of dehydrofluorinated PVDF film was found to be about 20 eV (Fig. 4) which is the same as that in all other condensed carbon structures [46]. The energy levels of the "skeleton" *sp*-hybridized orbitals make up the bottom region of the valence band with a binding energy of 5–20 eV. The characteristic feature of the spectrum is the existence of a σ-gap. The vertical solid lines in Fig. 4 show the gap width to be *ca.* 2.5 eV. The existence of a gap as well as of the maxima near its edges at 11 and 13.5 eV was interpreted as a consequence of the chain conformations and the limited length of the straight carbyne fragments of the chain [47] which is in conformity with a conclusion inferred from the AES and EELS results.

4.2.4. *X-Ray Emission Spectroscopy*

The π-electronic system is affected more than any other by deviation of the chains from straightness. This phenomenon can be qualitatively explained in view of the symmetry of π-electron wavefunctions in space. Indeed, the two π-subsystems are oriented relative to the carbon chain so that their deformation in any direction must at once result in the alteration of the wavefunction degree of overlap of one of the two π-electrons of a particular atom with that of an adjacent one. It was therefore interesting to analyze the shape of the x-ray emission (C K_α) characteristic spectrum which is dominated by the *p*-part of the valence band. The principal peak occurring at 279 eV in the spectrum of dehydrofluorinated PVDF film corresponds to the electronic states deep inside the valence band, 6 eV below the edge at 285 eV (the latter is presumably close to the energy of optical transitions from the top edge of the valence band) [42]. At still lower photon energies the spectrum intensity decreases rapidly. This effect is due to the dipole selection rules: because of them the experimental spectrum reflects only those electronic states which are characterized by *p*-type symmetry of the wavefunctions. The comparison of the experimental C K_α spectrum with spectra calculated theoretically for three different carbon-chain model structures (*viz.*, rectilinear, zigzag, and kinked models) showed the best agreement for the zigzag and kinked models [42]. The energy dependence of the density of *p*-states was obtained for these structures, which had the principal maximum at a binding energy of 6 eV. Being in agreement with the AES, EELS, and XPS data, this result was considered to provide further support to the existence of conformation features in the dehydrohalogenation products and, consequently, in the carbyne chains.

4.3. ^{13}C NUCLEAR MAGNETIC RESONANCE SPECTROSCOPY

The use of NMR spectroscopy provides a deeper understanding of the molecular structure of the dehydrohalogenation products since individual structural groups can be determined without problems. In particular, the chemical shifts of linear carbynoid fragments occur in a range not occupied, and therefore influenced by other resonances. Since the lengths of the fragments also control the chemical shift the NMR spectra yield additional information on the conformation of carbyne and carbynoid structures.

The CP MAS ^{13}C NMR spectrum of the starting PVDC shows two strong peaks at 63.3 and 85.3 ppm which are attributable, respectively, to methylene and dichloromethylene groups alternating in macromolecules of the original polymer (Fig. 5) [29]. The latter peak disappears completely from the spectra of dehydrochlorination products as the reaction proceeds, whereas the former one was found to be shifted downfield by 3.5 ppm, and broadened. The spectrum of the final product exhibits conspicuous features at 13.9 (strong, sharp), 28.8 (weak), 66.8 (strong, broad), 79.5 (weak), 133.9 (strong, very broad), 164.0 (shoulder), and 194.9 ppm (medium, broad) (Fig. 5) [29]. The features at 66.8 and 164.0 ppm are attributable to carbyne fragments, respectively, polyyne [48–50] and cumulene [48,51] structure, formed by exhaustive dehydrochlorination of PVDC, *i.e.*, the target reaction. The peak at 13.9 ppm can be assigned to methyl carbons in ethoxy-groups that arise in macromolecules of PVDC due to a side reaction of nucleophilic substitution of chlorine atoms by ethoxide species under alkaline conditions (Williamson-type reaction [52,53]). Nucleophilic addition to the multiple carbon-carbon bonds [54] is also possible in later

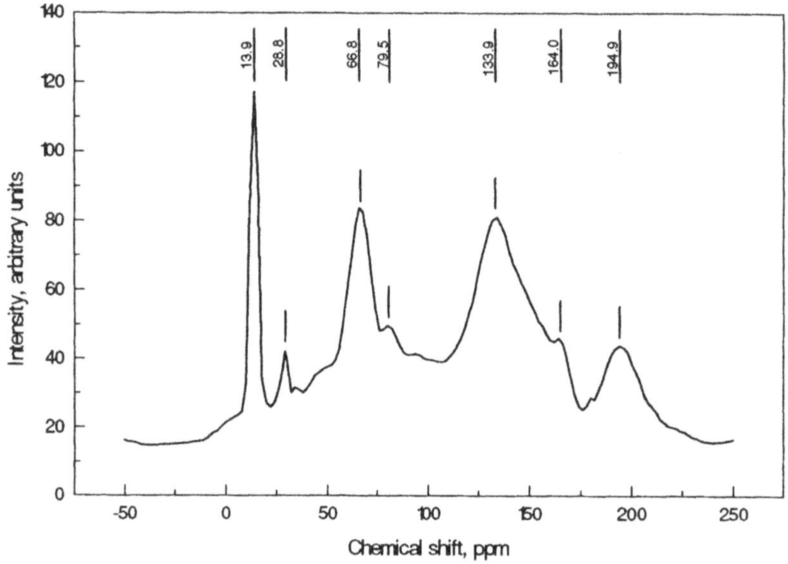

Figure 5. ^{13}C solid-state NMR spectrum of dehydrochlorinated PVDC powder.

stages of dehydrochlorination. It should be noted that oxymethylene moieties, $-O-CH_2-$, of the ethoxy groups may also contribute to the broad peak at 66.8 ppm. The broad peak at 194.9 ppm is due to carbonyl C=O groups arising in the reaction products through various mechanisms. For example, hydroxylated units formed *via* nucleophilic substitution of the chlorine atoms by hydroxyl groups can undergo further reactions of elimination and tautomeric transformation [19] (*Scheme 11*). Oxidation of multiple bonds in polyene, polyyne, or cumulene fragments is also feasible during synthesis or work-up procedures.

The low-intensity peak at 28.8 ppm can be assigned to methylene groups in the polymer chains as well as to methyl carbons in terminal ketomethyl (acetyl) groups, whereas the weak feature at 79.5 ppm cannot be unambiguously interpreted and has

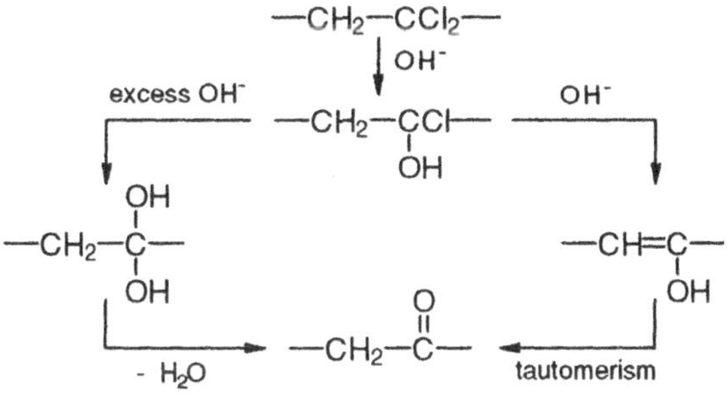

Scheme 11

yet to be dealt with. Similarly to mechanisms of alkali-induced degradation of poly(vinyl chloride) proposed by Schwenk *et al.* [55], the formation of terminal acetyl groups may occur owing to side reactions of the chain scission at the branching sites (*Scheme 12*, R=(–CH_2—CCl_2–)).

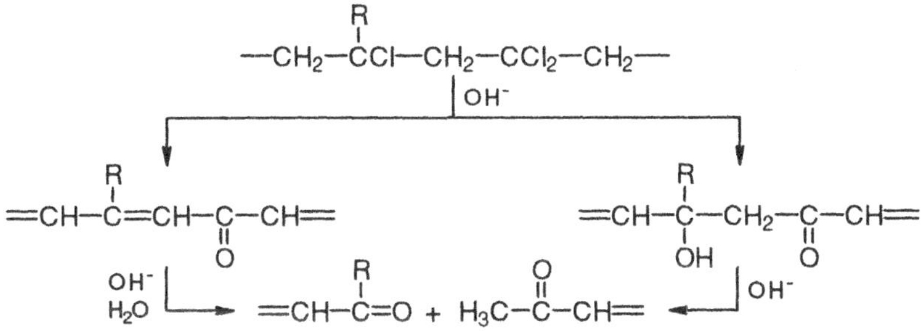

Scheme 12

Finally, the broad asymmetric peak at 133.9 ppm is attributable to various sp^2-carbon atoms in double bonds formed in the reaction products mainly because of the dehydrochlorination of defect structures arising in the macromolecules of the original PVDC due to side reactions of nucleophilic substitution as well as cross-linking. The cross-linking is supposed to involve a Diels-Alder mechanism in polyene sequences formed during the first stage of dehydrochlorination [14,18,56], although other mechanisms can also be expected to occur through breaking of multiple bonds in the reaction products (in particular, addition reactions of α,β-unsaturated ketones). It should be noted that signals from the carbon atoms in the cumulene fragments can also contribute to the peak at 133.9 ppm [57,58].

5. Conclusions

Based on the results described above, one can conclude that the interaction of halogen-containing polymers with basic reagents is a complex process involving dehydrochlorination as a target reaction along with side reactions of nucleophilic substitution and cross-linking as well as some secondary transformations of both intermediate and final products. The general route of this process for the case of PVDC and alcoholic alkali is outlined in *Scheme 13*.

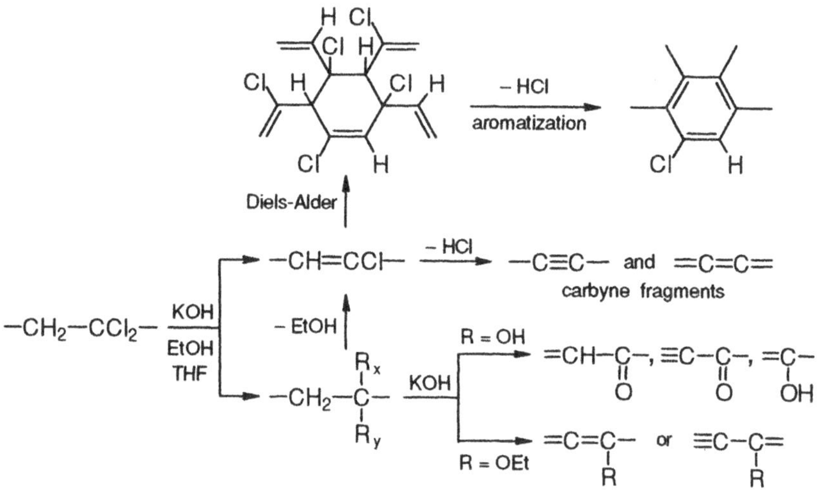

Scheme 13

The straightforward elimination of hydrogen halide results in the formation of carbyne fragments of both polyyne- and cumulene-type structures, whereas the side reactions produce various defects, *i.e.*, alkoxy and hydroxyl groups, carbonyl species, and cross-links randomly distributed within the polymer chains. It should be noted that the results obtained by NMR spectroscopy do not provide any evidence to confirm a previously proposed mechanism of the formation of the cumulene form of carbyne by alkali-induced dehydrohalogenation of poly(vinylidene halides) [2,3] inasmuch as both polyyne- and cumulene-type structures were found to be present in the reaction products. Moreover, taking into account an abundance of ethoxy groups in the products, one may assume that the mechanism of formation of double C=C-bonds *via* elimination of the alcohol molecules as suggested by Barton *et al.* [52] can also be involved in the overall process.

Based on the above considerations, it should be admitted that the final dehydrohalogenation products of halogen-containing polymers cannot be regarded as carbyne *sensu strictu* but rather as carbyne-containing or carbynoid polymeric materials. Considering the fact that the contribution of side reactions to the dehydrohalogenation process is quite noticeable, it can be concluded that the carbyne fragments in this material are rather short. This is in accord with evaluation of the length of linear carbyne fragments (*ca.* 8 carbon atoms) by Auger electron spectroscopy [40,41].

Since basic compounds presently used as dehydrohalogenating agents always possess some (often significant) nucleophilic reactivity, the complete suppression of the substitution side reactions seems to be rather unlikely, although a search for new dehydrohalogenating systems possessing high basicity but low nucleophilicity still remains a topical problem.

Another serious problem which makes almost hopeless the preparation of perfect carbyne structures (long-chain polyynes or cumulenes) is the presence of intrinsic defects in the original halogen-containing polymers. The polymers produced by radical polymerization always contain some defect structures such as "head-to-head" links and branching sites. For example, according to nuclear quadrupole resonance (NQR) spectroscopic data [59], the macromolecules of PVDC contain 12.5% of the "head-to-head" links which hamper exhaustive dehydrohalogenation (*Scheme 14*).

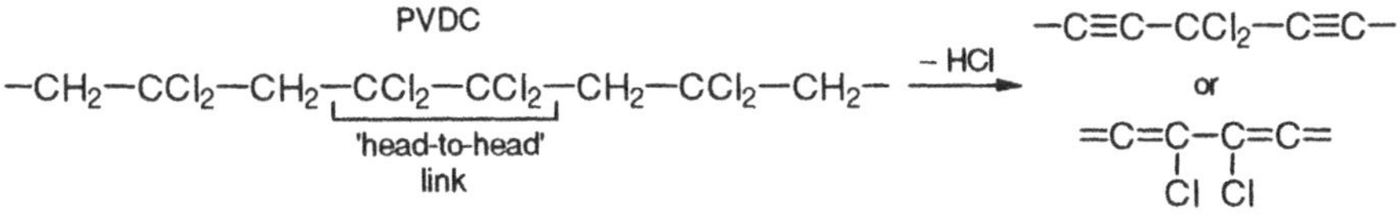

Scheme 14

Furthermore, poly(trichlorobutadiene)s contain 3,4-connected links along with predominating 1,4-links [12], that will afford allenic or acetylenic pendants during dehydrohalogenation [6] (*Scheme 15*).

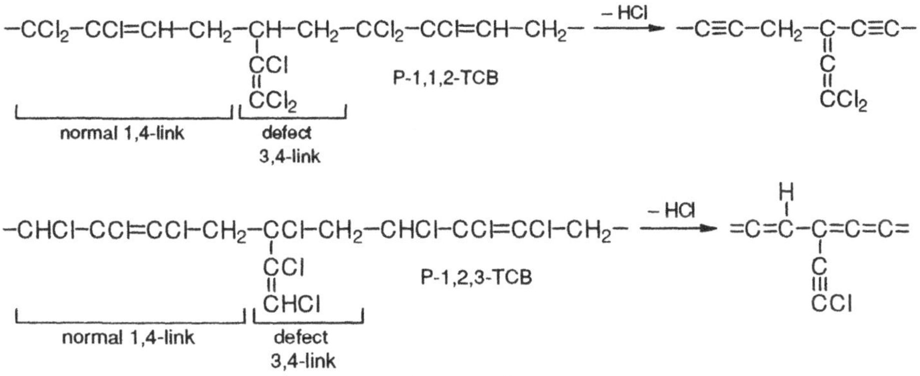

Scheme 15

Finally, poly(trichlorobutadiene)s are known to undergo allylic rearrangements with migration of allylic chlorine (1–4%) and hydrogen (3–10%) atoms under the action of solvents [60]. This accounts for remarkable similarity of the IR spectra of dehydrochlorinated P-1,1,2-TCB and P-1,2,3-TCB (Fig. 1).

The concurrent formation of polyyne- and cumulene-type fragments can be also accounted for by chemical reasons. For instance, only the cumulene structures can result from complete dehydrochlorination of the PVDC fragments located between ethoxylated carbon atoms, and only a triple bond can be formed at the carbon atom adjacent to the carbonyl group, provided that the elimination of hydrogen chloride is exhaustive. In these terms, however, it is not possible to take into account more complicated situations of different substituents at the same polymer fragment, odd number of the carbon atoms in a fragment linked to the carbonyl group, or the like.

Acknowledgement – I gratefully acknowledge the generous support by the Alexander-von-Humboldt Foundation under the auspices of its follow-up program.

6. References

1. Evsyukov, S.E., Kudryavtsev, Yu.P., and Korshak, Yu.V.: Chemical dehydrohalogenation of halogen-containing polymers, *Uspekhi Khimii* **60** (1991), 764–798 (in Russian) [*Russ. Chem. Rev.* **60** (1991), 373–390 (Engl. Transl.)].
2. Korshak, V.V., Kudryavtsev, Yu.P., Korshak, Yu.V., Evsyukov, S.E., Khvostov, V.V., Babaev, V.G., and Guseva, M.B.: Formation of β-carbyne by dehydrohalogenation, *Makromol. Chem., Rapid Commun.* **9** (1988), 135–140.
3. Evsyukov, S.E., Kudryavtsev, Yu.P., Korshak, Yu.V., Khvostov, V.V., Babaev, V.G., Guseva, M.B., and Korshak, V.V.: Synthesis of carbyne on the basis of poly(vinylidene halides), *Vysokomolek. Soedin., Ser. A* **31** (1989), 27–33 (in Russian) [*Polym. Sci. USSR* **31** (1989), 29–37 (Engl. Transl.)].
4. Akagi, K., Nishiguchi, M., Shirakawa, H., Furukawa, Y., and Harada, I.: One-dimensional conjugated carbyne – Synthesis and properties, *Synth. Met.* **17** (1987), 557–562.
5. Korshak, V.V., Kudryavtsev, Yu.P., Evsyukov, S.E., Korshak, Yu.V., Khvostov, V.V., Babaev, V.G., and Guseva, M.B.: Synthesis of carbyne on the basis of poly(vinyl chloride), *Doklady Akad. Nauk SSSR* **300** (1988), 889–892 (in Russian) [*Dokl. Phys. Chem.* **300** (1988), 510–512 (Engl. Transl.)].
6. Vointseva, I.I., Gilman, L.M., Kudryavtsev, Yu.P., Evsyukov, S.E., Pesin, L.A., Gribov, I.V., Moskvina, N.A., and Khvostov, V.V.: Chemical dehydrochlorination of polytrichlorobutadienes – A new route to carbynes, *Eur. Polym. J.* **32** (1996), 61–68 and 403.
7. Vointseva, I.I., Gilman, L.M., Kudryavtsev, Yu.P., Evsyukov, S.E., and Valetsky, P.M.: Synthesis of polymorphous modifications of carbyne by chemical dehydrochlorination of polytrichlorobutadiene isomers, *Vysokomolek. Soedin., Ser. A* **38** (1996), 1116–1121 (in Russian) [*Polym. Sci., Ser. A* **38** (1996), 707–711 (Engl. Transl.)].
8. Kudryavtsev, Yu.P., Sladkov, A.M., Aseev, Yu.G., Nedoshivin, Yu.N., Kasatochkin, V.I., and Korshak, V.V.: Investigation of properties and structure of carbyne, *Doklady Akad. Nauk SSSR* **158** (1964), 389–392 (in Russian).
9. Gordon, A.J. and Ford, R.A.: *The Chemist's Companion. A Handbook of Practical Data, Techniques, and References*, Wiley, New York, 1972.
10. Korshak, V.V. and Matveeva, N.G.: Macromolecular compounds. Report 56. Concerning the influence of stereochemical factors on the ability of halogen-substituted ethylenes to polymerize, *Izv. Akad. Nauk SSSR, Otdel. Khim. Nauk* No.2 (1953), 344–350 (in Russian).
11. *Chemical Encyclopaedia*, Vol. 1, Soviet Encyclopaedia Publ., Moscow, 1988, p. 105 (in Russian).
12. Vointseva, I.I., Klimentova, N.V., and Niazashvili, G.A.: Poly(1,1,2-trichlorobutadiene-1,3) and its compositions: Part I – Synthesis, structure and reactivity, *Int. J. Polym. Mater.* **29** (1995), 69–96.
13. Kudryavtsev, Yu.P., Evsyukov, S.E., and Babaev, V.G.: An efficient dehydrohalogenating system for poly(vinylidene fluoride), *Izv. Akad. Nauk, Ser. Khim.* (1992), 1223–1225 (in Russian) [*Bull. Russ. Acad. Sci., Div. Chem. Sci.* **41** (1992), 966–967 (Engl. Transl.)].
14. Yie-Shun, Ch., Jagur-Grodzinski, J, and Vofsi, D.: Dehydrochlorination of polyvinylidene chloride (PVDC) in the presence of crown ether as a phase transfer agent, *J. Polym. Sci.: Polym. Chem. Ed.* **23** (1985), 1193–1202.
15. Kise, H. and Ogata, H.: Phase transfer catalysis in dehydrofluorination of poly(vinylidene fluoride) by aqueous sodium hydroxide solutions, *J. Polym. Sci.: Polym. Chem. Ed.* **21** (1983), 3443–3451.
16. Kise, H. Ogata, H., and Nakata, M.: Chemical dehydrofluorination and electrical conductivity of poly(vinylidene fluoride) films, *Angew. Makromol. Chem.* **168** (1989), 205–216.
17. He, F.-F. and Kise, H.: Simultaneous dehydrochlorination and film formation of a vinylidene chloride copolymer by aqueous-organic interface reaction, *Makromol. Chem.* **186** (1985), 1395–1400.
18. He, F.-F. and Kise, H.: Dehydrochlorination of vinylidene chloride—vinyl chloride copolymer by aqueous sodium or potassium hydroxide solutions under two-phase conditions, *J. Polym. Sci.: Polym.*

Chem. Ed. **21** (1983), 1729–1740.
19. Pendleton, P., Vincent, B., and Hair, M.L.: Dehydrochlorination of monodisperse poly(vinylidene chloride) latex, *J. Colloid Interface Sci.* **80** (1981), 512–527.
20. Yamaguchi, T., Amagasa, M., Sasaki, E., Chiba, T., and Takama, M.: Dehydrochlorination reaction of polyvinylidene chloride in liquid ammonia, *Embi to Porima [Vinyls and Polymers]* **18**, No.5 (1978), 8–12.
21. Korshak, V.V., Kudryavtsev, Yu.P., Litovchenko, G.D., and Varfolomeeva, O.B.: A study of the dehydrochlorination reaction of halogen-containing compounds, *Doklady Akad. Nauk SSSR* **283** (1985), 1411–1414 (in Russian).
22. Kudryavtsev, Yu.P., Varfolomeeva, O.B., Litovchenko, G.D., and Sladkov, A.M.: A study of the dehydrochlorination reaction of poly(vinylidene chloride), *Izv. Akad. Nauk SSSR, Ser. Khim.* (1983), 195–197 (in Russian) [*Bull. Acad. Sci. USSR, Div. Chem. Sci.* **32** (1983), 170–172 (Engl. Transl.)].
23. Tsuchida, E., Shinohara, I., and Kambara, Sh.: Dehydrochlorination products from polyhalogen-containing polymers, *Kogyo Kagaku Zasshi [J. Chem. Soc. Jpn., Ind. Chem. Sect.]* **66** (1963), 1404–1405.
24. Tsuchida, E., Shih, Ch.-N., Shinohara, I., and Kambara, Sh.: Synthesis of a polymer chain having conjugated unsaturated bonds by dehydrochlorination of polyhalogen-containing polymers, *J. Polym. Sci.: Pt. A* **2** (1964), 3347–3354.
25. Komatsu, T., Nomura, M., Kakudate, Y., Fujiwara, S., and Heimann, R.B.: Characterization of dehydrochlorinated poly(vinylidene chloride) and the shock-compressed material, *Macromol. Chem. Phys.* **196** (1995), 3031–3040.
26. Dias, A.J. and McCarthy, Th.J.: Dehydrofluorination of poly(vinylidene fluoride) in dimethylformamide solution: Synthesis of an operationally soluble semiconducting polymer, *J. Polym. Sci.: Polym. Chem. Ed.* **23** (1985), 1057–1061.
27. Kijima, M., Toyabe, Ts., and Shirakawa, H.: Electrocatalytic dehydrofluorination of poly(vinylidene fluoride), *Chem. Lett.* (1995), 553–554.
28. Yamaguchi, T., Sasaki, E., and Chiba, K.: Dehydrochlorination reaction of poly(vinyl chloride-vinylidene chloride) copolymer in liquid ammonia, *Embi to Porima [Vinyls and Polymers]* **18**, No.8 (1978), 22–27.
29. Evsyukov, S.E., Paasch, S., Thomas, B., and Heimann, R.B.: Formation of carbynoid structures by chemical dehydrochlorination of poly(vinylidene chloride). A ^{13}C solid-state NMR study, *Ber. Bunsen-Ges. Phys. Chem.* **101** (1997), 837–841.
30. Korshak, V.V., Kudryavtsev, Yu.P., Evsyukov, S.E., Korshak, Yu.V., Guseva, M.B., Babaev, V.G., and Kostishko, B.M.: Infra-red spectra of carbyne, *Doklady Akad. Nauk SSSR* **298** (1988), 1421–1424 (in Russian) [*Dokl. Phys. Chem.* **298** (1988), 199–201 (Engl. Transl.)].
31. Heimann, R.B., Kleiman, J., and Salansky, N.M.: A unified structural approach to linear carbon polytypes, *Nature (London)* **306** (1983), 164–167.
32. Heimann, R.B., Kleiman, J., and Salansky, N.M.: Structural aspects and conformation of linear carbon polytypes (carbynes), *Carbon* **22** (1984), 147–156.
33. Kürti, J., Magyar, C., Balázs, A., and Rajczy, P.: Vibrational analysis for short carbon chains with alternating and cumulenic structure, *Synth. Met.* **71** (1995), 1865–1866.
34. Berdyugin, V.V., Kudryavtsev, Yu.P., Evsyukov, S.E., Korshak, Yu.V., Shorygin, P.P., and Korshak, V.V.: Resonance Raman scattering spectra of β-carbyne, *Doklady Akad. Nauk SSSR* **305** (1989), 362–365 (in Russian) [*Dokl. Phys. Chem.* **305** (1989), 217–219 (Engl. Transl.)].
35. Evsyukov, S.E.: Polymers with a system of multiple bonds on the basis of the dehydrohalogenation reaction, *Ph.D. Thesis*, A.N. Nesmeyanov Institute of Organoelement Compounds, Russian Academy of Sciences, Moscow, 1989 (in Russian).
36. Kastner, J., Kuzmany, H., Kavan, L., Dousek, F.P., and Kürti, J.: Reductive preparation of carbyne with high yield. An *in situ* Raman scattering study, *Macromolecules* **28** (1995), 344–353.
37. Korshak, V.V., Kudryavtsev, Yu.P., Khvostov, V.V., Guseva, M.B., and Babaev, V.G.: A study of electronic structure of carbyne by the Auger spectroscopy method, *Doklady Akad. Nauk SSSR* **280** (1985), 402–403 (in Russian).
38. Khvostov, V.V., Babaev, V.G., and Guseva, M.B.: Auger spectroscopy of amorphous carbon films, *Fiz. Tverd. Tela (Leningrad)* **27** (1985), 887–891 (in Russian).

39. Korshak, V.V., Kudryavtsev, Yu.P., Khvostov, V.V., Guseva, M.B., Babaev, V.G., and Rylova, O.Yu.: Experimental verification of novel structural hypothesis of carbyne, *Doklady Akad. Nauk SSSR* **293** (1987), 393–396 (in Russian).
40. Korshak, V.V., Kudryavtsev, Yu.P., Khvostov, V.V., Guseva, M.B., Babaev, V.G., and Rylova, O.Yu.: Electronic structure of carbynes studied by Auger and electron energy loss spectroscopy, *Carbon* **25** (1987), 735–738.
41. Kudryavtsev, Yu.P., Evsyukov, S.E., Guseva, M.B., Babaev, V.G., and Khvostov, V.V.: Carbyne – the third allotropic form of carbon, *Izv. Akad. Nauk, Ser. Khim.* No.3 (1993), 450–463 (in Russian) [*Russ. Chem. Bull.* **42** (1993), 399–413 (Engl. Transl.)].
42. Kudryavtsev, Yu.P., Baytinger, E.M., Kugeev, F.F., Korshak, Yu.V., and Evsyukov, S.E.: Electronic structure of carbyne studied by x-ray photoelectron spectroscopy and x-ray emission spectroscopy, *J. Electron Spectrosc. Relat. Phenom.* **50** (1990), 295–307.
43. *Electron Spectroscopy: Theory, Techniques and Applications*, ed. by Brundle, C.R. and Baker, A.D., Vols. 1–4, Academic Press, London, 1977.
44. Rice, M.J., Phillpot, S.R., Bishop, A.R., and Campbell, D.K.: Solitons, polarons, and phonons in the infinite polyyne chain, *Phys. Rev. B* **34** (1986), 4139–4149.
45. Streitwolf, H.W.: Physical properties of polyacetylene, *Phys. Stat. Solidi (B)* **127** (1985), 11–54.
46. Baytinger, E.M.: *Electronic Structure of Condensed Carbon*, Urals State University Publications, Sverdlovsk, 1988 (in Russian).
47. Brus, L.: Size dependent development of band structure in semiconductor crystallites, *New J. Chem.* **11** (1987), 123–127.
48. Kleinpeter, E. and Borsdorf, R.: *^{13}C-NMR-Spektroskopie in der Organischen Chemie*, Akademie-Verlag, Berlin, 1981.
49. Lagow, R.J., Kampa, J.J., Wei, H.-Ch., Battle, S.L., Genge, J.W., Laude, D.A., Harper, C.J., Bau, R., Stevens, R.C., Haw, J.F., and Munson, E.: Synthesis of linear acetylenic carbon: the "sp" carbon allotrope, *Science* **267** (1995), 362–367.
50. Bartik, T., Bartik, B., Brady, M., Dembinski, R., and Gladysz, J.A.: A step-growth approach to metal-capped one-dimensional carbon allotropes: Syntheses of C_{12}, C_{16}, and C_{20} μ-polyynediyl complexes, *Angew. Chem. Int. Ed. Engl.* **35** (1996), 414–417.
51. Touchard, D., Haquette, P., Daridor, A., Toupet, L., and Dixneuf, P.H.: First isolable pentatetraenylidene metal complex containing the $Ru{=}C{=}C{=}C{=}C{=}CPh_2$ assembly. A key intermediate to provide functional allenylidene complexes, *J. Am. Chem. Soc.* **116** (1994), 11157–11158.
52. Barton, S.S., Dacey, J.R., and Harrison, B.H.: Thermal and chemical dehydrohalogenation of Saran powder, *Polym. Prepr.* **12** (1971), 768–779.
53. Stark, J.G.: Organic name reactions –II, *Educ. Chem.* **4** (1967), 200–206.
54. Bohlmann, F. and Viehe, H.-G.: Polyacetylenverbindungen, IX. Mitteil.: Alkalische und saure Addition an Polyine, *Chem. Ber.* **88** (1955), 1017–1027.
55. Schwenk, U., Streitberger, H., and König, I.: Über die Spaltung von PVC mit Alkali in Tetrahydrofuran, *Angew. Makromol. Chem.* **47** (1975), 43–59.
56. Barton, S.S., Boulton, G., Harrison, B.H., and Kemp, W.: Study of the dehydrochlorination of polyvinylidene chloride by alcoholic base, using ultra-violet absorption spectroscopy, *Trans. Faraday Soc.* **67** (1971), 3534–3539.
57. Lin, J., Ijadi-Maghsoodi, S., Barton, Th.J., Meyer, R.K., Benner, R.E., Vardeny, Z.V., Wei, X., Smith, A.V., and Shinar, J.: First synthesis and study of cumulene-containing conjugated polymers as third-order nonlinear optical active materials, *Polym. Prepr.* **35** (1994), 832–833.
58. Lin, J., Ijadi-Maghsoodi, S., and Barton, Th.J.: Synthesis and study of cumulene-containing polymers with flexible blocks in the main chain, *Polym. Prepr.* **36** (1995), 499–500.
59. Semin, G.K., Robas, V.I., and Babushkina, T.A.: NQR in polymers, in: *Solid-State Radiospectroscopy*, Atomizdat, Moscow, 1967, pp. 218–220 (in Russian).
60. Lebedeva, T.L., Vointseva, I.I., Gil'man, L.M., Petrovskii, P.V., and Larina, T.A.: Solvent-induced allylic rearrangements in poly(trichlorobutadiene) chains, *Izv. Akad. Nauk, Ser. Khim.* (1997), 763–768 [*Russ. Chem. Bull.* **46** (1997), 732–738 (Engl. Transl.)].

3.2.2 PHOTO- AND LASER-INDUCED DEHYDROHALOGENATION OF POLYMERS

Akira YABE
National Institute of Materials and Chemical Research
Higashi 1-1, Tsukuba, Ibaraki 305-8565, Japan

Abstract - Photo-induced dehydrohalogenation of halogenated polymers produces a conjugated polyene and/or polyyne. The mechanism for dehydrohalogenation of poly(vinyl chloride) (PVC) was discussed in terms of the C-Cl bond cleavage and the following zip-dehydrochlorination. Irradiation of PVC, chlorinated PVC, and poly(vinylidene chloride) with an Ar ion laser or excimer lasers was carried out in order to synthesize longer conjugation polyene and/or polyyne. Irradiation with a KrF excimer laser in vacuum produced a polyyne structure with a conjugation length of about 30 carbon atoms.

1. Introduction

As one of the most promising methods to synthesize carbyne, dehydrohalogenation of halogenated polymer has been investigated by using a variety of transformations such as chemical, pyrolytic, photochemical, and electrochemical methods. In this chapter, a photochemical method will be discussed, in particular, using laser-induced reactions which have much progressed in this decade.

This method originated from the discovery of discoloration on poly(vinylchloride) to pale yellow on exposure to sunlight. Although this phenomenon of PVC is concerned with a troublesome problem in industrial applications, we would like to pay attention to the finding that it is attributed to the formation of conjugated double bonds by dehydrochlorination. Below, photochemical dehydrochlorination of PVC is reviewed as a preliminary study for the dehydrochlorination of higher-chlorinated polymers.

Based on many investigations on the photo-induced dehydrochlorination of PVC, chlorinated PVC (Cl-PVC) and poly(vinylidene chloride) (PVDC) were used as precursors for the synthesis of carbyne. The selection of polymer precursors is one of the most important subjects. Unfortunately, however, not many polymers were studied in the past. In the case of photo-induced reactions, dehydrohalogenation has been limited to elimination of hydrochloride in the previous works. The reasons can be explained by : (1) chlorinated polymers are easily accessible and fairly stable as precursors, and (2) the cleavage of the Cl-C bonds is effectively controlled by photochemical reactions.

R.B. Heimann et al. (eds.), Carbyne and Carbynoid Structures, 75–91.

$$-(CH_2\text{-}CHCl)_n\text{-} \xrightarrow{h\nu} -(\dot{C}H{=}CH)_{n-x}\text{-}CH\text{-}CH_2\text{-} \xrightarrow{heat} -(CH{=}CH)_n\text{-}CH_2\text{-}CHCl\text{-} \quad (n<20) \tag{1}$$

$$-(CH{=}CH)_n\text{-}\dot{C}H\text{-}CH_2\text{-} \quad (n=3) \tag{2}$$

Poly(tetrafluoroethylene) (PTFE), an important precursor for the chemical and electrochemical methods, is not an appropriate precursor because the cleavage of F-C bond does not proceed selectively upon photo-irradiation as compared to the cleavage of C-C bonds. However, the selective C-F bond cleavage by photochemical reaction is an interesting subject to be studied in the near future.

Another important subject is the control of the photochemical reaction. At present, a variety of lasers are available as a light source in photochemical reactions. In particular, excimer lasers, developed in the late-1970s to 80s, emit high quality ultraviolet photons and hence are interesting light sources for the control of the photochemical reaction. Nowadays excimer lasers have made possible many new photon processes such as microstructuring by excimer laser ablation and excimer-laser-induced surface modification in material science. Here, excimer laser-induced reactions of Cl-PVC and PVDC will be described. A new promising reaction for synthesis of carbyne has been advanced by excimer-laser-induced surface dehydrochlorination recently.

2. Photo-induced dehydrochlorination of PVC

Studies on photo-induced dehydrochlorination of PVC started from the viewpoints of its degradation and stability. Since PVC was applied in industry, it has been observed that PVC discolors to pale yellow on prolonged exposure to sunlight. The photochemistry of PVC has been widely investigated in order to clarify the mechanisms which are responsible for this discoloration. Here, we want to pay attention mainly to the results of dehydrochlorination to form conjugated carbon-carbon double bonds in attempt to synthesize conjugated carbon-carbon triple bonds.

To clarify the mechanism of discoloration of PVC, Nishijima *et al.* followed the change of PVC on light exposure by UV-vis spectroscopy [1]. Upon irradiation with a wavelength of 185 nm with a low pressure mercury lamp in vacuum, a new absorption peak at 280 nm appeared and increased with an additional tailing band in the longer wavelength by further irradiation (Fig. 1). On heating the UV irradiated sample above 80 ^{o}C, a broad absorption band with the maximum at 550 nm appeared (Fig. 2).

From these spectral results, the authors assumed that discoloration could be ascribed to the formation of relatively short polyene segments by photochemical dehydrochlorination. The "conjugation length (n)" of the 550 nm-polyene formed on heating was estimated to be up to 20 conjugated double bonds.

Moreover, they concluded that the polyene sequence was only propagated along the PVC chain when the chains were released. From further ESR study, it was concluded that the propagating species obtained by irradiation at 185 nm might be the polyenyl radical (UV max. around 325 nm) with about three conjugated double bonds [2].

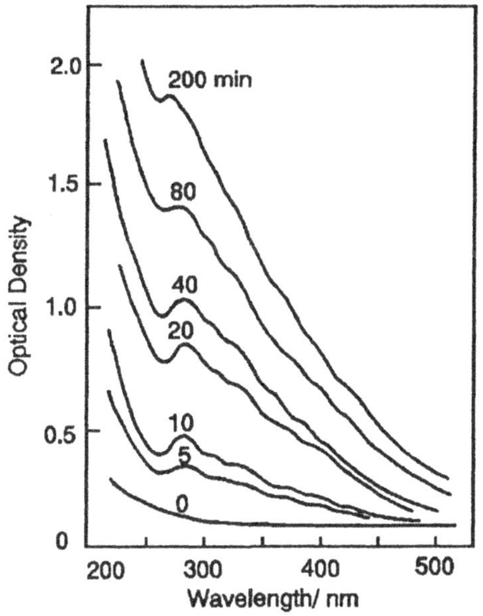

Figure 1. Changes of UV-vis absorption spectra of PVC on irradiation with a wavelength of 185 nm [1].

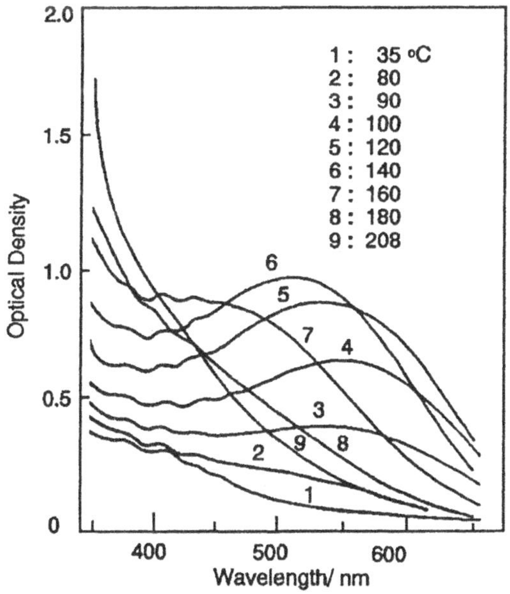

Figure 2. Changes of UV-vis absorption spectra of UV-irradiated PVC on heating [1].

Decker and Balandier studied the quantum yield of dehydrochlorination of PVC upon irradiation with a medium pressure mercury lamp and an argon ion laser [3,4].

In the case of an UV lamp, the quantum efficiency f was estimated to be 0.011 in a nitrogen atmosphere and 0.015 in pure oxygen. On the other hand, f is reduced by 50% when the degradation was induced by irradiation with an argon ion laser with wavelengths of 488 or 514 nm (Figs. 3 and 4).

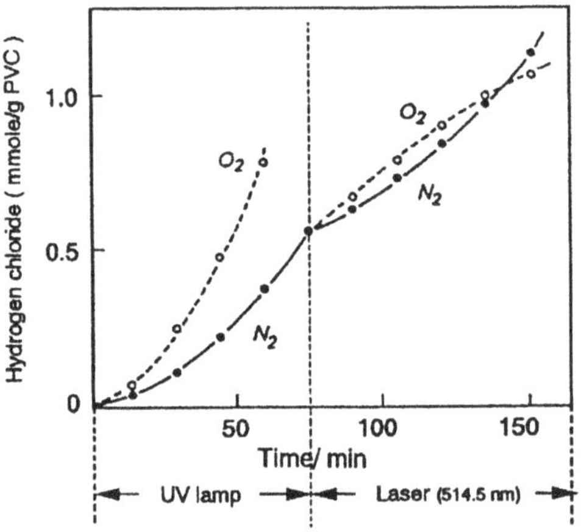

Figure 3. Kinetics of dehydrochlorination of PVC film in the UV and 514.5 nm photolysis in the presence of nitrogen or oxygen. Light intensity 1.5×10^{-7} einsteins s^{-1} [3].

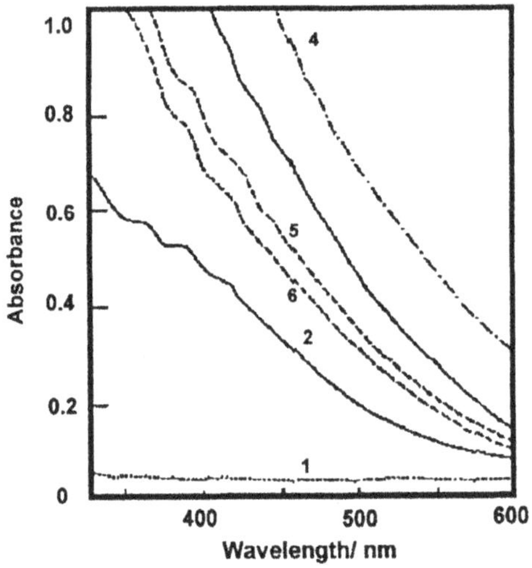

Figure 4. UV-vis spectra of PVC film ; (1): unirradiated, (2), (3): UV irradiated in nitrogen for 30 and 75 min, (4): UV irradiation for 75 min and laser for 60 min in nitrogen, (5), (6): UV irradiation in nitrogen for 75 min and laser irradiation in oxygen for 30 and 60 min [3].

$$\text{-(CH}_2\text{-CHCl)}_n\text{-} \longrightarrow \text{-(CH=CH)}_n\text{-} + n\,\text{HCl} \tag{3}$$

$$\phi_{HCl} = \frac{\text{Number of Hcl molecules evoloved per gram of PVC}}{\text{Number of quanta absorbed per gram of PVC}} \tag{4}$$

Moreover, they reported the formation of conjugated polyene sequences of various lengths up to about 20 by irradiation with the UV lamp for 75 min and further exposure with the 514.5 nm laser. In the case of laser irradiation under oxygen atmosphere, photobleaching of polyene with oxygen proceeded in quantum yield of 0.015. Concerning the mechanism of the promotion of polyene by further laser irradiation, they assumed the following reactions (5)-(11). The chlorine radical evolved in reaction (6) is expected to react with the polymer by abstracting a hydrogen atom, preferentially from a methylene group, either from the adjacent CH_2 of the polyenyl radical, thus increasing the polyene sequence length (cage reaction (6)),

$$\text{-(CH=CH)}_{14}\text{-} \xrightarrow{h\nu\,(514.5\text{ nm})} [\text{-(CH=CH)}_{14}\text{-CH(Cl)-CH}_2\text{-}]^* \tag{5}$$

$$[\text{-(CH=CH)}_{14}\text{-CH(Cl)-CH}_2\text{-}]^* \longrightarrow \text{-(CH+CH)}_{14}\text{-}\dot{\text{C}}\text{H-CH}_2\text{-} + \dot{\text{C}}\text{l} \tag{6}$$

$$\text{-(CH=CH)}_{14}\text{-}\dot{\text{C}}\text{H-CH}_2 + \dot{\text{C}}\text{l} \longrightarrow \text{-(CH=CH)}_{14}\text{-CH=CH-} + \text{HCl} \tag{7}$$

or from a methylene group of another chain, thus indicating a "*zip-dehydro chlorination*" process with the formation of new polyenes (reactions (8)-(11)):

$$\text{Cl} + \text{-CH}_2\text{-CHCl-CH}_2\text{-CHCl-} \longrightarrow \text{-}\dot{\text{C}}\text{H-CHCl-CH}_2\text{-CHCl-} + \text{HCl} \tag{8}$$

$$\text{-}\dot{\text{C}}\text{H-CHCl-CH}_2\text{-CHCl-} \longrightarrow \text{-CH=CH-CH}_2\text{-CHCl-} + \dot{\text{C}}\text{l} \tag{9}$$

$$\text{-CH=CH-CH}_2\text{-CHCl-} + \dot{\text{C}}\text{l} \longrightarrow \text{-CH=CH-}\dot{\text{C}}\text{H-CHCl-} + \text{HCl} \tag{10}$$

$$\text{-CH=CH-}\dot{\text{C}}\text{H-CHCl-} \longrightarrow \text{-CH=CH-CH=CH-} + \dot{\text{C}}\text{l} \longrightarrow \text{polyene} \tag{11}$$

This reaction scheme would explain the formation of polyenes with sequence lengths shorter than 14 observed during the 514.5 nm laser irradiation of PVC in nitrogen.

Wavelength effects on the discoloration of PVC were studied by Gardette and Lemaire [5]. Upon irradiation at short wavelength of 254 nm, a triene was formed and accumulated in the form of two photoisomers with absorptions at 270 and 284 nm.

These polyenes were photooxidizable and quickly disappeared on irradiation in the presence of oxygen. Irradiation at a long wavelength of 365 nm led to longer polyene sequences with conjugation lengths ranging from 2 to 10. However, their formation required prolonged exposure times, since the initial absorption of PVC was quite low at this wavelength. The triene with absorbance at 248 nm was shown to be a precursor of these polymers. The results are summarized in Scheme 1.

According to these mechanisms, polyene sequences are formed by successive and multiple photochemical excitations. Each excitation produces a direct scission of C-Cl bonds and the chlorine atom formed can abstract a neighboring hydrogen atom. Some chlorine atoms can escape from the intermediate surroundings and initiate an oxidation chain. Under the oxygen the radical in the polymer chain is terminated by the reactions shown in Scheme 2 .

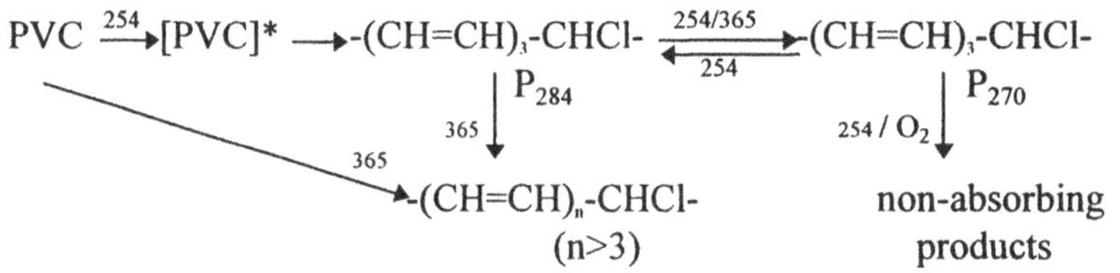

Scheme 1

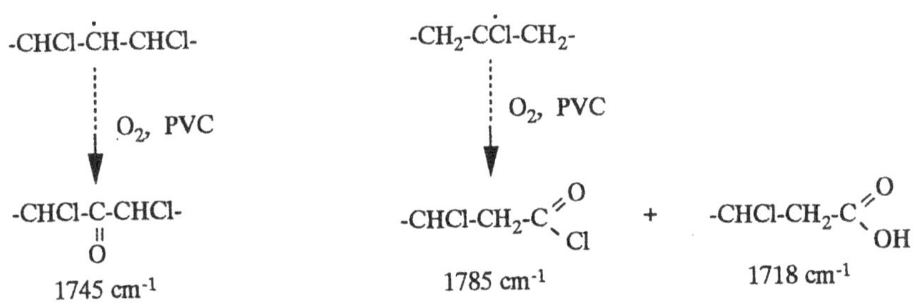

Scheme 2

The above-mentioned results can be summarized as follows:

(1) the first step is the C-Cl cleavage to form a Cl radical by shorter wavelength UV irradiation,

(2) heating and longer wavelength laser irradiation after UV irradiation is effective for the formation of longer polyene structures, and

(3) the mechanism for formation of new polyenes is explained by zip-dehydrochlorination.

3. Photo- and laser-induced dehydrochlorination of chlorinated-PVC (Cl-PVC)

3.1 UV lamp and Ar ion laser irradiation

Decker and Balandier showed a high instability of chlorinated PVC (Cl-PVC) against UV light [6]. They prepared 64.8 %-chlorinated PVC films by photochlorination of cast films of PVC. By UV exposure with a medium pressure mercury lamp (250-450 nm), the Cl-PVC underwent a fast dehydrochlorination process which proceeded ten times more efficiently than with PVC, leading to long chlorinated polyene sequences ($n > 30$) which were responsible for the rapid coloration of the irradiated polymer. Chain scission and cross-linking occurs simultaneously, both in the presence and in the absence of oxygen, indicating that UV radiation is able to cleave the polymer chains.
From the viewpoint of preparation of conductive polymers, Decker noticed the formation of long chain polyene sequences in the dehydrochlorination of Cl-PVC [7,8].

He intended to directly prepare conductive polymers by using Cl-PVC as a starting material that was shown to be a highly photosensitive polymer. Upon intense UV irradiation, it should *a priori* be possible to pursue the dehydrochlorination toward its ultimate stage in order to finally end up with a pure carbon polymer, according to the following Scheme 3:

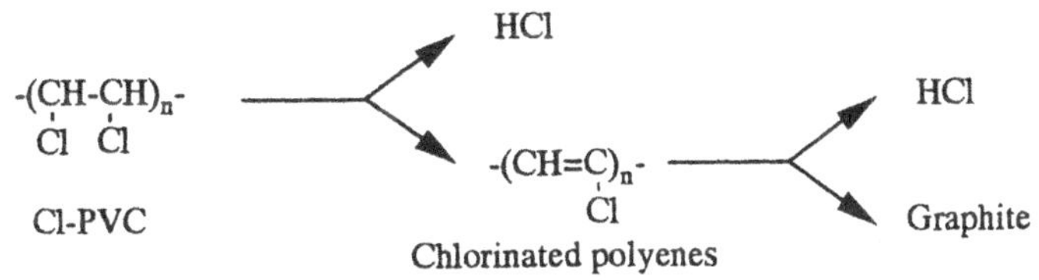

Scheme 3

In this work he used Cl-PVC films cast from Cl-PVC powder with a degree of conversion of 90 % (Cl 71 wt%). Since Cl-PVC is not totally transparent in the near-UV and visible range, it does not absorb at the 488 nm wavelength of an Ar ion laser. Therefore, Cl-PVC were first exposed to UV radiation with a medium pressure mercury lamp in order to produce strongly absorbing polyenes. This irradiation was carried out at room temperature in the absence of oxygen, thus preventing the formation of undesirable oxidation products.

When the Cl-PVC film (30 μm thick) was exposed to UV light with a wavelength above 250 nm, dehydrochlorination proceeded rapidly along the polymer chain leading to the formation of chlorinated polyene sequences according to the following set of reactions (Scheme 4):

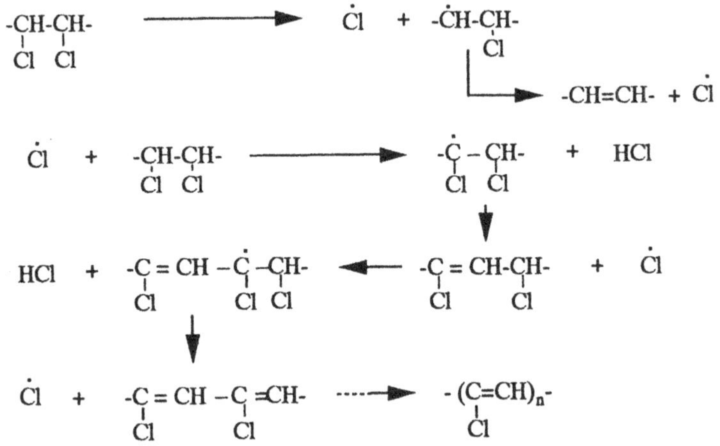

Scheme 4

The dehydrochlorination chain reaction was shown to proceed ten times more efficiently in Cl-PVC than in PVC, with a quantum yield of 0.12 mol photon^{-1} in the absence of oxygen. Figure 5 shows spectral changes (a) before and (b) after UV exposure for 15 min. Such a short irradiation time already appears to be sufficient to produce appreciable amounts of chlorinated polyenes with a sequence length of 12, which can absorb 488 nm photons corresponding to one of the argon ion laser lines. Upon prolonged UV exposure, strongly colored material was obtained.

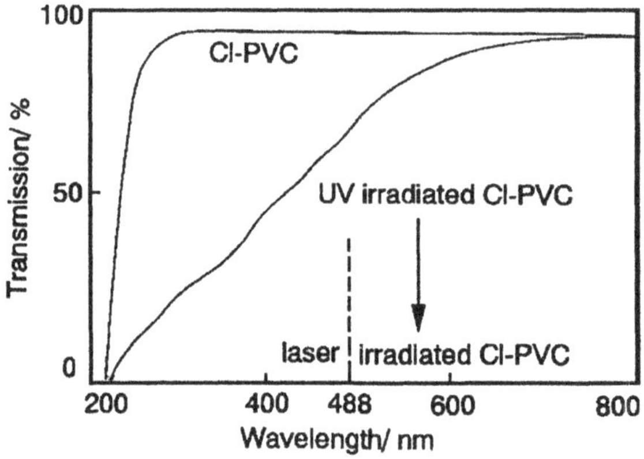

Figure 5. UV-vis absorption spectra of Cl-PVC film (30 μm thickness) (a), after irradiation at 250 nm for 15 min (b), and after subsequent laser irradiation at 488 nm in air for 0.1 s [7].

In order to pursue the dhydrochlorination a step further and transform the chlorinated polymers into a pure carbon material, UV degraded Cl-PVC films were exposed for a short time to the intense 488 nm beam of an argon ion laser. During laser irradiation for less than 0.1 s, large amounts of HCl were evolved and a black residue was produced that completely absorbed the UV-vis light (Fig. 5). This product was assumed to be graphite. The extent of the graphitization process was evaluated from the IR spectra (Fig. 6). All functional groups, in particular C-Cl bonds, essentially disappeared within 0.1 s of laser irradiation. The few remaining CH groups were probably due to the fact that the starting material still contained about 10 % of vinyl chloride units. It will degrade into $(CH{=}CH)_n$ structures which can no longer be decomposed upon UV-vis irradiation.

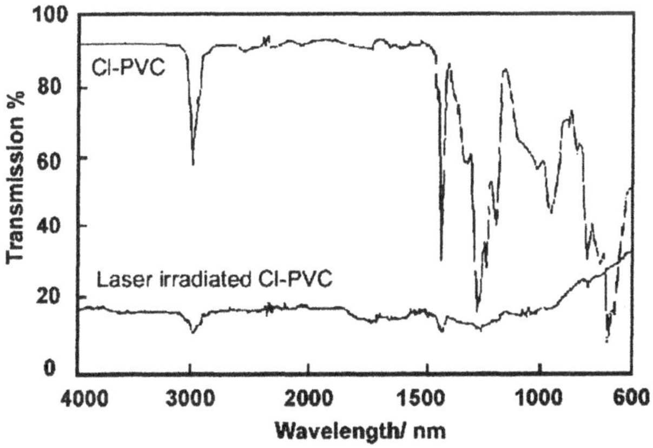

Figure 6. IR absorption spectra of Cl-PVC film (30 μm thickness) before and after UV irradiation at 250 nm and subsequent laser irradiation at 488 nm [7].

In addition, it was suggested that the dehydrochlorination during laser irradiation proceeded by zip-dehydrochlorination similar to that of PVC, as shown in the following scheme (Scheme 5). The quantum yield of carbonization appeared to be 3 to 12, much higher than unity, suggesting that carbonization occurred by a chain reaction process.

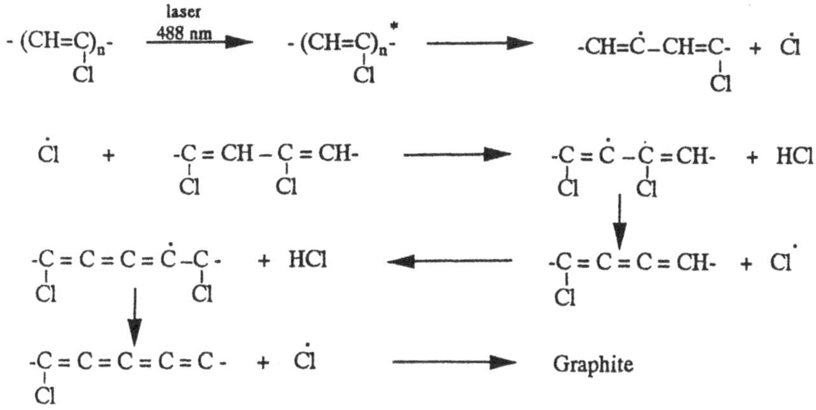

Scheme 5

Although the photochemical process plays a dominant role in the graphitization of Cl-PVC, the contribution of thermal effects must be also considered with regard to (1) the observed increase of the sample temperature which still remains below the softening point of the polymer and (2) the lack of any significant graphitization when conventional UV light was used.

In the above-mentioned papers by Decker *et al* [7, 8], the final carbonized product was graphite, and the formation of linear conjugated carbon bonds, i.e. polyene and polyyne were not mentioned by the authors. They did not consider the existence of linear carbon structures, even though they mentioned carbyne as final product in the reaction equation. Their work on laser-induced dehydrochlorination has stopped since the end of the 1980s.

3.2 EXCIMER LASER IRRADIATION

With the development of excimer lasers in 1980s, material processing by lasers has extended not only to metals but also to ceramics and polymers. In particular, the interaction of molecules with UV-laser beams can induce photochemical reactions through electronic excitation. In addition, excimer lasers emitting high intensity UV-photons are effective for photochemical reactions of condensed materials.

By using an excimer laser, Shimoyama *et al* studied the dehydrochlorination of Cl-PVC [9]. A Cl-PVC powder was used as a starting material, whose molecular weight and chlorine content were 37 000 and 64 wt-%, respectively. Cl-PVC films of 50 μm thickness were exposed to a KrF excimer laser (248 nm) and a low pressure mercury lamp (254 nm) as reference at room temperature in vacuum. Upon irradiation the absorbance at 300 nm increased and its absorption edge was shifted to longer wavelengths. Using the mercury lamp, the shift to longer wavelength was registered in 320 min (Fig. 7). On the other hand, the exposure to a KrF laser beam formed a new absorption band around 550 nm as a shoulder; the absorption edge was shifted to longer wavelengths (Fig. 8). Spectra of the laser-irradiated Cl-PVC film developed at about 20 000 pulses. At this time, the color of the sample was reddish black with a luster on the surface. The results suggest that the KrF irradiation is more effective for dehydrochlorination of Cl-PVC than the irradiation with a mercury lamp as shown in the UV-vis spectra.

The chemical analysis of laser-treated Cl-PVC films was carried out using X-ray photoelectron spectroscopy (XPS), Fourier-transform IR spectroscopy by the attenuated total reflection method (FTIR-ATR), and Raman spectroscopy. In the XPS spectrum (Fig. 9) of the film before irradiation, the $C(1_s)$ peak is composed of two components, corresponding to $-CH_2-$ and -CHCl-. The spectrum after irradiation consists of a single sharp component with a tail at high binding energy. This tail was ascribed to the formation of oxidation products at the surface, because no photoelectron peak could be detected in the $Cl(2_p)$ region. This assignment was suggested from the IR band at 1720 cm^{-1} as being due to carbonyl. On the other hand, the Raman spectrum gave peaks ascribable to C=C and C≡C as shown in Fig. 10. Strong peaks were observed at 1112, 1488 and 2188 cm^{-1}. The former two peaks were assigned to the structure of *trans*-polyacetylene, and another peak at 2188 cm^{-1} was assigned to the carbon triple bond.

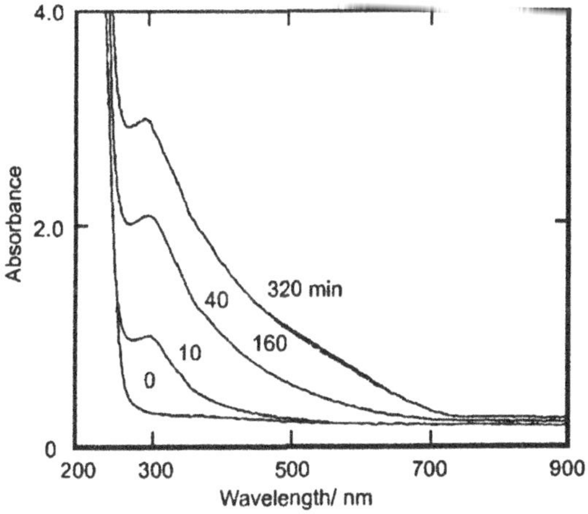

Figure 7. Changes in UV-vis spectra of Cl-PVC film irradiated with a low pressure mercury lamp [9].

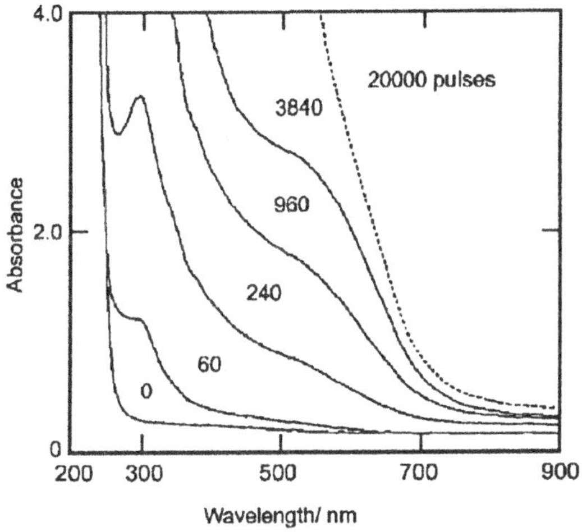

Figure 8. Changes in UV-vis spectra of Cl-PVC film irradiated with a KrF excimer laser in vacuum [9].

It has been reported by Furukawa *et a.l* [10] that a simple correlation exists between the number ($N_{C=C}$) of double bonds in *trans*-conjugation and the Raman shift ($\nu_{C=C\ stretching}$), and also between N and $n_{C\equiv C}$. From these correlation, Shimoyama *et al* [9] concluded that the laser irradiated Cl-PVC contained about 20 conjugated C=C bonds and about 8 conjugated C≡C bonds. The reason why short conjugated polyene and polyyne were produced although dehydrochlorination was almost complete as shown from the XPS results, was explained by the structural situation ascribed mainly to incomplete chlorination of the starting Cl-PVC (atomic ratio Cl/H = 0.5), as shown in Scheme 6.

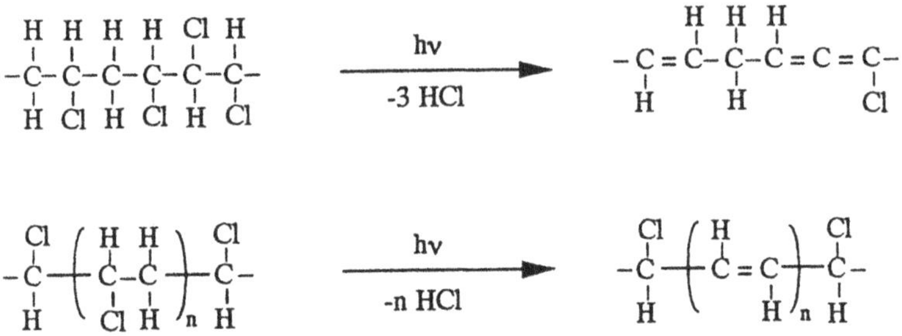

Scheme 6

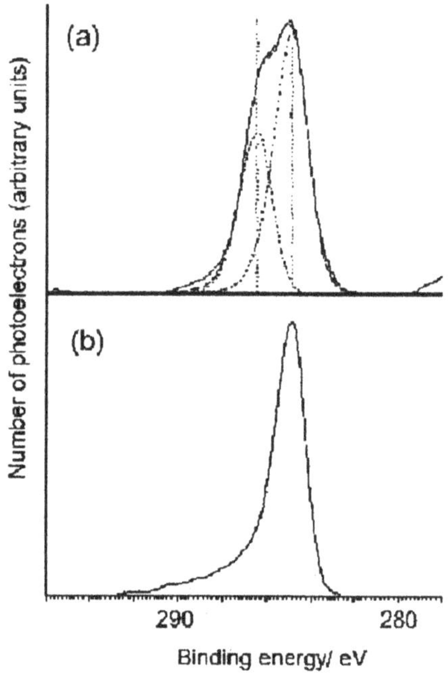

Figure 9. XPS spectra in the carbon region of Cl-PVC: (a) before irradiation, and (b) after irradiation with a KrF excimer laser of 20 000 pulses. Spectra were resolved into components using a Gaussian-Lorentzian function (dashed line) [9].

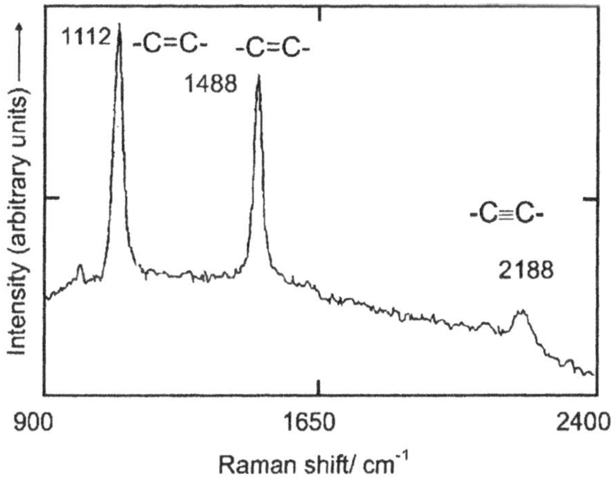

Figure 10. Raman spectrum of Cl-PVC film irradiated with a KrF excimer laser: excitation wavelength 647.1 nm [9].

4. Laser-induced dehydrochlorination of poly(vinylidene chloride) (PVDC)

Based on the preceding results of excimer laser-induced dehydrochlorination for Cl-PVC, Niino and Yabe used a higher chlorinated hydrocarbon polymer, poly(vinylidene chloride) (PVDC) as starting material for synthesis of polyyne with longer chain (Scheme 7) [11].

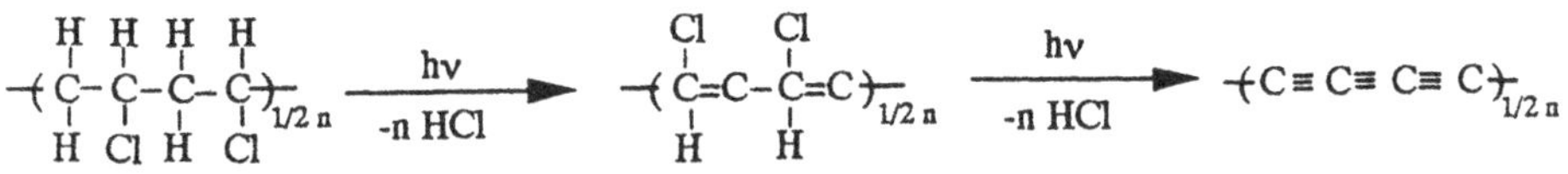

Scheme 7

PVDC films of approximate thickness of 100 μm were prepared from a boiling toluene solution of PVDC powder with M_w =100 000 and Mn =50 000. When the PVDC film was irradiated with a KrF excimer laser (248 nm) at a fluence of 30 mJ cm^{-1} $pulse^{-1}$ in a vacuum (10^{-4} Pa) at 90 °C, the film surface gradually changed from colorless to dark brown with increasing number of laser pulses. As shown in Fig. 11, new broad bands appeared in the visible and near IR region, indicating the formation of long conjugated chains of carbon multiple bonds in the PVDC film by photo-induced dehydrochlorination.

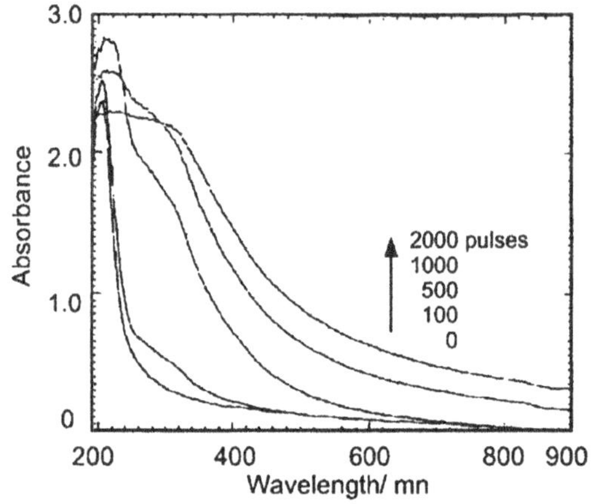

Figure 11. UV-vis absorption spectra of PVDC film irradiated with a KrF excimer laser with a fluence of 30 mJ cm^{-2} [11].

The dehydrochlorination of PVDC film was confirmed by analyses with XPS and quadrupole mass spectrometry (Q-MS). In the XPS measurement, it was observed that the atomic ratio of carbon / chlorine of the laser-irradiated surface changed to 1.00 : 0.015, whereas the original PVDC film had a ratio of 1:1 (Fig. 12).

The measurement with Q-MS detected gaseous molecules desorbed from the surface during laser irradiation in the vacuum chamber. The observed mass numbers of 36(intensity 3.8) and 38(1.0) were assigned to hydrochloride (HCl). Because no other peaks such as chlorine molecule and carbonized molecule were observed, it was concluded that the hydrochlorination of PVDC surface was the dominant reaction during KrF laser irradiation.

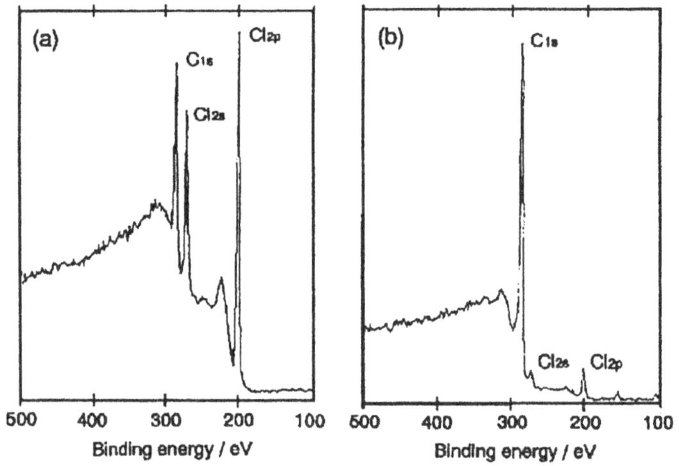

Figure 12. XPS spectra of PVDC film before (a) and after (b) KrF excimer laser irradiation with a fluence of 30 mJ cm^{-2} with 1000 pulses [11].

The laser-treated PVDC sample was measured by Raman spectroscopy. As shown in Fig. 13, three peaks were observed on a broad luminescent background, which were assigned to a triple bond (2125 cm^{-1}) and a *trans*-double bond (1485 and 1126 cm^{-1}). By using the correlation derived by Furukawa *et al* [10], 30 triple bonds and 10-25 *trans*-double bonds were estimated.

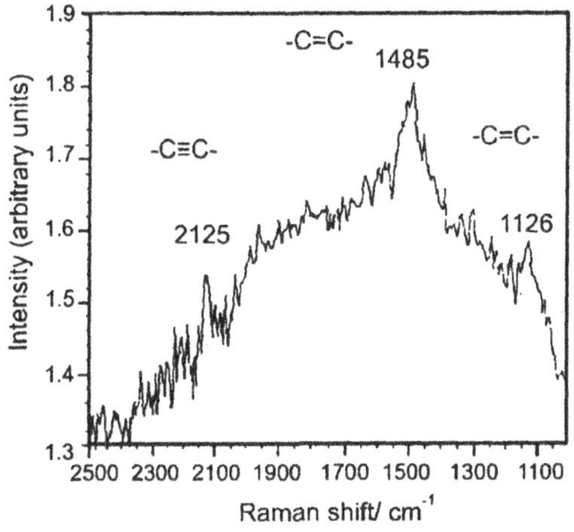

Figure 13. Raman spectrum of PVDC after KrF laser irradiation with a fluence of 30 mJ cm^{-2} with 21 000 pulses: excitation wavelength 1.06 mm of Nd:YAG laser [11].

Long-lived radical species in the PVDC sample were also observed by ESR spectroscopy. ESR spectra of PVCD powder after laser irradiation in a capillary tube. showed a broad peak with a peak-to-peak line width of ΔH_{pp}=2.1 mT belonging to the region of carbon radicals. Interestingly, further irradiation induced new carbon radicals with narrow widths (g=2.0035 and ΔH_{pp}=0.15 mT). The narrow peak was ascribed to carbon radicals on the conjugated bonds because unpaired electrons were localized on the polymer chain, being another evidence for the formation of conjugated bonds.

On the other hand, an ESR spectrum of the sample obtained by the thermolysis of PVDC powder at 200 °C in vacuum showed a broad singlet peak with ΔH_{pp} = 0.80 mT. This suggests that the thermolysis occurred at random to produce amorphous carbon with unsaturated chemical bonds.

In addition, Niino and Yabe [11] studied the effect of dehydrochlorination on light intensity. When the irradiation was carried out at the higher fluence of 40 mJ cm^{-1}, the Raman spectrum of the laser-treated film showed two broad peaks at 1597 and 1288 cm^{-1} which were attributed to an amorphous graphite structure. The more intense laser irradiation induced subsequent photo-induced reactions such as decomposition and rearrangement of the conjugated bonds to form amorphous graphite. On the other hand, the lamp-irradiated PVDC sample showed weaker absorption bands in the visible region than those of the laser-irradiated film (Fig. 11), indicating that less conjugated bonds with shorter lengths was formed. It is obvious that the laser irradiation with controlled fluence were more effective to form conjugated multiple bonds compared with the lamp irradiation.

5. Conclusion

It was shown that the dehydrochlorination of PVC plays an important role in the synthesis of conjugated polyyne. Advanced results were reported from the viewpoints of different starting materials and reaction control. The best results for carbynoid synthesis were obtained by excimer-laser-induced dehydrochlorination of PVDC, that resulted in a polyyne structure with around 30 conjugated chains. Our next important subject is to find ways to produce pure and longer-chained polyyne.

Although there is little doubt that polymers as starting materials will provide the most promising route to long-chained polyyne, we must carefully control how to prepare a starting sample. While it is difficult to prepare pure polyyne in the neat sample, it may be interesting to prepare it in inert media, by analogy to the use of alkalihalide media by Kavan *et al* [12, 13] and inert gas at low-temperature in the case of C_{18} synthesis by Diederich [14]. For this purpose, one needs to study dehydrochlorination of the starting polymer in a dispersed phase.

Photochemical reaction using excimer lasers was very effective for preventing graphitization as shown by Niino *et al* [9,11]. Moreover, their reaction control in vacuum and heating was also effective. However, the reaction did not give pure polyyne but fair amounts of amorphous carbon because of radical reactions in the condensed phase. As mentioned-above, it will be a key technology to prepare the starting material in a suitable reaction phase.

What is the best precursor for photo- and laser-induced dehydrochlorination or other dehydrohalogenation reaction? As far as we use conventional photochemical processes, it may be PVDC. However, PTFE may be expected to be a suitable precursors if we consider excitation of inner shell electrons by using a higher energy photon source. There may be a possibility for the selective bond cleavage of the C-F instead of the C-C bond. In other methods such as electrochemical and catalytic processes, defluorination of PTFE was very effective. PTFE has a high crystallinity which is preferable for synthesis of linear polyyne structures, *i.e.*, carbyne. Selective defluorination of PTFE may be an interesting future subject.

References

1. Nishijima, Y., Takadono, S., and Oku, O. (1966) Photochemistry of polyvinylchloride(I) Photochemical formation of polyene, *Rep. Prog. Polym. Phys. Jpn.* **9**, 497-500.
2. Yamamoto, Y., Yano, M., Nishijima, Y., Photochemistry of polyvinylchloride(I) Role of free radicals in photochemical formation of polyene, *Rep. Prog. Polym. Phys. Jpn.* **11**, 495-498.
3. Decker, C. and Balandier, M. (1981) Laser-induced degradation of polyvinyl chloride I: Quantum yield of dehydrochlorination, *J. Photochem.* **15**, 213-219.
4. Decker, C. and Balandier, M. (1981) Laser-induced degradation of polyvinyl chloride II: Oxygen bleaching of polyenes, *J. Photochem.* **15**, 221-228.
5. Gardette, J-L, and Lemaire, J. (1989) Wavelength effects on the discoloration and oxidation of poly(vinyl chloride), *Polym. Degradation Stability,* **25**, 293-306.
6. Decker, C. and Balandier, M. (1982) Etude de la stabilite a la lumiere du poly(chlorure de vinyle) chlore, *Makromol. Chem.* **183**, 1263-1278.

7. Decker, C. (1987) Synthesis of conductive polymers by laser irradiation of chlorinated PVC, *J. Polym. Sci. C. Polym. Lett.* **25**, 5-10.
8. Decker, C. (1988) Photochemical modifications of poly(vinyl chloride): conducting polymers and photostabilization, *ACS Symp. 364 : Chemical Reactions on Polymer,* 201-218.
9. Shimoyama, M., Niino, H., and Yabe, A. (1992) A KrF excimer laser induced dehydrochlorination of a chlorinated poly(vinyl chloride): preparation of conjugated polyene and polyyne, *Makromol. Chem.*, **193**, 569-574.
10. Furukawa, Y., Arakawa, T., Takeuchi, H., Harada, I., and Shirakawa, H.(1984) Vibrational spectroscopic stduy on partly hydrogenerated trans-polyacetylenes, *J. Chem. Phys.* **81**, 2907-2914.
11. Niino, H. and Yabe, A. (1998) Formation of conjugated polyene and polyyne structure by KrF-excimer-laser-induced dehydrochlorination of poly(vinylidene chloride) film, Submitted.
12. Kavan, L. (1991) *Electrochemical Carbonization of Fluoropolymers*, Marcel Dekker, New York, 69-171.
13. Kastner, J., Kuzmany, H., Kavan, L., Dousek, F.P., Kürti, J. (1995) Reductive preparation of carbyne with high yield. An in situ Raman Scattering study, Macromol., **28**, 344-353.
14. Diederich, F., Rubin, Y., Chapman, O.L., Goroff, N.S. (1994) Synthetic Routes to the Cyclo[*n*]carbons, *Helv. Chim. Acta,* **77**, 1441-1457

3.2.3. DEHYDROGENATION OF POLYACETYLENE AT HIGH STATIC PRESSURE

I.A. UDOD
Department of Chemistry,
M.V. Lomonosov Moscow State University,
119899 Moscow, Russia

Abstract – Interaction of polyacetylene with potassium at high pressure resulted in dehydrogenation of the former with retention of the original one-dimensional structure of the carbon skeleton. Semi-transparent, light brown crystals of carbyne were formed after removal of potassium hydride and the excess free potassium by thermolysis and acidolysis. A similar mechanism of the formation of crystalline carbyne was also observed during the reaction of a carbynoid material (the dehydrohalogenation product of poly(vinylidene halides)) with alkali metals. The structural data of intermediate compounds are identical in both cases. Some hypothetical models of a short carbon-chain structure of these compounds are discussed. A high pressure intercalation mechanism is proposed for the formation of linear carbon which includes complete isolation of polymer carbon chains by alkali metals.

1. Introduction

When considering the proposed chain-like structure of carbyne, it may appear at first glance that the methods of polymer-analogous transformations are most attractive for synthesizing linear carbon. Indeed, the simplest approach seem to take a polymer precursor and remove all side atoms and groups from its carbon skeleton. However, experiments have shown the reality, as it usually is, to be much more complicated than any theoretical prediction. Possibly, this is a consequence of the peculiar chemical nature of the subject of our interest.

Strictly speaking, the formation of linear carbon resulting from the elimination of different (usually, halogen-containing) groups should proceed step by step through a number of consecutive (activated) states. These are transitional polymer complexes containing large amounts of very reactive unsaturated chemical bonds. By this it is meant that the possibility of side reactions to occur is quite considerable, particularly, in the presence of oxygen-containing molecules. Indeed, those molecules are usually used as dehalogenating or dehydrohalogenating reagents [1–4]. Actually, recent ^{13}C-NMR data have shown that the carbynoid material synthesized by dehydrohalogenation of poly(vinylidene halide)[1] contains various bridging and terminal oxygen-based

[1] For details *cf.* Chapter 3.2.1 of this book.

R.B. Heimann et al. (eds.), Carbyne and Carbynoid Structures, 93–116.

groups along with double and triple C–C bonds in the carbon skeleton [5]. One of the limiting stages of heterogeneous polymer-analogous transformations is determined by the swelling of the polymer substrate. Perhaps, it is the presence of solvent molecules in the interchain space, that prevents carbynoid material to form numerous C–C cross-links, *i.e.* keeps essentially a one-dimensional structure of the carbon skeleton.

Heating to 600–700 °C is needed to completely remove the solvent molecules from a polymer sample [6]. One can hardly expect polyenyne fragments to survive in the macromolecules of carbynoid material under these conditions. Their extensive cross-linking can lead to profound chemical transformations in the structure of the carbon skeleton. The disappearance of the absorption band near 2100 cm^{-1} attributable to both cumulene and polyyne fragments in the IR spectra of the carbynoid obtained by dehydrohalogenation [6] might result from similar cross-linking processes. Thus, the preparative methods employing solution-based chemical reactions have limited possibilities for obtaining linear carbon in high yields and, particularly, in a crystalline state. Unfortunately, the lack of large carbyne crystals suitable for reliable x-ray analysis is the only reason for numerous discrepancies and controversies in the entire area of linear carbon.

In this case we should seriously consider the high pressure technique as a way to promote the transition from an amorphous to a crystalline state. It should be noted, however, that we cannot expect a complete disentangling and straightening of polymer molecules to occur under high pressure conditions, *i.e.* the transition from globular or fibrillar structures to a packing of carbon chains oriented along the same direction. Moreover, there is the likelihood of a detrimental effect of the high pressure. First, the compression could lead to a close approach of carbon atoms belonging to different carbon chains and hence the possibility of cross-linking would increase. The observed transition from α-carbyne (according to Refs. [7,8]) to the denser phase of β-carbyne upon high pressure treatment might be of the similar nature. Secondly, the rate of chemical reactions at high pressure usually slows down. Thus, the efficiency of the high pressure applied could be significant if the reaction would proceed with an essential volume decrease. To summarize, one can conclude that the ordering effect of the high pressure could manifest itself only if the carbon chains would be oriented and separated from each other by some chemical particles before the pressure treatment. Ideally, such particles (atoms or molecules) should be capable of an additional stabilization of linear carbon molecules.

Metals whose atoms possess appropriate size and do not form strong covalent bonds with carbon chains might be plausible candidates for such stabilizing particles. The weakest interaction, not affecting the type and character of chemical bonds in the original matrix, is realized in *intercalation compounds*. Carbon-based matrices (graphite, polyacetylene, and fullerene types) possessing a delocalized π-electron system are capable of forming such compounds with metals that have the lowest values of the ionization potential. According to my opinion, the character of the host-guest interactions arising in these cases is akin to weak chemical bonds in organic charge-transfer complexes. The linear structure predicted for carbyne also implies the presence of a system of delocalized π-electrons. This fact puts carbyne into the same group with the other carbon matrices mentioned above and suggests a promising approach to synthesize crystalline carbyne through the formation of such intercalation compounds.

These ideas were used as the basis in previous studies by Udod *et al.* [6,9–12]. Polyacetylene (PA) being a one-dimensional carbon-chain matrix capable of forming intercalation compounds with alkali metals was used as the first original substance in those experiments [6].

2. Polyacetylene Intercalation Compounds

The structure of polyacetylene intercalation compounds (PAIC) with alkali metals has been studied in sufficient detail [13–27]. Whilst structural data vary among samples obtained by different methods, it nevertheless is possible to display some particularities of polyacetylene structures doped with alkali metals (Fig. 1). First, the intercalation compounds are anisotropic: the PA chains run parallel to the *c*-axis as in the original matrix. The linear structure of the polyacetylene matrix determines the linear structure of the dopant metal 'channels'. The shortest distance between the metal and carbon atoms in a PA chain is close to a sum of the van-der-Waals radius of carbon and ionic radius of the metal. This rule remains true also for other types of intercalation compounds, *viz.*, graphite intercalation compounds, but does not involve full charge transfer from intercalated atoms to the carbon-based matrix. Otherwise, we should take into consideration an ionic radius of carbon rather than its van-der-Waals' radius. Presumably, this is a prerequisite for the realization of the host-guest interaction in the intercalation compounds. Thus, the coordination number of PA chains around the dopant metal atom depends on its size. Therefore, this number is 4 in PAICs with heavy alkali metals [17,18,23,25] (Fig. 1, *b*), whereas it is 3 in the lithium-intercalated PAIC [22] (Fig. 1, *c*). Sodium, probably, takes an intermediate position: it is capable of forming both an ordered hexagonal structure and a disordered one at higher degrees of intercalation in the Na–PA system [16]. As we can see from the two-dimensional structures shown in Fig. 1, the coordination number of PA chains around the metal 'channel' is determined just by the existence of the van-der-Waals interaction between them. The first stage of PAICs with heavy alkali metals, having an ordered structure commensurate with the original PA matrix, corresponds to the empirical formula $(C_4H_4)_2M$. In this case the M-M distance in a certain channel is close to 496 pm [25]. Further saturation, for example by electrochemical intercalation, might lead to disordering of the intercalation structure, probably resulting from co-intercalation of solvent molecules. In contrast to the electrochemical method, the intercalation from a vapour phase excludes the possibility of insertion of other substances. In this case a limiting composition of PAICs with heavy alkali metals could depend on the size of the doping metal. According to Billaud *et al.* [25], the post-intercalation into $(C_4H_4)_2M$ led to the disappearance of commensurability between the PA chains and the metal channels. The periodicity along the *c*-axis changed to depend on the size of metal atoms and was 398, 402 and 407 pm for K, Rb and Cs, respectively. The shortest M–M distance inside the channel (at least for heavy alkali metals) is less than that in a free metal state. Thus, we might assume the reality of the metal-metal interaction along the *c*-axis. In turn, the formation of a denser $(C_3H_3)_2M$ structure requires the approach of heavy alkali metal atoms to the distance up to ~370 pm. This is presumably impossible, at least under ambient pressure conditions. Actually, the attempt to obtain

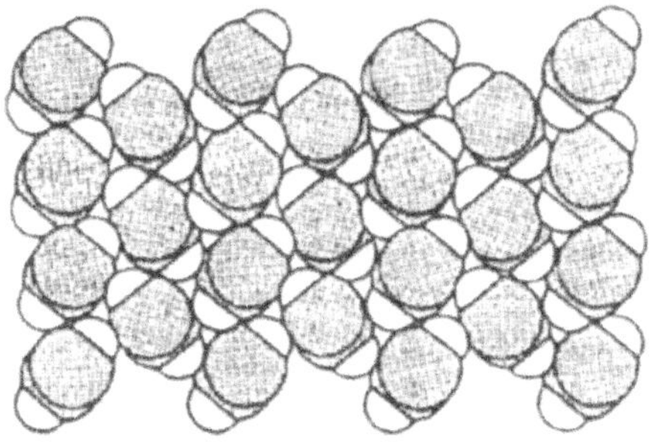

a

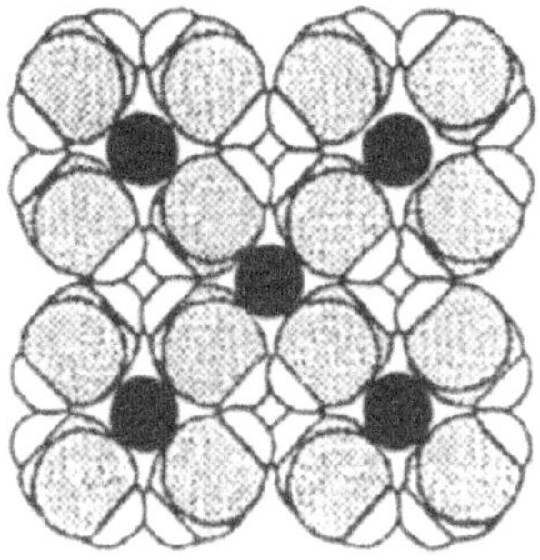

b

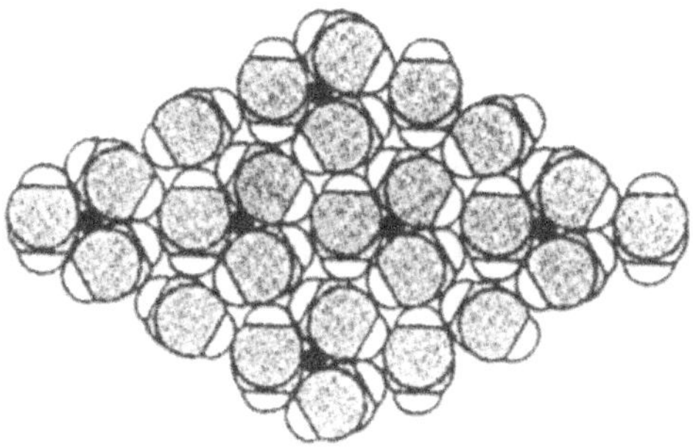

c

Figure 1. Two-dimensional structures of the original polyacetylene (*a*) and its intercalation compounds with heavy alkali metals (*b*) and lithium (*c*).

more saturated PAIC with cesium via the gas-phase intercalation at *ca.* 200 °C resulted in the unexpected formation of cesium hydride [25]. It is interesting to note that the CsH crystals were oriented with respect to the PA matrix structure. This fact might suggest that the reaction took place in the interchain space after the post-intercalation of excess cesium. The dehydrogenation of PA at a temperature substantially lower than that of the thermal elimination of hydrogen (the carbonization of PA proceeds moderately on heating just above 600–700 °C [28]) might result from the formation of fairly stable compounds of a newly formed carbon matrix with alkali metal. Unfortunately, this argument was not used by the authors [25].

The thermal treatment of PA in vacuum up to 1000 °C led to exhaustive dehydrogenation and the formation of an amorphous phase with basically sp^2-state of carbon atoms [28]. In turn, complete graphitization was observed only upon heating above 2500 °C. This fact underlines the difficulties to directly transform one-dimensional into two-dimensional structures of the carbon skeleton. The effect of the high pressure allows to essentially lower the temperatures of both the dehydrogenation reaction and the complete graphitization of the sample [6].

The DTA curves measured during the thermobaric treatment of polyacetylene have demonstrated a large endothermic effect at 400–500 °C and 20–30 kbar caused by the elimination of hydrogen and carbonization of the sample. At 800–1000 °C this process occurred almost completely: the content of carbon was found to be more than 98%. Moreover, in this case the absorption bands characteristic of C–H vibrations disappeared from the IR spectra. Broad lines with interplanar distances of 340, 208, and 169 pm, corresponding to poorly crystallized graphite, were observed in the x-ray diffraction pattern. Thus, the direct high-pressure treatment affects thermal dehydrogenation through the preferred formation of a denser, more ordered structure as a result of numerous interchain cross-links.

3. High-Pressure Studies of Polyacetylene+Alkali Metal Systems

The application of the high pressure technique turned out to be fairly efficient in studies of intercalated compounds of another carbon matrix, *viz.*, graphite [29–38]. A simple calculation of volume effect values of intercalation reactions points already to the considerable role high pressure plays in the synthesis of new superdense compounds (Table 1). The intercalation of alkali metals into graphite under such conditions leads to supersaturation of the carbon matrix with alkali metals to form intercalate layers denser than those in GICs obtained by conventional methods at ambient pressure. We might, therefore, expect that the use of high pressure for the intercalation of alkali metals into polyacetylene would be similarly efficient.

A study of the interaction of PA with some alkali metals has been performed by the DTA method *in situ* under high pressure in the 20-40 kbar range [6,10,11]. Typical DTA heating curves of a (PA+K) system are shown in Fig. 2. As can be seen from Fig. 2, solid-phase intercalation of potassium into PA did not occur (region *A*). Furthermore, an endothermic peak *B* corresponding to the melting of free potassium was observed. The absence of marked heat effects at higher temperatures (region *C*) might suggest the absence of significant intercalation of molten potassium as well. At first glance this result appears to be very strange. Actually, the intercalation of potassium into graphite under high pressure conditions starts immediately after metal melting. To explain the difficulties of the high-pressure intercalation of alkali metals

TABLE 1. Volume effects of intercalation reactions of alkali metals into graphite and interatomic distances in corresponding GICs

Metal, M	GICs	$\Delta V/V_0$, %[b]	$d_{M\text{-}M}$ in GICs, pm	$d_{M\text{-}M}$ in metal, pm
Li	C_6Li[a]	-21.1	426	310
	C_2Li	-44.0	246	
Na	-[a]			380
	C_2Na	-57.9	246	
K	C_8K[a]	-22.1	496	472
	C_6K	-33.4	426	
	C_4K	-48.4	330[c]	
Rb	C_8Rb[a]	-27.1	496	496
	C_6Rb	-38.7	426	
	$C_{4.3}Rb$	-51.1	340[c]	
Cs	C_8Cs[a]	-32.2	496	536
	C_4Cs	-58.1	330[c]	

[a] These compounds can be synthesized at the ambient pressure. Analogous GICs of sodium is unknown; [b] Calculations have been performed assuming that all GICs possessed a single-layer intercalate structure at least under high pressure conditions; [c] All M–M distances are thought to be equal. Thus, in some cases the position of the metal atoms above graphite hexagons is off-center.

in this case, we should consider the structural similarity of the original matrix and its intercalation compounds. While in the case of graphite the swelling of the carbon network along the *c*-axis alone could be sufficient for the intercalation to succeed, the intercalation of the metal atoms into the PA matrix would require some changes in the entire packing of polymer molecules, *i.e.*, their rearrangement (Fig. 1). The high-pressure conditions, in turn, decrease dramatically the mobility of PA chains. The region *D* (Fig. 2), being close to the region of thermal elimination of hydrogen from pristine PA, shows profound chemical transformations in the sample. The large exothermic effect might result from the formation of ionic hydride. To complete this reaction repeated heating-cooling in the 20–800 °C range was required. The amount of free potassium determined from the area under its melting peaks decreased gradually in each subsequent heating-cooling cycle until it completely disappeared. The DTA method allows to determine quantities of alkali metal reacted under high-pressure conditions. Indeed, if the content of potassium in the sample was increased to >1.7, then an excess of free metal was observed on DTA curves after completion of the reaction.

$$(CH)_n + 1.7nK \rightarrow nKH + (CK_{0.7})_n \quad (1)$$

Thus, as follows from equation (1), the composition of the metal-carbon phase at 20 kbar pressure is $CK_{0.7}$, *i.e.*, $C_{1.43}K$. This composition has the highest content of metal compared to known binary intercalation compounds based on carbon matrices, including even those synthesized using a high-pressure technique.

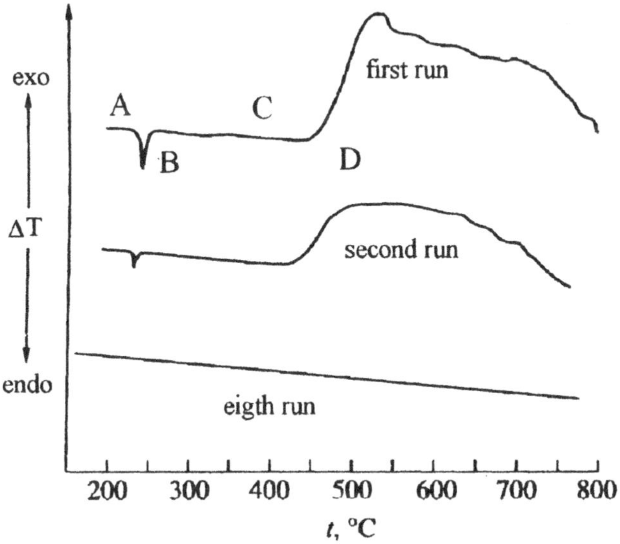

Figure 2. The interaction of polyacetylene with potassium under high-pressure conditions. DTA heating curves of a PA+K mixture for a ratio of CH:K = 1:1.7 (P = 20 kbar).

The x-ray diffraction data of the sample after thermobaric treatment confirmed the formation of KH. The interlayer spacings belonging to the carbon-metal phase could not be observed against the background of the strong and sharp reflections from the crystal lattice of ionic hydride. Nonetheless, we might assert that the compound with a composition of $C_{1.43}K$ does not present a well-ordered crystal phase as, for example, ionic carbide. The predominantly amorphous state of this carbon-metal phase could be due to its instability at ambient pressure. Actually, the superdense GICs with alkali metals are known to decompose after pressure relieve to release the metal surplus in a highly dispersed state, appearing amorphous in x-ray diffraction measurements [33,35].

The interaction of PA with lithium at 20–40 kbar and 200–900 °C proceeds with an exothermic effect. It does not begin upon the elimination of hydrogen (usually >500 °C), but immediately after lithium melting (Fig. 3). The maximum of lithium reacted with PA under those conditions corresponded to the ratio Li:(CH) = 2:1. The x-ray diffraction data showed the presence of two crystalline phases, *viz.*, lithium hydride and lithium carbide. Thus, lithium, being a carbide-forming metal, splits PA completely:

$$2(CH)_n + 4n\mathrm{Li} \rightarrow n\mathrm{Li}_2\mathrm{C}_2 + 2n\mathrm{LiH} \qquad (2)$$

The reaction of PA with sodium occurs similarly to that with potassium. The difference, probably, is in its incompleteness. The limiting metal content for the (PA+Na) reaction mixture has not been determined by the DTA method. Probably, the intercalation of sodium occurs somewhat slower and, therefore, cannot prevent some structural changes in the carbon matrix being formed. In turn, the lower reactivity of sodium during its intercalation into carbon matrices can be explained by its higher ionization potential compared to that of potassium.

Thus, a carbonaceous product was formed as a result of thermobaric treatment of PA with potassium, which contained impurities of potassium hydride and, probably, free metal phase partially deintercalated after pressure relieve. The removal of these impurities was possible by either thermal or chemical routes. The latter consists in the interaction of a sample with water or mineral acid solution, resulting in the formation of amorphous carbon soot containing up to 10% of hydrogen and oxygen. In this case partial hydrogenation and oxidation of the carbon skeleton might occur, owing to the formation of gaseous hydrogen and potassium hydroxide during aqueous treatment.

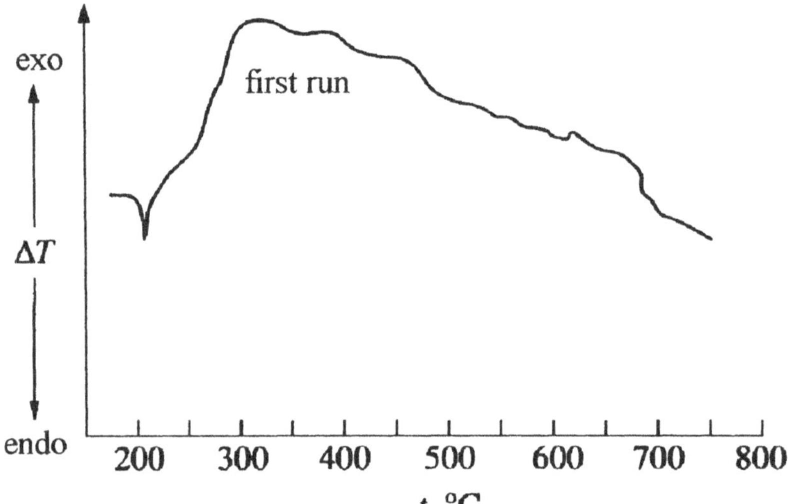

Figure 3. The interaction of polyacetylene with lithium under high-pressure conditions. DTA heating curves of a PA+Li mixture for a ratio of CH:Li = 1:2 (P = 20 kbar).

Therefore, the thermal approach was more suitable for our experimental task. In order to remove impurities, the sample was placed into a three-section ampoule (Fig. 4). The temperature of decomposition of KH exceeds 600 °C. We should also expect the decomposition of the metal-carbon compound to occur at this temperature. Actually, while free alkali metal condensed in the coldest part of the ampoule, a brown film consisting of carbon and potassium (according to x-ray microanalysis data, the content of oxygen was just a few percent) was formed in the second (central) part of the ampoule. The C/K ratio changed over a wide range along the whole length of this section. That is why we cannot consider any particular composition when discussing the possible structure of the film. It should be noted that a significant amount of the carbon material sublimated away, and that only a small part of it was transformed into non-volatile amorphous carbon.

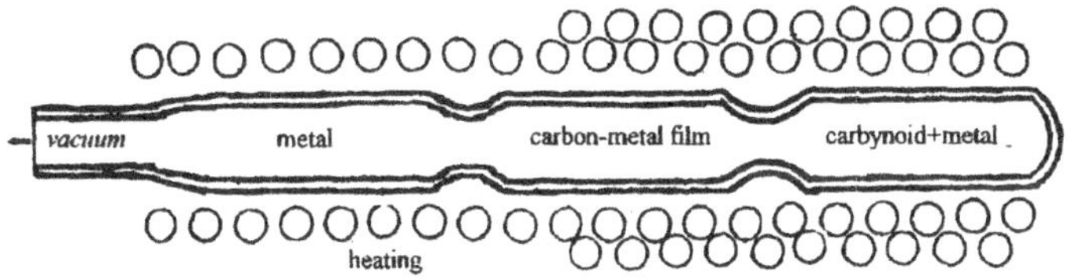

Figure 4. Three-section ampoule for the synthesis of carbon-metal films by a chemical transport reaction.

The acidic treatment of the film resulted in unexpected crystallization as a result of vigorous heat production accompanied by local burning. The electron microdiffraction data from these crystals revealed a well-ordered hexagonal structure with lattice parameters a = 516 pm[2] and c = 1.6 nm. Thus, investigations of the (PA+K) system under high pressure conditions disclosed a new method of synthesizing fully crystalline carbyne.

4. Interaction of Carbynoid Materials with Alkali Metals

Chemical reactions proceeding under the conditions described above touch upon several fields of chemistry such as high pressure chemistry, intercalation chemistry, polymer chemistry and also the most baffling part of carbon chemistry, 'carbyne chemistry'. That is why it is necessary to analyze all currently available data, which are few in number, if we are to understand the formation mechanism of carbyne via dehydrogenation of PA by potassium under high pressure. The chemical nature and the structure of the potassium-carbon film as an intermediate link in a sequence of chemical transformations leading to the formation of carbyne crystals was thought to be a key in solving this question.

In contrast to the interaction with acids, the reaction of the potassium-carbon film with water occurred markedly less energetic and led to its almost complete dissolution [6]. A dark-brown solution showing a considerable light dispersion (Tyndall effect) was formed. The formation of such a colloid solution might indicate the presence of hydrophobic surfactant molecules with a hydrophilic tail. Further acidification resulted in the precipitation of black-brown amorphous flakes. This is very unexpected because usually alkaline solutions are used as a coagulator. A bromine-water solution became colorless upon the interaction with just a small piece of the film. Moreover, the brominated derivatives had higher density and exfoliated completely in the aqueous solution. The presence of copper(+1) salts in the solution led to the precipitation of a

[2] The parameter a used in the present chapter has been chosen as the minimal value following from the diffraction data, being smaller than other reported values [39–44] by a factor of $3^{1/2}$.

black powder of, apparently, copper acetylides [6]. Taking into account all these facts, we can draw some preliminary conclusions:

(i) relatively light volatility at high temperatures and the capability of forming aqueous solutions (though colloid) attest to the non-polymeric structure of carbon particles in the potassium-carbon film;

(ii) the discoloration of a bromine-water solution and the formation of copper acetylides imply the presence of many unsaturated bonds and also terminal triple bonds.

The most simple explanation of the above facts relies on the assumption of a short-chain structure of carbon fragments in that substance. The stability of short carbon chains in a condensed state might be due to chemical bonding with metal atoms. In this case the formation of polymeric linear carbon molecules after acidolysis of the film could occur through the polymerization of short carbon chains.

These considerations formed the basis of the next stage of our investigations. According to a modern concept [5], the product of dehydrohalogenation of poly(vinylidene halides) is a typical carbynoid, *i.e.*, a carbon-rich material containing linear chain-like carbon fragments[3]. The carbynoids are usually amorphous because they can comprise, for example, some oxygen-containing groups forming both bridge and terminal bonds with carbon skeletons. The main idea of the next series of experiments was to isolate the linear carbon fragments from impurities by using a chemical transport reaction in order to attempt deliberate crystallization of carbyne. Sodium was chosen as a metal that could provide the required volatility of short carbon chains and their stability in a condensed state [9]. Indeed, sodium occupies an intermediate position between carbide-forming lithium and potassium capable of forming fairly stable intercalation compounds with carbon matrices.

The carbynoid precursor obtained by chemical dehydrochlorination of poly-(vinylidene chloride) is an infusible and non-volatile substance that does not markedly decompose up to 800–900 °C. The treatment of this carbynoid with sodium metal in vacuum resulted in the cleavage of polymer chains to form volatile low molecular-weight fragments followed by the deposition of a semi-transparent, dark-brown film. According to x-ray microanalysis data, the film consisted predominantly of carbon and sodium. The oxygen content was just a few percent. The C/Na ratio was not constant over the whole film area and changed under the electron beam irradiation towards an increase in the sodium content. Moreover, long exposure of the sample to the electron beam led to vaporization of carbon particles and, finally, emerging of a free sodium phase [9].

The sodium-carbon film oxidized rapidly in air. According to TEM data [9], the following phases were discovered: a cubic phase of sodium peroxide Na_2O_2 and a carbon phase, whose diffraction pattern contained only a diffuse ring with d = 437 pm (Fig. 5). The parameter a = 508 pm calculated under the assumption of a hexagonal symmetry, could be related to a multilayered chain-like structure proposed by Babaev *et al.* [42]. In contrast to the oxidation process, the reaction of the sodium-carbon film with water led to its dissolution. The gases released decolorized a bromine-water solution and reacted with an ammoniacal solution of a copper(+1) salt to form a black

[3] For details *cf.* Chapter 3.2.1 of this book.

precipitate, presumably of copper(+1) acetylides. Finally, the acidic treatment of the film resulted in the formation of carbyne crystals in quantitative yield similar to the previous experiment with a (PA+K) system after the high pressure treatment.

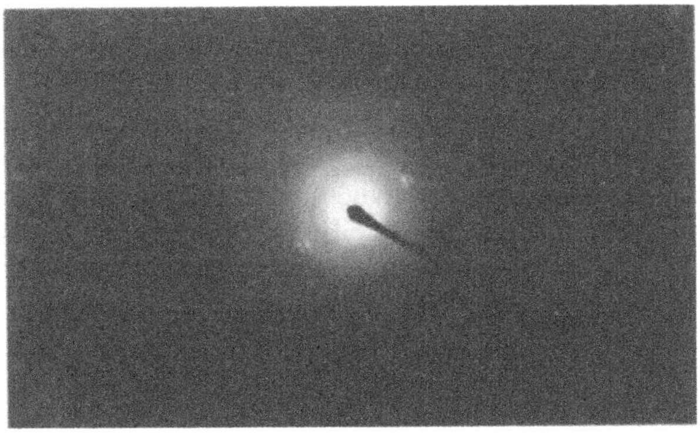

Figure 5. Electron diffraction pattern of the carbon-sodium film oxidized during exposure to air.

Thus, although the starting compounds in those experiments were different, as well as the conditions of the reaction between them and alkali metals, intermediate phases were formed in both cases with similar chemical properties. Analyzing these properties, one can assume that the films studied consist of short carbon chains. In my opinion, it is chemically warranted that the vaporization of short carbon chains with metal atoms at their ends results from destruction of the polymer chain induced by the alkali metal. According to mass spectroscopic measurements [45], similar molecules were observed during laser vaporization of a graphite–KOH mixture in vacuum. In our case the interaction of short carbon-chain organometallic compounds with water is much like the hydrolysis of metal acetylides and can be defined as hydrogenation of the terminal carbon atoms. Prolonged exposure of these compounds to air resulted in their oxidation and complete elimination of the alkali metal atoms from the carbon matrix. Apparently, the carbon phase being formed in this case retains predominantly a one-dimensional structure of its molecules, but with poorly ordered molecular packing. The reaction of the metal-carbon compounds with acids in turn results in oriented lengthening of carbon chains. Moreover, the energy released during this reaction, coupled with its high rate, is sufficient for intra- and inter-molecular transformations of carbon chains to occur along with capturing metal atoms in the vacancies being formed. The possible role of the metal atom impurities during crystallization of carbyne and structural models of the carbyne intercalation compounds are discussed in detail in Chapter 4.3 of this treatise.

5. Structural Models of Short Carbon-Chain Compounds with Metals

During discussion of the formation of linear carbon above, the structural aspect of the problem was omitted. We should realize the great importance of structural investigations not only for understanding of the chemical processes between alkali metals and polymeric carbon materials, but also for obtaining evidence for a one-dimensional structure of molecules in the carbyne crystals. That is why presently available structural data concerned with the short carbon-chain compounds will be discussed separately below.

Unfortunately, structural data concerning even diacetylene compounds with alkali metals are lacking. Of the binary compounds of that type, only alkali metal acetylides, MC≡CM (M= Li, Na, K), were investigated [46–48]. Sodium acetylide, Na_2C_2, has a distorted antifluorite structure [47], and its potassium analogue is isostructural [48]. The structure of solid alkali metal acetylides with diatomic anions C_2^{2-} and cations M^+ is a typical ionic lattice. The anion C_2^{2-} also occurs in such typical ionic structures as CaC_2, SrC_2, and BaC_2. It should be emphasized that the size of the cation is usually important for ionic structures of an A_2B type. That is why the stability of these acetylides decreases essentially from Li to Na to heavy alkali metals. Actually, in this case the possibility of screening the same charged particles by oppositely charged ones is weakened with increasing cation size.

Crystal lattices of a different kind are observed in alkali metal ethynides, MC≡CH (M = Na, K, Rb) [49,50] and propynides $MC{\equiv}CCH_3$ (M = Na, K) [49]. Their layered structure consists of linear molecules forming a two-dimensional packing (Fig. 6). In this case the formation of C_2^{2-} (as a basis of the above-mentioned crystal lattices) becomes impossible, and molecular fragments keep their original linear structure. Such linear orientation of the M–C bond can testify to the emergence of some covalent contribution. In other words, the metal in these compounds might be present in a non-ionic state. This can explain the coordination around alkali metal atoms of four C≡C bonds.

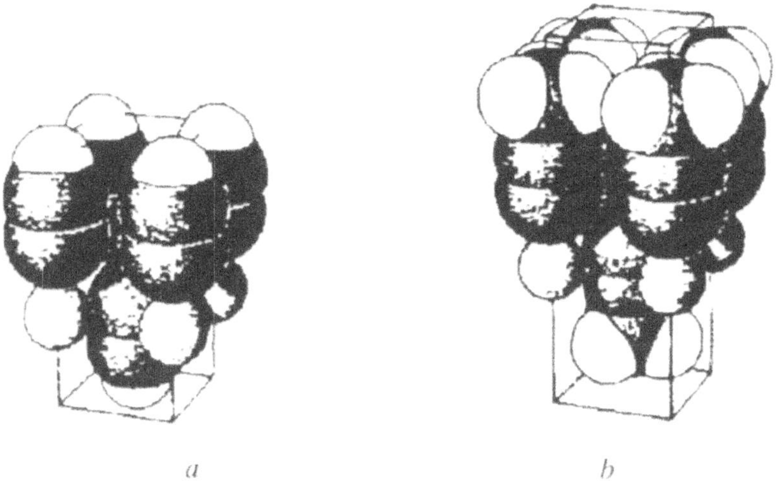

Figure 6. Layered structures of MC≡CH, where M = Na, K, Rb and Cs (a) and $MC{\equiv}CCH_3$, where M = Na, K (b).

In reality, the alkali metals form chemical bonds with the carbon matrix possessing a system of delocalized π-electrons (graphite, polyacetylene, fullerene, nanotubes, and others). Such substances are usually referred to as intercalation compounds. The metal atoms can be entirely isolated (as in fullerenes doped with heavy alkali metals [51–57]) or build up one-dimensional channels (doped polyacetylene [13–27]), two-dimensional layers (graphite intercalation compounds [29–38]) or helical structures (doped nanotubes [58]). The type of the intercalated sub-lattice is determined by the structure of the starting matrix. It is important to emphasize that metal atoms can form in such compounds, probably, very weak but quite real M–M bonds [59].

Similarly layered structures, in which the fragment M–O–C is linear, have been observed in some alkali metal alkoxides with a non-branched hydrocarbon tail [60–63]. In all these cases the molecular fragments have only one terminal metal atom.

The closest known structures to the substances of our interest are acetylenediolates of alkali metals. The formation of black and extremely reactive 'potassium carbonyl' by reacting CO with molten potassium in a 1:1 molar ratio was reported as far back as the last century [64]. Originally, the product was supposed to be a potassium derivative of hexahydroxybenzene, $K_6C_6O_6$, which can be formed by careful hydrolysis. Subsequently, it was shown that under mild conditions (for example, when introducing CO into an ammoniacal solution of alkali metals) a dimer was formed, according to reaction (3):

$$2M + 2CO \rightarrow MOC{\equiv}COM \qquad (3)$$

It is very interesting that the stability of a linear OC≡CO fragment increases significantly with an increase in the size of alkali metal atoms [65–67]. The compounds of dihydroxyacetylene with alkali metals crystallize tetragonally and form isotropic layered structures (Fig. 7). Each oxygen atom is surrounded by five potassium atoms (a tetragonal pyramid with an O atom at the center of its base). The same environment occurs in potassium methoxide. The most important structural element is a linear anion $^{-}OC{\equiv}CO^{-}$ with approximate interatomic distances of C–C = 120 pm and C–O = 128 pm. The latter distance is close to the normal double C=O bond in carbonyls (120 pm) rather than to an ordinary C–O bond, for example, in $KOCH_3$ (145 pm) (Fig. 6*b*), whereas the length of a C–C bond corresponds satisfactorily to that of the cumulated bonds in carbon suboxide C_3O_2 (Table 2). Probably, there is an intermediate state (between $^{-}O{-}C{\equiv}C{-}O^{-}$ and O=C=C=O) of chemical bonding in linear molecules. The M–O bond length (266 pm) in a linear MOC≡COM molecule is shorter than the other four M–O distances (277 pm), and this difference is greater in methoxides than in acetylenediolates (*cf.* $KOCH_3$ and KOC≡COK, Table 2). The main difference between these two types of substance lies in the presence in the MOC≡COM compounds of non-screened alkali metal atoms forming, probably, diatomic metal clusters. Indeed, the M–M distances in acetylenediolates are shorter than those in free metals (for instance, $d_{K\text{-}K}$ in KOC≡COK is 364 pm [65], while it is 472 pm in free potassium). The layered structure of all compounds mentioned above might, in turn, be determined by the linear structure of their particles. Actually, molecules with higher degrees of spatial branching usually form spherical and cyclic aggregates [68–71]. For example, the structure of alkali metal *tert*-butoxides can consist of tetramers of a cubane type.

The layered structures are known to possess a spatial anisotropy, *i.e.* there is a certain direction in which the density of layers is markedly higher than in other directions. In contrast to usual layered structures consisting of 2D-polymer molecules (*e.g.*, graphite) the structures built from linear molecules are characterized by strong *00l* and *hk0* reflections. However, the intensities of *hkl* reflections are usually very weak. As is evident from Fig. 7, the value of parameter *c* in acetylenediolates is determined by the length of MOC≡COM fragments.

TABLE 2. Interatomic distances in some carbon-containing compounds

Compound	d(C–O), pm	d(C–C), pm	d(K–O), pm
$H_2C=O$	120.9	–	–
C≡O	113.1	–	–
O=C=O	116.3	–	–
O=C=C=C=O	116.0	128.0	–
KOC≡COK	128.0	120.0	266.0 (277.0)
$KOCH_3$	145.0	–	260.0 (280.0)

In our case, there is no chemical factor that would be able to control the destruction process of the starting polymer. In other words, the lengths of short carbon chains formed can vary. This assumption allows one to explain why the composition of the metal-carbon films is not constant in different parts of the same film. Generally, the length variability in linear fragments should lead to an incommensurate parameter *c* and, therefore, to the absence of *00l* and *hkl* interplanar spacings in the diffraction pattern.

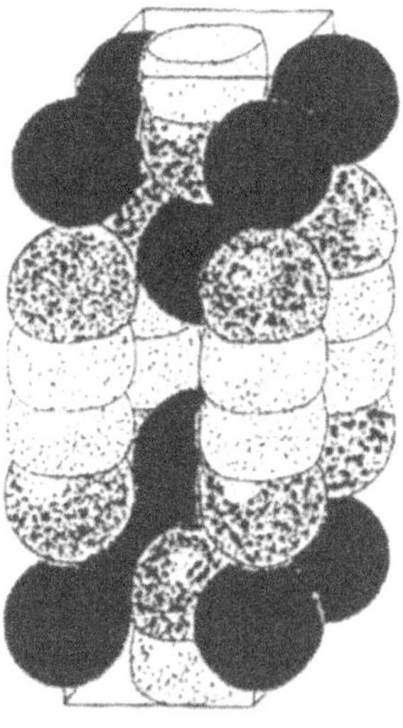

Figure 7. Structure of acetylenediolates (MOC≡COM), where M = Na, K, Rb, and Cs.

The texture pattern of layered crystals should be such of an orientation perpendicular to the electron beam axis. This is why the *hk0* section of the reciprocal lattice is usually observed by TEM diffraction. In order to obtain *hkl* lines, the crystal must be rotated. The rotation of all carbon films with sodium and potassium investigated so far did not give rise to the appearance of additional lines in the diffraction pattern, while the lines being present weakened to the point of their complete disappearance. Obviously, this fact suggests that the diffraction pattern observed contained only *hk0* reflections and that the parameter c was absent in those organometallic compounds constituting metal-carbon film.

In my opinion, the most important experimental fact obtained by electron diffraction is the confirmation of the identity of structures of metal-carbon films (with both sodium and potassium) synthesized by high-pressure dehydrogenation of PA in the presence of alkali metal and those obtained by simple interaction of carbynoid material with alkali metals [6,9–12]. Therefore, the mechanism of formation of a carbyne crystal during acidic treatment of the metal-carbon films in both cases is definitely the same. Indeed, the intermediate compounds in both cases were short carbon-chain compounds with alkali metals.

The majority of particles obtained by interaction of the carbynoid material with sodium exhibited diffraction lines as shown in Table 3. According to Udod *et al.* [9], these lines correspond to the *hk0* reflections of a two-dimensional tetragonal lattice with parameter a = 340 pm. The absence of *100*, *210*, *etc.* reflections of the tetragonal symmetry was explained by the proposed two-layer structure of the sodium-carbon compound (I)[4] (Fig. 8*a*). As discussed above, the two-layer structure is characteristic of a number of alkali organometallic compounds, especially acetylenediolates. It is noteworthy that in the model shown in Fig. 8*a* the shortest distance between the metal atom and an adjacent carbon chain in the same layer is the sum of the van-der-Waals radius of *sp*-carbon and the ionic radius of sodium (150 + 90 = 240 pm). The analogous compound with potassium also conforms to a tetragonal symmetry with the C–K distance of 287 pm. At the same time, the C–K distance expected in the potassium-carbon compound (I) is 150 + 133 = 283 pm, being close to the experimental value. This distance in potassium acetylenediolate (KOC≡COK) is 277 pm. All these compounds possess a tetragonal symmetry determined by the interaction of metal with adjacent carbon or oxygen-carbon chains and van-der-Waals interaction between them. The increase of the tetragonal parameter to a = 400 pm (Table 3) observed in sodium-carbon compound (II) might result from substantial overlapping of the upper and lower layers of the two-layer structure. As can be seen from Fig. 8*b*, the distance between carbon chains is 283 pm. Thus, the contribution of the van-der-Waals interaction between carbon atoms to a total energy of the crystal should increase significantly in the metal-carbon compounds with longer chains. Hence, the formation of a dense hexagonal structure becomes possible in potassium-carbon compound (II) (Fig. 8*c*). The value of the hexagonal parameter a (the distance between carbon chains) is 294 pm, close to the van-der-Waals diameter of *sp*-carbon atoms [42].

---

[4] Since the resulting lattice corresponds to a tetragonal body-centered cell the reflection conditions $h+k+l = 2n$ are not fulfilled for *100* and *210*, i.e. they will be extinct (Eds. comment).

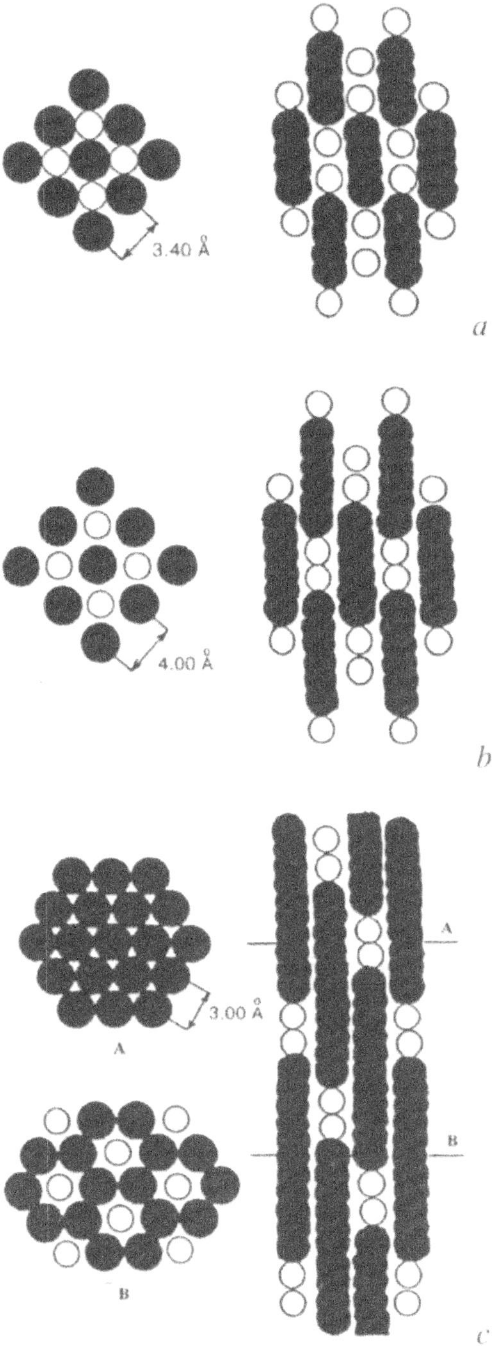

Figure 8. Hypothetical models of short carbon-chain organometallic compounds with different average lengths of the carbon molecules and the size of alkali metal.

TABLE 3. Interplanar distances observed in the sodium-carbon film synthesized by Udod *et al.* [9]

d_{exp}, pm	C/Na *hk0*	(I) d_{calc}, pm	C/Na *hk0*	(II) d_{calc}, pm
288	–	–	*110*	289
241	*110*	240	–	–
199	–	–	*200*	204
170	*200*	169	–	–
147	–	–	*220*	145
125	–	–	*310*	129
99.7	–	–	*400*	99.5
91.6	–	–	*420*	91.2

Normally, in some areas the potassium-carbon films [10,11] contain three phases: potassium-carbon compound (I) with short chains, potassium-carbon compound (II) with medium-length chains, and crystalline carbyne with polymer chains (Table 4). This fact confirms the conjecture that the formation of carbyne crystals in our case initially occurs through assembling organometallic compounds of short carbon chains with alkali metals.

Thus, we can distinguish two basic types of chemical interaction between carbon and metallic atoms in the short-chain alkali metal compounds:

– chemical bonds between the metal and carbon atoms are, probably, of ionic type (similarly to alkali metal acetylides);

– chemical bonds are formed by partial charge transfer from the alkali metal atom to π-orbitals of carbon chains (much like in graphite intercalation compounds or in doped PA).

TABLE 4. Interplanar distances of the three phases present in a potassium-carbon film, synthesized by Udod *et al.* [10]

d_{exp}, pm	C/K-I *hk0*	(tetragon.) d_{calc}, pm	C/K-II *hk0*	(hexagon.) d_{calc}, pm	carbyne *hk0*	- crystal d_{calc}, pm
460	–	–	–	–	*100*	460
293	*110*	291	–	–	–	–
252	–	–	*100*	255	–	–
208	*200*	206	–	–	–	–
152	–	–	–	–	*300*	153
146	*220*	146	*110*	148	–	–
129	*310*	130	*200*	127	–	–
105	*400*	103	–	–	–	–
95	*330*	97	–	–	–	–
85	–	–	*210*	85	–	–

Eventually, the contribution of the van-der-Waals interaction between carbon atoms to a total energy of the crystal in longer chain compounds increases, whereas the contribution of the metal-carbon interaction should decrease. One can expect that the van-der-Waals interaction is of vital importance to the stability of carbyne crystal with polymer-like carbon chains[5].

6. Possible Mechanism of Linear Carbon Formation during High Pressure Treatment of a PA+K System

As mentioned above, the retention of the linear structure of the original carbon skeleton during high pressure dehydrogenation of polyacetylene is conditioned by the presence of alkali metal atoms. The formation of alkali metal hydride was also observed during intercalation of alkali metals into polyacetylene at ambient pressure [25]. However, this reaction occurs as a result of the interaction of gaseous hydrogen evolved with free alkali metal, and the carbon phase is transformed at the temperature of hydrogen elimination into an amorphous one with predominant sp^2-hybridization state of the carbon atoms. The carbon skeleton can be prevented from cross-linking by isolating the carbon chains formed from each other by the intercalation of metal atoms. It is important to mention that all known intercalated compounds are unstable within the temperature region of hydrogen thermal elimination from PA. Therefore, the role of high pressure conditions might be to increase the thermal stability of intermediate intercalation phases. Actually, the deintercalation of alkali metals would take place with increasing volume (Table 1).

As discussed above, the high-pressure intercalation of alkali metals into graphite resulted in new superdense GICs. The composition of the superdense compounds synthesized was determined by different methods. In my view, the most valuable data were obtained by methods that allow investigation of the composition *in situ*, *i.e.* directly under high pressure conditions. The shortest distance between metal atoms in a superdense monolayer can be calculated on the assumption that they are being packed in a closest sphere arrangement. The basis for such an assumption lies in the tendency of a chemical system to occupy the smallest volume possible under thermobaric treatment. This fact excludes the formation of vacancies and low coordination bonds in the intercalate layer. This might be proven indirectly by the high rate of the intercalation-deintercalation cycle [30,32,36,38].

The compression of alkali metal atoms in the interlayer space of graphite can be achieved at pressure essentially lower than in the free metal state. According to Mordkovich [72], such additional 'compression' of metal atoms results from specific influence of the graphite matrix. In the case of superdense sodium GIC (C_2Na) the essential decrease in the M–M distance can be explained by centering of sodium atoms above all carbon hexagons [37], whereas the commensurability between these two sub-lattices is probably absent in superdense GICs of heavy alkali metals (C_4M compounds) [11]. Table 1 shows possible values of the M–M distances in the intercalated alkali metal layers, that are of particular interest for understanding the chemical processes

[5] For details see Chapter 4.3 of this book.

occurring in the PA+alkali metal systems during thermobaric treatment. Perhaps, we might observe similar compression also of metal atoms in the intercalated channels in comparison with the free metal state due to the action of the PA matrix.

As visible in Table 1, the existence of $(C_3H_3)_2M$ at high-pressure conditions is not unlikely (the M–M distance in this case should be *ca.* 370 pm). The formation of structures substantially more saturated with alkali metal could be possible when removing hydrogen atoms. In this case the cylindrical symmetry of polymer molecules would appear. Therefore, during retention of the original tetragonal symmetry in general, the formation of an additional channel of alkali metal becomes possible (Fig. 9*a*). The composition of such a tetragonal structure (Fig. 9*b*) should be close to C_3M. Further saturation of the carbon phase with alkali metal might occur due to a rearrangement in packing of polymer molecules resulting in a change from the tetragonal to hexagonal symmetry (Fig. 9*c*). It should be noted that the composition of this hypothetical intermediate compound should be close to C_3M_2, correlating satisfactorily with the experimental value of $C_{0.7}K$ or $C_{1.43}K$ (eq. 2) determined by DTA measurements directly under high-pressure conditions. As may be seen from Fig. 9*c*, the carbon chains are isolated from each other entirely by alkali metal atoms that form their own three-dimensional sub-lattice. Thus, the formation of one-dimensional carbon molecules under 'compression' conditions is possible only when there is complete isolation by alkali metal atoms. The metal atoms do not form strong chemical bonds with these chains, but, probably, stabilize them additionally. The nature of the chemical interaction between the polymer carbon chains and alkali metal atoms in such a high-pressure phase is very similar to that in superdense GICs with alkali metals.

The proposed mechanism of the formation of carbyne during thermobaric treatment of the PA+K system shows the possibility in principle to transform sp^2-carbon atoms of PA into the *sp*-state under high-pressure conditions only when carbon chains are completely isolated from each other. Otherwise, the energy released by the formation of new single C–C bonds would be sufficient to induce extensive interchain cross-linking. On the other hand, direct transition from a one-dimensional state to a well-ordered two-dimensional structure does not occur, at least not at the temperature of hydrogen elimination from polyacetylene. The cleavage of all original C–C bonds as well as those formed intermediately requires significantly higher temperatures. Thus, in order to understand and thoughtfully plan chemical transformations in a carbon phase, one should consider first all possible kinetic mechanisms. The reaction pathway, that appears to be most preferable from a thermodynamic point of view, often cannot be realized in practical situations because of a kinetic hindrance.

This fact is clearly demonstrated by the formation of fullerenes and nanotubes [73,74], which are essentially less stable than 'normal' carbon forms, *viz.*, graphite and diamond.

In connection with the aforesaid, one needs to mention a recent study by Yamada *et al.* [75] concerning a possible mechanism of carbyne destruction under electron beam irradiation. According to the authors' opinion, the intermediate substance in this case is rhombohedral graphite. In my view, a situation where the intermediate substance (graphite) is thermodynamically more stable than the starting one, and, moreover, the endproduct appears to be unlikely. Furthermore, graphite under similar conditions is

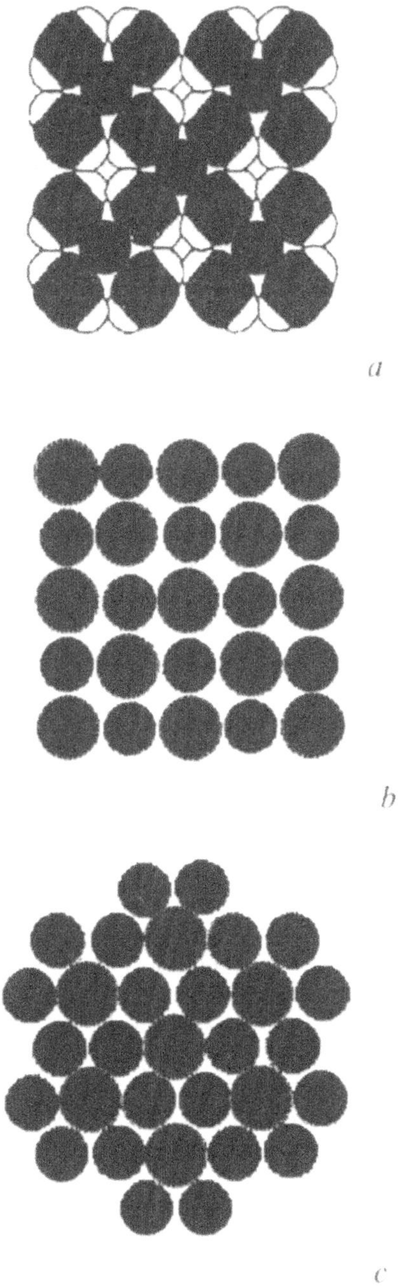

Figure 9. Possible structures of intermediate compounds obtained under high-pressure conditions: (*a*) polyacetylene doped with heavy alkali metal; (*b*) tetragonal linear carbon–metal compound, (*c*) hexagonal linear carbon–metal compound. Grey and black circles represent, respectively, carbon and metal atoms.

fairly stable, at least it keeps its 2D structure. The fact of identifying graphite by *hkl* diffraction spacings such as 208.0 and 146.4 pm is even more unusual. The absence of *hk0* lines possessing much stronger intensities than *hkl* lines in the diffraction pattern from polycrystalline graphite [75] should attest to the incorrectness of the proposed hypothesis. The values of the interplanar distances reported rather correspond to a tetragonal symmetry because 208.0:146.4 » $2^{1/2}$. These values are very close to our data shown in Table 4. That is why there is good reason to believe that intermediate substances in the destruction of crystalline carbyne under electron beam irradiation are short carbon-chain structures, probably stabilized by impurity atoms[6]. Thus, the electron beam-induced destruction of carbyne leading eventually to the transformation of the entire linear structure into amorphous carbon begins with scissions in the polymer chains. Short chains, being less stable and more mobile particles, in turn can transform into amorphous carbon as a result of extensive cross-linking.

7. References

1. Akagi, K., Nishiguchi, M., Shirakawa, H., Furukawa, Y., and Harada, I.: One-dimensional conjugated carbyne - synthesis and properties, *Synth. Met.* **17** (1987), 557–562.
2. Korshak, V.V., Kudryavtsev, Yu.P., Korshak, Yu.V., Evsyukov S.E., Khvostov, V.V., Babaev, V.G., and Guseva, M.B.: Formation of β-carbyne by dehydrohalogenation, *Macromol.Chem., Rapid Commun.* **9** (1988), 135–140.
3. Evsyukov S.E., Kudryavtsev, Yu.P., Korshak, Yu.V., Khvostov, V.V., Babaev, V.G., Guseva, M.B., and Korshak, V.V.:Synthesis of carbyne on the basis of polyvinylidene halides, *Polymer Science USSR* **31** (1989), 29–37.
4. Vointseva, I.I., Gil'man, L.M., Kudryavtsev, Yu.P., Evsyukov, S.E., and Valetskii, P.M.: Synthesis of polymorphic modification of carbyne by dehydrochlorination of poly(trichlorobutadiene) isomers, *Polym. Sci., Ser A* **38** (1996), 707–710.
5. Evsyukov S.E., Paasch, S., Thomas, B., and Heimann, R.B.: Formation of carbynoid structures by chemical dehydrohalogenation of poly(vinylidene chloride). A ^{13}C solid-state NMR study, *Ber. Bunsen-Ges. Phys. Chem.* **101** (1997), 837–841.
6. Udod, I.A., Shchurik, V.I., Bulychev, B.M., Sirotinkin, S.P., Guseva, M.B., Babaev, V.G., Kudryavtsev, Yu.P., and Evsyukov, S.E.: Formation of carbyne in the interaction of polyacetylene with potassium under high pressure quasi-hydrostatic pressure, *J. Mater. Chem.* **3** (1993), 413–416.
7. Kasatochkin, V.I., Shterenberg, L.E., Kazakov, M.E., Slesarev, V.N., and Belousova, L.V.: The thermal transformation of carbyne under pressure, *Dokl. Akad. Nauk SSSR* **209** (1973), 388–391 (in Russian).
8. Kasatochkin, V.I., Korshak, V.V., Kudryavtsev, Yu.P., Sladkov, A.M., and Shterenberg, L.E.: On crystalline structure of carbyne, *Carbon* **11** (1973), 70–72.
9. Udod, I.A., Bulychev, B.M., Sirotinkin, S.P., Babaev, V.G., and Kudryavtsev, Yu.P.: An interaction of amorphous carbyne with sodium: the formation of carbyne intercalation compounds, *Synth. Met.* **60** (1993), 57–62.
10. Udod, I.A., Bulychev, B.M., Babaev, V.G., and Kudryavtsev, Yu.P.: Formation of carbyne intercalation compounds with alkali metals, *Zh. Neorg. Khim.* **41** (1996), 382–387 (in Russian).
11. Udod, I.A.: Synthesis and structure of carbyne and graphite intercalation compounds with sodium and potassium, *Ph.D. Thesis*, Moscow State University, Moscow, 1994 (in Russian).
12. Bulychev, B.M. and Udod, I.A.: Linear carbon (carbyne): approaches to synthesis, identification of structure and intercalation, *Ross. Khim. Zh.* **39** (1995), 9–18 (in Russian).
13. Baughman, R.H., Murthy, N.S., and Miller, G.G.: The structure of metallic complexes of polyacetylene with alkali metals, *J. Chem. Phys.* **79** (1983), 515–520.

[6] The presence of impurity atoms as one of necessary conditions for successful crystallization of carbyne is discussed, among others, in Chapter 4.3 of this book.

14. Shacklette, L.M., Murthy, N.S., and Baughman, R.H.: New structural phases of polymer battery anode materials: alkali-metal-doped polyacetylene and polyphenylene, *Mol. Cryst. Liq. Cryst.* **121** (1985), 201–209.
15. Baughman, R.H., Shacklette, L.M., Murthy, N.S., Miller, G.G., and Elsenbaumer, R.L.: The evolution of structure during the alkali metal doping of polyacetylene and poly(*p*-pheylene), *Mol. Cryst. Liq. Cryst.* **118** (1985), 253–261.
16. Winokur, M., Moon, Y.B., Heeger, A.J., Barker, J., Bott, D.C., and Shirakawa, H.: X-ray scattering from sodium-doped polyacetylene: incommensurate-commensurate and order-disorder transformation, *Phys. Rev. Lett.* **58** (1987), 2329–2332.
17. Murthy, N.S., Shacklette, L.M., and Baughman, R.H.: Effect of charge transfer on chain dimension in *trans*-polyacetylene, *J.Chem.Phys.* **87** (1987), 2346–2348.
18. Billaud, D., Ghanbaja, J., and Coulon, C.: Heavy alkali metal doping of (CH)*x*: new data, *Synth. Met.* **17** (1987), 497–502.
19. Letner, O., Kahlert, M., Leising, G., Fink, J., and Fritzsche H.: Doping induced structural changes of oriented trans-polyacetylene, *Synth.Met.* **28** (1989) D225–D229.
20. Murthy, N.S., Shacklette, L.M., and Baughman, R.H.: Structures and staging in polyacetylene charge-transfer complexes, *Solid State Commun.* **72** (1989), 267–270.
21. Djurado, D., Fischer, J.E., Heiney, P.A., Ma, J., Coustel, N., and Bernier, P.: Staging in potassium-doped polyacetylene: in situ x-ray diffraction, *Synth. Met.* **34** (1989), 683–688.
22. Murthy, N.S., Shacklette, L.M., and Baughman, R.H.: Structure of lithium-doped polyacetylene, *Phys. Rev. B* **40** (1989), 12550–12553.
23. Ma, J., Djurado, D., Fischer, J.E., Coustel, N., and Bernier, P.: X-ray structure of Cs-doped polyacetylene, *Phys. Rev. B* **41** (1990), 2971–2975.
24. Billaud, D., Saldi, F., Ghanbaja, J., Begin, D., and Lelaurain, M.: Structural variations of oriented polyacetylene upon intercalation with selected electron donating and accepting species, *Synth. Met.* **35** (1990), 113–121.
25. Saldi, F., Lelaurain, M., and Billaud, D.: Evolution of the tridimensional organization of alkali metal doped polyacetylene complexes $(M_yCH)_x$ with the doping level for M = Na, K, Rb, Cs, *Solid State Commun.* **76** (1990), 595–599.
26. Murthy, N.S., Shacklette, L.M., and Baughman, R.H.: Structural models for alkali metal complexes of polyacetylene, *Phys. Rev. B* **41** (1990), 3708–3718.
27. Saldi, F., Lelaurain, M., and Billaud, D.: Structural disorders in $(M_yCH)_x$ compounds (M = K, Rb, Cs): *Mater. Sci. Forum* **91-93** (1992), 363–368.
28. Chuvyrov, A.N., Lebedev, Y.A., Kornilov, V.N., Salimgareeva, V.N., Sannikova, N.S., and Leplyanin, O.V.: Structure and conductivity of new carbonaceous films obtained by polyacetylene pyrolysis, *Synth. Met.* **41–43** (1991), 1443–1446.
29. Semenenko, K.N., Avdeev, V.V., and Mordkovich, V.Z.: Intercalation of alkali metal into graphite under high pressure, *Dokl. Akad. Nauk SSSR* **271** (1983), 1402–1405 (in Russian).
30. Semenenko, K.N., Avdeev, V.V., and Mordkovich, V.Z.: A study of reactions of graphite with alkali metals under high pressures by quantitative differential thermal analysis, *Zh. Neorg. Khim.* **29** (1984), 2236–2239 (in Russian).
31. Nalimova, V.A., Udod, I.A., Avdeev, V.V., and Semenenko, K.N.: Synthesis and compressibility of lithium graphite intercalation compounds of the first stage, *Zh. Obshch. Khim.* **60** (1990), 868–871 (in Russian).
32. Nalimova, V.A., Avdeev, V.V., and Semenenko, K.N.: New alkali metal-graphite intercalation compounds at high pressures, *Mater. Sci. Forum* **91–93** (1992), 11–16.
33. Belash, I.T., Bronnikov, A.D., Zharikov, O.V., and Palnichenko, A.V.: On the superconductivity of graphite intercalation compounds with sodium, *Solid State Commun.* **64** (1989), 1445–1447.
34. Belash, I.T., Zharikov, O.V., and Palnichenko, A.V.: Synthesis, stability and structure of GIC with Li, Na and K, *Synth. Met.* **34** (1989), 47–52.
35. Syassen, K., Sonnenschein, R., Hanfland, M., and Beister, H.J.: Graphite graphite intercalation compounds under pressure: Raman modes, optical reflectivity and phase changes, *Synth. Met.* **34** (1989), 293–306.
36. Nalimova, V.A., Chepurko, S.N., Avdeev, V.V., and Semenenko, K.N.: Intercalation in the graphite-rubidium system under high pressure, *Synth. Met.* **40** (1991), 267–273.
37. Udod, I.A., Orman, H.B., and Genchel V.K.: The sodium-graphite system under high pressure conditions: the comparison with the lithium-graphite system, *Carbon* **32** (1994), 101–106.
38. Avdeev, V.V., Nalimova, V.A., and Semenenko, K.N.: High-pressure synthesis of the graphite-cesium intercalation compounds, *Dokl. Akad. Nauk SSSR* **304** (1989), 111–113 (in Russian).
39. El Goresy, A. and Donnay, G.: A new allotropic form of carbon from the Ries Crater, *Science* **161** (1968), 363–364.

40. Heimann, R.B., Kleiman, J., and Salansky, N.M.: A unified structural approach to linear carbon polytypes, *Nature* **306** (1983), 164–167.
41. Heimann, R.B., Kleiman, J., and Salansky, N.M.: Structural aspects and conformation of linear carbon polytypes (carbynes), *Carbon* **22** (1984), 147–156.
42. Kudryavtsev, Yu.P., Evsyukov, S.E., Babaev, V.G., Guseva, M.B., Khvostov, V.V. and Krechko, L.M.: Oriented carbyne layers, *Carbon* **30** (1992), 213–221.
43. Guseva, M.B., Babaev, V.G., Babina, V.M., and Khvostov, V.V.: Formation of oriented nitrogen doped carbon films, *JEPT Lett.* **62** (1995), 715–718.
44. Ando, Y.: A carbyne observed in a marketing microgrid, *Carbon* **33** (1995), 171–175.
45. Rohlfing, E.A., Cox, D.M., and Kaldor, A.: Production and characterization of supersonic carbon cluster beams, *J. Chem. Phys.* **81** (1984), 3322–3330.
46. Juza, R., Wehle, V., and Schuster H.-U.: Zur Kenntnis des Lithiumacetylids, *Z. Anorg. Allg. Chem.* **352** (1967), 252–257.
47. Atoji, M.: Crystal structure determination of disodium acetylide, Na_2C_2, by neutron diffraction, *J. Chem. Phys.* **60** (1974), 3324–3327.
48. Foeppl, H.: Strukturuntersuchungen an Alkalicarbiden und -disulfiden, *Angew. Chem.* **70** (1958), 401.
49. Weiss, E. and Plass, H.: Die Kristallstrukturen der Alkalihydrogenacetylide $MC{\equiv}CH$ (M= Na, K, Rb) und -propinylide $MC{\equiv}CCH_3$ (M = Na, K), *Chem. Ber.* **101** (1968), 2947–2955.
50. Atoji, M.: Neutron structure of monosodium acetylide, NaC_2H, at 293 and 5 K, *J. Chem. Phys.* **56** (1972), 4947–4951.
51. Haddon, R.C., Hebard, A.F., Rosseinsky, M.J., Murphy, D.W., Duclos, S.J., Lyons, K.B., Miller, B., Rosamilla, J.M., Fleming, R.M., Kortan A.R., Glarum, S.H., Makhija, A.V., Muller, A.J., Elck, R.H., Zahurak, S.M., Tycko, R., Dabbagh, G., and Thiel F.A.: Conducting films of C_{60} and C_{70} by alkali-metal doping, *Nature* **350** (1991), 320–322.
52. Hebard, A.F., Rosseinsky, M.J., Haddon, R.C., Murphy, D.W., Glarum, S.H., Palstra, T.T.M., Ramirez, A.P., and Kortan A.R.: Supercondactivity at 18 K in potassium-doped C_{60}, *Nature* **350** (1991), 600–601.
53. Stephens, P.W, Mihaly, L., Lee, P.L., Whetten, R.L., Huang, S.-M., Kaner, R., Deiderich, F., and Holczer, K.: Structure of single phase superconducting K_3C_{60}, *Nature* **351** (1991), 632–634.
54. Holczer, K., Klein, O., Huang, S.-M., Kaner, R., Fu, K.-J., Whetten, R.L., and Diederich, F.: Alkali-fulleride superconductors: synthesis, composition and diamagnetic shielding, *Science* **252** (1991), 632–634.
55. Fleming, R.M., Ramires, A.P., Rosseinsky, M.J., Murphy, D.W., Haddon, R.C., Zahurak, S.M., and Makhija, A.V.: Relation of structure and superconducting transition in A_3C_{60}, *Nature* **352** (1991), 787–788.
56. Tanigaki, K., Ebbesen, T.W., Saito, S., Mizuki, J., Tai, J.S., Kubo, Y., and Kuroshima, S.: Superconductivity at 33 K in $Cs_xRb_yC_{60}$, *Nature* **352** (1991), 222–223.
57. Kelty, S.P., Chen, C.-C., and Lieber, Ch.: Superconductivity at 30 K in caesium-doped C_{60}, *Nature* **352** (1991), 223–225.
58. Mordkovich, V.Z., Baxendale, M., Yoshimura, S., and Chang R.P.H.: Intercalation into carbon nanotubes, *Carbon* **34** (1996), 1301–1303.
59. Udod, I.A.: Sodium-graphite intercalation compound of the first stage: two-dimentional structure and stability, *Synth. Met.* **88** (1997), 127–131.
60. Weiss, E.: Die Kristallstruktur des Kaliummethylats, *Helv. Chim. Acta* **46** (1963), 2051–2054.
61. Weiss, E.: Die Kristallstruktur des Natriummethylats, *Z. Anorg. Allg. Chem.* **332** (1964), 197–203.
62. Weiss, E.and Alsdorf, H.: Die Kristallstrukturen des Kalium-, Rubidium- and Casiummethylats, *Z. Anorg. Allg. Chem.* **372** (1970), 206–213.
63. Greiser, T. and Weiss, E.: Kristallstructur der Alkali-isopropoxide des Kaliums, Rubidiums und Caesiums, *Chem. Ber.* **112** (1979), 844–848.
64. Buchner, W. and Weiss, E.: Zur Kenntnis der sogenannten "alkalicarbonyle" IV. Über die Reaktion von geschmolzenem Kalium mit Kohlenmonoxid, *Helv. Chim. Acta* **47** (1964), 1415–1423.
65. Weiss, E. and Buchner, W.: The structure of the so-called "alkali carbonyls", *Angew. Chem. Int. Ed. Engl.* **3** (1964), 152.
66. Weiss, E. and Buchner, W.: Zur Kenntnis der sogenannten "Akalicarbonyle" I. Die Kristallstruktur des Kalium-acetylendiolats, $KOC{\equiv}COK$, *Helv. Chim. Acta* **46** (1963), 1121–1127.
67. Weiss, E. and Buchner, W.: Zur Kenntnis der sogenannten "Alkalicarbonyle" II. Die Kristallstruktur des Rubidium- und Caesium-acetylendiolats, $RbOC{\equiv}CORb$ und $CsOC{\equiv}COCs$, *Z. Anorg. Allg. Chem.* **330** (1964), 251–258.
68. Weiss, E., Alsdorf, H., and Kuhr, H.: Structure of alkali metal t-butoxides, *Angew. Chem. Int. Ed. Engl.* **6** (1967), 801–802.

69. Weiss, E., Alsdorf, H., Kuhr, H., and Grutzmacher H.-F.: Rontgengraphische, NMR- und Massenspektrometrische Untersuchungen der *tert.*-Butylate des Kaliums, Rubidiums und Casiums, *Chem. Ber.* **101** (1968), 3777–3786.
70. Greiser, T. and Weiss E.: Kristallstructur des Natrium-*tert.*-butoxids, $[(CH_3)_3CO)Na]_9$ x $[(CH_3)_3CO)Na]_6$, ein neuer Strukturtyp mit nonameren und hexameren Assoziaten, *Chem.Ber.* **110** (1977), 3388–3396.
71. Davies, J.E., Koff, J., and Weiss, E.: Structure of sodium *tert*-butoxide: a re-refinement, *Acta Cryst.* **B38** (1982), 2251–2253.
72. Mordkovich, V.Z.: Model for constitution of graphite intercalation compounds, *Synth.Met.* **63** (1994), 1–6.
73. Kroto, H.W., Heath, J.R., O'Brien, S.C., Curl, R.F., and Smalley, R.E.: C_{60}: Buckminsterfullerene, *Nature* **318** (1985) 162–163.
74. Iijima, S.: Helical microtubules of graphitic carbon, *Nature* **354** (1991), 56–58.
75. Yamada, K., Burkhard G., Tanabe T., and Sawaoka A.B.: An in-situ observation of amorphization of carbyne particle under electron-beam irradiation, *Carbon* **34** (1996), 1601–1602.

3.3. Pyrolytic methods

3.3.1. DECOMPOSITION OF HYDROCARBONS

A. SOKOŁOWSKA and A. OLSZYNA
Faculty of Materials Science, Warsaw University of Technology, Narbutta 85, 02-524 Warsaw, Poland

Abstract – The chapter contains a survey of methods based on the pyrolysis of carbon-containing organic vapours that are suitable for the production of carbyne and/or chaoite. It has been shown that the synthesis of the $\sigma sp\pi p^2$ phases can only occur when the process is activated. The agents suitable for the activation include: diamond surfaces, electrons, ions and absorbed photons. The additional nonmechanical energy delivered to the system should be smaller than the energy necessary for the formation of the σsp^3 phase. It is being proposed to describe the mechanism of the formation of the $\sigma sp\pi p^2$ phases in terms of a chemical reaction with the participation of an 'active complex'.

1. Introduction. Pyrolysis of Volatile Hydrocarbons

Low-molecular weight hydrocarbons undergo pyrolysis when subjected to heating under normal conditions, according to the reactions:

$$CH_4 \longrightarrow H + CH_3^*$$
$$CH_3^* \longrightarrow C_s + 3H$$
$$CH_4 \longrightarrow C_s + 2H_2 \qquad \Delta H_0 = 17.88 \text{ kcal/mol}$$

A solid product of this reaction is carbon with the $\sigma sp^2\pi p$ configuration. Depending on the process temperature and the quenching rate of the products, carbon appears in various forms with different particle sizes and textures, from soot up to pyrographite.

2. Epitaxial Growth during Pyrolysis

Numerous investigators, among the first Deryagin *et al.* [1] and Eversole *et al.* [2], have shown that the high temperature pyrolysis of hydrocarbons at a diamond substrate in the presence of excessive hydrogen or a nonstoichiometric amount of oxygen results at 1000 K in a homoepitaxy of carbon in the form of diamond. Based on the concept first proposed by Spitsyn [3], Bachman [4] and Frenklach [5], a particular phase equilibrium system was suggested in which, under the conditions of pyrolysis carried

R.B. Heimann et al. (eds.), Carbyne and Carbynoid Structures, 117–131.

out in the presence of other gases active chemically towards graphite, diamond is a thermodynamically stable phase. In this process, a fundamental role is being played by the adsorption of atomic hydrogen H on the substrate surface and the kinetics of hydrogenation of graphite.

This theory, however, was not supported unequivocally by any experiment in which the pyrolysis of hydrocarbons would yield metastable carbon without any activation through additional energy or the contribution of a diamond substrate. Deryagin *et al.* [6] have proposed a general relationship determining the conditions for metastable carbon phases to form:

$$J_{ch.c.}/J_{ph.c.} \approx p_{\Sigma} \exp\,(\Delta G_T{}^{o} - E + Q/RT) \qquad (1)$$

where: J - crystallization rate, ch.c. - 'chemical crystallization', ph.c. - 'physical crystallization', $\Delta G_T{}^{o}$ - enthalpy change of the chemical reaction, E - activation energy, Q -evaporation heat.

However, this expression was based on the tenuous assumption that the value of the supersaturation $k\mathrm{T}\cdot\ln(p/p_o)$ is identical for graphite and the (metastable) diamond phase. So it seems that one factor influencing the phase composition of the products of pyrolysis may be the nature of the substrate on which crystallization takes place. On graphite substrates and on other non-carbon substrates, without additional activation only graphite layers can be grown during CVD processes.

Fedoseev *et al.* [7] proved that at 1200 K from carbonaceous gas diluted with hydrogen a mixture of various carbon forms (including α-carbyne) can crystallize on a diamond substrate. The authors explain this characteristic phenomenon with the specific influence of the diamond surface. Simultaneous presence of four forms of carbon indicates the non-equilibrium state of the whole system. It also appears that the energy state of carbon atoms forming carbyne is closer to that of the diamond than for graphite. And what is more striking, the $\sigma sp\pi p^2$ electron hybridization in carbyne seems to be more akin to the $\sigma sp^2\pi p$ hybridization in graphite than to the σsp^3 hybridization in diamond.

3. Activated Pyrolysis

An additional non-thermal activation of a gaseous phase containing hydrocarbons, can be afforded by

– electric energy which, at the initial step, increases the kinetic energy of the electrons,

– electromagnetic energy whose initial action results in the absorption of a photon by the gas particles.

Depending on the amount and kind of energy introduced into the system, the activated gas contains a number of excited particles, ions and free electrons which can essentially affect the energy state of the valence electrons of the synthesis product. When we deal with carbon, this product will be metastable phases with σsp^3 configuration (diamond) or $\sigma sp\pi p^2$ configuration (chaoite and α-carbyne).

The kind of particles of elevated energy, their number and their modes of collision decide whether the critical nuclei formed will be those of graphite or those of another carbon phase. By way of example of collisions described by the equations given below, the role played by the negative ions present in the ionized gas during formation of a metastable phase may be considered [8]:

$$A^- + B \rightarrow AB^* + e^- \quad (2)$$

$$A^- + B^- \rightarrow AB^* + e^- \quad (2a)$$

The emission of an electron is associated with a release of energy from the system, thus stabilizing the excited particle.

Studies of the low-pressure synthesis of diamond by the activated CVD method, rapidly increasing in number for the past 20 years, often lead to the synthesis of chaoite or carbyne as the only product or a by-product that accompanies diamond or graphite.

Elkaseh [9] investigated the pyrolysis of a volatile carbon compound with the aim of determining how the various factors that activate the vapour particles affect the phase composition of the carbon obtained. The effects of the individual factors were examined separately.

Ethanol was subjected to pyrolysis at ambient pressure. The conditions under which the process was carried out were selected so as to avoid the formation of a plasma. The lowest substrate temperature at which carbon precipitated was 900 K. In one of its variants, the synthesis proceeded under the action of temperature alone. In the other variant, the reaction was carried out at the same temperature and, additionally, with a the pulsed laser beam ($h\nu$ = 3.3 eV, power 50 kW). The third variant involved the temperature and electrons that were field-emitted from an edge maintained at a voltage of ≤12 kV.

The carbon layers obtained were nanocrystalline. The only method that permitted their identification was electron diffraction. Under the action of the temperature alone, graphite crystallized. When using the temperature and photons as well as the temperature and electrons, chaoite was deposited. When the voltage applied to the edges was equal to or exceeded 10 kV, the chaoite crystallized in the form of minute spherules positioned on the edge and, further on its extension, like a chain arranged along the electric field lines.

Hence, chaoite already forms at a relatively low temperature, a relatively low photon energy, and a rather low kinetic electron energy. It forms through a specific chemical reaction in which the excitation of the reactants does not affect the rate and direction of the reaction but has an effect upon the electron configuration of the product. We can expect that an environment which contains a mixture of ions, electrons and excited particles, *i.e.*, a plasma, will be an exceptionally effective medium to synthesize metastable carbon phases.

An important question pertains to the difference between conditions necessary for the formation of carbon with σsp^3 configuration, *i.e.*, diamond, and for the formation of carbon with $\sigma sp\pi p^2$ configuration, *i.e.*, chaoite.

3.1. PLASMA ASSISTED HYDROCARBON PYROLYSIS

Recently CVD techniques have been extensively used in surface engineering for fabrication of different types of coatings. CVD stands for Chemical Vapor Deposition and refers a group of processes, during which products of chemical reactions in the gas phase are being deposited on various substrates. Deposition of these products leads to the formation of films only if the state of the substrate ensures conditions that allow the growing material to attain its state of thermodynamic stability.

Pyrolysis of hydrocarbons can be regarded as a good example of such a CVD process. Activation of the reaction with a plasma produced in the gaseous environment, PA-CVD (*i.e.*, Plasma Assisted CVD) is one of the most popular methods of fabrication of diamond or diamond-like carbon (DLC) layers. As experiments on optimization of PA-CVD of diamond films show, these processes frequently result in the formation of chaoite and α-carbyne with $\sigma sp\pi p^2$ electron configuration.

In surface engineering the most commonly exploited type of plasma is non-isothermal low-pressure plasma generated in a 13.56 MHz (radio-frequency) electromagnetic field. A scheme of typical apparatus for r.f. PA CVD of carbon films is shown in Fig. 1.

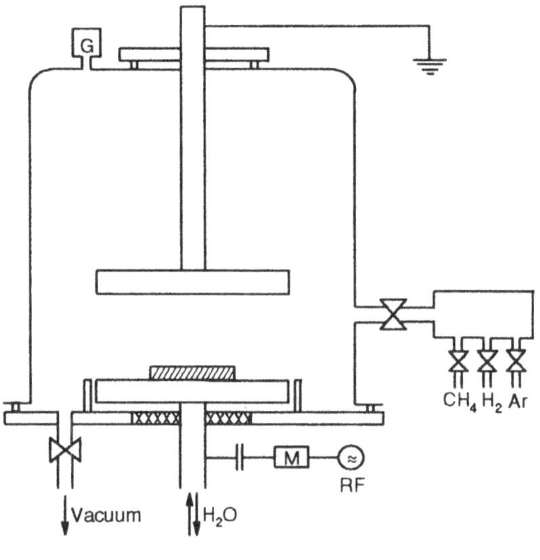

Figure 1. Apparatus of r.f. PA CVD process (radio-frequency 13.56 Mhz). RF – radio frequency generator, M – matching system, G – gauge.

The plasma generator is capacitively coupled with an inner non-grounded electrode at which surface films are being grown. Due to the fact that at so high a frequency of all particles present only electrons are able to preserve their translatory motion, on the inner electrode a negative self-bias potential V appears. Its value depends on the plasma state:

$$V \cong -kT_e \ln \frac{T_e m_j}{T_j m_e} \tag{3}$$

The self-bias potential describes the mean kinetic energy of the electrons present in the plasma. It can be controlled by gas pressure and the power of the generator.

In a series of experiments [10] during which a constant pressure (130 Pa) of methane atmosphere was maintained whereas the self-bias potential of the inner electrode was changed, it was possible to determine the conditions that allow the formation of carbyne. In Fig. 2, the dependence between the *V* value and the phase composition of the product of the r.f. plasma assisted-pyrolysis is presented. The carbyne phase forms at an ion energy < 100 eV. It should be noted here that so called 'hard amorphous carbon films', frequently obtained during processes of r.f. plasma assisted pyrolysis of hydrocarbons also contain small amounts of a carbon phase of *sp* electron configuration ($sp^3:sp^2:sp$ = 68:30:2) [11].

The fact that self-bias potential value, reflecting mean kinetic energy of plasma electrons, is also a measure of the energy lost by these particles in the course of their collisions [11] seems to suggest that formation of crystal nuclei or clusters of chaoite as well as of α-carbyne requires larger amounts of energy than is necessary to form graphite nuclei and lesser than is necessary to form diamond nuclei.

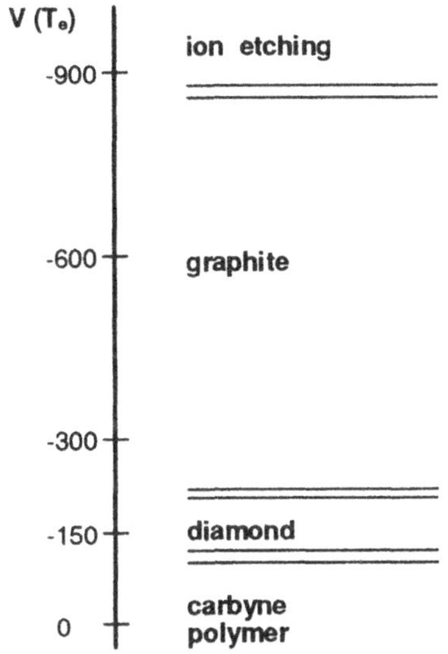

Figure 2. Carbon phases deposited on a non-grounded electrode in a r.f. PA–CVD process depending on self-bias potential V.

For fabrication of metastable phases of carbon, plasma created by means of a 2.4 GHz electromagnetic field (MW plasma assisted-CVD) is also applied. Synthesis of diamond films occurring under conditions of of MW PA-CVD under reduced pressure is considered to be the method yielding the chemically and physically purest diamond layers [12].

For production of highly dispersed carbon aerosols containing carbyne crystals there has been utilized a specific kind of MW (microwave) PA-CVD method [13]. A schematic block diagram of apparatus used for this purpose is shown in Fig. 3.

During this process low-molecular weight hydrocarbons are introduced into the plasma beam, generated in nitrogen by a high frequency electromagnetic field. As a result a carbon aerosol is formed which subsequently is cooled down by jet of neutral gas. The thermodynamic temperature of the plasma beam is controlled by the nitrogen flow as well as by the power of the electromagnetic field. In the experiments this temperature ranged between 2400 K and 4400 K. Carbon-containing gases introduced into the plasma could be diluted with neutral gases. The carbyne content in the pyrolysis product was observed to increase with increasing degree of dilution while simultaneously decreasing the plasma temperature to 3200 K. The authors attributed this effect to the establishment of thermal conditions at which the stable form of carbon is carbyne [14].

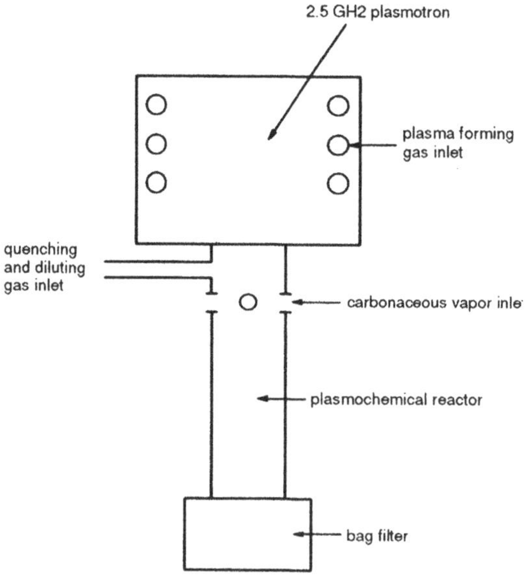

Figure 3. Block diagram of the apparatus of UHF assisted-CVD.

It appears, however, that the role of ions in plasma-chemical processes should not be overlooked. Also, the role of plasma should not be considered only as a 'heating medium'.

A specific type of plasma used for obtaining nanocrystalline diamond films is an impulse plasma [15,16]. A schematic diagram of apparatus serving for IPA (Impulse Plasma Assisted)-CVD process is shown in Fig. 4.

Crystallization of material takes place during movement of non-isothermal plasma packets with supersonic speed, which are produced by a high-current impulse discharge occurring in the space between the coaxial electrodes. The plasma-chemical crystallization process is determined by the energy released in a single discharge impulse and by the gas pressure. At an energies of around 1 kJ the excitation temperature within the interelectrode area was approximately 10^4 K whereas the thermodynamic temperature of plasma was 2000 K.

By means of the IPA-CVD process at ambient pressure a highly dispersed chaoite aerosol with nanometer-sized particles was produced from hydrocarbons [15].

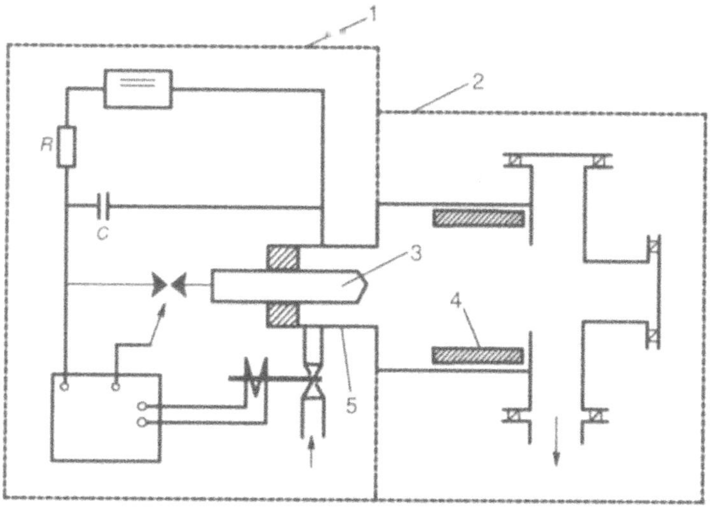

Figure 4. Apparatus for the IPA-CVD process; 1 – electric supply of impulse discharge, 2 – reaction chamber, 3 – internal electrode, 4 – substrate, 5 – external electrode.

The phase composition of carbon layers obtained under low pressure conditions (~20 Pa) by the use of IPA-CVD was dependent in a characteristic way on the local composition of the plasmoid, which was non-homogeneous and had the form of a multilayer cylinder [17]. The temperature of the substrate was stable during the whole process and did not exceed 600 K. At the pressure of 30 Pa in some regions of the deposited films formation of chaoite microcrystals was observed. Their location was strictly associated with the local concentration of carbon ions (Fig. 5).

Lowering the pressure back to 20 Pa resulted in formation of only diamond, lonsdaleite and graphite (Fig. 6) in spite of the fact that the distribution of carbon ion concentration did not significantly change. This indicates that an extension of the mean free path lengths and increasing plasma non-isothermality, the latter resulting in a raise of electron energies), promote diamond crystallization. Chaoite formation occurred in areas where high concentration of ions of a certain moderate energy range was observed.

Comparing results of the investigations of various plasma assisted hydrocarbon pyrolysis processes some general principles ruling the formation of chaoite and/or α-carbyne can be recognized which are independent on the plasma type. They are:
– presence of carbon ions;
– gas pressure ranging from ambient down to 30 Pa;
– a specific range of ion energies of values higher than those necessary for graphite formation and lower than those required for diamond crystallization.

3.2. ELECTRON ASSISTED HYDROCARBON PYROLYSIS

The general literature concerning the effect of an electric field and charged particles upon the supersaturation and phase composition of the material being formed during homogeneous nucleation includes the works by Thompson [18], Kashchiev [19], Rusanov [20] and Sokolowski [21].

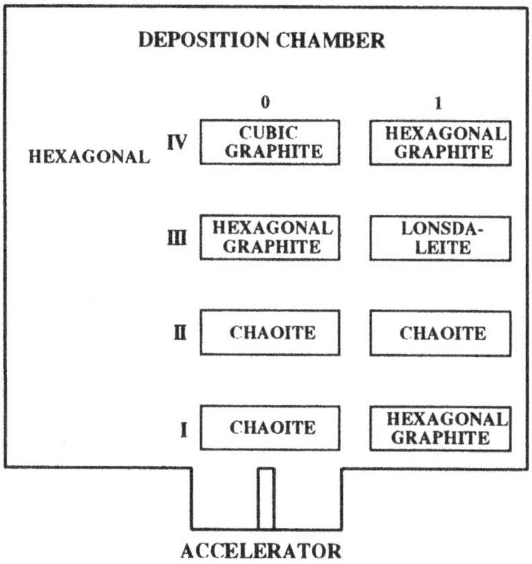

Figure 5. Spatial distribution of the identified carbon phases (p = 0.3 hPa, U = 4 kV).

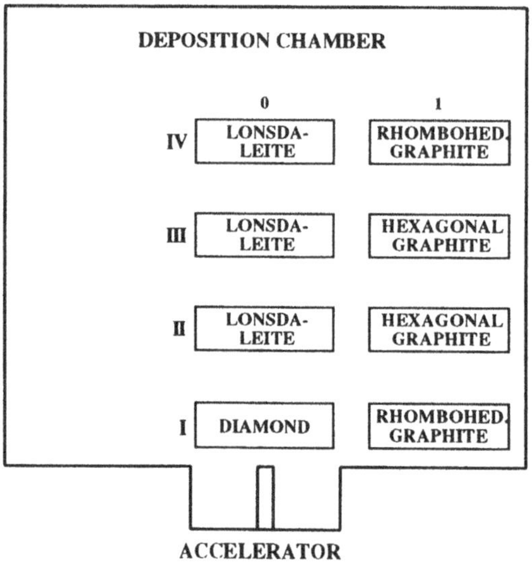

Figure 6. Spatial distribution of the identified carbon phases (p = 0.2 hPa, U = 4 kV).

In the electron-assisted CVD processes (EA-CVD), the electrons are delivered from a metallic edge by autoemission due to a corona discharge [22] or emitted thermally from a metallic wire (HF-CVD) [23]. In the zone where the chemical reactions proceed, inelastic collisions of electron-atom and electron-particle types take place, which may induce the processes of ionization

$$e^- + A \rightarrow A^+ + 2e^-, \qquad (4a)$$

excitation

$$e^- + A \rightarrow A^*, \qquad (4b)$$

and dissociation

$$e^- + AB \rightarrow A + B^- \quad \text{or} \quad e^- + AB \rightarrow e^- + A^+ + B^- \qquad (4c)$$

due only to collisions with electrons. When a neutral CH_4 particle collides with a free electron whose energy exceeds 8.5 eV, the processes may proceed according to the formulae:

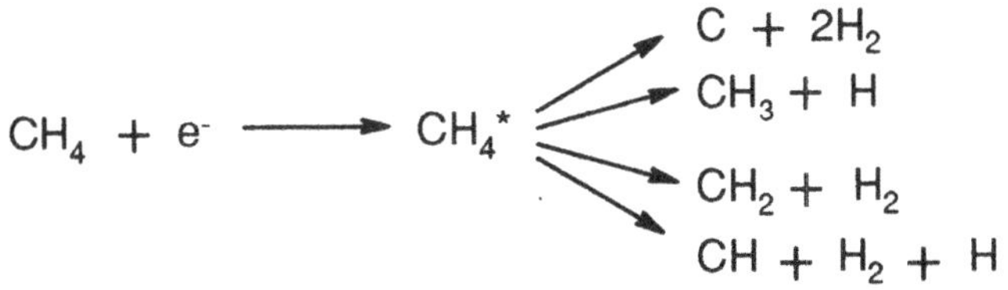

The energy gained by the electrons in the electric field depends on their free path lengths and on the strength of the electric field; as the values of these quantities increase, the energy increases.

The effect of electrons upon the phase composition of the carbon layers that crystallized in a pyrolysis process with C_2H_5OH vapor as a carbon source and a corona discharge as an electron source was studied by Olszyna *et al.* (Fig. 7) [24]. The experimental conditions were selected so as to avoid the formation of plasma. The substrate temperature was 1000 K. These investigators found that the allotropic carbon phases obtained by the free electron assisted-CVD technique (pyrolysis) differ from those produced by a non-activated CVD process. The EA-CVD process yields chaoite, whereas non-activated CVD yields graphite.

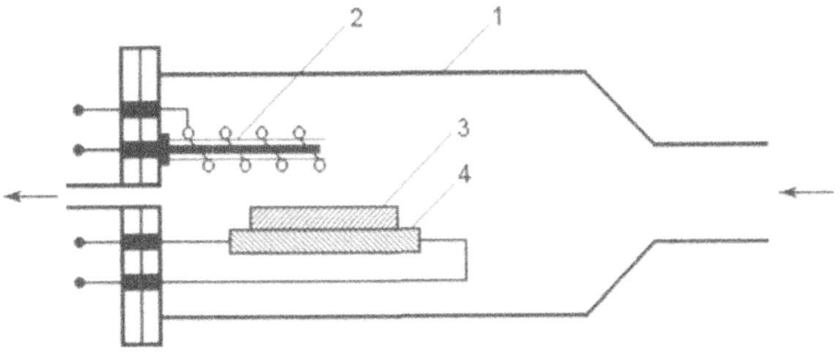

Figure 7. Experimental setup of the EA-CVD process of carbon layers; 1 – reaction chamber; 2 – source of field-emitted electrons; 3 – substrate; 4 – heater.

Digilov *et al.* [25] reported on the pyrolysis of C_4H_{10} using a pyrolysis process activated by electrons emitted thermally from a tungsten wire at a temperature of 2500 K. Analyses by XPS, Auger spectroscopy and TEM showed that on a substrate at 1000 K the crystalline layer is composed of a mixture of diamond and α-carbyne.

Although the literature about electron assisted-pyrolysis is scarce, one can ascertain their characteristic similarity to the plasma assisted-pyrolysis. They are:
– activation with charged particles (electrons) of a certain moderate energy range, and
– ambient pressure.

3.3. PHOTON ASSISTED HYDROCARBON PYROLYSIS

The effect of a photon field of high power density on the crystallization process is being examined in two aspects:
i) in the CVD processes as the means of photochemical activation of the gas (absorption of the photons by the reactants and crystallization products, in particular the absorption accompanied by excitation of electrons, atoms and molecules) and
ii) as a factor that elevates the temperature of the substrate.

Like electrons, photons can be considered the carriers of non-mechanical work [26]. Hence collisions with photons introduce non-mechanical work into a system. However, from the point of view of materials synthesis the excitation due to collisions with electrons differs essentially from the excitation due the collisions with photons. This is, in particular, with regard to the quantum principles of the conservation of energy and momentum. During effective collision with a photon, the photon must annihilate, whereas the target must have an energy structure that is able to absorb the entire quantum of energy. As the mass of a photon is negligibly small, no momentum exchange occurs in a collision with it. This is why the photon assisted synthesis may have a peculiar character.

In the case of photon activated-pyrolysis of low-molecular weight hydrocarbons a difficulty arises concerning the fact that photon absorption of these vapors begins from $h\nu > 7$ eV ($\lambda < 165$ nm). However, a light source of these wavelengths is hard to apply. Nevertheless, the synthesis of metastable forms of carbon under these conditions has been published.

Fedoseev *et al.* [27] irradiated hydrocarbon vapors with an IR laser ($\lambda = 10.6$ μm, power density 10 W/cm^2) in a 'cool gas - hot substrate' system. Depending on the conditions under which the synthesis proceeded (T,p) and on the crystallographic orientation of the substrate, the authors obtained whiskers of various shapes identified as graphite or carbyne. Carbyne crystallized on PAN fibers subjected to dehydrogenation simultaneously with the deposition process. The temperature ranged from 1000 K to 2300 K and the pressure from 1 to 500 Pa.

Digilov *et al.* [25], using the pyrolysis of C_4H_{10} at a temperature of 2800 K and irradiation both the gas and the substrate with a CO_2 laser beam ($\lambda = 10.6$ μm, power 0.08 kW), deposited carbon layers on a high-strength PAN-based fibre (Fig. 8). As shown by x-ray examinations, they obtained the two phases: diamond and α-carbyne.

It is interesting to note that irradiation with an IR beam of a wavelength such that the radiation was not absorbed by the hydrocarbon vapors led to a specific photochemical synthesis that yielded *sp*-bonds and not sp^2-bonds. This may be explained in terms of a pure thermal effect due to substrate heating.

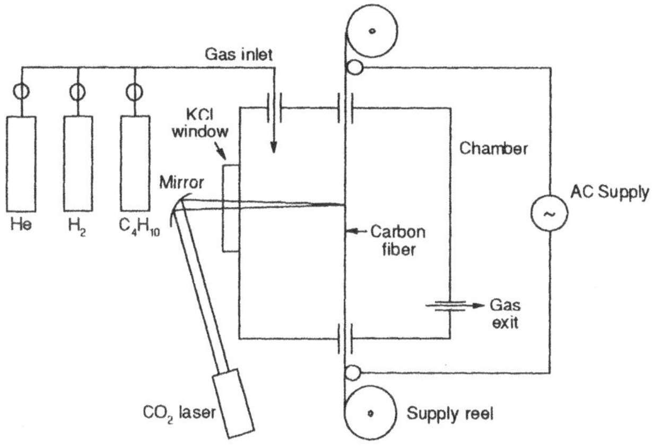

Figure 8. Diagram of the experimental setup to coat fibers with carbon by a PhA-CVD process.

Fedoseev *et al.* [28,29] reported on the homogeneous crystallization of metastable carbon phases at a high degree of supersaturation. They used a CO_2 laser beam with the wavelength $\lambda = 10.6$ µm and a power density ranging from 1500 to 2000 W/cm^2. The sources of carbon were liquid hydrocarbons (such as octane, decane and dodecane) that were delivered in the form of drops through a capillary. With a laser power density of 1500 W/cm^2 the authors obtained a highly dispersed C_8 powder with a structure very similar to that of chaoite. Much better crystallized grains ($\phi < 1$ µm) of carbyne and diamond were obtained at a power density of 2000 W/cm^2.

Fedoseev *et al.* explained these results as a consequence of the high supersaturation. The homogeneous nucleation rate J at a pressure P and temperature T follows the relationship:

$$J = [\alpha P^2/(2\pi m kT)^{0.5}\sigma]\cdot(U/kT)\cdot\exp(-U/kT) \qquad (5)$$

where U is the work required to produce a critical nucleus by homogeneous nucleation, α is the condensation coefficient, σ is the surface energy and m is the mass of the atom (molecule).

The nucleation rate has the maximum:

$$J_{max} = [\alpha P^2/(2\pi m kT)^{0.5}\sigma]\cdot\exp(-U/kT)_{max} \qquad (6)$$

and is evidently dependent on the surface energy of the nucleus.

Nucleation in a carbon system at high degrees of supersaturation and at high temperatures can be accompanied by the formation of drops of molten carbon which crystallize to form metastable phases when rapidly cooled. These small particles appear to be strongly compressed at the Laplace pressure $p = 2\sigma/r$, where 'r' is the radius of curvature of the nucleus. Taking for example $\sigma = 5\cdot10^3$ ergs/cm^2 and $r = 10^{-7}$ cm, we obtain p = $60 \cdot 10^3$ atm. This pressure favors the synthesis of high-pressure phases [26]. But there is no evidence that carbyne is a 'high-pressure' phase. So this reasoning might explain the formation of diamond but not that of carbyne.

Sokolowska *et al.* [30,31] irradiated propane vapor using an UV nitrogen laser operated in a pulse mode at a wavelength of λ = 337.1 nm (hv = 3.68 eV) and a power of 150 kW (Fig. 9). At room temperature, they obtained chaoite that crystallized at the substrate surface.

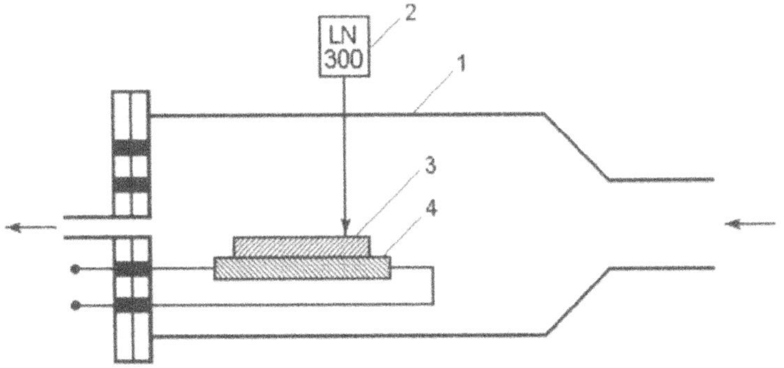

Figure 9. Experimental setup of PhA-CVD process of carbon layers: 1 – reaction chamber; 2 – pulsed nitrogen laser LN 300; 3 – substrate; 4 – heater.

Also, the crystallization of carbon layers in a cold gas (C_2H_5OH) – hot (1000 K) substrate system by irradiation with a pulsed UV beam of 3.68 eV and 75 kW power or with a halogen lamp with a power of 1000 W and a filament temperature corresponding to λ_{max} = 800 nm led to the formation of chaoite [30]. The same pyrolysis process without any additional excitation resulted in graphite being crystallized.

Under all these conditions the influence of heat on the crystallization process can be excluded. The energy of the photons is low and still the '*sp*'-phase forms. Hence we can infer that photons even those not absorbed by gas molecules and not heating the substrate, can influence the metastable phase formation.

The specific photon activated-pyrolysis of volatile carbonaceous compounds may be explained by a process occurring in the very carbon nucleus. The nucleus (or cluster) according to the phase rule is one of graphite configuration. The band model of graphite (Fig. 10) possess a half-filled π-band about 5 eV wide [32] of metallic character. The valence electrons in graphite can absorb all photons with various energies applied in the experiments and occupy energetically higher empty levels producing in this way a new different electron configuration.

4. Conclusions

Chaoite and α-carbyne, *i.e.*, allotropic carbon phases with a specific atomic structure imposed by the stiffening effect of the π-bonds, have until recently been an undesirable side product of the synthesis of diamond. Now it is known that these allotropic forms also have useful properties and thus methods of producing carbyne in its pure form are of interest. The pyrolysis of carbonaceous volatile compounds seems to be a relatively easy and inexpensive technique for producing the $\sigma sp\pi p^2$ phases.

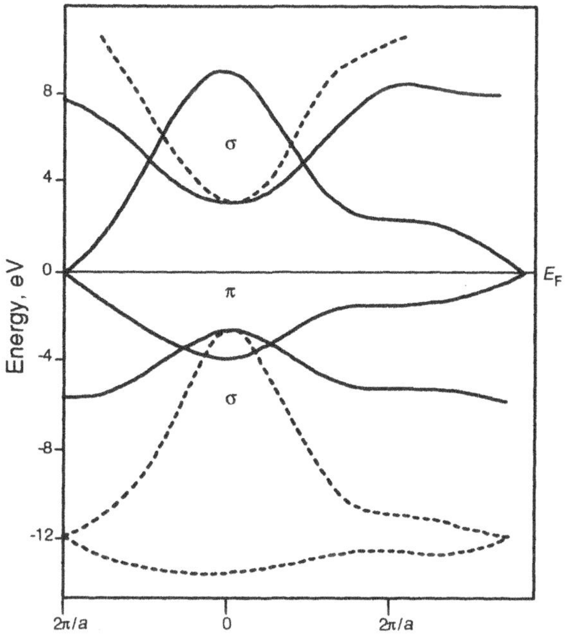

Figure 10. Energy band model of graphite; σ – the band of σsp^2 electrons, π – the band of πp electrons.

From the review of the state-of-the-art we can infer that all kinds of appropriately activated pyrolyses reactions, such as the pyrolysis activated by the diamond substrate, by charged particles and by photons, are suitable for producing carbyne. In most general terms, to specify the optimum conditions necessary for carbyne to form, we can formulate the following requirements:

- the kind of the volatile carbonaceous compound used is of no importance;
- the kind of plasma is not important;
- the gas pressure should exceed 30 Pa (with an aerosol – ambient pressure);
- the substrate temperature > 900 K;
- the gas temperature = 2000 K;
- the energy of the ions < 100 eV;
- the irradiating light should have a high power density within a wavelength range from infrared to near ultraviolet;
- the substrate material – diamond.

With the conditions specified above we can attempt to describe the mechanism of the formation of the $\sigma sp\pi p^2$-phases during the pyrolysis reaction (*i.e.*, during a CVD process). Carbon is here the product of a chemical reaction and thus in describing its crystallization process we should adopt adequate formulae, *e.g.* those proposed by Deryagin *et al.* [6]:

$$AB_g \rightarrow A_s + B_g \tag{7}$$

where the subscript 'g' denotes the gas, and the subscript 's' the solid, and

$$I = \text{const}\cdot(p_{AS} - p_{AB})\cdot\exp(-E/RT) \tag{8}$$

where 'p' is the pressure of the vapors, E is the activation energy of the chemical reaction, and I is the rate of crystallization due to the chemical reaction.

This description of the 'chemical crystallization' clearly indicates the significance of the activation energy E in the synthesis of the low-pressure phase whose electron configuration, however, differs from typical; the term P_{AS} is only significant for the crystallization of high-pressure phases where it should satisfy the condition $p_{Diam} < p_{Graph}$, which is inconsistent with the typical phase equilibrium rule.

It appears that the general requirements formulated above can be reduced to the requirement of the appropriate selection of the way in which the energy E_1 (necessary to form the $\sigma sp^2\pi p$-confiuration) is changed into the energy E_2 (necessary to form the $\sigma sp\pi p^2$-configuration). The most simple method of achieving this is a reaction with the participation of an 'active complex'. For example, if the direct reaction is

$$AB \rightarrow A_{Graph} + B, \tag{9a}$$

the reaction involving an active complex is

$$AB \xrightarrow{C} ABC \longrightarrow A_{Carb} + B + C \tag{9b}$$

where C is an atom or a particle, but may also be an electron, ion or photon.

This description [32] of the synthesis of carbyne explains well the role played by the activation in the process of pyrolysis.

In systems where the formation of an 'active complex' is hardly possible, *e.g.* when the photon of a given energy cannot be absorbed, another concept [31] seems to be convincing, namely that the graphite cluster being formed elevates the state of its π-electrons by absorbing a sufficient number of photons of arbitrary energy.

5. References

1. Spitsyn, B.V. and Deryagin, B.V.: Method of building up diamond facets, *USSR Pat. 339134*, (1956) [*Chem. Abstr.* **93** (1980), 97748f].
2. Eversole, W.G.: Synthesis of diamond by deposition on seed crystals, *US Pat. 3030187*, (1959) [*Chem. Abstr.* **57** (1962), 4136d].
3. Spitsyn, B.V., Bouilov, L.L., and Derjaguin, B.V.: Diamond and diamond-like films: deposition from the vapour phase, structure and properties, *Prog. Cryst. Growth Charact.* **17** (1988), 79–170.
4. Bachmann, P.K., Leers, D., and Lydtin, H.: Towards a general concept of diamond chemical vapor deposition, *Diamond Relat. Mater.* **1** (1991), 1–12.

5. Frenklach, M., Kematick, R., Huang, D., Howard, W., Spear, K.E., Phelps, A.W., and Koba, R.: Homogeneous nucleation of diamond powder in the gas phase, *J. Appl. Phys.* **66** (1989), 395–399.
6. Spitsyn, B.V., Bouilov, L.L., and Deryagin, B.V.: Vapor growth of diamond on diamond and other surfaces, *J. Cryst. Growth* **52** (1981), 219–226.
7. Fedoseev, D.V., Deryagin, B.V., Varnin, V.P., and Vnukov, S.P.: On the polymorphism in carbon and boron nitride systems, *Carbon* **16** (1978), 191–193.
8. Sokolowski, M.: Influence of the pulse plasma chemical content on the crystallization of diamond under conditions of its thermodynamic instability, *J. Cryst. Growth* **54** (1981), 519–522.
9. Elkaseh, A.A.: Effect of electrons and photons on the creation of metastable forms of carbon and boron nitride, *Ph.D. Thesis*, Warsaw University of Technology, Warsaw, 1995.
10. Mitura, S.: The role of electrons in the process of low-pressure synthesis of diamond, *Ph.D. Thesis*, Lodz University of Technology, Lodz, 1992.
11. Dischler, B., Bubenzer, A. and Koidl, P.: Bonding in hydrogenated hard carbon studied by optical spectroscopy, *Solid State Commun.* **48** (1983), 105–108.
12. Morrish, A.A. and Pehrsson, P.E.: Effects of surface pretreatments on nucleation and growth of diamond films on a variety of substrates, *Appl. Phys. Lett.* **59** (1991), 417–419.
13. Nikulin, Yu.N., Melnichenko, V.M., Kudryavtsev, Yu.P., and Sladkov, A.M.: Plasmochemical synthesis of carbyne, *Izv. Akad. Nauk SSSR, Ser. Khim.* (9) (1981), 1943–1948 (in Russian) [*Bull. Acad. Sci USSR, Div. Chem. Sci.* **30** (1981), 1593–1597 (Engl. Transl.)].
14. Whittaker, A.G.: Carbon: a new view of its high-temperature behavior, *Science* **200** (1978), 763–764.
15. Sokolowski, M., Sokolowska, A., Gokieli, B., Michalski, A., Rusek, A., and Romanowski, Z.: Reactive pulse plasma crystallization of diamond and diamond-like carbon, *J. Cryst. Growth* **47** (1979), 421–426.
16. Sokolowska, A., Olszyna, A., Michalski, A., and Zdunek, K.: Diamond layers deposited from pulse plasma, *Surf. Coat. Technol.* **47** (1991), 144–155.
17. Walkowicz, J.: Simplified spectral diagnostics of pulse plasma for control of diamond-like layers deposition process, in *Proc. 11th Int. Symp. on Plasma Chemistry*, 1993, Vol. 4, pp. 1656–1661.
18. Thompson, J.J.: *Elektrizitätsdurchgang in Gasen*, B.G.Teubner, Berlin, 1906.
19. Kashchiev, D.: Nucleation in external electric electric field, J. Cryst. Growth **13/14** (1972), 128–130.
20. Rusanov, A.I.: Thermodynamic theory of nucleation on charged perticles, *J. Coll. Interface Sci.* **68** (1979), 32–47.
21. Sokolowski, M., Sokolowska, A., Michalski, A., and Ziólkowski, Z.: Synthesis of Al_2O_3 from the vapour in an external electric field, J. Cryst. Growth **52** (1981), 274–278.
22. Loeb, L.B.: *Electrical Coronas - their Basic Physical Mechanisms*, University of California Press, Berkeley, 1965.
23. Matsumoto, S., Sato, Y., Tsutsumi, M., and Setaka, N.: Growth of diamond particles from methane-hydrogen gas, *J. Mater. Sci.* **17** (1982) 3106–3112.
24. Olszyna, A., Elkaseh, A.A., and Sokolowska, A.: A comparison of electro- and photon assisted CVD deposition of carbon and BN, in *Proc. 11th Int. Symp. on Plasma Chemistry*, 1993, Vol. 3, pp. 517–521.
25. Digilov, M., Vilcev, K., Zhikharev, A., Kovalev, A., Maluchkov, O., Stemin, S., Ilizarov, L., and Kostikov, V.: Deposition of carbon-based coatings on the surface of carbon fibers, *Carbon* **31** (1993), 365–371.
26. Sokolowska, A., Michalski, A., Olszyna, A., and Zdunek, K.: *Niekonwencjonalne Srodki Syntezy*, PWN, Warszawa, 1990.
27. Fedoseev, D.V., Deryagin, B.V., Varshavskaya, I.G., and Lavrent'ev, A.V.: The growth of graphite whiskers, *Dokl. Akad. Nauk SSSR* **221** (1975), 149-152 (in Russian) [*Dokl. Phys. Chem.* **221** (1975), 229–231 (Engl. Transl.)].
28. Fedoseev, D.V., Deryagin, B.V., Lavrent'ev, A.V., and Varshavskaya, I.G.: Homogeneous formation of a diamond in the gaseous phase, *Pis'ma Zh. Eksp. Teor. Fiz.* **32** (1980), 7–9 (in Russian) [*JETP Lett.* **32** (1980), 5–7 (Engl. Transl.)].
29. Fedoseev, D.V., Deryagin, B.V., Varshavskaya, I.G., Lavrent'ev, A.V., and Matveev, V.V.: Homogeneous formation of metastable carbon phases at high degrees of supersaturation, *Zh. Eksp. Teor. Fiz.* **80** (1981), 413–419 (in Russian) [*Sov. Phys. JETP* **53** (1981), 210–213 (Engl. Transl.)].
30. Sokolowska, A., Michalski, A., and Wronikowski, M.: UV Photon assisted CVD of carbon, *Proc. 6th Eur. Conf. on Chemical Vapour Deposition*, ed. by Porat, R., Iscar Ltd., Nahariya, Israel, 1987, 423–427.
31. Sokolowska, A. and Wronikowski, M.: Effect of free electrons and photons upon the crystallization process of material layers under monotropic transformations, *Arch. Nauki Mater.* **10** (1989), 189–198 (in Russian).
32. Coulson, C.A. and Taylor, R.: Studies in graphite and related compounds. I. Electronic band structure in graphite, Proc. Phys. Soc. **A65** (1952), 815–825.

3.3.2. PYROLYSIS OF ORGANIC POLYMERS

S.E. EVSYUKOV
A.N. Nesmeyanov Institute of Organoelement Compounds,
Russian Academy of Sciences,
117813 Moscow, Russia

Abstract – The possibility of the formation of carbyne during pyrolytic decomposition of organic polymers has been considered. The formation of carbyne should generally depend on experimental conditions (temperature, heating rate, heat exchange, environment) as well as on the chemical structure of the original polymers. The mechanisms of its formation should be different in linear aliphatic polymers and in aromatic thermosetting resins.

1. Introduction

The formation of carbyne and carbynoid structures during pyrolysis of organic polymers is likely the most vague area of all in the field of linear carbon allotropes. Studies reporting on such phenomena are few in number, and their interpretations (if any) are very ambiguous. Therefore, it is truly a difficult task to summarize the fragmentary data available in the literature to date. All data on observations of carbyne and carbynoid structures in pyrolysed organic polymers can be quite arbitrarily separated, according to the number of publications encountered, into three groups: pyrolysis of (i) polyacrylonitrile; (ii) glassy carbons (pyrolysed thermosetting resins); and (iii) other polymers. The two former groups include several publications each, whereas the later includes the only studies that ever appeared (to my best knowledge, of course) on the pyrolysis of other polymeric materials.

2. Pyrolysis of Poly(acrylonitrile)

In 1970, Kasatochkin and Kargin [1] proposed, for the first time, an idea on possible formation of carbon chains during the pyrolysis of oriented poly(acrylonitrile) (PAN) fibers. The pyrolysis of PAN is known to result in the formation of ordered graphite-like ribbons being started from the intramolecular polymerization of cyano-groups to form azo-acenic strips [2]. The authors, however, assumed that the insertion of firmly bound nitrogen atoms between the carbon chains through inter-molecular cross-linking in early stages (T < 250°C) may prevent the chains from being transformed to the ordered structure of fused aromatic layers. The cross-linked fragments were supposed to transform into linear carbon chains of polyyne or cumulene structure by losing nitrogen at higher temperatures (*Scheme 1*).

R.B. Heimann et al. (eds.), Carbyne and Carbynoid Structures, 133–138.

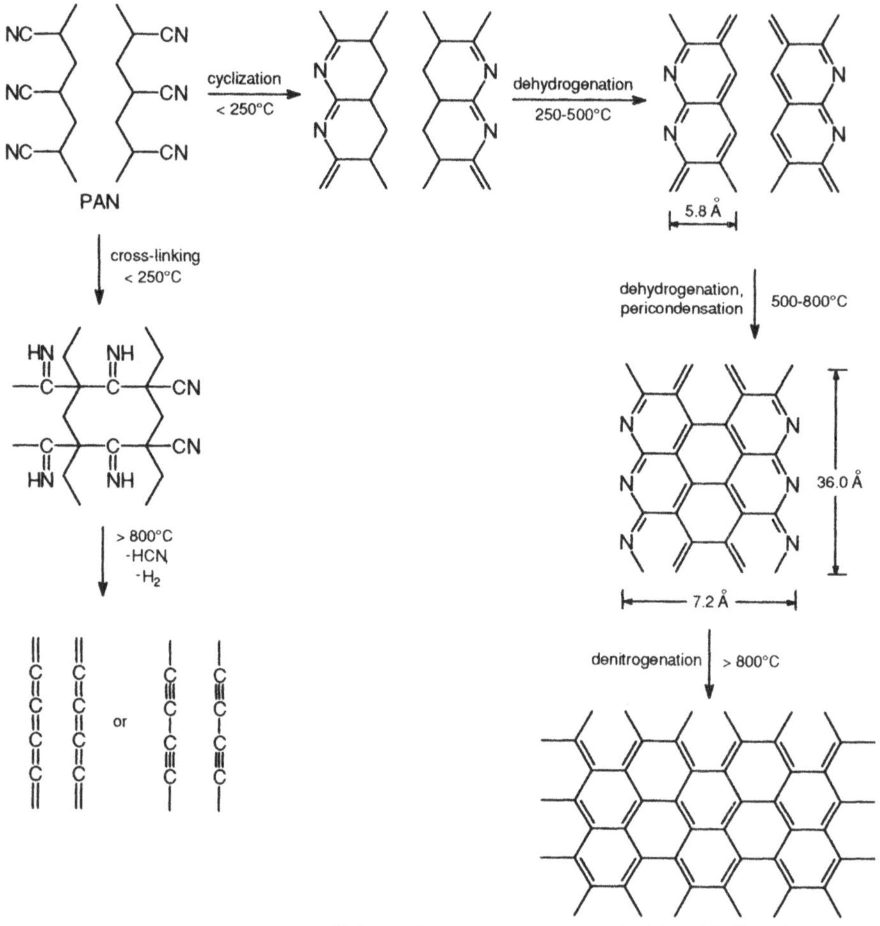

Scheme 1

This conjecture was then confirmed experimentally on the basis of x-ray analysis by calculating the function of the radial electron density distribution in the PAN-based carbon [3]. The interatomic C–C distances of 0.128 nm characteristic of *sp*-hybridized bonds in the carbon chains were found in the fibers prepared by carbonization at 1000°C of a partially oxidized PAN fiber.

Subsequently, Kasatochkin *et al.* [4] revealed crystalline microinclusions of carbyne in dispersed carbon fibers obtained by carbonization at 2750°C of a partially oxidized PAN fiber. The interplanar spacings derived from the TEM microdiffraction pattern coincided with those of carbyne synthesized by laser-induced vaporization of graphite as well as oxidative dehydropolycondensation of acetylene[1]. Selective oxidation of amorphous and graphite-like phases by prolonged boiling of the dispersed PAN-based carbon fibers in a 1:1 HNO_3/H_2SO_4 mixture allowed the isolation of a crystalline carbyne phase in low yield as a white powder [5]. The lattice parameters of the

[1] For details *cf.* Chapter 3.1.1 of this book.

hexagonal crystals were found to be a=0.892 nm and c=1.536 nm [4], although some crystals with a=0.911 nm were also observed [5].

In 1991, Kugeev *et al.* [6] studied the structure of the PAN-based carbon fibers by x-ray photoelectron spectroscopy (XPS) and confirmed the presence of the chain-like species along with graphite-like components. Noteworthy was the effect of high-temperature annealing on the structure of the fibers. While the rearrangement (ordering) processes inside the graphite-like crystallites are dominant in the 1000–2000°C range, further annealing up to 3100°C results in cross-linking of the one-dimensional species and graphitization of the entire fiber [6].

It is interesting to note that in some CVD[2] studies utilizing the PAN-based carbon fibres as a substrate for the deposition of carbon, formation of carbyne was observed [7,8]. In these cases the carbyne species present in the original carbon fibers might serve as nuclei for the epitaxial growth of carbyne crystals.

3. Carbyne in Glassy Carbons

Glassy (vitreous) carbon is normally produced by pyrolysis of aromatic and heteroaromatic thermosetting resins such as phenol-formaldehyde and furan resins [9]. Therefore, studies dealing with the observation of carbyne and/or carbynoid structures in glassy carbons are very pertinent to the present brief review.

Whittaker and Tooper [10,11] studied the structure of various glassy (vitreous) carbons with electron diffraction. Although most of the particles obtained by powdering the samples were amorphous and hence gave diffuse ring patterns, single-crystal patterns also occurred with a mean frequency of *ca.* 1 in 10, which varied significantly among the samples. However, the real 'concentration' of crystalline regions was difficult to estimate by the electron diffraction measurements, and none of the samples showed x-ray reflections of the carbon forms observed during the TEM measurements. Generally, the frequency of the occurence of particles with single-crystal patterns was reported to increase with increasing curing temperature of the samples. Actually, almost all known crystalline forms of carbon were observed. Apart from graphite (single crystals and twins), diamond and lonsdaleite (ring patterns), several polymorphs of carbyne were also found. Single-crystal pattern of chaoite and carbon VI were found with a frequency exceeding that of graphite in all glassy carbon samples studied. The size of the crystallites was estimated to be 0.1 to several microns (on the average, 1÷2 μm). Based on their observations, the authors concluded that vitreous carbon may consist of crystalline regions in which carbon is present in all its possible allotropic forms. These crystallites are connected by a three-dimensional random network of a variety of C–C bonds, probably mostly graphite-like, which forms the greater portion of the bulk of glassy carbon [10,11].

Baker and Bragg [12,13] studied the temperature dependence of the electrical conductivity of glassy carbon heat-treated in the 1200–2700°C range. An empirical formula (1) consistent with experimental data was suggested to describe the dependence of the conductivity σ on the temperature T.

---

[2] Chemical Vapour Deposition; for details *cf.* Chapter 3.3.1 of this book.

$$\sigma = A + B\cdot\exp(-CT^{-1/4}) - DT^{-1/2} \tag{1}$$

The first and largest term was attributed to metallic conduction and was thought to be affected by scattering from crystallite boundaries. The second term was ascribed to variable range-hopping of carriers between localized states. Finally, the third term represented as a negative correction term of the Kondo logarithmic form was attributed to a one-dimensional transport and, hence, to the presence of one-dimensional structures in glassy carbon. While the constants A, B, and C were insensitive to the heat-treatment temperature, the constant D decreased with increasing temperature until it disappeared at about 2200°C. It was found that the negative correction component has an inverse square root temperature dependence, compatible with the theory for the low-temperature correction for one-dimensional metallic filaments [12]. The authors pointed out that the microstructure of glassy carbon, as examined in lattice TEM images, the Porod-law dependence obtained in small-angle scattering, as well as the decrease in the d_{002} interlayer spacing above 2200°C measured by x-ray diffraction, support this conjecture [13]. However, no further details were given.

Similarly, when studying the influence of the annealing conditions on the XPS spectra of glassy carbon, Pesin *et al.* [14] observed a decrease of the contribution of the chain-like components to the imaginary part of the dielectric function with increasing heat treatment temperature. This result agrees with that obtained in previous studies of PAN-based fibers, and strongly suggests cross-linking of the chain-like carbon structures at elevated temperatures.

4. Pyrolysis of Other Polymers

Recently, Bianconi *et al.* [15] reported on an exciting method for the preparation of diamond-like carbon *via* pyrolysis of soluble organic precursor synthesized by dehalogenative polymerization of α,α,α-trichlorotoluene[3] in an ultrasonically generated emulsion of a sodium/potassium alloy in terahydrofuran (*Scheme 2*) [15,16].

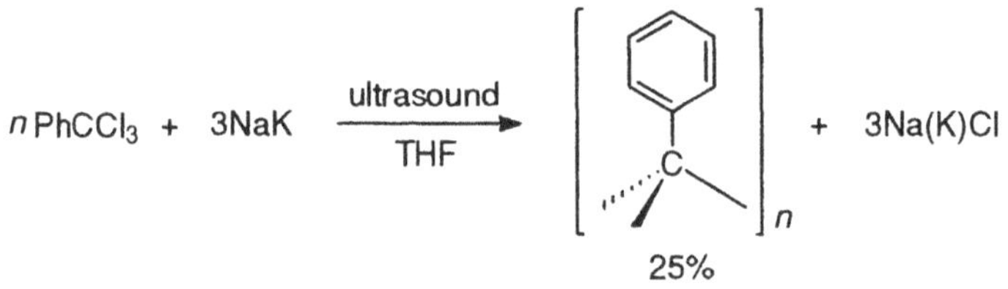

Scheme 2

This procedure was then improved by Rieke and Chen [17], who used highly reactive calcium ('Rieke calcium') produced by reacting lithium biphenylide with CaI_2 [4] to obtain the same product in 46% yield.

[3] Co- and polymerization of 1,1,1-trichloroethane and chloroform was also reported to be possible [16].

[4] $SrBr_2$ and BaI_2 were also used [17].

The precursor is highly branched oligomer[5] called by authors 'poly(phenyl-carbyne)', probably according to the IUPAC-adopted term for carbon-centered organic triradicals [18]. This name, however, has nothing to do with linear carbon allotropes and can be misleading in terms of the present book. The pyrolysis of the precursor at 1000 or 1600°C in argon resulted in the formation of diamond-like carbon containing minor inclusions of α- and β-carbyne [15]. The pyrolysis of poly(diphenylacetylene) perfomed under the same conditions yielded poorly crystallized graphite and chaoite.

Fialkov *et al.* [19] studied the structure and paramagnetic properties of residual carbon obtained by thermal decomposition of fluorocarbon. The latter was synthesized by fluorination of viscose fibers carbonized at 2500°C. The decomposition at 650–1000°C occurred in a burst-like manner to produce mostly amorphous carbon. An x-ray study revealed a line with interplanar spacing d_{002} = 0.404 nm, whereas analogous spacing in original carbon fibers was found to be 0.344–0.380 nm, characteristic of turbostratic carbon. The increased d_{002} value was explained by the contribution of linear carbon chains formed during the defluorination. The concentration of paramagnetic species was found to reach $5 \cdot 10^{20}$ spin/g, as inferred from ESR spectra. The two lines observed in the ESR spectra at low temperatures suggested the existence of two carbon phases. The authors believed that these were carbyne- and graphite-like phases, but no further structural evidence was given [19].

Finally, Kasatochkin *et al.* [20] reported on a surprising transformation of polyadamantan into β-carbyne. The authors studied the behaviour of polyadamantane[6] under thermobaric conditions, supposing that the diamond-like core of adamantane would predetermine the formation of diamond. Unexpectedly, it was found instead that polyadamantane heated at 1800°C and 9 kbar for 5 min transformed into β-carbyne and graphite. This result provides one more reason to question the assignment of the β-carbyne phase to the family of linear carbon allotropes. Taking into account the fact that β-carbyne was reported to form during thermobaric treatment of α-carbyne prepared by oxidative dehydropolycondensation of acetylene, one can suggest this phase may be in effect a partially (but regularly) cross-linked carbynoid structure[7].

5. Conclusions

Based on the fragmentary data described above one is led to conclude that there is no clear-cut 'recipe' to produce carbyne by pyrolytic decomposition of organic polymers, and that the observations reported so far are mostly accidental. It is evident that the pyrogenetic formation of carbyne should generally depend on experimental conditions (temperature, heating rate, heat exchange, environment) as well as on the chemical structure of original polymers, and that all these factors are mutually interconnected. Two completely different mechanisms of pyrogenetic formation of carbyne can be envisaged. The first including elimination of side substituents while retaining a linear

---

[5] The authors referred to their product as 'network polymer', which appears to be incorrect because the notion of a network structure is totally incompatible with the solubility observed.

[6] Polyadamantane was prepared *via* coupling of brominated adamantane by treating with sodium.

[7] *cf.* Chapter 5.1 of this book.

chain structure can occur during the pyrolysis of linear aliphatic polymers (*e.g.*, PAN). The formation of random cross-links preventing the polymer chains from extensive cyclization and aromatization can be also involved in this case. A different mechanism should take place in aromatic and heteroaromatic thermosetting resins. The formation of linear carbon chains in this case could occur through a profound degradation of the polymer skeleton including both elimination and ring-opening reactions.

Acknowledgement – I gratefully acknowledge the generous support by the Alexander-von-Humboldt Foundation under the auspices of its follow-up program.

6. References

1. Kasatochkin, V.I. and Kargin, V.A.: Thermal transformation of oriented polyacrylonitrile, *Dokl. Akad. Nauk SSSR* **191** (1970), 1084–1087 (in Russian) [*Dokl. Phys. Chem.* **191** (1970), 303–306 (Engl. Transl.)].
2. Gupta, A.K., Paliwal, D.K., and Bajaj, P.: Acrylic precursors for carbon fibers, *J. Macromol. Sci. – Rev. Macromol. Chem. Phys.* **C31** (1991), 1–89.
3. Khomenko, A.A., Smirnov, Yu.E., Sosedov, V.P., and Kasatochkin, V.I.: Interatomic bonds in carbonized polyacrylonitrile fiber, *Dokl. Akad. Nauk SSSR* **197** (1971), 1074–1076 (in Russian).
4. Kasatochkin, V.I., Smutkina, Z.S., Kazakov, M.E., Radimov, N.P., Nabatnikov, A.P., and Yares'ko, T.D.: Molecular structure of an anisotropic carbon fiber, *Dokl. Akad. Nauk SSSR* **205** (1972), 1090–1092 (in Russian).
5. Kasatochkin, V.I., Mel'nichenko, V.M., and Élizen, V.M.: Electron microdiffraction of carbyne single crystals, *Vysokomol. Soed., Ser. A* **17** (1975), 1883–1887 (in Russian) [*Polym. Sci. USSR* **17** (1975), 2167–2173 (Engl. Transl.)].
6. Kugeev, F.F., Baitinger, E.M., Teterin, Yu.A., and Gagarin, S.G.: Structure of carbon fibers from by x-ray photoelectron spectroscopic data, *Khim. Tverd. Topliva* (3) (1991), 120–125 (in Russian).
7. Fedoseev, D.V., Deryagin, B.V., Varshavskaya, I.G., and Lavrent'ev, A.V.: The growth of graphite whiskers, *Dokl. Akad. Nauk SSSR* **221** (1975), 149-152 (in Russian) [*Dokl. Phys. Chem.* **221** (1975), 229-231 (Engl. Transl.)].
8. Digilov, M., Vilcev, K., Zhikharev, A., Kovalev, A., Maluchkov, O., Sternin, S., Ilizarov, L., and Kostikov, V.: Deposition of carbon-based coatings on the surface of carbon fibers, *Carbon* **31** (1993), 365-371.
9. Otani, S. and Oya, A.: Glass-like carbons, *Mater. Sci. Technol.* **9** (1991), 549–572.
10. Whittaker, A.G. and Tooper, B.: Single-crystal diffraction patterns of several carbon forms from glassy carbon, *Extended Abstr. and Program 11th Biennial Conf. on Carbon*, June 4–8, 1973, Gatlinburg, Tennessee, USA, 1973, pp. 184–186.
11. Whittaker, A.G. and Tooper, B.: Single-crystal diffraction patterns from vitreous carbon, *J. Am. Ceram. Soc.* **57** (1974), 443–446.
12. Baker, D.F. and Bragg, R.H.: The electrical conductivity and Hall effect of glassy carbon, *J. Non-Crystalline Solids* **58** (1983), 57–69.
13. Baker, D.F. and Bragg, R.H.: One-dimensionality in glassy carbon, *Phys. Rev. B* **28** (1983), 2219–2221.
14. Pesin, L.A., Baitinger, E.M., Kuznetsov, V.L., and Sokolov, O.B.: Peculiarities of x-ray photoelectron spectra of core electrons in glassy carbon, *Fiz. Tverd. Tela* **35** (1993), 2262–2266.
15. Visscher, G.T., Nesting, D.C., Badding, J.V., and Bianconi, P.A.: Poly(phenylcarbyne): a polymer precursor to diamond-like carbon, *Science* **260** (1993), 1496–1499.
16. Visscher, G.T. and Bianconi, P.A.: Synthesis and characterization of polycarbynes, a new class of carbon-based network polymers, *J. Am. Chem. Soc.* **116** (1994), 1805–1811.
17. Rieke, R.D. and Chen, T.-A.: Facile synthesis of poly(phenylcarbyne): a precursor for diamondlike carbon, *Chem. Mater.* **6** (1994), 576–577.
18. Glossary of Terms Used in Physical Organic Chemistry, *Pure Appl. Chem.* **55** (1983), 1281–1371.
19. Fialkov, A.S., Dobryakov, S.N., Khorkhorin, A.V., Tyan, L.S., Polyakova, N.V., Kuznetsov, Yu.N., and Yurkovskii, I.M.: Concerning structural features of residual carbon obtained by thermal decomposition of fluorocarbon, *Zh. Strukt. Khim.* **31** (3) (1990), 48–52 (in Russian).
20. Kasatochkin, V.I., Sterenberg, L.E., Kazakov, M.E., Slesarev, V.N., and Belousova, L.V.: Thermal transformation of carbyne under pressure, *Dokl. Akad. Nauk SSSR* **209** (1973), 388-391 (in Russian).

3.4. Phase transformation of carbon materials

3.4.1. Condensation of carbon vapour

3.4.1.1 RESISTIVE HEATING AND LASER IRRADIATION

R.B.HEIMANN
Department of Mineralogy, Freiberg University of Mining and Technology, D09596 Freiberg, Germany

Abstract - Vapourization of carbon by utilization of resistive heating, electric arc discharge, magnetron sputtering or high-energy laser irradiation, and subsequent condensation of the vapour by quenching on a cooled substrate is being frequently employed to obtain carbyne and carbynoid films of various structures. The films generally consist of a matrix of amorphous carbon and more or less well ordered carbyne microcrystals.

1. Introduction

The existence of long linear chain-like molecules in a hot carbon vapour has been theoretically explained by Pitzer and Clementi [1] based on earlier mass spectroscopic studies [2-7] that established the presence of C_1, C_2, C_3 and C_5 species. Measurements of the vapour pressure by Basset [6] showed that it reaches one atmosphere already below 4,000°K suggesting additional, longer chained vapour species to be present.[1] A long chain of carbon atoms with localized double or triple bonds is unstable as compared to graphite. However, complete resonance or conjugation such that each bond is axially symmetric yields a large stabilization. Calculations of molecular orbital energies for C_n linear molecules indicated that chains with an odd number of carbon atoms have a closed shell structure and lower energies than even species [1]. Pitzer and Clementi [1] also proposed that liquid carbon consists of essentially infinite linear chains. This view was supported by Sladkov and Kydryavtsev [10] and Whittaker [11] who determined experimentally the optical properties of liquid carbon to be those of carbyne [11a]. Hence he concluded that the solid carbon phase that melts at 3,800°K (see footnote 1) is presumably ß-carbyne or carbon VI [12,13], not graphite.
The existence of long linear carbon chains in the vapour state was also confirmed by spectroscopic analyses [12,14,15], laser Raman studies [16,17], ion microprobe mass analyses [18], and time-of-flight mass spectrometry [19,20]. Therefore, the crucial molecular structures leading eventually to condensation of carbyne appear to be present already in the carbon vapour phase. The formation of linear chains, for example, by

[1] The melting point of graphite has now been established to be close to 5,000°K at ambient pressure [8,9].

R.B. Heimann et al. (eds.), Carbyne and Carbynoid Structures, 139–148.

resistive heating or laser ablation of graphitic carbon can be seen as the result of breaking of single bonds in 'meta'-position in a graphene sheet and shifting of electrons towards adjacent double bonds (Fig. 1a). Hence, isolated polyyne-type chains parallel <10.0> of graphite will be formed that are, in principle, straight [12]. An analoguous mechanism was conjectured for the formation of cumulene-type chains by breaking carbon single bonds in 'para'-position to generate chains parallel <11.0> (Fig. 1b) [21].

During quenching of the carbon vapour the polymerization [22] and subsequent crystallization must be sufficiently slow to afford carbyne formation by heterogeneous nucleation. On the other hand, the reaction must be fast enough to prevent high-temperature annealing of the carbyne structure that would result in deposition of graphite-like structures.

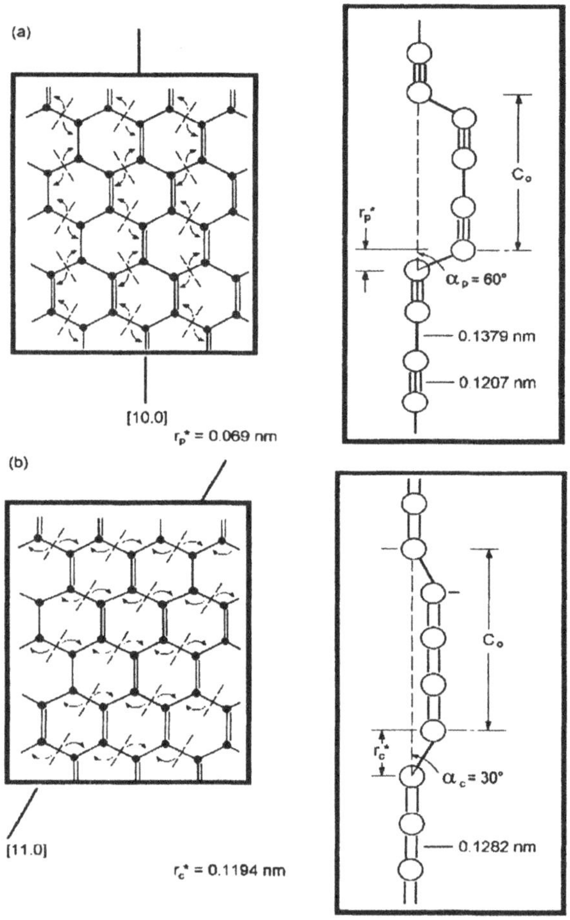

Figure 1. Model of formation of linear carbon chains by breaking single bonds in a graphene latice in (a) 'meta'-position to form chains parallel to <10.0> [12] and (b) 'para'-position to form chains parallel to <11.0> [21].

2. Resistive Heating of Graphite

Much work was carried out between 1969 and 1985 studying the high-temperature, low-pressure behaviour of carbon by a research group around Arthur Greenville Whittaker of The Aerospace Corporation, El Segundo, California [23-27]. During resistive heating of small bars of pyrolytic graphite (2x3x40 mm) to temperatures between 2,700 and 3,000°K in a low pressure argon atmosphere (10^{-4} Torr), a silvery white coating was formed whose electron diffraction pattern revealed hexagonally arranged spots from which the lattice constants $a_0 = 894.5 \pm 0.7$ pm, $c_0 = 1407.1 \pm 1.1$ pm were calculated [26]. This materials was coined 'white carbon' and was subsequently found to be identical to chaoite, i.e. carbyne with n = 11 [21] (n = number of carbon atoms in the chain) discovered earlier in shock-fused graphitic gneisses of the Ries crater, Nördlingen, Bavaria, Germany [28]. This experiment effectively disproved the original assumption by El Goresy and Donnay [28] that chaoite is a high-pressure polymorph of carbon. In addition to chaoite, α-SiC was formed in Whittaker's experiments. Although the 'white carbon' synthesized contained about five times as much silicon as the natural chaoite, there was no discernible effect of the silicon content on the numerical value of the a_0-parameter. Hence it was concluded that it is unlikely that silicon may be directly involved in the formation of carbyne.

Along with chaoite, a second carbyne polytype (carbon VI) was formed on graphite under free-vapourization conditions at low pressure (10^{-4} to 760 Torr) and temperatures above about 2,550°K [23]. It was pointed out by the authors that this polytypic carbyne phase could be structurally related to hexagonal diamond (lonsdaleite) in that a_0 for carbon VI (533 ± 5 pm) is rather close to $2 \cdot a_0$ for lonsdaleite (252 pm), and c_0 for carbon VI (1224 ± 18 pm) is very close to $3 \cdot c_0$ for lonsdaleite (412 pm). However, some similarities also exist with the chaoite structure in that a_0 of the larger hexagonal unit cell for carbon VI ($a_0 \cdot \sqrt{3} = 923$ pm) resembles $a_0 = 894.5 \pm 0.7$ pm of chaoite, and c_0 of carbon VI (1224 ± 18 pm) is reasonably close to c_0 of chaoite (1407 ± 1 pm). Apart from the somewhat numerological aspects of these statements, the presumed space groups of both chaoite and carbon VI of P3 or $P3_{1,2}$, point to materials with interesting optical and electrical characteristics that include optical activity, the linear electrooptical effect, and piezoelectric properties in hydrostatic, compressional, and torsional modes. It should be added that the possible relationships of carbyne to crystalline chiral poly(diacetylene) carbon as recently suggested by Baughman *et al* [29] (cp. Chapter 4.2) may add other potential properties to this list including non-vanishing second order non-linear optical susceptibility.

3. High-Energy Laser Irradiation

The carbyne forms deposited by condensation of a carbon vapour plume produced by laser ablation of graphite are essentially identical to those described above. A carbon phase analogous to chaoite was obtained by irradiation of pyrolytic graphite with a pulsed Nd:YAG laser (laser energy: 250-500 J, impulse duration: 10^{-3} s) [30].

Graphitic and vitreous carbon samples, shrouded with argon or helium near atmospheric pressure, were irradiated at power densities up to 25 kW/cm² with focused 10.6 μm wavelength radiation from a continuous CO_2-N_2-He laser [31]. Craters were formed in the samples that were lined and surrounded by a light-reflecting white deposit consisting of chaoite and carbon VI as confirmed by electron diffraction. Beyond a threshold power density of about 15 kW/cm² chaoite was always formed associated with the laser crater. Detailed investigations of cross-sections of the craters by SEM showed features highly suggestive of the presence of a liquid phase from which a crystalline deposit was formed coating the more or less unaffected graphite surface. These coatings were, in some cases, layered and showed a strong incidence of carbyne polytypes such as chaoite and carbon VI. Thus there is a suggestion that carbyne may actually crystallize from a carbon liquid, a contention that should be approached with extreme caution since it contradicts, at least at this time, almost all published ideas about the low pressure region of the carbon P,T-phase diagram (0-0.3 GPa) as shown for example, in Bundy *et al* [9]. Although evidence has been reported that the low pressure carbon liquid is electrically insulating [32] and changes at higher pressure >0.2 GPa at 5,000°K to a conducting liquid [8], latest more rigorous experiments do not support the existence of a non-conducting liquid carbon phase below a line 0.2 GPa/5,000°K to 0.1 GPa/6,000°K subdividing the phase field of liquid carbon [9].

It should be noted that the nature of the irradiated target seems to be less important for carbyne formation than the intensity of the laser beam impinging on the substrate surface. For example, high intensity laser irradiation of chlorinated poly(vinylchloride), leading initially to graphitization with concurrent formation of chlorinated polyene sequences by a zip-dehydrochlorination mechanism, results on exceeding a threshold laser fluence in the formation of a white residue that consists of 'ashes' [33]. Although these 'ashes' were not characterized by the author, we conjecture that 'white carbon' (chaoite) may have been formed in accord with results obtained by other research groups (cp. Chapter 3.1.2).

Additional work performed by Kasatochkin's group [34] revealed a host of novel carbon forms with a wide range of a_0-parameters as determined by electron diffraction studies: ß-carbyne (a_0 = 824 pm), chaoite (a_0 = 894 pm), carbon VI (a_0 = 923 pm), carbon VIII (a_0 = 911 pm) and α-carbyne (a_0 = 892 pm). In addition, electron diffraction signatures of two other phases with a_0 = 880 pm and 902 pm were recorded. These phases were later found in carbonaceous chondrites and tentatively named carbon XI and XII by Whittaker *et al* [35] based only on their c-axis electron diffraction pattern.

An attempt to utilize carbyne as a hard, transparent optical coating as well as for antireflection or optical corrective films was made by Whittaker [36]. Gaseous carbon produced by laser heating of a rapidly rotating graphite rod to temperatures of about 3,000°K was quenched in an oxygen containing atmosphere (100-200 μTorr) and deposited onto various substrates. The carbyne polytype deposited from carbon gas quenched from a temperature of between 2,600 and 2,800°K is a chaoite thin film (0.1-2.5 μm) characterized by its hardness greater than boron carbide, chemical stability, high refractive index, and good adhesion to metals, polymers, glass, sapphire, calcium fluoride and potassium chloride. As discussed in more detail in Chapter 7.1 it is puzzling how a material that is essentially bonded by van der Waals-interaction could exhibit such a high hardness. It is conceivable that instead of carbyne a highly cross-linked material was formed with sp^3-hybridization akin to diamond or diamond-like carbon (DLC).

We are at this time still far from being able to explain in detail the chemical processes occuring in the laser ablation plume and during condensation of the carbon vapour at the substrate surface. While the emission of small carbon clusters C_n ($1<n<3$) from a laser-irradiated graphite surface appears to be the predominant mechanism it was assumed by Kokai *et al* [20] that the growth of larger clusters ($5<n<19$) results from the occurence of various ion-neutral reactions that may be promoted by a magnetically induced change of the cyclotron radii of charged species. However, such a change is even more pronounced for electrons. Therefore it is more likely that the enhancement of ion-neutral reactions is due to an increase of ionic species resulting from the collisional ionization of initially neutral species through a confinement of electrons. Figure 2 shows that among the species in the plume, C_3 must play an important role since collision-induced reactions of lighter species result, in the presence of a magnetic field, in the buildup of heavier species up to C_{19}^+. In particular, linear C_{11}^+ clusters are predominantly present in the laser ablation plume [19,37]. This may be able to explain the occurence of chaoite, thought to contain a repeating unit of n = 11 carbon atoms [21] as the main product of laser irradiation of pyrolytic graphite [30,31,36].

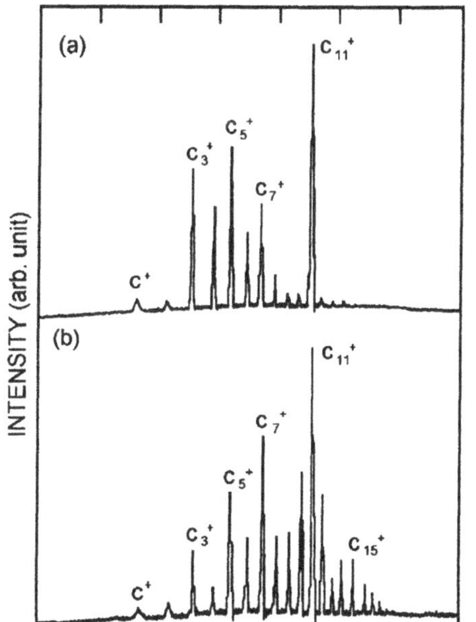

Figure 2. Time-of-flight mass spectra for positive ionic species observed during laser ablation of graphite at a laser fluence of 1.3 J/cm² in the presence of magnetic fields of (a) 0.03-0.05 T and (b) 0.08-0.10 T [20].

4. Ion Sputtering

Sputter deposition of carbon from graphite targets in argon at various d.c. powers produced films whose structure could be identified as amorphous carbon with inclusions of either chaoite or α-carbyne [38] (cp. Chapter 4.4). In addition, crystallites were formed whose electron diffraction pattern resembled diamond but with forbidden reflections present that suggested the formation of a superstructure by ordered incorporation of argon. Although such thin films (8.5 -50 nm) were not very hard, Knoop indentation microhardness measurements and scratch testing showed that they did improve the scratch resistance of float glass by about a factor of three. X-ray photoelectron (XPS) and Auger electron (AES) spectroscopy performed on the films presumably reflected only the electronic structure of the amorphous carbon, not that of the carbyne or diamond superstructure inclusions because of their extremely small crystal size. Similar results were obtained for carbon films produced by fast atom bombardment (FAB) using butane and butane-argon gas mixtures [39] (cp. Chapter 4.4).

Recently highly oriented thin films of chaoite were produced by d.c. magnetron sputtering of graphite in th presence of a hot tungsten filament as a thermal radiation source [40]. By varying the three process parameters filament temperature (R.T. to 2,000°C), substrate (Vycor glass) temperature (300 to 700°C) and argon sputter gas pressure (130 to 150 mTorr), the crystallinity and orientation of the chaoite microcrystals could be optimized. The x-ray diffraction pattern obtained from the films showed that the intensities of the (220) and (600) interplanar spacings increased very strongly with increasing filament temperature, *i.e.* thermal electron emission (Fig. 3a), strongly with the substrate temperature above 400°C (Fig. 3b) and rather weakly with the pressure of the argon used as sputtering gas (Fig. 3c). The reflection high-energy electron diffraction (RHEED) pattern indicates the preferred (220) orientation of chaoite (Fig. 4). Field emission scanning electronmicroscopy (FE-SEM) micrographs show chaoite microcrystals of about 60 nm size agglomerated to clusters of about 750 nm (Fig. 5).

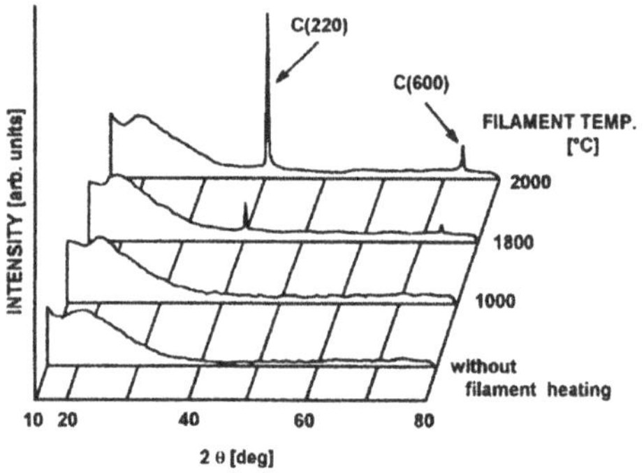

Figure 3. X-ray diffraction results on carbon thin films produced by d.c. magnetron sputtering of graphite in the presence of a thermal radiation source (hot tungsten filament) [40].
(a) Variation of filament temperature (substrate temperature: 600°C, argon flow rate: 4.0 sccm, total pressure: 130 mTorr).

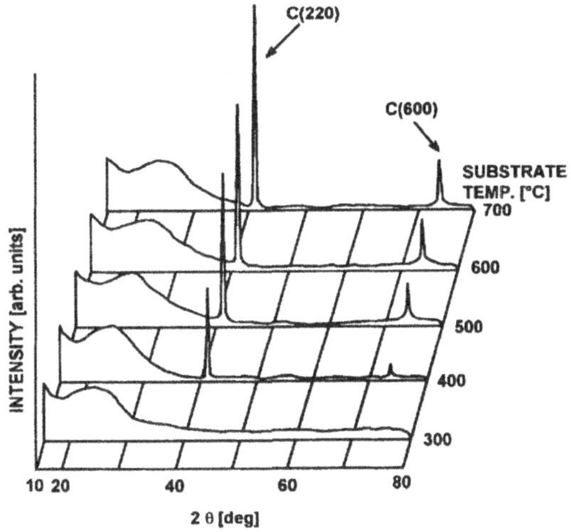

(b) Variation of substrate temperature (filament temperature: 2,000°C, argon flow rate: 4.0 sccm, total pressure: 130 mTorr).

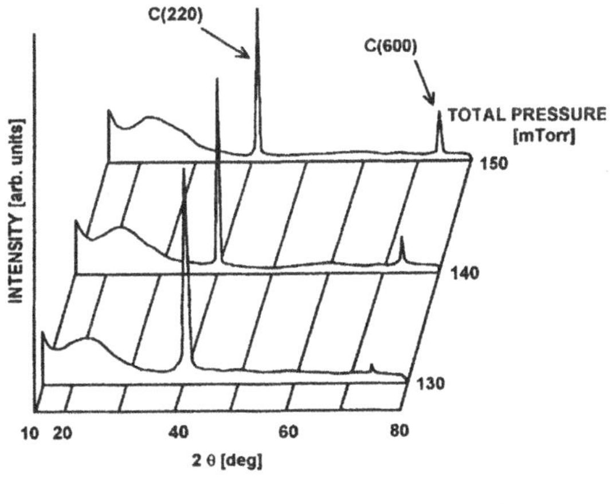

(c) Variation of total gas pressure (filament temperature: 2,000°C, substrate temperature: 600°C, argon flow rate: 4.0 sccm).

This result is to be compared with previous work that yielded, by magnetron sputtering, only amorphous carbon films with very few if any crystalline inclusions [41,42]. Apparently, thermal electron emission from the hot tungsten filament played a crucial role in the process of crystallization and the attainment of a preferred orientation of carbyne.

Figure 4. RHEED pattern of carbon film (filament temperature: 2,000°C, substrate temperature: 600°C, argon flow rate: 4.0 sccm, total gas pressure: 130 mTorr) [40]. Courtesy: Y.Onuma, Shinshu University, Nagano, Japan.

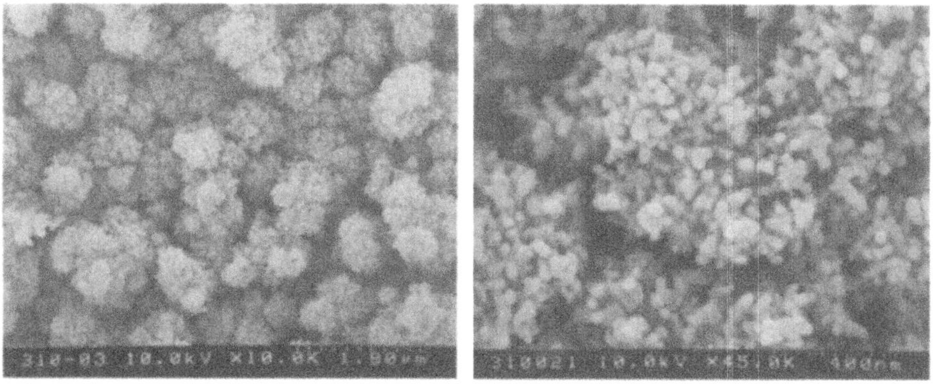

Figure 5. FE-SEM micrographs of carbon film obtained under the conditions shown in Fig. 4 [40]. Courtesy: Y.Onuma, Shinshu University, Nagano, Japan.

5. Conclusion

The deposition of carbon from a gaseous phase obtained by resistive heating, laser ablation or ion sputtering is, at this time, the only conceivable technique to produce mixed amorphous and microcrystalline carbyne films as opposed to those obtained by chemical or physical dehydrochlorination of halogenated polymers including PVDC and PTFE that inevitably results in polyene/polyyne-type carbynoid hybrid structures. In particular, d.c. magnetron sputtering in the presence of a thermal radiation source yields well-oriented, highly crystalline carbyne films of a quality that is expected to be of interest for application in electronic devices. Indeed, preliminary measurements revealed very high electrical conductivity and a positive thermal coefficient of resistance [40]. It would be of great interest to determine whether this electrical conductivity is akin to a process mediated by a charge density wave or a soliton mechanism as suggested earlier [43,44].

6. References

1. Pitzer, K.S. and Clementi, E. (1959) Large molecules in carbon vapor, *J.Am.Chem.Soc.* **81**, 4477-4485.
2. Chupka, W.A. and Inghram, M.G. (1954) Direct determination of the heat of sublimation of carbon, *J.Chem.Phys.* **22**, 1472.
3. Honig, R.E. (1954) Mass spectrometric study of the molecular sublimation of graphite, *J.Chem.Phys.* **22**, 126-131.
4. Chupka, W.A. and Inghram, M.G. (1955) Direct determination of the heat of sublimation of carbon with the mass spectrometer, *J.Phys.Chem.* **59**, 100-104.
5. Brewer, L. and Engelke, J.L. (1962) Spectrum of C_3, *J.Chem.Phys.* **36**(4), 992- 998.
6. Basset, J. (1942) Experimental realization of the melting of graphite at argon ressures up to 11,500 kg/sq.cm. Statement of the triple point and drawing of a preliminary diagram of the solid, liquid and gaseous state of carbon, *Brennstoffchemie* **23**, 127-135.
7. Kertesz, M., Koller, J. and Azman, A. (1978) Ab initio Hartree-Fock crystal orbital studies: II. Energy bands of an infinite carbon chain, *J.Chem.Phys.* **68**, 2779-2782.
8. Baitin, A.V., Lebedev, A.A., Romanenko, S.V., Senchenko, V.N. and Scheindlin, M.A. (1990) The melting point and optical properties of solid and liquid carbon at pressure of up to 2 kbar, *High Temp.High.Press.* **22**(2), 157-170.
9. Bundy, F.P., Bassett, W.A., Weathers, M.S., Hemley, R.J., Mao, H.K. and Goncharov, A.F. (1996) The pressure-temperature phase and transformation diagram for carbon; updated through 1994, *Carbon* **34**(2) 141-153.
10. Sladkov, A.M. and Kudryavtsev, Yu.P. (1969) Diamond, graphite, carbyne- allotropic forms of carbon, *Priroda* **5**, 37-44 (in Russian).
11. Whittaker, A.G. and Kintner, P.L. (1981) Carbon vapor pressure in the range 3450 to 4500 K and evidence for melting at 3800 K. Report.Mater.Sci.Lab., The Aerospace Corporation, El Segundo, Ca 90245.

11a. Wang, C.Z. et al. (1993) Phys. Rev.B 47, 14835.

12. Whittaker, A.G. (1978) Carbon: A new view of its high-temperature behavior, *Science* **200**, 763-764.
13. Whittaker, A.G. and Kintner, P.L. (1977) Comments on the solid-liquid-vapor triple point of carbon, *High Temp.Sci.* **9**, 71-72.
14. Whittaker, A.G. (1979) Carbon's high-temperature behavior, *Science* **203**, 1266.
15. Whittaker, A.G. (1985) Carbyne forms of carbon: Evidence for their existence, *Science* **229**, 485-486.
16. Nakamizo, M., Kammereck, R. and Walker, P.L. (1974) Laser Raman studies on carbons, *Carbon* **12**, 259-267.
17. Nakamizo, M. and Kammereck, R. (1973) Laser Raman sudies on carbons, Abstr. 11th Biennial Conf. on Carbon, Catlinburg, TN., June 4-8, p. 187.
18. Stuckey, W.K. and Whittaker, A.G. (1971) The identification of carbon allotroes by ion microprobe mass analysis, Abstr. 10th Biennial Conf. on Carbon, Lehigh University, Bethlehem, PA, June 27-July 2, p. 278.
19. Heath, J.R., Zhang, Q., O'Brian, S.C., Curl, R.F., Kroto, H.W. and Smalley, R.E. (1987) The formation of long carbon chain molecules during laser vaporization of graphite, *J.Am.Chem.Soc.* **109**, 359-363.

20. Kokai, F., Koga, Y and Heimann, R.B. (1995) Magnetic field enhanced growth of carbon clusters in the laser ablation plume of graphite, *Appl.Surf.Sci.* **96-98**, 261-266.
21. Heimann, R.B., Kleiman, J. and Salansky, N.M. (1983) A unified structural approach to linear carbon polytypes, *Nature* **306**, 164-167.
22. Whittaker, A.G. and Evangelides, J.S. (1977) Influence of carbon polymorphs on ablation of graphitic materials, Abstr. 13th Biennial Conf. on Carbon, Irvine, CA, July 18-22, p.410.
23. Whittaker, A.G. and Wolten, G.M. (1972) Carbon: A suggested new hexagonal crystal form, *Science* **178**, 54-56.
24. Whittaker, A.G. and Kintner, P.L. (1976) Particle emission and related morphological changes occurring during the sublimation of graphitic carbons, *Carbon* **14**, 257-265.
25. Whittaker, A.G. and Kintner, P.L. (1985) Carbon: Analysis of spherules and splats formed from the liquid state and of the forms produced by quenching gas and solid, *Carbon*, **23**, 255-262.
26. Whittaker, A.G. and Kintner, P.L. (1969) Carbon: Observations on the new allotropic form, *Science* **165**, 589-591.
27. Whittaker, A.G. (1973) Chaoite, U.S.Patent 3,733,394 (May 15, 1973).
28. El Goresy, A. and Donnay, G. (1968) A new allotropic form of carbon from the Ries crater, *Science* **161**, 363-364.
29. Baughman, R.H., Galvao, D.S., Cui, C. and Dantas, S.O. (1997) Hinged and chiral polydiacetylene carbon crystals, *Chem.Phys.Letters* **269**, 356-364.
30. Kasatochkin, V.I., Kazakov, M.E., Savranski, V.V., Nabatnikov, A.P. and Radimov, N.P. (1971) Synthesis of new allotropic carbon forms from graphite, *Dokl.Akad.Nauk SSSR* **201**(5), 1104-1105 (in Russian).
31. Nelson, L.S., Whittaker, A.G. and Tooper, B. (1972) The formation of new polymorphs of carbon and fluid flow patterns by irradiating solid carbon with a CO_2 laser, *High Temp.Sci.* **4**, 445-477.
32. Jones, M.T. (1958) Nat.Carbon Res.Lab.Report PRG.
33. Decker, C. (1987) Synthesis of conductive polymers by laser irradiation of chlorinated PVC, *J.Polymer Sci., Part C, Polymer Letters*, **25**, 5-10.
34. Kasatochkin, V.I., Savranski, V.V., Smirnov, B.N. and Mel'nichenko, V.M. (1974) Investigation of carbyne condensed from carbon vapor, *Dokl.Akad.Nauk SSSR* **217**(4), 796-799.
35. Whittaker, A.G., Watts, E.J., Lewis, R.S. and Anders, E. (1980) Carbynes: Carriers of primordial noble gases in meteorites, *Science* **206**, 1512-1514.
36. Whittaker, A.G. (1981) Chaoite coating process, U.S.Patent 4,248,999 (Feb 3, 1981).
37. Rohlfing, E,A., Cox, D.M. and Kaldor, A. (1984) Production and characterization of supersonic carbon cluster beams, *J.Chem.Phys.* **81**, 3322-3330.
38. Fitzgerald, A.G., Simpson, M., Dederski, G.A., Moir, P.A., Matthews, A. and Tither, D. (1988) Triode technology in the sputter deposition of carbon films, *Carbon* **26**(2), 229-234.
39. Fitzgerald, A.G., Simpson, M., Dederski, G.A., Storey, B.E., Moir, P.A. and Tither, D. (1988) Carbon films produced by fast atom bombardment, *Carbon* **26**(4), 547-552.
40. Onuma, Y., Kawai, Y., Nakao, M. and Kamimura, K. (1996) Highly oriented crystalline carbon films obtained by DC magnetron sputtering with a hot filament, *Jpn.J.Appl.Phys.* **35**, L1354-1356.,
41. Kudryavtsev, Yu.P., Evsyukov, S.E., Babaev, V.G., Guseva, M.B., Khvostov, V.V. and Krechko, L.M. (1992) Oriented carbyne layers, *Carbon* **30**(2), 213-221.
42. Khvostov, V.V., Guseva, M.B., Babaev, V.G. and Rylova, O.Yu. (1986) Auger-spectroscopy studies of the electronic structure of amorphous carbon films, *Surf.Sci.* **169**, L253-L258.
43. Rice, M.J., Phillpot, S.R., Bishop, A.R. and Campbell, D.K. (1986) Solitons, polarons, and phonons in the infinite polyyne chain, *Phys.Rev.B* **34**(6), 4139-4149.
44. Heimann, R.B. (1994) Linear finite carbon chains (carbyne): their role during dynamic transformation of graphite to diamond, and their geometric and electronic structure, *Diamond Rel.Mater.* **3**, 1151-1157.

3.4.1.2 CONDENSATION OF CARBON VAPOR OBTAINED BY ELECTRICAL ARC DISCHARGE

Quasi One Dimensional Carbon Crystal "Carbolite"

S. TANUMA
College of Science and Engineering, Iwaki Meisei University
Chuodai, Iwaki, Fukushima 970 Japan

Abstract - A quasi one-dimensional carbon crystal was synthesized by means of a carbon arc. The high temperature carbon clusters were quenched on a copper or sapphire plate to obtain a quasi one-dimensional crystal, i.e., a trigonal array of linear carbon chains. The chains, consisting of four carbon atoms, have a distance of 344 pm; the intra-atomic distance along the chain is 138 pm which is the average value of the alternating C≡C bond length and C-C bond length of a polyyne chain. The specific density of the crystal was estimated to be 1.46 g/cm^3. The stability of the crystal was checked by specific heat and x-ray diffraction measurements. A complicated irreversible change in the specific heat was found above 350 K, and small and large changes were found in the x-ray diffraction patterns at 350 K and 600 K, respectively. Intercalation of alkali and halogen species was successfully performed.

1. Introduction

Carbon has the largest variety of allotropic forms compared to almost all other elements. Diamond forms a three-dimensional structure, graphite has a quasi-two-dimensional structure and fullerene has a ball-like quasi-zero dimensional structure. It is important to explore a quasi-one-dimensional carbon.

Kasatochkin *et al.* [1,2] performed dehydrogenation experiments on acetylene and obtained a hexagonal aggregate of macromolecular chains. They named such quasi-one-dimensional carbon "carbyne" in 1973. El Goresy and Donnay [3] discovered an allotropic form of carbon in 1968 in the Ries crater in Bavaria which was a material similar to that produced by Kasatochkin *et al.* Whittaker and Kintner [4] also made a similar white carbon in 1969 by Joule heating of a graphite tip up to a temperature of more than 2550 K in argon atmosphere. Almost all materials in such previous reports had too small a sample size in order to investigate such solid state properties as electrical conductivity and others.

Our first aim was to grow carbyne crystals of sufficient sizes to investigate the thermal and electrical properties. We used a method similar to Joule heating of a

R.B. Heimann et al. (eds.), Carbyne and Carbynoid Structures, 149–158.

graphite tip applied by Whittaker and Kinter, but failed to grow carbyne. Next we adopted a technique to quench the high temperature carbon arc and obtained a somewhat different kind of allotropic form of a quasi-one dimensional carbon whose sample size was considerably larger than that of the above mentioned carbyne samples. This crystal should obey a kind of carbynoid structure.

2. Syntheses and Structures

As mentioned above, we first adopted Joule-type heating of a graphite tip to a temperature higher than 2600°C, followed by a rapid cooling to room temperature. The length of the tip was 1 cm and its thickness was 1-2 mm. Although the actual temperature was 2700-2800°C measured by a pyrometer, no carbyne crystal was obtained after cooling. Even if carbyne could be generated at high temperature, the rate of transformation of graphite to carbyne is very slow compared to that of the reverse transformation, and therefore the rate of cooling for obtaining carbyne crystals at room temperature should be higher than 10,000 K/s which was not achieved by a natural cooling by cutting the heating current. There should be a genuinly special method of cooling.

We therefore tried another method with the same aim, that is, a method of quenching the high temperature carbon gas on a metallic plate at room temperature. The material obtained was not any of already known carbynes, but rather a new kind of allotropic form of carbon, still consisting of quasi-one-dimensional carbon chains.

The carbon gas was generated by arcing between a pair of sharpened tips of carbon rods of 6 mm diameter as illustrated in Fig.1a. The electric current was about 600 A generated by two car batteries connected in series. The carbon rods were set in a stainless steel pot covered with a seal ring so that it could be evacuated or filled with argon or hydrogen gas (method 1). The substrate was a 0.2 mm thick copper plate or a 1 mm thick sapphire plate; the widths and lengths were 2-3 cm each.These two plates were set in a symmetric position with respect to the aperture of the carbon tips, 3 to 5 mm away from the tips. Such a short distance from the center of the arc resulted in a high density carbon gas in the vicinity of the substrate plates, and further led to a high temperature gradient which was the driving force for intensive gas condensation at the substrate. The rate of carbon film deposition was 10 to 100 μm/s. Carbon gas did impact the substrate plates almost perpendicularly with a high temperature gradient of about 1,000°C/mm. This situation is thought to be advantageous for growing of carbon filaments at a right angle to the substrate surface that is further supported by the rapid extraction of heat from the carbon layer by the thermally well conducting substrate.

During condensation of carbon gas, linear carbon clusters may play a prominent role. Rohlfing *et al.* [5] investigated a non-equilibrium carbon gas which was generated by laser irradiation, and found that linear clusters consisting of 2 to 30 atoms were stable at high temperatures. Such linear carbon clusters supposedly collide with each other to produce larger clusters of complicated form and, finally, fullerenes when the clusters move away from the center of the laser beam. Hence the position of the substrate should be set near the arc center in order to make linear chains arrive at the substrate surface.

Figure 1b is an illustration of a revised apparatus (method 2). Two parallel circular copper discs accommodate about twenty curved copper substrates put between them. By rotating the discs by an appropriate angle, the curved substrate plates are successively moved into a suitable position at the center of the carbon tips. This reduces the time required to mount the samples (Fig.1a).

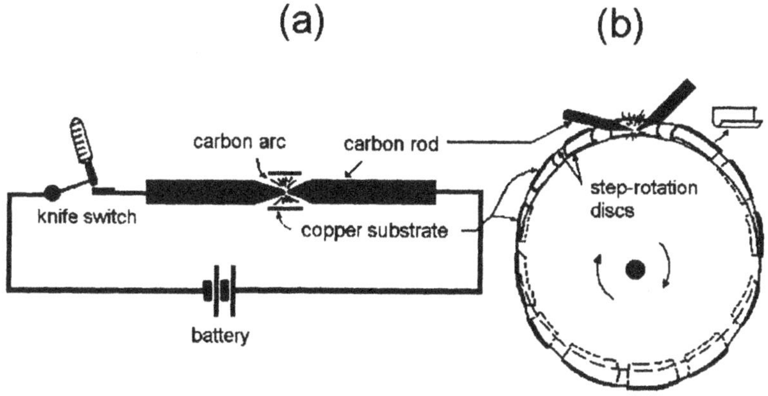

Figure 1. Quenching methods of a high temperature carbon gas generated by arcing between a pair of carbon tips. (a) Single shot method. (b) Multi-shots method using rotating discs.

In method 1, an atmosphere of 0.5 torr argon gas was used. In method 2, a mixture of 4 torr hydrogen gas and 0.5 torr argon gas was used mainly for the purpose of stabilizing the arc.

After arcing, the film-like deposit of carbon quenched on the copper plate was collected by dissolving the copper in nitric acid and picking up the material using a stainless mesh, rinsing with pure water, and drying.

The powder-like samples prepared were investigated by optical microscopy in transmitted light (Figs.2 a and 2b). The sample shown in Figure 2a was slightly orange-colored and the dark parts are due to a superposition of two specimens. The sample shown in Fig.2b is colorless and translucent. Its opaque part is caused by a curling of the edges of the plates.

Figure 3 shows an SEM picture which suggests a parallel columnar arrangement of carbon chains. The thickness of the sample is several micrometers.

The specific weight was calculated to be 1.46 g/cm³ from the structural data in the way described below. On the other hand, the experimental determination of the specific weight was done using the method of floating the specimen in the middle of a liquid column of suitably diluted sulphuric acid solution.

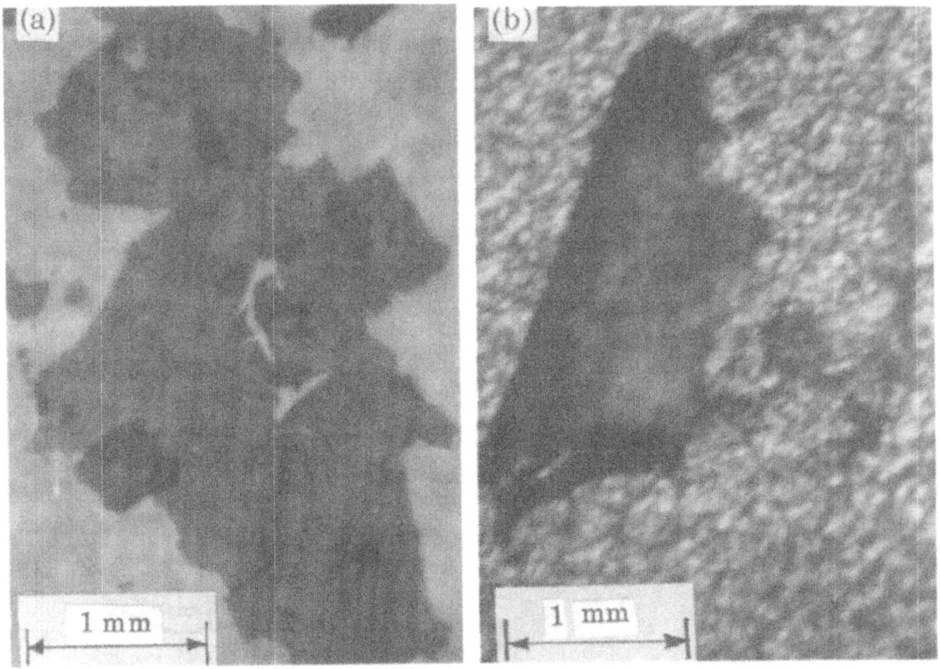

Figure 2 Optical micrographs of carbolite crystals. (a) Slightly orange-colored thin platy crystal synthesized in argon gas,, (b) Translucent colorless crystal synthesized in a mixed hydrogen and argon gas..

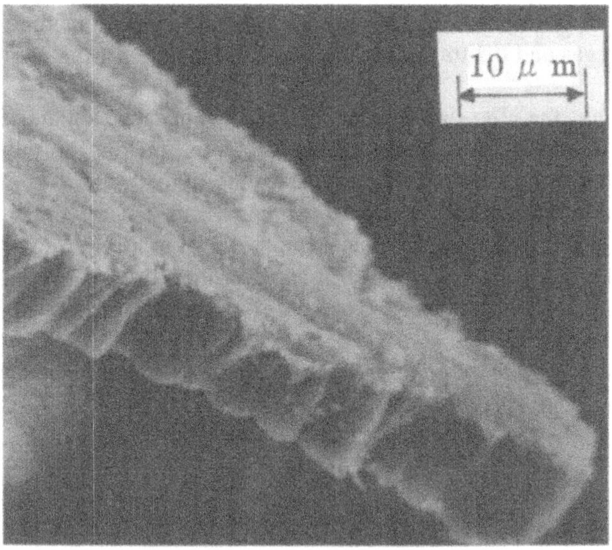

Figure 3 SEM picture of a specimen grown in argon gas according to method 1.

The result has shown a value of 1.45 ± 0.05 g/cm³. This value is much smaller than the values of 2.25 and 3.51 of graphite and diamond, respectively, and also smaller than the values of 2.68 of α-carbyne and 3.13 of β-carbyne (in g/cm³) [6]. The material has been tentatively coined "carbolite" owing to its extremely low specific weight. The materials synthesized in argon gas and in a mixed gas of hydrogen and argon were called carbolite I and carbolite II, respectively.

The chemical composition of the carbolite samples was analyzed by an x-ray micro-analyzer (JED-2001) for elements with atomic numbers larger than 11 (Na); no impurities were found within the sensitivity of 1 % of carbon. Carbolite II has been found by chemical analysis to include hydrogen on the same order of magnitude as that of carbon[1].

To investigate the crystal structure of carbolite, the deposit was crushed and spread on a vitreous quartz plate which gives no x-ray diffraction line in a θ-2θscan. Using characteristic CuKα radiation, such scans were taken. Figures 4a and b show typical scans for respectively, carbolite I and II specimens. Table 1 gives the lattice spacings for a carbolite I crystal. The values of d(obs) were obtained from experimental diffraction peaks, and the values of d(calc) were calculated by indexing the peaks for a hexagonal lattice with a unit cell size of a_0 = 1193 ± 3 pm and c_0 = 1062 ± 3 pm. The agreement between the observed and calculated spacings is good. We should mention that the diffraction lines are rather broad. The coherence length estimated from the line width of the largest diffraction line in Figure 4a is as short as 20 nm. Such a low value of the coherence length can be explained as the result of the very high speed of quenching of carbon gas.

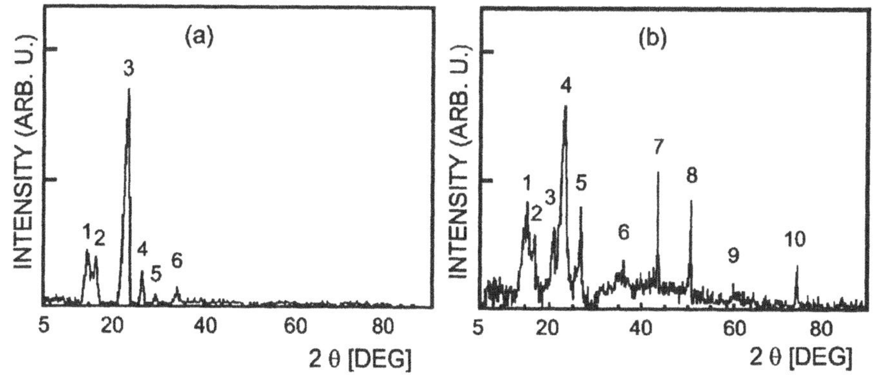

Figure 4. X-ray (Cu Kα) diffraction (θ-2θ scan) pattern of crystals of (a) carbolite I and (b) carbolite II. The numbers at the diffraction peaks refer to Tables 1 and 2.

[1] Hence, 'carbolite II' should be considered a crystalline hydrocarbon rather than a carbyne or carbynoid material (Comment by the editor (RBH)). See also Rietmeijer, F.J.M.: Are crystalline C-(H-O-N) carbons the elusive meteoritic carbyne?, *Meteoritics* **28** (1993), 242-245.

These values are not in agreement with the reported values of a_0 and c_0 of carbynes which are listed in Table 1 of reference [6], e.g., a_0 and c_0 values of hexagonal α-carbyne and β-carbyne that are, respectively 894 pm and 1536 pm, and 824 pm and 768 pm. Hence carbolite should be classified as a new polytype of carbyne or a carbynoid structure. The mechanism of crystallization of carbon gas will be clarified in a future study.

TABLE 1. Lattice spacings of carbolite I grown in argon gas

Line no.	hkl	d(obs) [nm]	d(cal) [nm]	d(obs)-d(cal) [nm]	Intensity (arb.units)
1	11.0	0.5964	0.5964	0	24
2	00.2	0.5313	0.5310	0.0003	20
3	12.0	0.3903	0.3903	0	100
4	10.3	0.3352	0.3349	0.0003	13
5	11.3	0.3044	0.3044	0	5
6	12.3	0.2618	0.2623	-0.0005	9

Figure 4b shows a θ-2θscan of a specimen of carbolite II grown in mixed hydrogen and argon gas. Table 2 gives the corresponding d(obs) and d(calc) values. Since the extinction rule of (-h+k+1= 3N) holds, the crystal lattice may belong to the rhombohedral system, and the value of $a_0/\sqrt{3}$ = 673 pm has been chosen for the value of a_0(rhomb) of the rhombohedral cell. The size of c_0(rhomb) is 1568 pm. The rhombohedral edge length is 852 pm and the rhombohedral angle is 86.34°. The possible crystal structures of carbolite I and II are illustrated in Figures 5 a-d. Microscopically, carbon chains are directed along the c-axis which is perpendicular to the substrate surface. The positions of the chains are assumed to coincide with the dots illustrated in Figures 5 a and c.

TABLE 2. Lattice spacings of carbolite II crystals synthesized in a hydrogen-argon gas mixture.

Line no.	hkl (hex)	hkl (rhomb)	d(obs) [nm]	d(calc) [nm]	d(ob)-d(cal) [nm]	Intensity (a.u.)
1	11.0	10-1	0.5813	0.5813	0	52
2	00.3	111	0.5226	0.5226	0	32
3	20.2	200	0.4226	0.4245	-0.0019	37
4	11.3	2-10	0.3882	0.3892	-0.0010	100
5	30.0	211	0.3358	0.3366	-0.0008	50
6	40.1	3-1-1	0.2491	0.2493	-0.0002	23
7	31.5	410	0.2087	0.2089	-0.0002	70
8	51.1	4-1-2	0.1806	0.1802	0.0004	56
9	43.4	51-2	0.1537	0.1529	0.0008	14
10	63.0	51-4	0.1277	0.1272	0.0005	23

The ratio of c_0 of carbolite I and II is 1.062:1.568, very close to 2:3. This suggests that the unit length of the chain is approximately 0.5 nm. Hence two 4-membered unit chains in series are included in the hexagonal unit cell, and three unit chains in series are included in the rhombohedral unit cell, as shown in Figure 5. The interchain connections are made by kinks which are indicated as slanted bonds in Figure 5.

A trigonal array of the chains perpendicular to the direction of the chain is also shown in the figure. A reasonable assumption of the length of chain is that the unit chain is made up by four atoms. Then the mean C-C distances for carbolite I and II are calculated to be 132.8 pm and 130.7 pm, respectively. Under this assumption, twelve chains are included in each unit cell section of carbolite I and II. Thus the interchain distances are calculated to be 344.3 pm and 336.6 pm for carbolite I and II, respectively. These values have been confirmed by θ-2θ scan of a sample quenched on a sapphire plate with such a configuration that the incident x-ray beam was parallel to the substrate surface, i.e., perpendicular to the carbon chains. A diffraction peak was found which indicated the interchain distance to be 336 pm [7].

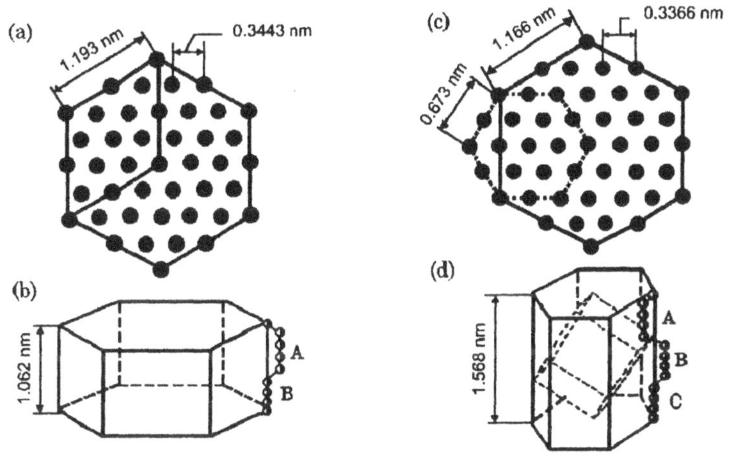

Figure 5. Crystal structures of carbolites I and II. (a) and (b) show the hexagonal lattice of carbolite I, and (c) and (d) show the rhombohedral lattice of carbolite II.

The appearance of as many as twelve chains contained in the sectional area of a unit cell should underscore the complexity of the mutual arrangements of the kinks, although the pattern of the arrangement of kinks is not yet known.

3. Nature of Bonds and Electrical Transport

The value of the interchain distance of 336 pm is close to the interlayer distance of 335.4 pm between the honeycomb planes of graphite. The C-C distances of 132.8 pm and 130.7 pm, respectively, of carbolites I and II are close to the C-C distance of 142.1 pm in the honeycomb plane of graphite. Thus the structure of carbolite I and II are commensurate with that of graphite.

The specific weight has been calculated to be 1.46 g/cm³ from the above structural data, and this value accords very well with the above mentioned experimental value of 1.45 ± 0.005 g/cm³ that is the smallest value among all carbon allotropes observed so far. The presence of many long interchain bonds, that is, a high density of π-bonds results in the low specific weight of carbolite. The interatomic bonds along a chain are of σ-type similar to the interatomic bonds in the honeycomb plane of graphite.

The observation by a microscope showed transparency suggesting that the crystals are electrical insulators. The one-dimensional chain of carbon atoms should be subject to a Peierls transition, resulting in a band gap, and the carbon-carbon linkages should be of polyyne type, i.e. alternating single bond and triple bonds, -C≡C-C≡C-. In a preliminary observation of the optical absorption spectrum, the band gap was estimated to be about 2 eV. A more detailed discussion is included in Ref. [8].

Electrical resistances were measured on samples of carbolite I. The samples were crystals as-quenched on sapphire plates of 10 mm square. The four leads were 0.15 mm thick gold wires glued by silver paste to the sample surface. The current flowed perpendicular to the carbon chains, so the resistance was measured parallel to the c-plane. The temperature dependence of the resistance is shown in Figures 6 a and b on a logarithmic scale. The excitation energies were estimated to be respectvely, 0.12 eV, 0.31 eV and 0.43 eV, in the temperature ranges below 140 K, around 300 K and above 370 K. These energies suggest the position of various impurity levels in the band gap to be around 2 eV.

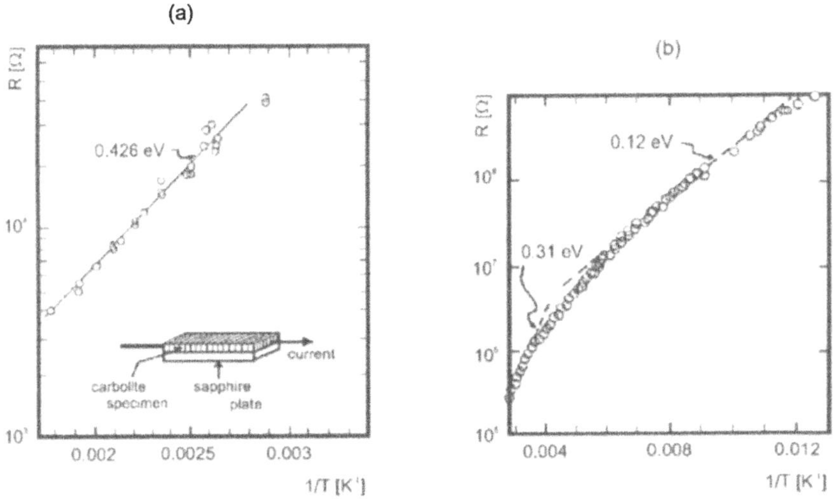

Figure 6. Temperature dependences of the electrical resistance of carbolite I in the high temperature region of 300 K ~ 523 K (a) and in the low temperature region of 77 K ~ 350 K (b).

The interchain force is considered to be of Van der Waals-type. The Van der Waals interplanar space of graphite accommodates very many kinds of intercalated species; similarly, the interchain spaces of carbolite could have a potential for intercalation. It has been shown that alkali metal atoms such as Na, K, Rb and Cs as well as I_2 molecules could be intercalated. On the other hand, Br_2 and ICl molecules have been found to dissolve carbolite.

Figures 7 a and b show examples of the logarithmic dependence of the resistance on the reciprocal temperature for the intercalation of K and I_2, respectively. Both relations display a peculiar feature: the high resistance plateaus, still one to two orders of magnitude lower than the resistance value of non-intercalated crystals, persist for a rather long time after the initiation of the intercalation reaction. At the end of the plateaus, the resistances of the specimens decrease rapidly to the minimum value by as much as six orders of magnitude.

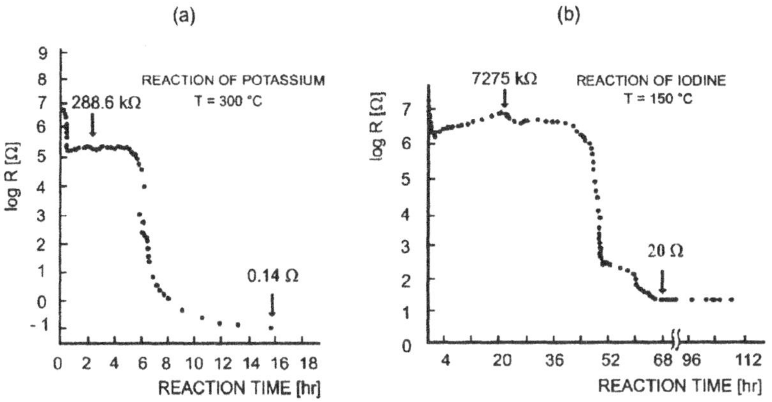

Figure 7. Changes of resistance of carbolite I depending on the reaction time for intercalation. (a) Intercalation of potassium atoms. (b) Intercalation of iodine molecules.

The stability of carbolite I has been examined by Palnichenko *et al* [9] by means of temperature-dependent x-ray diffraction and specific heat measurements. The diffraction patterns of θ-2θ scans at temperatures of 300 K and 450 K are almost identical with those shown in Figure 4 a. The diffraction pattern at 600 K is different from that at lower temperature, with broad the peaks indicating amorphization. A large anomaly starts at about 350 K in the specific heat curve and ends at about 700 K. The begin of the anomaly is not clearly seen in the x-ray scan at temperatures below 600 K, but the crystal may be metastable at temperatures above 350 K.

4. Acknowledgments

The author expresses his cordial thanks to Dr. Andrei Palnichenko of the Institute of Solid State Physics, The Russian Academy of Sciences, for his effective cooperation in this work when he was visiting with the College of Science and Engineering, Iwaki Meisei University.

5. References

1. Kasatochkin, V.I., Popov, N.M., Sladkov, A.M., Kudryavtsev,Yu P. and Korshak,V.V.:*Dokl. Akad. Nauk. SSSR* . **177** (1967), 358-360.
2. Kasatochkin, V.I., Korshak, V.V., Kudryavtsev, Yu P., Sladkov, A.M. and Sterenberg, I.E.: On crystalline structure of carbyne, *Carbon* **11** (1972), 70-72.
3 El Gorsey, A. and Donney, G,:.A new allotropic form of carbon from the Ries crater, *Science* **161**, (1969),363-364.
4. Whittaker,A. G. and Kinter, P.I,: Observations on the new allotropic form, *Carbon* **165** (1969), 589-591.
5. Rohlfing, E.A.,Cox, D.M. and Kaldor, A. Production and characterization of supersonic carbon cluster beams *J.Chem. Phys.* **81** (1983), 3322-3330.
6. Heimann, R.B., Kleiman, J.K. and Salansky, N.M.: A unified structural approach to linear carbon polytypes. *Nature* **306** (1983), 164-167.
7. Ishizawa, Y. and Tanuma, S.: to be published.
8. Tanuma, S. and Palnichenko A. ; Synthesis of low density carbon "carbolite" by quenching of carbon gas, *J.Mater. Res.* **10** (1995), 1120-1125.
9. Palnichenko, A.V., Gurov, A.V., Kopylov, V.., Kusano, K., Tanuma, S. and Salamatov, E.I.: Specific heat of a one-dimensional carbon phase synthesized by condensation of high-temperature carbon gas, *Phys. Rev. B* (1997), in the press.

3.4.2. ION-ASSISTED CONDENSATION OF CARBON

V.G. BABAEV and M.B. GUSEVA
Department of Physics,
M.V. Lomonosov Moscow State University,
119899 Moscow, Russia

Abstract – The chapter reports on a new, promising method for preparing oriented films of linear chain-like carbon. The method is based on ion-assisted condensation of carbon utilizing bombardment of a growing film by low-energy inert gas ions. The carbon flux produced by ion sputtering of graphite contains carbon chains. Ion-induced electrical charging of the chains prevents them from interchain cross-linking and contributes to the formation of solitons and kinks that divide the carbon chains into linear fragments and, at the same time, stabilize the structure of chains. The ion irradiation applied during the epitaxial condensation of carbon vapor on substrates decreases the temperature of heteroepitaxial growth and facilitates alignment of carbon chains. A crystalline material coined Tetracarbon™ based on two-dimensionally ordered carbon chains prepared by ion-assisted condensation will be described.

1. Introduction

Ion-assisted condensation occupies a special position among the various methods to obtain carbyne inasmuch as this method allows the preparation of carbyne films with a controllable degree of ordering – from amorphous to single-crystal layers. Furthermore, carbynes of a required polytype, *i.e.*, with controlled crystal-lattice parameters can be prepared by this method. Finally, films of other allotropic forms of carbon, including metastable ones, can be obtained by varying the ion-irradiation parameters. Prospects for creating layered structures, such as superlattices, on the basis of various carbon modifications in a single technological process are thereby within reach. It was shown by our research group [1-4] that during preparation of thin carbon films in a vacuum under the conditions of either irradiation simultaneously with condensation or alternate irradiation by slow ions of inert gases, stimulation of phase transitions can be observed along with the formation of metastable phases [1–4]. The yield of this process is maximized under a regime of resonance charge exchange (neutralization) interaction of ions with the amorphous condensate. The resonance charge exchange between ion electron levels and a surface (filled with electrons) is observed in selective ranges of ion energies.

R.B. Heimann et al. (eds.), Carbyne and Carbynoid Structures, 159–171.

2. General Principles of Ion-Assisted Condensation of Carbon

During ion-assisted condensation in a high vacuum (10^{-7} Torr) the carbon and inert gas ion (Ar^+) fluxes impinge on the substrate either simultaneously or alternately. The carbon flow is produced by thermal vaporization or ion sputtering of a graphite target. The energy of the ions impinging on the substrate varies within the range of 1.5–2 keV, the Ar^+ ion current density (j) at the substrate is typically 10 mA/cm^2, and the rate of film growth is very slow, typically ~1 nm/min. The thickness of the films produced under these processing conditions ranges from 20 to 100 nm. The carbon films prepared by alternate condensation and ion irradiation are generally amorphous, with their structure and properties depending greatly on the irradiation conditions. The specific dependence of the electric conductivity of the films on the energy of the irradiating ions (Fig. 1) is particularly revealing in this respect. As shown, the resistivity (R) of the films varies non-monotonically with the Ar^+ ion energy, and its minimum and maximum values differ by several orders of magnitude.

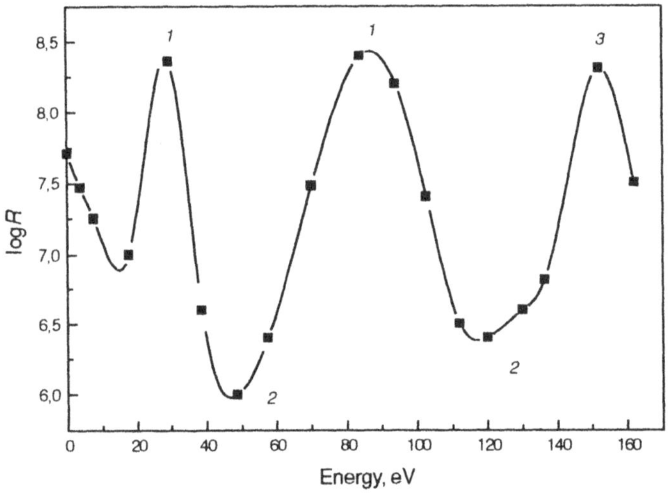

Figure 1. Dependence of the electrical resistivity (R) of a carbon film on the energy of irradiating ions (E). *1, 2, 3* – extremal values of the resistivity.

Other electrophysical properties of the films vary similarly, including the optical density and the forbidden gap E_g, from 0.5 eV at the maxima to 2 eV at the minima of the electric conductivity. A series of electron microscopic and spectroscopic (Auger, IR, Raman) investigations[1] has established that in amorphous carbon films (α-C) prepared at E = 50 and 125 eV (see Fig. 1) the structural elements with a graphite-type short-range order predominate. In the films prepared at E = 30, 90, and 150 eV (α-C

[1] For details see Chapters 6.1 and 6.3 of this book.

films with improved dielectric properties), structural elements predominate whose short-range atomic order is arranged as in graphite and/or diamond. Thus, ion irradiation of growing carbon films can stipulate phase transitions, and this effect is a non-monotonic function of the energy of the ions. The effect of ion irradiation can be adequately explained in terms of the quasi-resonance charge exchange model [3]. As previously shown by theoretical calculations, atomic fluctuations corresponding to metastable phases can be formed [5] in a graphite cluster at energies of Ar^+ ions as low as a few electron volts. They are unstable, but can become stable by capturing a hole, *i.e.*, an ionized C^+ state formed as a result of charge exchange of Ar^+ ions with the carbon condensate. From the macroscopic point of view, resonance charge exchange transfers the system from a C–C term to a C–C^+ term, along which contraction occurs. An important point is that the ionized C^+ state is a localized hole due to the disordered state of the amorphous system. This hole subsequently creates fluctuonic states and plays the role of a inhibitor for the new phase being formed. The time dependence of the charge-exchange process of Ar^+ with a discrete energy level of solid has a sinusoidal character yielding the oscillations in energy coordinates. The ion energy determines the 'ion surface' interaction. Annealing of carbon films with a varying degrees of short-range order leads to the crystallization of carbon in various allotropic states – graphite, diamond, carbyne. The carbon films obtained under the conditions of simultaneous condensation and irradiation may contain both amorphous and oriented phases (microcrystalline inclusions). Microcrystalline inclusions of 10–100 nm size appear in these films at $E > 30$ eV, and their diffraction pattern corresponds to a hexagonal structure, which may be identified as carbyne [1,2]. The concentration of carbyne crystallites depends on the current density j of the ions impinging on the substrate during the process of film growth, and it increases as j is increased. Variation of the energy within the range studied does not have an appreciable effect on the density of the crystalline inclusions. However, an increase in the energy of the ions causes a very significant change in the parameter a of the carbyne crystal lattice: from 1.03 nm at $E_{Ar} = 30$ eV up to 0.831 nm at $E_{Ar} = 150$ eV. Similar changes [6] in the parameter a are observed after high-temperature annealing of the carbyne single crystals formed in an amorphous matrix during the crystallization process at 600°C. Initially, the crystals have the shape of low-contrast, flat, acute-angled particles. When the annealing time and temperature are increased, they grow rapidly and reach a size of 1 μm at 650°C. Their microdiffraction pattern contains point reflections corresponding to a hexagonal structure with a parameter $a = 0.901$ nm. Successive increase of the annealing temperature results in changes of the parameters of the carbyne cell. Initially, the a value decreases with increasing temperature from 0.901 nm at 700°C to 0.840 nm at 750°C, and then it increases back up to 0.901 nm at 800°C. Subsequent annealing at 850°C results in the phase transition of carbyne to amorphous carbon.

The introduction of nitrogen into the vacuum chamber during the ion assisted condensation brings about a significant increase in the yield of the carbyne crystals [7]. Figure 2a shows a comparatively large (1 μm)[2] single crystal of carbyne prepared by

[2] Larger crystals (~ 10 μm) of the same habit were also observed.

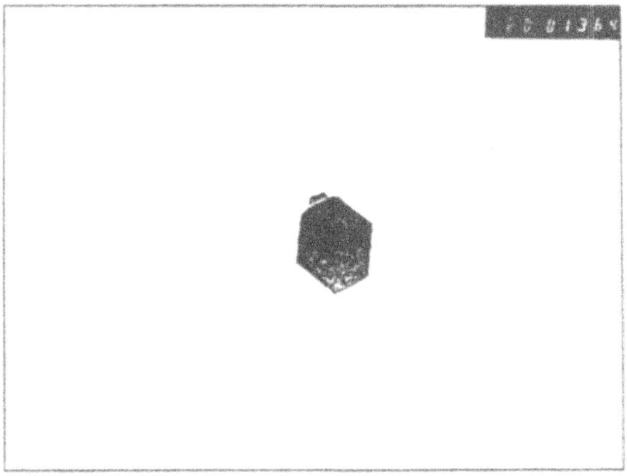

a

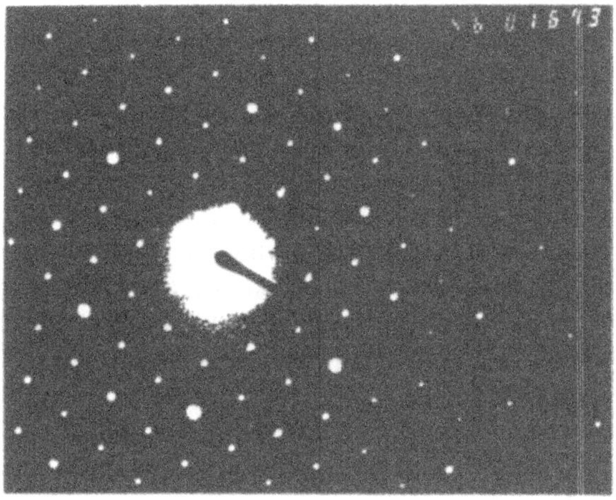

b

Figure 2. Single crystal of nitrogen-doped carbyne obtained by ion-assisted deposition (*a*) and its electron diffraction pattern (*b*).

ion-assisted deposition of carbon at $T = 150°C$ in an Ar N atmosphere. The corresponding Patterson function (Fig. 3*a*) and projection of the crystal potential on a (*00.1*) basal plane (Fig. 3*b*) are also shown. Based on the crystal potential distribution, a two-layered lattice can be constructed that is supposed to be characteristic of the structure of that crystal. The first layer is a close-packed hexagonal lattice with carbon chains oriented normally to the projection plane [8]. The second layer contains regular vacancies[3]. The addition of nitrogen markedly stimulates the growth of linear carbon chains by polycondensation of =C=N– radicals with *sp*-type of hybridisation of carbon atoms. At higher temperatures, desorption of the N atoms from the film surface will occur, yielding purely carbyne crystals.

3. Preparation of Carbyne Films

The main method for preparing single-crystal films is epitaxial growth. However, this method requires high temperatures and is inapplicable in the case of metastable carbyne. Single-crystal films of carbyne [7,8] were prepared by the technique of ion-assisted deposition based on inelastic interactions of low-energy ions with a solid that activates oriented growth of carbon molecules on its surface. This results, in particular, in a considerable decrease of the crystal growth temperature that suppresses the formation of cross-links between carbon chains. The ion irradiation is accompanied by surface charging, and additional possibilities appear when the chain-like carbon is being formed: the deformation of the carbon chain caused by electronic excitation corresponds to the generation of a soliton in accordance with predictions by Rice *et al.* [9]. The soliton[4] arises, in particular, from the insertion of a charge into the chain, presumably at a kink site, and increases the energetic advantage of the linear conformation, as confirmed by theoretical calculations [9].

3.1. 2D-ORIENTED LINEAR CARBON – A NEW PROMISING MATERIAL

The two-dimensionally ordered linear chain-like carbon was originally prepared by the ion-assisted deposition method [8] and was later coined Tetracarbon™ [10]. The basic idea was to use ion-assisted epitaxial condensation of carbon clusters on orienting surfaces to prepare ordered layers of linear carbon, while the carbon clusters are produced by ion sputtering of graphite. The mass-spectroscopic (SIMS) studies showed that carbon flux obtained by ion sputtering of a graphite target with a low-energy ion beam ($E = 5$ keV) consists of linear carbon chains C_n with $n = 1$–8 (Fig. 4). This follows from the specific intensity distribution of carbon clusters during ion sputtering of graphite. As established by SIMS [11], the concentration of linear carbon clusters C_n in a carbon vapour varies periodically as a function of the cluster size with a period n =2, which is in agreement with the dependence of the linear carbon chain stability on n as predicted theoretically by Pitzer and Clementi [12]. During the condensation, the Ar^+ ion beam irradiating the growing carbon film has an energy of 1.5 keV, which

---

[3] *cf.* Fig. 14 in Chapter 4.2 of this book.

[4] For a more general account on soliton formation in carbon chains *cf.* Chapter 6.2.

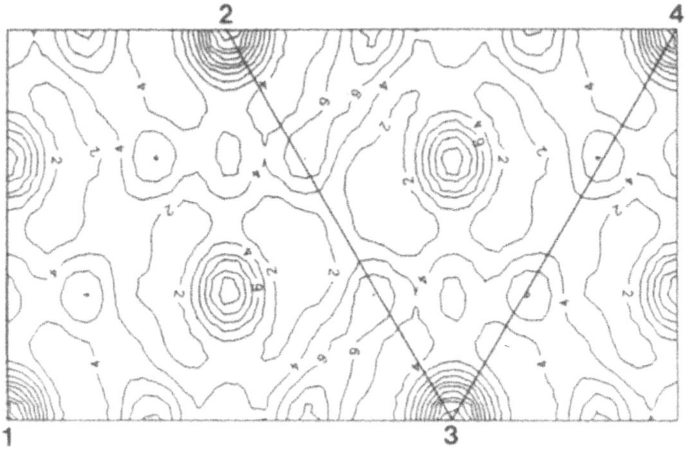

a

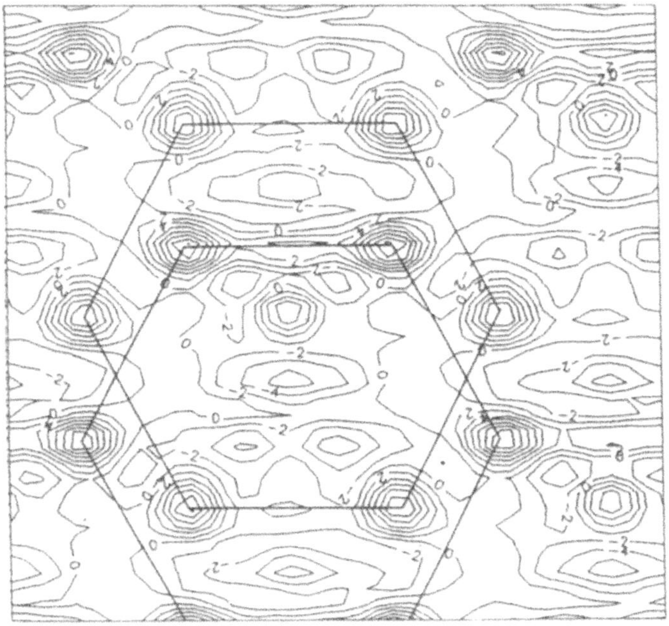

b

Figure 3. The Patterson function of the nitrogen-doped single crystal of carbyne (a), and the distribution of crystal potential in its unit cell (b).

corresponds to the maximum cross-section for neutralization of Ar^+ ions on the carbon chain-like clusters. This situation is favourable for low temperature deposition of a single-crystal film in the temperature range 50–180 °C which prevents the formation of inter-chain cross-links. The following parameters are normally used to prepare oriented linear-chain carbon [8]: a deposition rate of 1 nm/min; an ion current density of the beam irradiating the growing film of 10 mA/cm^2; and a residual gas pressure of $P = 10^{-7}$ Torr. The thickness of films thus prepared ranges from 10 to 100 nm. High-purity graphite (99.99 %) is used as a target. Representative electron diffraction pattern of oriented carbon films are shown in Figure 5. This diffraction pattern consists of a single ring or six bright reflections located on the same ring. The absence of higher-order reflections in the diffraction pattern is accounted for by a layered structure of the film and by small random displacements of the layers constituting the film. It should be noted that similar diffraction pattern are characteristic for multilayered films prepared by the Langmuir-Blodgett (LB) method in which the chain axes are oriented normally to the surface of the film. This normal orientation of carbon chains can also be proven by tilting a diagnostic electron beam relative to the direction perpendicular to the film surface. In this case only two reflections can observed in the diffraction pattern that are located parallel to the tilting axis. Such a pattern corresponds to the diffraction by parallel chains.

Presently, Tetracarbon™ films are also prepared on large areas with good adhesion by the method of ion-beam assisted deposition. The films have been distinguished due to their specific atomic and electronic structure that was studied by various spectroscopic methods.

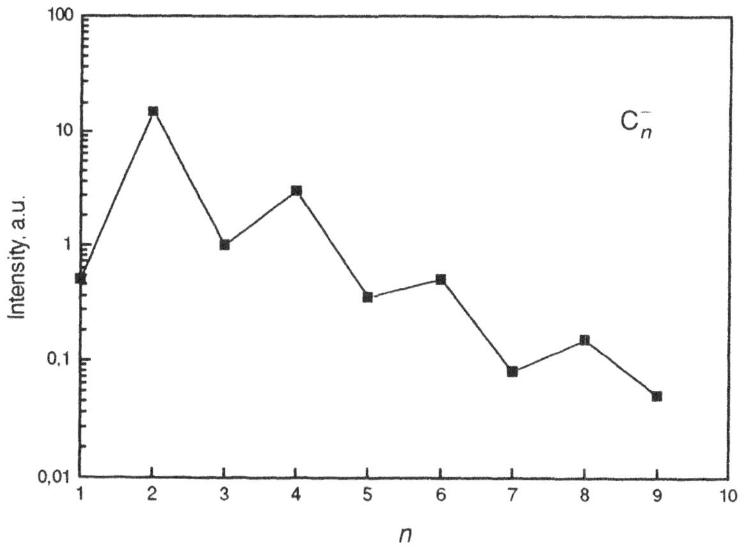

Figure 4. Mass spectrum of sputtered C_n^- ions.

3.1.1. *Atomic Structure*

As evident from Figure 5, there is only one bright narrow diffraction ring (Fig. 5*a*) or six point reflections equally spaced on this ring (Fig. 5*b*) in the diffraction pattern of Tetracarbon™. The interplanar distance $d_{10.0}$, as inferred from the diameter of the diffraction ring, ranges from 0.415 to 0.437 nm in different Tetracarbon™ samples. If the electron beam is tilted with respect to the film surface, then two arcs are observed in the diffraction pattern instead of the ring (Fig. 5*c*). The arcs are positioned symmetrically around the ring centre. This feature provides evidence for a one-dimensional texture of the film that is not characteristic of any known carbon form but is a typical feature of some LB films. The LB films consist of chain-like organic molecules closely packed into hexagonally arranged layers with an orientation of the molecules normal to the surface. These layers are randomly shifted with respect to each other.

3.1.2. *Electronic Structure*

The electronic structure of Tetracarbon™ was studied by means of Auger spectroscopy. The Auger spectra showed that the films consist of carbon with a purity exceeding 99%. The comparison of the KVV carbon Auger line shape of Tetracarbon™ with that of other carbon forms (diamond, graphite, turbostratic graphite) revealed that amongst all carbon forms, graphite exhibits the most pronounced shift of the high-energy edge of the KVV line in its Auger spectrum. The high-energy edge of turbostratic graphite (point line) is very close to that of graphite, *i.e.*, it is shifted relative to graphite by 0.3 eV to lower energy values. The KVV high-energy edges of both Tetracarbon™ and diamond are even more shifted toward lower energies. Compared to the KVV high-energy edge of graphite, they are shifted by a value of 5 eV.

Even after a small dose of ion beam etching the shape of the Tetracarbon™ KVV line becomes similar to that of graphite. The density of valence states (DOVS) was calculated from the experimental Auger line shape of Tetracarbon™ film by self-deconvolution of the carbon KVV line [13,14]. The DOVS curve of Tetracarbon™ has some specific features, *viz.*, the maxima at 16.5, 18 and 26 eV that distinguish it from diamond. These maxima correspond to the top and bottom of the π- and σ-subbands of linear carbon chains of cumulene type. In addition, the maximum at 13 eV provides evidence for the presence of *sp*-bonds as well. In general, the Auger spectroscopic data support the conclusion that the structure of Tetracarbon™ includes linear carbon chains with *sp*-bonds as distinguishing features. The optical gap width of Tetracarbon™ was determined from the optical absorption edge to be $\Delta E = 1.6$ eV.

3.1.3. *Atomic Model of Tetracarbon™*

Figure 6 shows a structural model of a multilayered Tetracarbon™ film. Each layer consists of carbon atom chains in an *sp*-hybridization state, densely packed into a hexagonal lattice and randomly shifted relative to each other. The distance between chains ranges from 0.490 to 0.503 nm in different Tetracarbon™ samples. In other words, Tetracarbon™ is constituted of identical chains with kinks oriented randomly relative to their axis. Hence, the offset angle has an arbitrary value. Since the chains are densely packed, the kinks of adjacent chains correlate and are positioned in the

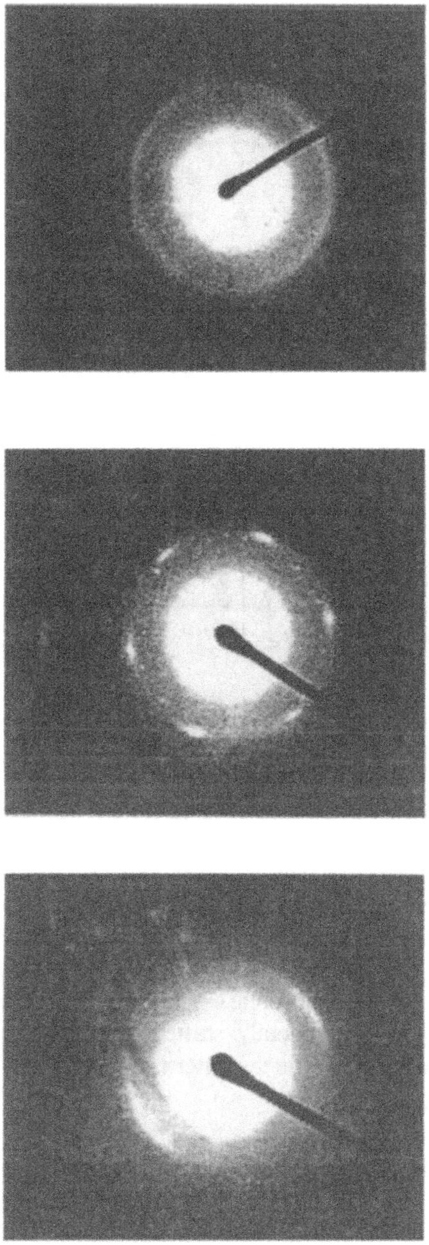

a

b

c

Figure 5. Electron diffraction patterns from oriented crystalline Tetracarbon™ films: *a* – normal incidence of the electron beam; *b* – highly ordered Tetracarbon™; *c* – tilted incidence of the electron beam.

same plane and point in the same direction[5].

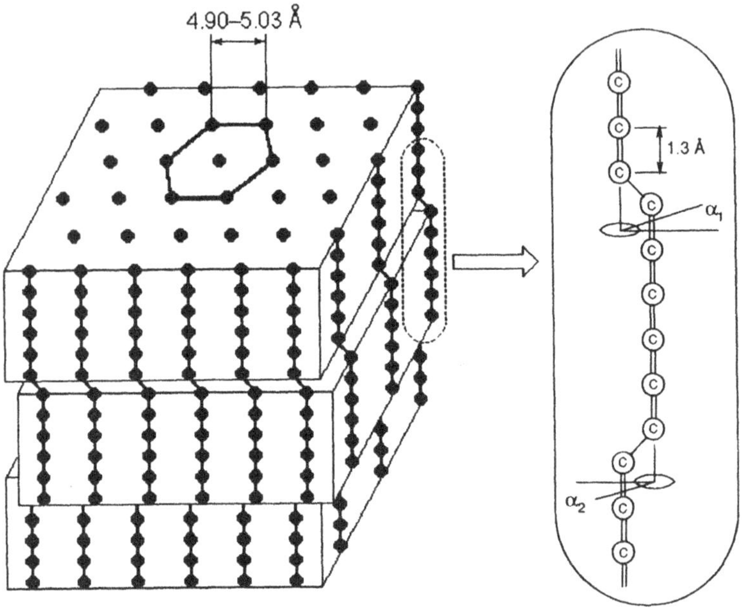

Figure 6. Structural model of a multilayered Tetracarbon™ film.

3.1.4. *Atomic Force Microscopy*

An oriented linear-chain carbon film of 4 nm thickness was studied by atomic force microscopy (AFM) using a Femtoscan 001 microscope. Figure 7 shows an image of the film surface in a height mode. The image clearly demonstrates a hexagonal array of carbon atoms. The lattice parameter a was determined to be 0.486 nm. There is obviously good conformity between the hexagonal parameters obtained by AFM and TEM methods. The crystal structure of oriented chain-like carbon films differs substantially from that of carbyne single crystals reported so far. The main difference is that the former consists of randomly shifted, two-dimensionally ordered layers of closely packed carbon chains, while the latter is a three-dimensional lattice of regularly positioned carbon chains.

3.1.5. *Raman Spectroscopy*

The data obtained by Raman spectroscopy are also consistent with the linear chain structure of Tetracarbon™. Besides a maximum near 1500 cm^{-1}, the Raman spectrum of the Tetracarbon™ film exhibits two additional maxima at 2100 and 2300 cm^{-1}, attributable to stretching vibrations of the *sp*-type of bonds in linear carbon chains. The vibrational modes were also calculated using the equation for long linear carbon chains,

[5] *cf.* Chapter 4.2 of this book.

$$\nu = \frac{1}{2}\left(\frac{k_1}{\pi^2 m}\right)^{\frac{1}{2}} \sqrt{1 + a^2 + (1 + a^2 + 2a\cos(\frac{2\pi n}{N}))} \qquad (1)$$

where N is the number of atoms in the chain (N = 20), m is the atomic mass of the C atom, a is the ratio of the force constant in a single bond (k_2) to that in the triple bond (k_1) [15][6]. The comparison between calculated and experimental stretching modes clearly showed that Tetracarbon™ film consists of long linear carbon chains with *sp*-hybridization of the carbon atoms. The calculated vibrational spectra show that in the presence of kinks in the carbon chain there is a band of stretching vibrational modes in the 1500–1600 cm^{-1} range.

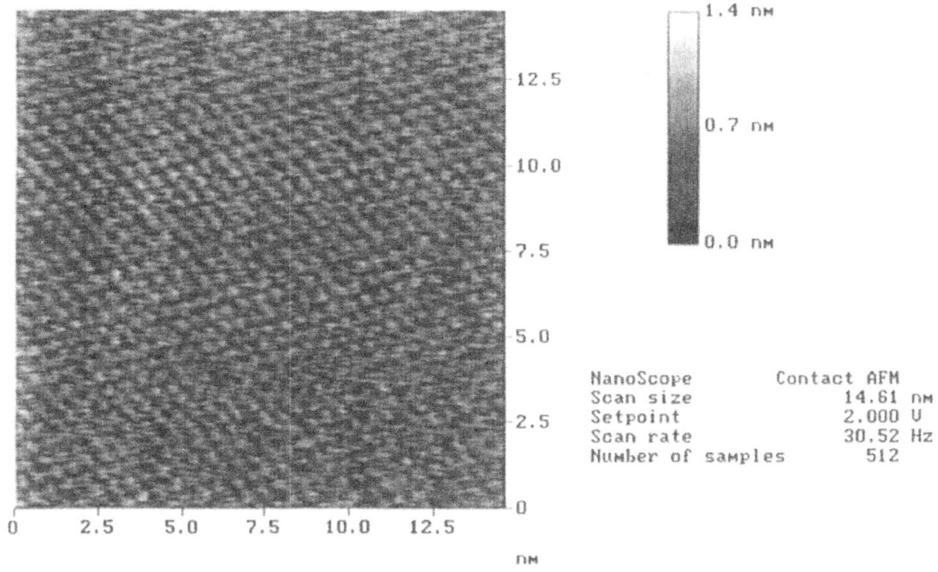

Figure 7. Atomic force microscopy image of Tetracarbon™ film of 4 nm thickness

It should be noted that Raman studies of linear chain carbon require special precautions because of its instability under laser irradiation. As a result of this instability, rapid degradation of the atomic structure of a sample occurs. Thus, it is difficult to observe a characteristic vibration band of linear carbon chain. The degradation of linear chain carbon under a laser beam increases intensities of bands at 1580 and 1360 cm^{-1} attributable to sp^2- and sp^3-type defects in carbon chains, and decreases intensity of a peak at 2100 cm^{-1} characteristic of linear carbon chains.

Summarizing the results of TEM, AFM, Auger and Raman spectroscopic studies, we can conclude that Tetracarbon™ films comprise predominantly long carbon chains of *sp*-hybridized carbon atoms, which are closely packed into two-dimensional

---

[6] This dependency is being discussed in more detailes in Chapter 6.1 of this book.

hexagonal layers that are randomly displaced relative to another[7]. The distance between the carbon chains was found to be 0.480–0.507 nm. In other words, Tetracarbon™ thus consists of nearly identical kinked[8] carbon chains with kinks oriented randomly relative to the chain axis; the angle α shown in Figure 6 is an arbitrary value along the chain, and identical for chains arranged in a layer. Since the chains are close-packed, the kinks of the neighbouring chains should be correlated and positioned in the same plane.

3.1.6. *Electrical Properties*

The Tetracarbon™ films reveal a marked anisotropy of their properties. In general, Tetracarbon™, being a semiconductor, occupies an intermediate position between graphite (conductor) and diamond (insulator). The conductivity of Tetracarbon™ films obeys an activation-type temperature dependence at high temperatures. At lower temperatures their conductivity is in agreement with a hopping model. The activation energy of conductivity is temperature-dependent. When the film prepared by ion assisted sputtering of graphite was brought into contact with crystalline silicon of *n,p*-type, the formation of a cut-off heterojunction with an extremely small back current and a high value of the photoelectromotive force (~0.6 eV) was observed. Thus, the Tetracarbon™ films were shown to be promising materials for application in electronics and electrical engineering. The activation energy is 20 meV which corresponds well to the soliton activation energy in one-dimensional system (as in polyacetylene) and does not contradict the predicted soliton conductivity in linear chain carbon.

4. Conclusion

Ion assisted condensation of carbon appears to be a promising method for depositing linear chain carbon, including two-dimensionally ordered coatings as well as microcrystals with variable crystal lattice parameters. The ion-assisted technique is realized under conditions of pulsed carbon plasma deposition. Carbon coatings of *ca.* 10×10 cm^2 area can thus be prepared with a deposition rate 100 nm/min on surfaces of any desirable shape and of different materials (metals, dielectrics, ceramics, *etc.*) with good adhesion [10].

Using the ion-assisted condensation method, films can be produced possessing a set of useful properties such as high hardness approaching that of diamond, anisotropic electrical conductivity and, finally, unique biocompatibility, useful for application in medicine[9] [10].

Acknowledgement – The authors are indebted to Dr. I. Yaminskii for the AFM studies of the Tetracarbon™ films.

[7] Editor's comment: This statistical nature has been addressed previously by considering a paracrystalline nature of the carbyne structure (see Heimann, R.B., *Carbon*, **35**(1997), 1669-71).

[8] *cf.* Chapter 4.2 of this book.

[9] *cf.* Chapter 7.2 of this book.

5. References

1. Guseva, M.B., Babaev, V.G., Nikiforova, N.N., and Savchenko, N.F.: Alteration in structure and properties of carbon films under conditions of irradiation by slow ions, *Vopr. Atom. Nauki Tekhn., Ser.: Fiz. Radiats. Povrezhdenii i Radiats. Materialoved.* (2 [25]) (1983), 92–95 (in Russian)
2. Savchenko, N.F., Guseva, M.B., Babaev, V.G., and Palatnik, L.S.: Structural modification of carbon films by the ion irradiation, *Poverkhnost'. Fiz., Khim., Mekh.* (6) (1985), 106–111 (in Russian).
3. Guseva, M.B., Savchenko, N.F., and Babaev, V.G.: The influence of ion irradiation on the structure and properties of amorphous carbon films, *Radiat. Eff. Lett. Sect.* **87** (1986), 215–224.
4. Palatnik, L.S., Guseva, M.B., Babaev, V.G., Savchenko, N.F., and Fal'ko, I.I.: Concerning γ-carbon, *Zh. Eksp. Teor. Fiz.* **87** (1984), 914–917 (in Russian).
5. Guseva M.B. and Babaev V.G.: Ion stimulation on the surface of solids by low energy ions, in *Ion Implantation in Semiconductors and Other Materials*, Vil'nyus, 1985, pp. 263–275.
6. Guseva, M.B., Savchenko, N.F., and Babaev, V.G.: New structural data concerning carbyne, *Dokl. Akad. Nauk SSSR* **283** (1985), 1336–1339 (in Russian).
7. Guseva, M.B., Babaev, V.G., Babina, V.M., and Khvostov, V.V.: Formation of oriented nitrogen-doped carbon films, *Pis'ma Zh. Eksp. Teor. Fiz.* **62** (1995), 698–701 (in Russian) [*JETP Lett.* **62** (1995), 715–718 (Engl. Transl.)].
8. Kudryavtsev, Yu.P., Evsyukov, S.E., Babaev, V.G., Guseva, M.B., Khvostov, V.V., and Krechko, L.M.: Oriented carbyne layers, *Carbon* **30** (1992), 213–221.
9. Rice, M.J., Phillpot, S.R., Bishop, A.R., and Campbell, D.K.: Solitons, polarons, and phonons in the infinite polyyne chain, *Phys. Rev. B* **34** (1986), 4139–4149.
10. Guseva, M.B., Novikov, N.D., Babaev, V.G., Adamyan, A.A., and Lavygin, I.A.: Tetracarbon, *PCT Int. Appl.* WO 97 25078, (1997) [*Chem. Abstr.* **127** (1997), 152988g].
11. Nikiforova, N.N. and Guseva, M.B.: Mass spectrometric analysis of the carbon flux by ion sputtering of graphite and diamond targets, in: *Secondary Ion and Ion-Photon Emission: Proc. All-Union Conf.*, Khar'kov, USSR, 1983 pp. 30–32 (in Russian).
12. Pitzer, K.S. and Clementi, E.: Large molecules in carbon vapor, *J. Am. Chem. Soc.* **81** (1959), 4477–4485.
13. Khvostov, V.V., Babaev, V.G., and Guseva, M.B.: Auger spectroscopy of amorphous carbon films, *Fiz. Tverd. Tela* **27** (1985), 887–891 (in Russian).
14. Khvostov, V.V., Guseva, M.B., Babaev, V.G., and Rylova, O.Yu.: Auger-spectroscopy studies of the electronic structure of amorphous carbon films, *Surf. Sci.* **169** (1986), L253–L258.
15. Weltner, W. and McLeod, D.: Spectroscopy of carbon vapor condensed in rare-gas matrices at 4°K. III, *J. Chem. Phys.* **45** (1966), 3096–3105.

3.4.3 DYNAMIC PRESSURE SYNTHESIS OF CARBYNES

J. I. KLEIMAN[1], K. YAMADA[2], A. B. SAWAOKA[3] and R.B. HEIMANN[4]

[1]*Integrity Testing Laboratory Inc./University of Toronto Institute for Aerospace Studies, 4925 Dufferin Street, North York, Ontario, M3H 5T6, Canada;*

[2]*Department of Chemistry, The National Defense Academy, Hashirimizu, Yokosuka 239, Japan;* [3]*Center for Ceramics Research, Tokyo Institute of Technology, Midori, Yokohama 227, Japan;*

[4]*Freiberg University of Mining and Technology, D 09596 Freiberg, Federal Republic of Germany*

Abstract - A review is presented of the high pressure - high temperature investigations of carbynes and their role in phase transformation processes of carbon.

1. Introduction

In recent years a considerable amount of work was performed to investigate the high pressure-high temperature behaviour of carbynes and their precursors. This chapter will attempt to summarize these activities.

Shock synthesis of carbyne from graphite was performed for the first time by Litvinova *et al.* [1]. Thereafter, several carbyne forms have been shock-synthesized from carbon materials using different shock wave methods [2-5]. In nature, carbyne which seems to have been formed by dynamic compression was discovered in shock-fused graphitic gneisses [6] and in carbonaceous meteorites [7-9].

From comprehensive reviews of the pressure-temperature phase and transformation diagram of carbon [10, 11] it is evident that the low pressure-high temperature region of the diagram where the carbynes appear to exist, is still not resolved. It is not even known yet whether there exists a true graphite/melt/vapour triple point. A number of authors [12-14] proposed the existence of carbynes as transformation product of graphite at temperatures above 2500 K and low pressures. According to Whittaker [14, 15] the compound that actually melts is carbon VI [16], and the triple point solid/liquid/vapour is at 3800 K and 2 104 Pa [17] (see Fig. 4, Chapter 7.1.).

More recently, a revised P-T diagram was discussed [11] which is shown in Figure 1. The major features are the graphite (and carbyne?)/vapour line, the liquid/vapour line

R.B. Heimann et al. (eds.), Carbyne and Carbynoid Structures, 173–187.

and the graphite (carbyne?) melting line. Evidence has been reported that the low pressure liquid is electrically non-conducting whereas the higher pressure liquid is conducting. Indeed, a phase boundary between the two kinds of liquids has been suggested, as illustrated in the figure. The liquid/vapour line should end at a critical point which has been estimated to be roughly in the vicinity of 0.2 GPa/6800 K [11]. At high temperatures below the melting point there is some evidence for transformations from graphite to other solid phases which are proposed to be carbyne.

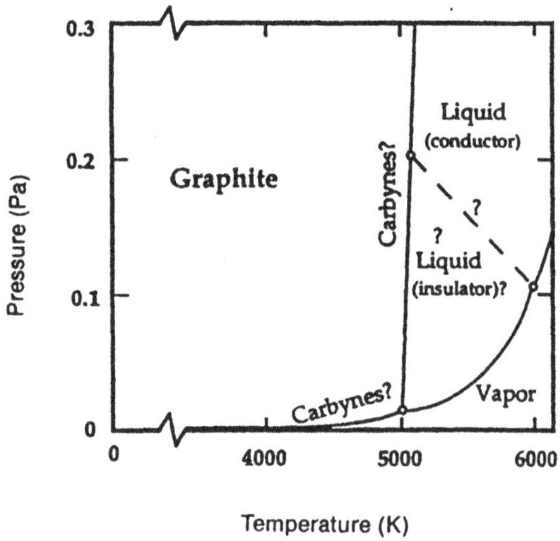

Figure 1. P, T phase diagram for the lower pressure region (according to Ref. [11]). The question marks following "carbynes" indicate that the existence of carbyne solid phases in this P, T region is controversial. The question marks in the liquid phase region indicate that the newest more rigorous experimental results do not support the earlier experimental interpretation that a non-conducting liquid phase of carbon exists.

2. Experimental Techniques Used in High P-T Carbyne Synthesis

A number of different experimental set-ups were used in high pressure-high temperature synthesis of carbynes. A brief account of some of them will be given below.

A combined system consisting of an implosion chamber and connected to a flash-heating system was used in a research program on synthesis of diamonds [2, 10] in which a number of carbyne forms were reported to have been formed. The implosion chamber was an improved version of that utilized by Glass and co-workers [18-20]. As shown in Figure 2, the 20-cm diameter cavity of the hemispherical chamber rested on a massive supporting frame with its axis in a vertical position. Two hydraulic jacks (H) connected to a compressor could lift or lower the upper part of the chamber in about 60 seconds. Four studs (G), threaded at the ends, guided the upper part of the chamber to the lower plate. During the experiments, four large nuts (L) threaded onto the studs held the two parts of the chamber together in addition to the hydraulic jacks.

All manipulations of the vacuum and gas systems were performed remotely. Figure 3 shows the principal of operation of the chamber. During the experiment, the chamber was first filled with a stoichiometric mixture of hydrogen and oxygen to a total pressure of 2.8 MPa. The gases were then detonated by exploding a small nickel wire located at the centre of the hemisphere (Figure 3a) through a capacitor discharge. A detonation wave was created that propagated outward to the PETN explosive liner [19, 20]. The explosive was instantly and simultaneously detonated (Figure 3b) upon reflection of the detonation wave and a strong, stable, focused implosion wave converged inwards and was reflected at the centre as a shock wave (Figure 3c), leaving behind a high-pressure and high-temperature region (Figure 3d). The implosion also produced a complex shock-wave rarefaction-wave system in the sample holder containing the material.

The flash-heating system consisted of a 160 μF capacitor bank connected in series to the sample by means of a spark gap and a low inductance/low-loss transmission line. The capacitors could be charged to a maximum 20 kV and held then an energy of 32 kJ. The capacitors could be discharged in 250 μs (90% of their energy was discharged within 50-100 μs [2].

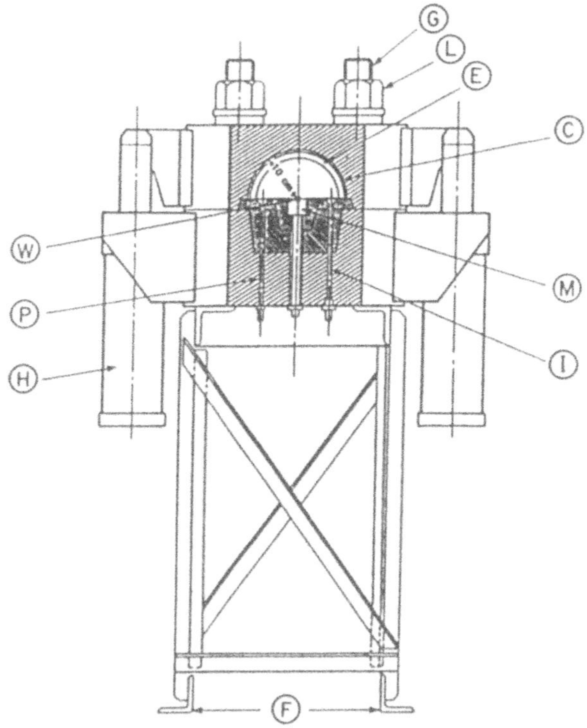

Figure 2. Schematic view of an explosion-driven 20 cm diameter hemispherical implosion chamber. E-PETN explosive; F-Support frame; G-Guide studs; H-Hydraulic jacks; I-Ignition wire; L-Lock nuts; M-Sample material holder; P-High-pressure gas inlet; W-Exploding wire; C-Copper liner

The peak current through the sample could reach 100 kA at 20 kV. A wide range of P,T, conditions was covered in experiments with the Flash-Heating Hemispherical Implosion System (FHHIS).

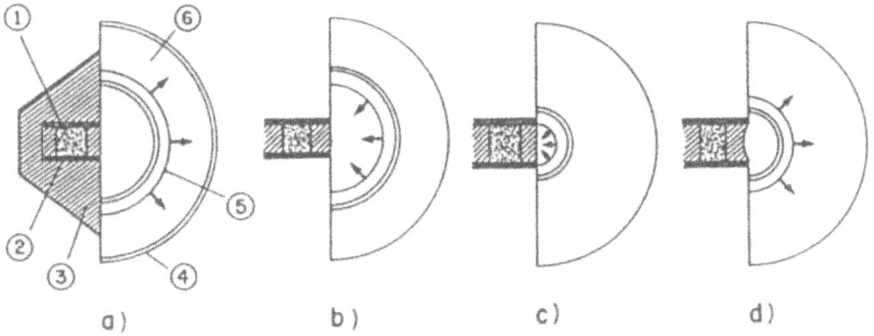

Figure 3. Schematic illustration of the principle of operation of the hemispherical implosion chamber. (a) Outgoing detonation wave of oxygen/hydrogen mixture: (1) sample material; (2) sample holder with plugs; (3) supporting crucible; (4) copper shell lined with PETN explosive; (5) detonation front; (6) H_2-O_2 mixture. (b) Ignition of explosive material. (c) Converging shock wave. (d) Reflection of shock wave at upper plug surface.

In experiments with shock compaction of diamond-powder mixtures, an impacting iron-flyer plate assembly was used in a set-up frequently called "mousetrap" assembly. The assembly was placed in the momentum trap type recovery fixture illustrated in Figure 4. The plane shock wave is first introduced into the sample capsule by impact of the 3.2 mm-thick iron flyer plate onto one end of the assembly at the velocity of 1.6 km/s or 2.3 km/s. The shock pressure in this experiment was calculated using the two-dimensional AUTODYNE 2D computer code, which applies the Eulerian processor [21]. After the shock treatment, the recovered stainless steel capsule was cut along its central axis with a silicon carbide blade, and then washed in an acetone bath with ultrasonic vibration in order to remove any contamination. The phases present in the shocked samples and the morphologies of the constituent particles were examined by high resolution electron microscopy (HREM). For examination by HREM, the powders were simply scraped from the shock compact onto a holey carbon film mounted on Cu HREM grids. In this manner, very fine particles can be easily sampled from the compact in a narrow area on the HREM grid without any damage. For the original powder, HREM grids were rubbed into the powder and the excess powder was blown off.

In a different series of experiments, formation of carbynes at dynamic shock loadings was studied by rapid sublimation and condensation of amorphous carbon black by means of a cylindrically converging shock wave technique. Structural details of individual carbynes and graphite particles produced in the shocked sample powders were examined using transmission electron microscopy. Moreover, numerical simulation which defines the shock compression conditions within the powder compacts was carried out.

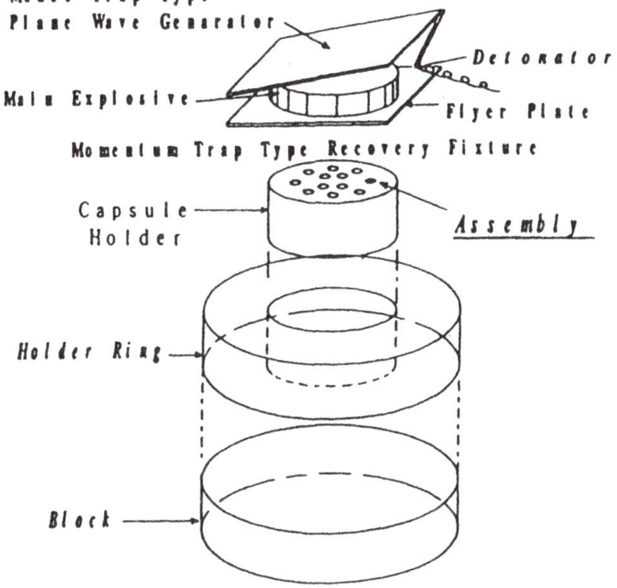

Figure 4. Diagram of a mousetrap plane-wave generator and a sample recovery system.

Experiments were performed with amorphous acetylene black starting material (>99% pure, 45 nm average particle size, supplied by the Denki Kagaku Co. Ltd., Japan), packed into a sample chamber of a cylindrical container, as shown in Fig. 5, to form a cylinder 40 mm in height and 10 mm in diameter. The initial density of the sample was 0.061 g/cm^3; about 3% of the density was determined by the pycnometer method. High explosive, trimethylene trinitramine (80 g) was used. The containers were recovered after the explosive loading and the shocked samples were removed from the container using a lathe.

Figure 6 shows a vertical section of the shocked container. The substance subjected to the high thermal pressure expanded adiabatically. As a result a conical cavity was formed along the central axis of the container. Different forms of crystalline particles of carbon were observed on the inner wall of the cavity. Most of them were sheet-like or spherical particles. Analysis of diffraction patterns from these particles confirmed their graphitic nature. Spherical particles were found to be composed of concentric graphitic layers parallel to the surface of the particle [5].

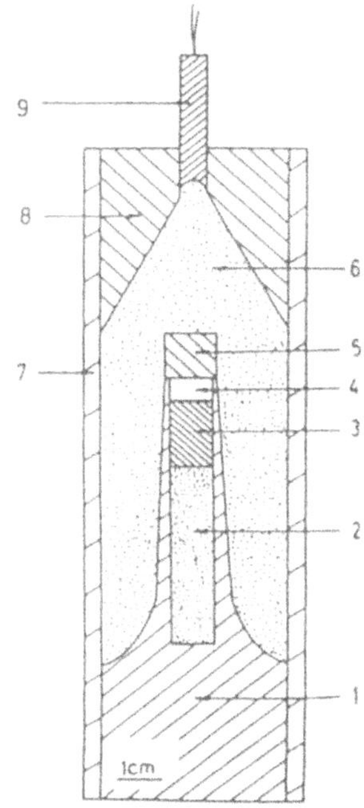

Figure 5. Schematic cross-sectional view apparatus (after [5]); 1-sample container of steel; 2-unshocked acetylene black powder; 3-plug of steel; 4-free space; 5-thick disk of brass; 6-explosive charge; 7-plastic tube of polyvinyl chloride; 8-wooden plug; 9-electric detonator

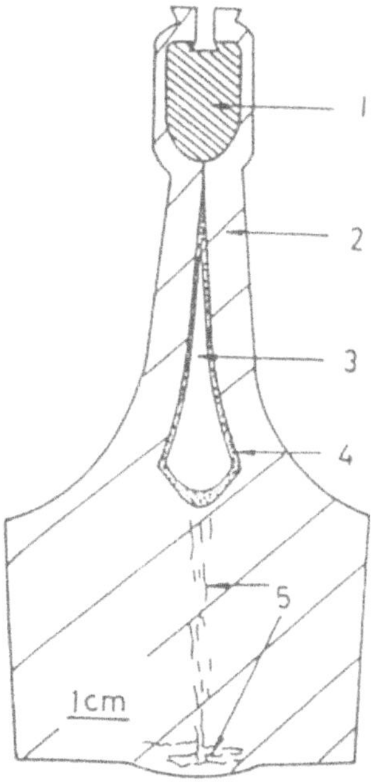

Figure 6. Vertical cross-section of the shocked container (after [5]); 1-steel plug; 2-sample container of steel; 3-conical cavity; 4-amorphous acetylene black; 5-fractures.

3. Experimental Results

In experiments with the FHHIS several polymorphic carbon forms (diamond, “tetragonally” crystallized diamond [2]), and differently ordered structural variants of graphite (turbostratic graphite or non-graphite carbon [22]) were identified.

Besides these compounds, many grains were found in the shock-reacted graphite, closely resembling those reported in the literature and ascribed to the carbyne family. Several grains were investigated with a scanning transmission electron microscope (STEM). Figures 7 and 8 show two representative grains. They have almost identical diffraction patterns (see Tables I and II) which reveal the strongest spacing of chaoite (“white carbon”) with l=0 [23]. Their EDX (energy-dispersive X-ray analysis) patterns, however, are quite different. In Figure 7c, the only element present is copper, which can be ascribed to the mounting grid material.

During the analysis of different parts of the grain in Figure 8, it was found that EDX patterns were varying (Figures 8c and d), while the diffraction pattern remained always the same (Figure 8b).

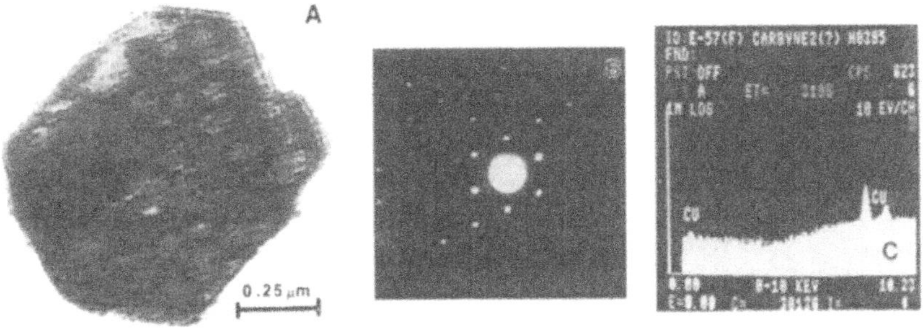

Figure 7. Scanning transmission electron microscopy (STEM) analysis of a grain with a chaoite-like diffraction pattern. (a) Transmission electron micrograph; (b) EDX analysis; (c) electron diffraction pattern

TABLE I. Electron diffraction data as obtained from the grain shown in Fig. 7

d (nm) observed	0.445	0.2597	0.2225	0.1679	0.1486	0.1333
d (nm) calculated	0.447	0.2578	0.2232	0.1688	0.1488	0.1289

The presence of different elements found in the EDX analyses of some grains with carbyne-type diffraction patterns could therefore be the result of a contamination by various elements. Moreover, contamination of electron microscope grids with airborne particles such as talc or asbestos often abundant in laboratory environment, were claimed to contribute to diverging analytical results [24].

TABLE II. Electron diffraction data as obtained from the grain shown in Fig. 8.

d (nm) observed	0.445	0.254	0.2209	0.1676	0.148	0.1286	0.112
d (nm) calculated	0.447	0.2578	0.2232	0.1688	0.1488	0.128	90.1116

In a different study an investigation was conducted on the microstructure of carbyne crystals formed from carbon materials (diamond and pyrolytic graphite) by two types of shock-wave techniques described in the preceding section. Based on the study, mechanisms for shock-induced transformation of the carbon materials to carbyne were developed from information obtained from the microstructure of carbyne crystals.

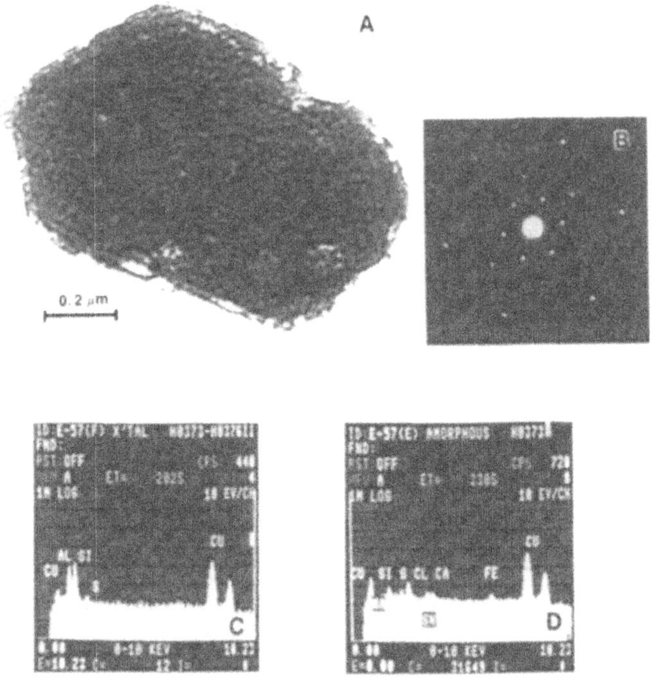

Figure 8. STEM analysis of a grain with chaoite-like diffraction pattern. (a) Transmission electron micrograph; (b) electron diffraction pattern; (c, d) EDX analysis of two different parts of the grain.

Fine particles of carbyne were found on the surface of diamond particles in the maximum pressure region (Figure 9) using the rod-in-cylinder method described above (Figures 5 and 6). The maximum pressures of 56 GPa occurred in the place where the oblique shock waves met.

The electron diffraction pattern (see insert in Figure 9a) from the particles showed that the d-spacings are in fairly good agreement with the reported values of the α-modification of carbyne [23, 25]. The evaluated lattice parameters were a_0 = 892 pm and c_0 =1536 pm. The HREM image in Fig. 9(b) reveals that the particles have facets and that the particle size is in the range of 2-8 nm.

In a series of experiments, an oriented pyrolytic graphite cube (purity: commercial grade) of side 5 mm was shock-compressed by a planar shock wave created by the mousetrap technique (see Figure 4). The sample was annealed in an argon flow at 3300 K for 1hr in a carbon resistance furnace and placed in a copper capsule. In this case, the carbon layers in the sample were perpendicular to the capsule axis.

Subsequently, the copper capsule was placed inside a cylinder of 2024 aluminum alloy. A 3.2 mm thick iron flyer-plate explosively driven at 2.0 km/s impacted onto the sample holder. A maximum pressure of 15 GPa was calculated [21]. All carbyne particles produced in this experiment were platy single crystals (Fig. 10a), in the size range of 2-15 μm. The corresponding electron diffraction pattern (see insert in Fig. 10a) represents the basal plane of the reciprocal lattice of a crystal of the hexagonal system. The calculated interplanar spacings are given in Table III. The a_0-value of the hexagonal cell parameter was found to be 893 pm and the plate-like particle was identified as α-carbyne. Figure 10b depicts a HREM image of the particle shown in Figure 10a whose lattice fringes have a spacing of 447 pm. If the above α-carbyne particles were formed by regraphitization of shock-formed polycrystalline carbon allotropes, they must be polycrystalline.

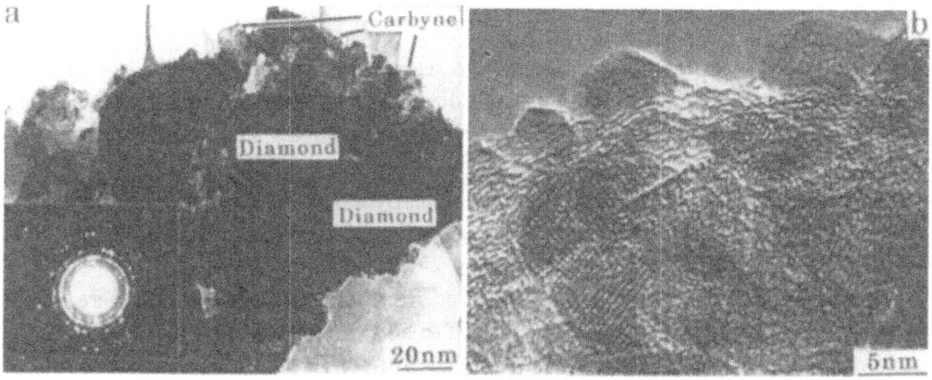

Figure 9. (a) Low-magnification HREM image of diamond surrounded by α-carbyne. The insert shows the electron diffraction pattern; (b) lattice fringes image of (a).

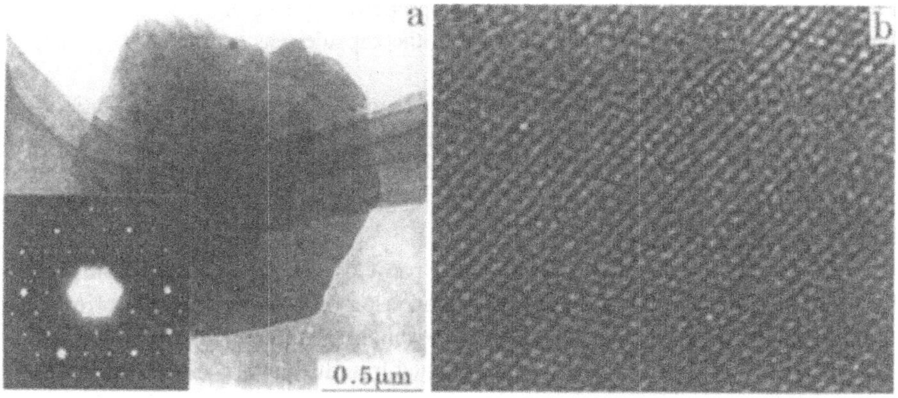

Figure 10. (a) Low-magnification HREM image of a platelet particle of α-carbyne. The insert shows the electron diffraction pattern; (b) lattice image of (0001) plane of (a).

TABLE III. Electron diffraction data for α-carbyne along the [0001] direction.

Observed interplanar spacings d (nm)		
hkil	Present Work	Ref. 5
1120	0.4470	0.4460
3030	0.2570	0.2560
2240	0.2235	0.2230
4150	0.1687	0.1686
3360	0.1488	0.1489
0660	0.1290	0.1288
2570	0.1240	0.1237

In addition to the carbyne particles, the occurrence of several other polymorphic carbon forms such as lonsdaleite, i-carbon, diamond, *etc.* was confirmed. These carbon allotropes were irregularly shaped polycrystalline particles consisting of grains of several to several tens of nanometers in size.

Additional evidence of the formation of carbyne-like structures was obtained in shock-compression experiments in a "mousetrap" type explosive plane-wave generator [3] similar to one shown in Fig. 4.

Specimens of glassy carbon were prepared from furfural resins, which were placed in a silica-glass boat, heated in N_2 at 5°C/min to 500°C and held for 3 hrs. The resulting carbon precursors were again heat-treated at 1600°C for 30 min in a graphite boat.

The glassy carbon obtained was ground and passed through a 300-mesh sieve. The powders were mixed with 300-mesh Cu powders (weight ratio 4:96) and were mechanically compacted in a stainless steel capsule at 392 MPa to form a disk 4.0 mm thick and 12 mm in diameter with a density of 6.02 g/cm^3.

The disk was shock-compressed by an explosive "mousetrap" plane-wave generator. An iron projectile plate (3.2 mm thick) was propelled at a velocity of 3.6 km/s against the capsule. The impact pressure induced in the capsule was estimated to be 100 GPa using the impedance matching method.

The shock-compressed disk specimen was immersed in dilute HNO_3 solution for 24 h to dissolve the copper matrix. The remaining black powder was washed with distilled water and alcohol, then oxidized for several hours with a hot solution of concentrated HNO_3 containing a small amount of sodium chlorate to dissolve unconverted carbon. Dark silver-grey powders were obtained.

The diffraction patterns of the obtained particles were examined by transmission electron diffraction. Many of the particles gave ring diffraction patterns corresponding to cubic diamond reflection. Occasionally, hexagonal symmetry patterns, consisting of spots, were also obtained. The d-values obtained from the pattern are shown in Table IV along with those of the chaoite reported by Whittaker and Kintner [23]. These values indicated close agreement with the reflection l = 0 produced by chaoite.

TABLE IV. Electron diffraction data for chaoite along [00 1] zone direction.

Observed interplanar spacings d (nm)	
Present Work	Ref. 23
0.4464	0.4467
0.2560	0.2571
0.2294	0.2234
0.1710	0.1681
0.1530	0.1496
0.1256	0.1283
0.1240	0.1238
0.1132	0.1118
0.1047	0.1080
0.0978	0.0975
0.0879	0.0899
0.0856	0.0865

The average value obtained for the hexagonal cell dimension a_0 was 893,7 ± 4 pm. The value was within the experimental error of the reported value, a_0 = 894,8 ± 1 or 893,7 ± 3,5 pm. The hexagonal symmetry shows that the electron beam was parallel to the c axis of chaoite.

Unexpectedly, a pattern corresponding to a zone direction other than [00.1] was found. The rectangular array was indexed as chaoite along the [12.0] zone direction. The d-values observed from the pattern and corresponding indices are given in Table V. The agreement between the observed and reported values of chaoite is within 1% experimental error.

TABLE V. Electron diffraction data for chaoite along [12.0] zone direction.

	Observed interplanar spacings d (nm)	
hkl	Present Work	Ref. 23
002	0.4470	0.4460
210	0.2570	0.2560
212	0.2235	0.2230
004	0.1687	0.1686
214	0.1488	0.1489
006	0.1290	0.1288
216	0.1240	0.1237

4. Role of Carbynes in Shock-Induced Phase Transformation in Carbon

In a basic study aimed at resolving the role of carbynes in the transformation mechanisms of graphite-to-diamond [26], carbyne precursor samples were treated with shock waves of various intensities. The carbyne samples used in the investigation were prepared according to a procedure similar to the Shirakawa method [27] since the trans-elimination of HCl should result in only polyyne- but not in polycumulene-type chain segments. The latter are thought to be less useful for low energy transformation routes. Results of these experiments can be found in Chapter 7.1. of this book.

Additional evidence of diamond formation in these shock experiments was obtained from an EELS study of the materials before and after shock treatment. Figure 11 shows the plasmon energy range of two EELS spectra from an unshocked carbyne precursor (a) and shocked to 5 GPa (b).

For comparison, spectra of pyrolytic graphite (c) and diamond powder (d) in the same regions are shown as well.

To account for the shift in the position of the σ-plasmon in the EELS spectra, it was proposed that the carbyne chains include a mix of sp-sp^2 hybridized atoms (see Chapter 6.4.). The co-existence of carbon atoms in different types of hybridization was proposed for a number of structures. The presence of sp^3-sp layer-chain carbons was suggested in a study on the behavior of the abnormal reflection of d = 336 pm in a number of carbon substances [28]. It is also established that the fullerenes are built from a mix of sp^2-sp^3 carbon atoms.

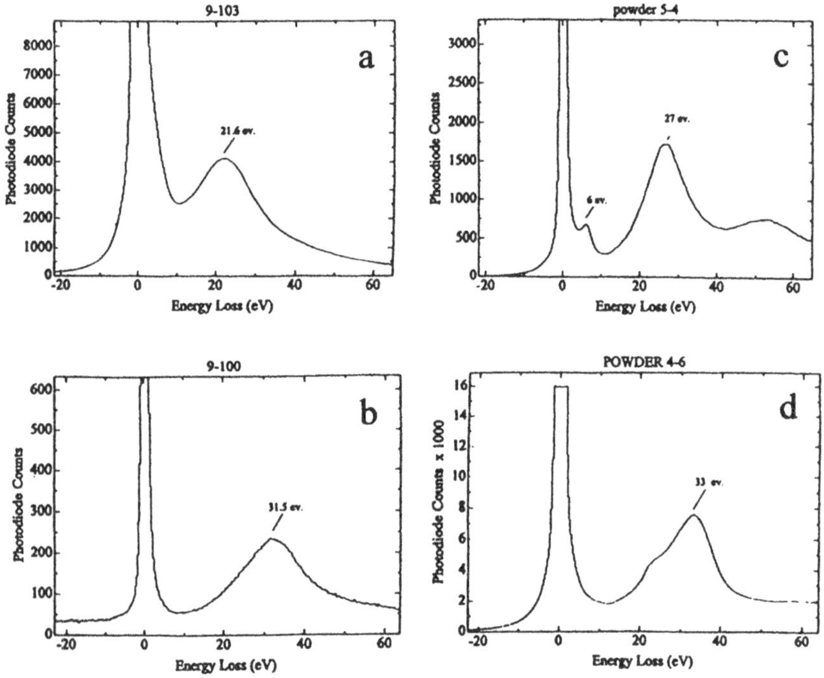

Figure 11. EELS spectra of the low energy plasmon region of the carbyne sample before shock treatment (a) and after a 5 GPa shock treatment (b). (c) - EELS of a pyrolytic graphite sample; (d) - EELS of a 2 μm diamond powder sample.

The formation of carbyne has been confirmed in studies on intercalation of metals into polyacetylene under quasi-hydrostatic pressure of 20-40 kbar [29], and in experiments on intercalation of carbyne prepared by chemical dehydrochlorination of poly(vinylidene chloride) (PVDC) powder mixed with potassium hydroxide in ethanol and THF [30] using a procedure reported in [31].

Adapting the kinked-chain model [32] to the results of the analysis conducted on the products of intercalation, a structure of linear carbon chains with intercalated Na atoms (CIC-Na) was proposed to explain the observed diffraction pattern shown in Table VI.

The length of the straight line part of the carbyne molecule was established to be 690 pm, corresponding to a five-atom chain (see Figure 7 of Chapter 4.3.).

Combining the kink-chain model [32] with the possible presence of intercalated metal atoms [29, 30] and the presence of sp-sp^2 atoms in such a structure, an "intuitive" unit cell of a carbyne material was proposed recently (see Chapter 6.4.).

TABLE VI. Interplanar distances corresponding to hk0 reflexes of a single crystal CIC-Na [29, 30].

hk0	dexp	dcalc
100	4.544	4.539
110	2.630	2.621
20	2.263	2.269
210	1.716	1.715
300	1.515	1.513
220	1.312	1.310
310	1.263	1.259
400	1.134	1.135
320	1.041	1.041
410	0.989	0.990
500	0.909	0.907
330	0.870	0.872
420	0.861	0.857
510	0.813	0.814
600	0.753	0.754
430	0.748	0.745
520	0.724	0.727
610	0.688	0.690

5. Conclusions

During study of the dynamic synthesis of carbyne from diamond powder and oriented pyrolytic graphite, the following informations were obtained:

When diamond powder was shock-compressed by a strong oblique shock wave, polycrystalline particles of α-carbyne and chaoite, respectively were formed by condensation of gaseous carbon. On the other hand, when the oriented pyrolytic graphite was shock-compressed by a weak planar shock wave parallel to the basal plane, occurrence of single-crystals of α-carbyne was confirmed whose origin was inferred to be due to Whittaker's formation mechanism. This mechanism involves bond-splitting within a graphene layer to form either conjugated triple-bonded carbon chains [10.0] [14] or cumulated double-bonded carbon chains parallel [11.0] [32].

6. References

1. Litvinova, V.A. and Cherkinskaya, K.T.: Phase transformations in pyrographite and glassy carbon on shock impact, *Khim. Tverd. Topliva* **3** (1976), 159-164.
2. Kleiman, J.I., Heimann, R.B., Hawken, D. and Salansky, N.M.: Shock compression and flach heating of graphite/metal mixtures at temperatures up to 3200 K and pressures up to 25 GPa, *J. Appl. Phys.* **56** (1984), 1440-1454.
3. Setaka, N. and Sekikawa, Y.: Chaoite, a new allotropic form of carbon, produced by shock compression, *J. Amer. Ceram. Soc.* **63** (1980), 238-239.
4. Sekine, T., Akashi, M., Setaka, N. and Kondo, K.: *J. Mater. Sci.* **22** (1987), 159.
5. Yamada, K., Kunishige, H. and Sawaoka, A.B.: Formation process of carbyne produced by shock compression, *Naturwiss.* **78** (1991,) 450.
6. El. Goresy, A. and Donnay, G.: A new allotropic form of carbon from the Ries crater, *Science* **161** (1968), 363-364.
7. Vdovikin, G.P.: Forms of carbon in the new Havero Ureilite of Finland, *Meteoritics* **7** (1972), 547-552.
8. Hayatsu, R., Scott, R.G., Studier, N.H., Lewis, R.S. and Anders, E.: Carbynes in meteorites, detection, low-temperature origin, and implications for interstellar molecules, *Science* **209** (1980), 1515- 1518
9. Whittaker, A.G, Watts, E.J., Lewis, R.S. and Anders, E.: Carbynes: carriers of primordial noble gases in meteorites, *Science* **209** (1980), 1512-1514.
10. Heimann, R.B. and Kleiman, J.I.: Shock-Induced Growth of Superhard Materials, in: Crystals II, Springer-Verlag Berlin, Heidelberg, (1988) 1-74.
11. Bundy, F.P., Basset, W.A., Weathers, M.S., Hemley, R.J., Maoand, H.K. and Goncharov, A.F.: The pressure-temperature phase and transformation diagram for carbon: updated through 1994, *Carbon* **34(2)** (1996), 141-153.
12. Deriagin, B.V., Fedoseev, D.V., Varnin, V.P. and Vnukov, S.P.: The nature of metastable phases of carbon, *Nature* **269** (1977), 398-399.
13. Kasatochkin, V.I., Savranskii, V.V., Smirnov, B.N. and Mel'nichenko, V.M.: An investigation of carbyne condensed from carbon vapours, *Dokl. Akad. Nauk SSSR* **217(4)** (1974), 796-799.
14. Whittaker, A.G.: *Science* **200** (1978), 763.
15. Whittaker, A.G. and Kintner, P.L.: Carbon: analysis of spherules and splats formed from the liquid state and of the forms produced by quenching gas and solid, *Carbon* **23(3)** (1985), 255-262.
16. Whittaker, A.G. and Wolten, G.M.: Carbon: a suggested new hexagonal crystal form, *Science* **178** (1972), 54-56.
17. Whittaker, A.G., Kintner, P.L., Nelson, L.S. and Richardson, N.: Aerospace Corporation Report SD-TR-81-60, NTIS No. A17106134, Sept.1/81
18. Flagg, R.F. and Glass, I.I., *Phys. Fluids* **11** (1968), 2282.
19. Glass, I.I.: Progress in *Aerospace Science* **13** (1972), 223.
20. Glass, I.I. and Sharma, S.P.: *AIAA J.* **14** (1976), 402.
21. AUTODYN, Compendium of Papers Rev. 2, pp. 1-6, Century Dynamic Inc., Oakland CA (1991).
22. Mering, J.: *Acta Cryst.* **2** (1949), 371.
23. Whittaker, A.G. and Kintner, P.L.: Carbon: observation on the new allotropic form, *Science* **165** (1969), 589-561.

24. Smith, P.P.K. and Buseck, P.R.: Carbyne forms of carbon: do they exist?, *Science* **216** (1982), 984-986.
25. Kasatochkin, V.I., Korshak, V.V., Kudryavisev, Yu.P., Sladkov, A.M. and Sterenberg, I.E.: On crystalline structure of carbyne, *Carbon* **11**(1973), 70-72.
26. Heimann, R.B., Fujiwara, S., Kakudate, Y., Koga, Y., Komatsu, T. and Nomura, M.: A new carbon form obtained by ewear shock compression of carbyne, *Carbon* **33(6)** (1995), 859-863.
27. Akagi, K., Nishigushi, M., Shirakawa, H., Furukawa Y. and Harada, I.: *Synth. Metals* **17** (1987), 557.
28. Mel'nichenko, V.M., Nikulin, Yu.N. and Sladkov, A.M.: Layer-chain carbons, *Carbon* **23**(1) (1985) 3-7.
29. Udod, I.A., Shchurik, V.I., Bulychev, B.M., Sizotinkin, S.P., Guseva, M.B., Babaev, V.G., Kudryavtsev, Yu.P. and Evsyukov, S.E.: Formation of carbyne in the interaction of polyacetylene with potassium under high quasi-hydrostatic pressure, *J. Mat. Chem.* **3(4)** (1993), 413-416.
30. Udod, I.A., Bulychev, B.M., Sirstinkin, S.P., Babaev, V.G. and Kudryavtsev, Yu.P.: An interaction of amorphous carbyne with sodium: the formation of carbyne intercalation compounds, *Synthetic Metals* **60** (1993), 57-62.
31. Korshak, V.V., Kudryavtsev, Yu.P., Korshak, Yu.V., Evsyukov, S.E., Khvostov, V.V., Babaev, V.G. and Guseva, M.B.: Macromol. Chem., *Rapid Commun.* **9** (1988), 135.
32. Heimann, R.B., Kleiman, J.I. and Salansky, N.M.: Structural aspects and confirmation of linear carbon polytypes (carbynes), *Carbon* **22** (1984), 147-155.

3.5. ELECTROCHEMICAL METHODS

L. KAVAN
J. Heyrovský Institute of Physical Chemistry,
Dolejškova 3, CZ-182 23 Prague 8, Czech Republic

Abstract. Electrochemical carbonization is a useful synthetic strategy towards unusual carbon structures including carbyne-like materials. The best products were prepared by cathodic defluorination of poly(tetrafluoroethylene) and some other perhalo-n-alkanes. Owing to high overpotentials of electrocarbonization, classical electrochemical processes in liquid media are sensitive to breakdown reactions which are triggered by protic impurities in electrolyte solutions. This problem is avoided if the reaction is carried out in a solid-state cell with alkali metal amalgam anode. Electrochemically synthesized carbyne-like materials exhibit spontaneous n-doping which may stabilize the carbyne structure by electrostatic charging and inserting of counterions between the *sp*-carbon chains. Although no defined "electrochemical carbyne" was prepared yet, electrosynthesis of a reasonably pure elemental carbon containing some amount of *sp*-bonded (carbynoid) moieties has been convincingly demonstrated.

1. Introduction

Electrochemical methods are essential in large-scale production of certain elements, e.g. halogens, alkali metals and aluminium. Carbon is a useful electrode material in electrochemistry [1-4], but electrochemical synthesis of elemental carbon [5,6] is still considered as curiosity among the common methods which are mostly based on pyrolysis of organic precursors. Nevertheless, electrochemistry may offer interesting pathways towards peculiar carbons, such as carbon nanotubes and onions [7-9] and, presumably, also carbyne [5,6,10]. The aim of this chapter is to scrutinize the salient issues of electrochemical synthesis of carbyne-like materials in general, while special attention will be paid to cathodic dehalogenation of polymers and application of alkali metal amalgams. Alternative electrosynthetic strategies, based on polycondenzation of low-molecular-weight carbon halides, are further discussed in more detail by Kijima and Shirakawa in chapter 3.1.2.

Elemental carbon is accessible either by anodic oxidation or by cathodic reduction of a suitable precursor, CY_n (Y denotes general substituent(s)) [5]:

$$CY_n \xrightarrow{\pm ne^-} C + nY^{\mp} \qquad (1)$$

R.B. Heimann et al. (eds.), Carbyne and Carbynoid Structures, 189–214.

Standard electrochemical potentials, E_0 of the reaction (1) equal:

$$E_0 = -\frac{\Delta G_0}{nF} \tag{2}$$

(G_0 is the Gibbs free energy, n the number of electrons and F the Faraday constant).

They can be calculated from the free energies of the CY_n formation assuming a hypothetical galvanic cell with a standard hydrogen electrode and the actual oxidation-reduction electrode corresponding to the reaction (1). Examples of E_0 values are collected in Table 1 for several typical precursors of electrochemical carbon. Table 1 demonstrates that some electrochemical carbonization reactions are thermodynamically very favored.

TABLE 1. Thermodynamic data on selected electrochemical carbonization processes at 25°C

Precursor	State	Products	$-\Delta G_0$ [kJ/mol]	E_0 [V]	Carbyne formation
$CH_3 - CH_3$	g	$C + 3\,H_2$	-7.86	-0.06	no data
$CH_2 = CH_2$	g	$C + 2\,H_2$	-68.17	-0.18	no data
$CH \equiv CH$	g	$C + H_2$	-209.2	-1.08	YES
CF_4	g	$C + 4\,HF$	236.65	0.56	no data
$CF_3 - CF_3$	g	$2\,C + 6\,HF$	459.57	0.79	YES
$(CF_2)_n$; PTFE	s	$2\,C + 4\,HF$	193.18	1.00	YES
CCl_4	l	$C + 4\,HCl$	459.77	1.19	NO
CO_2	g	$C + 2\,H_2O$	80.90	0.21	NO

For instance, perhalogenated hydrocarbons, e.g. CCl_4 or poly(tetrafluoroethylene) (PTFE) can be formally considered strong oxidants, and unsaturated hydrocarbons, e.g. acetylene, strong reductants (Table 1). However, practical redox carbonizations of these species are often slow and complicated by side reactions [5,10].

Table 2 surveys the electrochemical carbonization processes yielding carbyne-like materials. Since the relevant carbynoid products are often characterized by IR and/or Raman spectroscopy, Table 2 collects also the frequencies of the CC stretch characteristic for *sp*-carbon chains. For a more detailed treatment of IR and Raman spectra of carbyne see chapter 6.1.

2. General Aspects of Electrochemical Carbonization

The CY bond breaking, promoted by electron transfer (for a review see Ref. [31]) proceeds already at low (room) temperatures. Consequently, the structure and properties of electrochemical carbons may differ considerably from those of usual pyrolytic carbons [5,6]. The low-temperature electrochemical carbonization seems to be favorable also to produce carbyne-like materials. Various chemical and electrochemical pathways towards carbyne have been briefly reviewed by Kijima [10].

TABLE 2. Examples of electrochemical carbonization processes yielding carbyne-like materials.

Precursor	Working electrode	Potential [V] (Ref. electrode)	Electrolyte (solution)	ν (CC), cm^{-1}	Refs.
(Anodic processes)					
H-C≡C-H	Pt	1.5 (SCE)	$LiClO_4$ + DMF	2116 (IR)	[11]
(Cathodic processes)					
PTFE	GC, Fe	-2.6 (SCE)	NBu_4ClO_4 + DMF	2170 (IR)	[12-15]
PTFE	Cu-composite	1 (Li/Li^+)	NBu_4ClO_4 + THF	2177 (Ra)	
				2059 (IR)	[16]
PTFE	Mg	-	LiCl + $FeCl_2$ + THF	2100-2200 (IR, Ra)	[17]
PCTFE	C-composite	1.5 (Li/Li^+)	$LiBF_4$ + γ-butyrolactone	1750 (IR)	[18,19]
PVDF	Pt	galvanost.	NBu_4ClO_4 + DMF	2116 (IR)	[20]
I-C≡C-I	Pt	-2 (SCE)	NBu_4I + DMF	2050 (IR)	
				2150 (Ra)	[21,22]
HCBD	Pt	-2 (SCE)	NBu_4ClO_4 + DMF	1979 (IR)	
(Amalgam/cathodic processes)					
PTFE	C-as grown	0.91[a] (Li/Li^+)	C-MF (solid)	[b]1996-2045 (Ra)	[25,26]
[c] C_xF_{2x+2}	*ditto*	*ditto*	*ditto*	[b]2080-2122 (Ra)	[26,27]
PCTFE	*ditto*	*ditto*	*ditto*	[b]2074-2110 (Ra)	[26]
FEP	*ditto*	*ditto*	*ditto*	2100 (IR)	[28-30]
				2050 (Ra)	[26]

a: thermodynamic value for Li-amalgam [5]
b: values vary depending on the alkali metal type (M)
c: x = 6, 9, 20, 24; Raman frequencies vary both with x and M

Bu: butyl
DMF: dimethylformamide
FEP: poly(tetrafluoroethylene-*co*-hexafluoropropene)
GC: glass-like carbon
HCBD: hexachlorobuta-1,3-diene
PCTFE: poly(chlorotrifluoroethylene)
PTFE: poly(tetrafluoroethylene)
PVDF: polyvinylidene fluoride
SCE: saturated calomel electrode
THF: tetrahydrofuran

Electrons in reaction (1) are supplied either through redox active molecules (chemical carbonization) or through an electrode, which contacts an ionically conducting solid or liquid phase and the precursor (electrochemical carbonization). A sharp distinction between "chemical" and "electrochemical" processes cannot be drawn. This conclusion is supported by two principal arguments:

1. In certain cases, the carbonization might proceed *via* electrochemical generation of reactants, typically organic radical anions. These radical anions may have a sufficiently negative redox potential, that they can, subsequently, convert CY_n to elemental carbon by a chemical reaction (*ec* mechanism).

2. Some reactions of CY_n, normally regarded as chemical processes, are, in fact, electrochemical ones. This situation typically occurs if the precursor, CY_n and the second reactant are isolated from a direct "chemical contact" by a solid layer of carbonaceous products.

The fact that some virtually "chemical carbonizations" proceed *via* an electrochemical mechanism, has been first demonstrated by Jansta and Dousek [32-34] for the reduction of poly(tetrafluoroethylene) (PTFE) with liquid alkali metal amalgams, M(Hg), M = Li, Na, K. Since this reaction also seems to have a unique position in the synthesis of carbyne-like materials [5,25,35], it will be discussed in more detail in the next section.

2.1. ELECTROCARBONIZATIONS BY ALKALI METAL AMALGAMS

If a carefully outgased PTFE foil is contacted with liquid alkali metal amalgams, M(Hg) in vacuum, a spontaneous blackening of the PTFE surface occurs. This blackening is apparent already at room temperature, while its rate increases considerably at elevated temperatures. The reaction yields a fully defluorinated polymeric carbon in a mixture with a stoichiometric amount of alkali metal fluoride (MF, MF/C = 2) [32-34].

The reaction is initiated by a fast chemical reduction, which produces the first monolayers of C-MF. The C-MF layer is compact, non-porous and impermeable to the amalgam. However, it exhibits both electronic [36-38] and ionic (M^+) [39] conductivities, i.e. a short-circuited galvanic cell is formed:

$$M(Hg) \mid C\text{-}MF \mid PTFE$$

in which the C-MF layer acts simultaneously as a solid electrolyte, electron conductor and electrode. During the self-discharge, the PTFE carbonization takes place at the carbon (C-MF) cathode, which is continuously renewed *in-situ* as the C-MF layer grows. The cathodic process consists of splitting of C-F bonds, while the fluoride anions formed subsequently recombine with M^+ cations that also passed through the C-MF layer.

If we neglect the effects of n-doping (see section 3.5. - 5), the cell reaction is:

$$(\text{-}CF_2\text{-}) + 2\ M(Hg)_x \rightarrow (\text{-}C\text{-}) + 2\ MF + 2x\ Hg \quad (3)$$

The thermodynamic cell voltage, ΔE equals:

$$\Delta E = \frac{\Delta G_{PTFE} - 2\Delta G_{MF} + 2F(E_a - E_0) + 2RT\ln c}{2F} \quad (4)$$

(E_0 denotes standard redox potential of alkali metal (M/M^+), E_a standard redox potential of the amalgam, $M(Hg)/M^+$ and c amalgam concentration). Eq. 4 translates (for E in Volts and M = Li) into:

$$E = 3.251 + \frac{RT}{F}\ln c \quad (5)$$

The current (I) flowing through the cell (cf. Fig. 1) can be expressed according to Faraday's law as:

$$I = \frac{2S\,L_0\,\rho F}{t\,M} \quad (6)$$

where S is the area of the PTFE/amalgam interface, L_0 is the thickness of carbonized PTFE layer, ρ is the density of PTFE, t is the time, M is the molar mass of CF_2 unit and $a = L/L_0$ (L is the thickness of the formed C-MF layer). The current I obeys Ohm's law, assuming a series combination of ionic (σ_i) and electronic (σ_e) conductivities of C-MF (cf. Fig. 1):

$$I = \frac{S\,\Delta E\,\sigma_e\,\sigma_i}{L(\sigma_e + \sigma_i)} \quad (7)$$

Combining Eqs. 6 and 7 leads to the kinetic equation:

$$L_0 = \sqrt{\frac{\Delta E\,M\sigma_e\sigma_i\,t}{2aF\rho(\sigma_e + \sigma_i)}} = k \cdot t^{1/2} \quad (8)$$

where k is a rate constant. The experimentally found values of the rate constants, k of fluoropolymers carbonized by amalgams are summarized in Table 3. For comparison, estimated values for some other carbonization processes are also shown. A more precise mathematical model, proposed by Micka *et al.* [30] considers (i) the chemical initialization to produce rapidly a thin layer L_1:

$$L_0 = L_1 + kt^{1/2} \tag{9}$$

and (ii) the electrical conductivity of C-MF to be a function of time [30]. However, differences between the two models (Eqs. 8 and 9) are beyond the usual errors of kinetic measurements. The depicted model of a spontaneously formed local galvanic cell resembles that of the corrosion of metals. Hence, the term "electrochemical corrosion of PTFE by alkali metal amalgams" also appears [32-34].

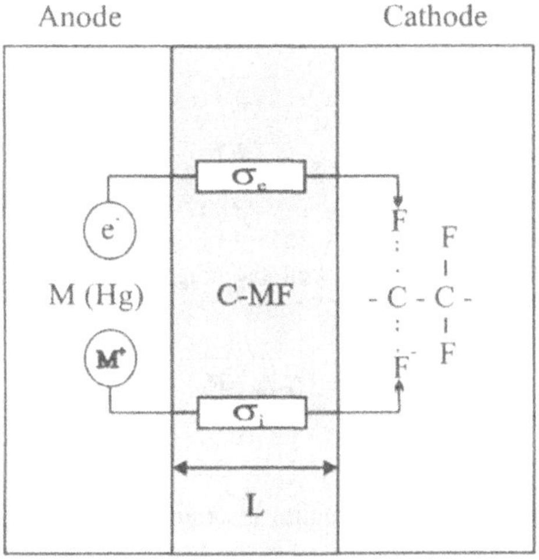

Figure 1. Scheme of the processes occurring at the interface between liquid alkali metal amalgam, M(Hg) and solid poly(tetrafluoroethylene).

The electrochemical scheme is applicable also for the carbonization by amalgams of various other fluoropolymers [5,39,43] and low-molecular-weight perhalogenated hydrocarbons [26,27] (cf. Tables 2 and 3). The kinetic model (Eqs. 4-9) is valid also for liquid or gaseous precursors. However, these systems are sensitive to accidental cracking of the C-MF layer; the precursor subsequently permeates through cracks which causes uncontrolled kinetics acceleration [5,26,27].

Compared to alternative electrochemical carbonizations, there are three main advantages of the amalgam carbonization:

(i) The defluorination is quantitative, independent of temperature (down to sub-room temperatures) [39], amalgam concentration and reaction time (L_0 values) [34,39].

(ii) Any common impurities (e.g. H, O from solvents, electrolytes or air) are safely excluded because the reaction proceeds through a "dry contact" of PTFE and amalgam, usually in a high vacuum.

(iii) The reduction rate is well controlled (Eq. 8) by only two variables: temperature and the type of alkali metal [34,39] (cf. Table 3).

TABLE 3. Electrochemical carbonization of fluoropolymers. Values of rate constant, k (Eq. 8)

Precursor	Reactant/ react. medium	Temperature [°C]	k [nm/s$^{1/2}$]	Refs.
PTFE	Li amalgam	10	27.6	[39]
PTFE	Li amalgam	25	47.5	[33,34]
PTFE	Li amalgam	50	109	[33,34]
PTFE	Li amalgam	100	352	[33,34]
PTFE	Na amalgam	25	2.7	[33,34]
PTFE	Na amalgam	50	3.7	[33,34]
PTFE	Na amalgam	100	17	[33,34]
PTFE	K amalgam	25	0.1	[39]
PTFE	K amalgam	50	0.2	[39]
PTFE	K amalgam	100	3.7	[39]
PTFE	Pt cathode/DMF+NEt_4BF_4	r.t.	120[a]	[40]
PTFE	Pt cathode/DMF+NBu_4BF_4	r.t.	70[a]	[40,41]
PTFE	Pt cathode/DMF+$NOct_4BF_4$	r.t.	30[a]	[40]
PTFE	benzoin dianion in Me_2SO	50	2[a]	[42]
PCTFE	Li amalgam	10	55	[39]
PCTFE	Li amalgam	25	124	[39]
PCTFE	Li amalgam	100	1550	[39]
Nafion® 117	Li amalgam	25	15.3	[43]
Nafion® 117	Li amalgam	25	115	[43]
PTFE-PPVE	Li amalgam	25	38.3	[39]
PTFE-PPVE	Li amalgam	100	360	[39]
FEP	Li amalgam	25	47.0	[28,30]
FEP	Na amalgam	25	2.7	[30]
FEP	K amalgam	25	0.1	[30]
FEP	sodium dihydronaphthylide/THF	0	0.4[a]	[44]

[a] Values estimated from literature experimental data; Me = methyl, Et = ethyl, Bu = n-butyl, Oct = n-octyl

PTFE-PPVE = poly(tetrafluoroethylene-co-perfluoropropylvinylether)
r.t. = room temperature
Nafion = sulfonated fluoropolymer from duPont (see Ref. [43] for a chemical formula)
(For explanation of the other symbols see Tab. 2)

The similarity between the PTFE/amalgam reaction (Fig. 1) and classical electrochemical reduction of PTFE at metal cathodes (section 3.5. - 2.2.) was first addressed by Barker *et al.* [41]. On the other hand, any "chemical" defluorination of PTFE must, in principle, proceed electrochemically as well (Fig. 1 and Eqs. 3-9) provided the C-MF is not permeable to the reactants. McCarthy *et al.* have discussed this mechanism for the reaction of PTFE with benzoin dianion [42] and poly(tetrafluoroethylene-co-hexafluoropropene) (FEP) with sodium dihydronaphthylide [44]. Their experimental L-t plots are similar to the corresponding plots for "PTFE/metal-cathode" [40,41] and "PTFE/amalgam" [33,34] which permits the rate constants *k* to be estimated (Table 3).

Generally, the electrochemical route of carbonization requires the existence of a compact and electrically conducting (both for electrons and ions) interfacial layer of the carbonaceous product. This situation occurs almost ideally at the interface of fluoropolymers with diluted alkali metal amalgams [5]. If the interfacial layer is mechanically damaged and the pores are transparent for the reactants, both electrochemical and chemical carbonization are at play.

2.2. ELECTROCHEMICAL CARBONIZATION IN LIQUID MEDIA

Electrocarbonization may also be realized by "classical" electrochemistry whereby the charge transfer reaction occurs at an interface of metal electrode and liquid electrolyte solution. Fig. 2 displays schematically the reductive carbonization of PTFE at a metal cathode [45].

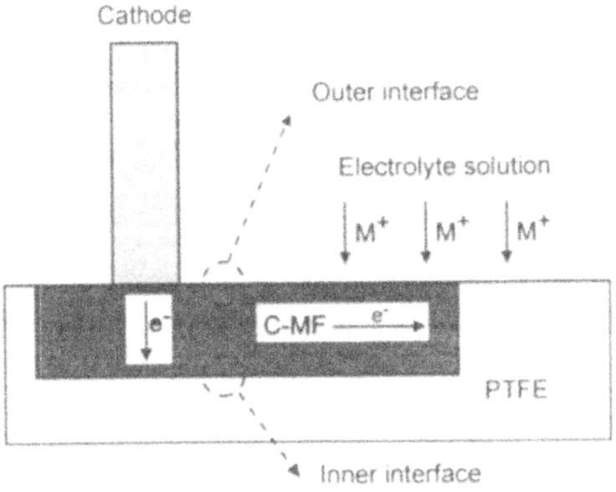

Figure 2. Scheme of the processes occurring at the contact of metal cathode with poly(tetrafluoroethylene) (PTFE) in a liquid electrolyte solution.

The reduction proceeds at a three-phase interface: PTFE/electronic conductor/ionic conductor. (Practical cells with large proportion of three-phase contacts are described in Ref. [45]). Electrons from the cathode propagate through the growing C-MF phase, while we can distinguish two interfaces:

(i) *Outer interface* is in contact with liquid electrolyte solution. The carbonized area propagates rapidly along this interface, since the transport of ions (M^+) in liquid phase is fast. Hence, the rate limiting process is the interfacial charge transfer reaction [40,46].

(ii) *Inner interface* is equivalent to the interface in the PTFE/amalgam reaction. The kinetics is controlled by the ionic conductivity, σ_i (Eq. 8), i.e. the measured rate constants, k, are expected to be similar for a given electrolyte cation, M^+ (cf. Table 3). If the electrolyte solution permeates accidental cracks in C-MF, the model (Eq. 8) is perturbed and k values increase randomly.

3. Anodic processes

Oxidative carbonization processes are mostly connected to early works on surface enhanced Raman spectroscopy on silver electrode [47]. The carbon layer is prepared by photoassisted oxidation of various organic precursors, while the products usually show graphite-like Raman spectra with no signs of carbyne-like ordering. Only occasionally, a weak (unintepreted) Raman band at *ca.* 2200 cm^{-1} [48] appears, which may be characteristic for *sp*-carbon links.

A more pronounced formation of carbyne *via* anodic oxidation was reported by Shirakawa et al. [11]. The oxidation of acetylene on a Pt anode at 1.5 V *vs.* SCE in the presence of CuCl and N,N,N',N'-tetramethylethylenediamine was interpreted as a polycondensation towards polyyne (cf. chapter 3.1):

$$\text{n H-C}\equiv\text{C-H} \xrightarrow[\text{Cu cat.}]{-2\text{n}(e^- + H^+)} (\text{-C}\equiv\text{C-})_n \qquad (10)$$

The product of reaction (10) shows a weak IR band supporting this interpretation ($\nu_{C\equiv C}$ at 2116 cm^{-1}), but its deeper characterization is still missing. Analogous catalytic polycondensation reactions of acetylene or butadiyne [49] (called Hay-, Glaser- or Eglinton reaction) [10,50] occur by the action of oxygen, [10,50,51] iodine, [52] and some other mild oxidants, [53-55]. The products are, however, poorly defined.

4. Cathodic processes

Cathodic reactions are more frequently used to produce carbyne-like materials. They are based either on cathodic dehalogenation of perhalo-n-alkanes or dehydrohalogenation of poly(vinylidene halides). The electrochemical carbon-halogen bond fission leads primarily to a halide anion and a radical intermediate which triggers subsequent protonization, dimerization, coupling, nucleophilic attack, etc. (cf. Scheme 1 in chapter 3.1.2 [31,56]). Even a very weak acid can serve as source of protons, e.g., dimethylformamide or acetonitrile, which are thus converted to $CON(CH_3)_2^-$ or CH_2CN^- [57].

For production of "electrochemical carbyne", the intermediate protonization is detrimental. It blocks further carbonization, since hydrocarbons are no more reducible to elemental carbon. (Reductive splitting of C-H bonds to C + H^- occurs only at drastic conditions, e.g. by action of potassium at 800°C and 20-40 kbar [58]). Synthesis of electrochemical carbyne requires, therefore, to work in perfectly aprotic media. This condition is almost ideally fulfilled in the case of amalgam reduction (see section 3.5. - 2.1.)

4.1 LOW-MOLECULAR-WEIGHT CARBON HALIDES

Reductive carbonization of CCl_4, CBr_4 and CI_4 with mercury was first reported by Tammann in 1921 [59]. The author had speculated about the formation of a new carbon allotrope, other than cubic diamond or graphite. Analogously, Holcombe *et al.* [60,61] have claimed a synthesis of "carbon VI", "chaoite" and "carbyne" as well as lonsdaleite (hexagonal diamond) *via* the reduction of carbon tetrafluoride with Si, SiC and CH_3SiCl_3. A quantitative low-temperature carbonization of CCl_4 by sodium amalgam was first reported by Fetkenheuer in 1923 [62]. On the other hand, CF_4 and C_2F_6 do not react markedly with alkali metal amalgams at 25-150°C, contrasting the behavior of higher perfluoro-n-alkanes [26].

A "classical" electroreduction of carbon tetrahalides, CX_4 at a mercury cathode in dimethylacetamide or acetonitrile usually yields dihalocarbene, $:CX_2$ or $:CX_2^{2-}$, which eventually convert to CH_2X_2 [57]. A complete carbonization of CCl_4 to carbonaceous fibers at Ni, Cu, Ag, Pt and Fe cathodes in acetonitrile or dichloromethane was reported by Sailor *et al.* [63]. However, the product was a disordered, sp^2 bonded carbon with no signs of carbynoid ordering. Substituted carbon tetrachlorides, $RCCl_3$ (R = H, methyl, phenyl) are reducible by Na/K alloy [64,65] or $M(biphenyl)_2$ (M = Ca, Ba, Sr) [66] only to RC, since the C-H and C-C bonds are inert against reductive splitting. The product, called poly(alkylcarbyne), represents an interesting three dimensional sp^3 carbon network [64,65,67,68]:

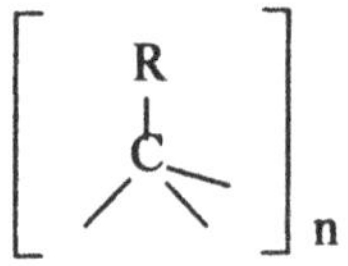

(Note a terminological inconsistency in the name of the above species and "carbyne" as a hypothetical *sp*-carbon allotrope). Pyrolysis of $(RC)_n$ yields predominantly sp^3 (diamond-like) carbons [64,65].

Hexachlorobuta-1,3-diene (HCBD) was reduced at glass-like carbon, Pt, Ni, Cu and In_2O_3/SnO_2 cathodes in NBu_4BF_4/CH_3CN [69-72] or NBu_4ClO_4/dimethylformamide [10,23,24]. Shirakawa et al. [10,23,24] have found that cathodic dechlorination of HCBD yields a promising carbynoid product, formulated as n-doped polymer with *sp*- and *sp*^{2} links:

$$[-(C\equiv C)_{10}-(-\underset{\vdots}{C}=)_6]^{1.5-}$$

Analogous material was also prepared by cathodic reduction of diiodoacetylene on a platinum electrode in the same medium [10,21,22]. Both these products are discussed in more detail in chapter 3.1.2.

Carbyne-like materials are also prepared by reductive defluorination of perfluoro-n-alkanes [26,27]. A quantitative defluorination of C_xF_{2x+2} (x = 6, 9, 20, 24) and perfluorobenzene has been achieved by alkali metal (Li, Na, K) amalgams [26,27]. The amalgam carbonization is conditioned by an electrical contact between the product (C-MF, growing on top of the M(Hg) surface) and the precursor (fluorocarbon). This is achieved by continued apposition of the precursor from liquid or gaseous phases [26].

4.2. HALOGENATED POLYMERS

Since the C-C bonds are inert against reductive splitting, electrochemical carbonization of perhalo-n-alkanes should ideally produce polyyne of the same conjugation length as the polymerization degree of the precursor (n):

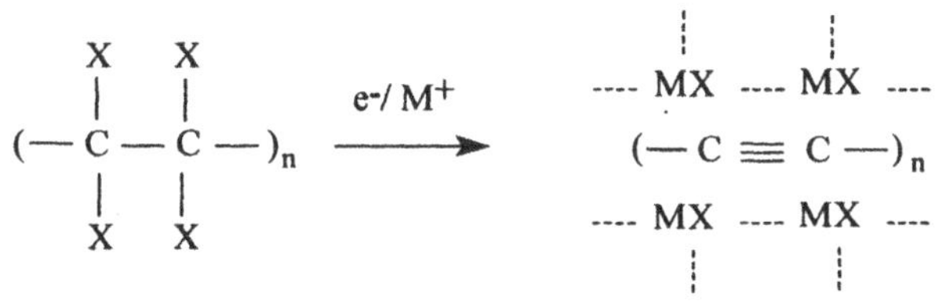

(11)

For steric reasons, all known perhalogenated polymers are fluoropolymers, although they may also contain a limited number of other atoms, such as oxygen, sulfur, other halogens, etc. [73]). The presence of certain heteroatoms (oxygen in etheric groups, sulfonic groups) decreases the reaction rate, but the C-F bond splitting is still quantitative [39,43]. This contrasts the behavior of partly halogenated polymers which usually show poor electrochemical activity (section 3.5. - 4.2.1.).

4.2.1 *Poly(vinylidene halides)*

Poly(vinylidene fluoride) (PVDF), and poly(vinylidene chloride) (PVDC):

$$-(CH_2-CX_2)-,\ X = F, Cl$$

similarly as poly(vinyl chloride) and poly(ethylene) do not react with amalgams even at 150°C [39] and they are also inert against direct electrolysis in liquid electrolyte solutions [20,74]. This is caused by the presence of non-reducible C-H bonds, which block the formation of conjugated carbon-carbon bonds and the growth of a conductive carbon skeleton (except for drastic conditions [58]; cf. section 3.5. - 4).

Poly(vinylhalides) are, however, chemically dehydrohalogenated in alkaline media [75-77] and these reactions are considered as a promising source of carbyne-like materials (cf. chapter 3.1.):

$$-CH_2-CX_2- \xrightarrow{-HX} -CH=CX- \xrightarrow{-HX} -C\equiv C- \qquad (12)$$

Analogous reactions proceed with a copolymer of vinylidene fluoride and tetrafluoroethylene [78], and poly(trichlorobutadienes) [79]. The dehydrohalogenation (Eq. 12) can be also driven electrochemically. Shirakawa *et al.* [20] have found that reductive carbonization of PVDF dissolved in dimethylformamide + NBu_4ClO_4 takes place at a Pt cathode, provided t-butanol (BuOH) is added to the electrolyte solution. The catalytic cycle is schematically depicted as follows:

$$BuOH \xrightarrow{e^- \text{(Pt cathode)}} BuO^- \qquad (13a)$$

$$-CH_2-CF_2 \xrightarrow[-(F^- + BuOH)]{BuO^-} -CH=CF- \xrightarrow[-(F^- + BuOH)]{BuO^-} -C\equiv C- \qquad (13b)$$

The product contains (besides some residual F and H) also inserted NBu_4^+, i.e. it is naturally n-doped by electrochemical overreduction (cf. section 3.5. - 5). It has been formulated as a copolymer of polyyne and fluoropolyacetylene:

$$Bu_4N^+ [-(C\equiv C)_{1.6}-(CH=CF)_{5.1}]^-$$

There seems to be a significant difference between chemical (Eq. 12) and indirect electrochemical (Eqs. 13a, b) dehydrohalogenation of poly(vinylidene halides), as the former reaction does not lead to n-doping (cf. section 3.5. - 5). Even in the absence of BuOH, reactive nucleophiles, $[(CH_3)_2NCHO]^-$ or $(CH_3)_2N^-$ can be generated by pre-electrolysis of pure electrolyte solutions [74]. They react with PVDF to give eventually carbon (cf. Eq. 13b). At the same conditions, PTFE shows no reactivity [74].

4.2.2 *Other partly halogenated polymers*

Copolymers of tetrafluoroethylene and vinylfluoride are reducible to carbon (or, presumably, to carbynoid materials) by indirect electrolysis analogous to that of pure PVDF [74]. As expected, the copolymers show intermediate electrochemical activity between that of pure PVDF and PTFE [74].

Poly(vinyl chloride), $(\text{-CH}_2\text{-CHCl-})_n$ is electrochemically dehydrochlorinated only to polyacetylene, $(\text{-CH=CH-})_n$ [80-83]. The reaction is again triggered by reactive nucleophiles, generated by a cathodic breakdown of the electrolyte solvent, i.e. acetonitrile or dimethylformamide.

Chlorinated polyacetylene [84] and chlorinated poly(vinyl chloride) [85-87] (both formulated as $(\text{CHCl})_n$) are carbonizable by dehydrochlorination in alkaline media [84] or photochemically [85-87], while formation of carbyne was sometimes observed [84,87].

Poly(tetrafluoroethylene-co-perfluoropropylvinylether) and perfluoropropenoxide are quantitatively carbonizable by amalgams [39]. The same holds for sulfonated fluoropolymers, Nafion® 117, Nafion® 125 and Nafion® 901 (duPont's trade mark) [43]. Sulfonic groups in Nafions are not reduced, which allows a synthesis of sulfonated carbon materials [43]. However, formation of carbyne from these precursors was not yet noted [39,43].

4.2.3. *Poly(tetrafluoroethylene)*

The best studied precursors of electrochemical carbyne are poly(perhalo-alkanes) *viz.* PCTFE and especially PTFE. Despite their well-known chemical and thermal stability, these polymers are thermodynamically unstable against reductive dehalogenation (cf. Table 1), and they should be carbonizable even with mild reductants. All these reactions can, at least theoretically, be regarded as electrochemical processes (cf. section 3.5. - 2).

Most of the carbonization studies were carried out with as-received commercial poly(tetrafluoroethylene) samples (duPont's Teflon®) usually in the form of foils. Mechanically oriented foils [26,46] and highly oriented films prepared by friction deposition [88] were also tested [89]. The latter seem to be particularly interesting precursors of electrochemical carbyne, as they show a perfect orientation of individual PTFE macromolecules [90-94]. Whether or not this orientation is also preserved in the follow-up electrochemical carbons has not been confirmed yet [89,95]. Some anisotropy was apparent on electrochemical carbyne produced from stretch-oriented foils [26,46].

The first reports about reductive carbonization of PTFE appeared in 1957-8 as several patents granted to duPont and 3M aimed at the promotion of adhesion and bondability of Teflon [96,97]. The carbonization was carried out either by the action of solvated electrons (ammonia solutions of alkali metals) or by organic radical anions, typically sodium dihydronaphthylide ($NaC_{10}H_8$).

The defluorination of PTFE with ammonia solutions of alkali metals [98,99] and Mg [100] is incomplete, and the resulting carbons also contain C-H bonds since NH_3 acts as an efficient proton donor. At certain conditions, this reaction may even yield polyethylene [101]:

$$(\text{-}CF_2\text{-}) + 4\ Li + 2\ NH_3 \rightarrow (\text{-}CH_2\text{-}) + 2\ LiF + 2\ LiNH_2 \quad (14)$$

The reactant is prepared by dissolving alkali metals in ammonia or (in the case of Mg) electrochemically [100]. Analogously, dihydronaphthylide is prepared from an alkali metal and naphthalene in tetrahydrofuran, but it can be also generated electrochemically in a solution of naphthalene in dimethylformamide + NBu_4BF_4 or NBu_4ClO_4 at -2.45 to -2.6 V *vs.* SCE [12,102].

Since the PTFE reduction regenerates the starting naphthalene (15a), the reaction propagates catalytically in the presence of sub-stoichiometric amounts of naphthalene:

$$C_{10}H_8^- + 0.5\ (\text{-}CF_2\text{-}) \rightarrow 0.5\ (\text{-}C\text{-}) + F^- + C_{10}H_8 \quad (15a)$$

$$C_{10}H_8 + e^- \rightarrow C_{10}H_8^- \quad (15b)$$

The catalytic cycle is driven (in step 15b) either electrochemically [12,102] or through the reaction with excess alkali metal. A similar scheme is applicable to reactions with radical anions ($\bullet R^-$) derived from benzoin [42,103], pyrene, benzanthracene, phenanthrene, triphenyl, benzpyrene, 9,10-dimethylanthracene, acenaphthalene, fluoranthene, perylene, biphenyl, benzophenone, and triphenylborane [104,105]. The reaction rate increases with decreasing redox potential of the $R/\bullet R^-$ couple and also in the sequence K < Na < Li [104,105].

The product's structure is far from being explored, and, even with optimized materials, the presence of carbyne is not strongly supported. For instance, the product made by the action of benzoin dianion was identified by one group [42] as a cross-linked polymeric carbon with short carbynoid (- C ≡ C -) segments, single and double bonds, as well as F, H, and small amount of O. However, a different group [106] has reported that the same reaction gives only low yields of carbon, the major product being trans-polyacetylene, $(CH{=}CH)_n$.

Another example of indirect electrocarbonization of PTFE *via* electron mediators was reported by Matsui et al. [17]. In this case, the PTFE foil was reduced in the electrolyte solution by highly-reactive Mg^+ species originating from anodic dissolution of a Mg electrode in tetrahydrofuran medium. The material produced showed several well-resolved carbyne-related Raman bands in the region of 2100-2200 cm^{-1} [17].

A direct cathodic carbonization of PTFE was first described by Gabano *et al.* [107]. The reaction was carried out with a composite cathode made by pressing PTFE powder with a conducting binder (graphite or carbon black). The electrolyte solutions were $LiClO_4$ or $LiBF_4$ in tetrahydrofuran, dioxolan, dimethoxyethane, or diethylene glycol and lithium metal served as an anode. The cathode delivered roughly the theoretical faradaic charge referred to a complete defluorination of PTFE (1.07 Ah/g). Quantitative defluorination in a PTFE/graphite composite was also confirmed with poly(ethylene oxide)-based solid electrolyte at 1-1.5 V *vs.* M/M^+ (M = Li, Na) and 58-82°C [108]. In some cases, the experimentally found carbonization charge capacity of PTFE (1.2 Ah/g [109]) even exceeded the theoretical value for a full defluorination (1.07 Ah/g). This points to the effect of n-doping (see section 3.5. - 5 for details).

The carbonization in a Li / PTFE / graphite system may consist of two parallel reactions: besides the direct cathodic carbonization of PTFE at three-phase interfaces (Eq. 11, Fig. 2), there is a second possible reaction triggered by the intermediate formation of Li/graphite intercalate, Li_mC (or even metallic Li) that subsequently reduces PTFE chemically [109]:

$$Li_mC + x(\text{-}CF_2\text{-}) \rightarrow Li_{m-2x}C + x(\text{-}C\text{-}) + 2x\,LiF \qquad (16)$$

PTFE was suggested as a cathode material for primary lithium batteries [107,108]. However, electrocarbonization of PTFE may also cause a capacity loss of lithium-ion battery with PTFE-bonded graphite electrodes [109].

The structure of carbon made in the PTFE composite electrodes [107-109] was not yet explored systematically. A first evidence that this process yields a carbyne-like material was obtained only recently by Oya et al. [16] (cf. Table 3). The carbyne formation is perturbed by proton-transfer from electrolyte cations (NBu_4^+) which, on the other hand, accelerates the PTFE defluorination [16].

Reductive carbonization of a pure PTFE foil (without conductive additives) on a lead-antimony cathode in a solution of NBu_4BF_4 + dimethylformamide was first reported by Brewis *et al.* [102]. The same reaction at a platinum or glass-like carbon electrode was further studied in detail by this [40,41,46,110,111] and other groups [12-15,45,74,112,113]. Coulometric experiments indicate a complete defluorination of PTFE, i.e. a consumption of 2.05-2.45 electrons per ($-CF_2-$) group [40,41,111]. (A small superstoichiometric electron consumption is again due to n-doping; cf. section 3.5. - 5). With increasing size of the electrolyte cation (NR_4^+ ; R = Et, Bu, Oct) the carbonization rate drops [40] (Table 2). However, the onset potential of the PTFE reduction surprisingly shifts to less negative values in the same series [13,114].

The carbonized area on top of a PTFE foil propagates in all directions from the contact point (cf. Fig. 2). Its shape depends on the preparation history of the PTFE sample: the electrochemical propagation proceeds rapidly along the chains in mechanically oriented PTFE foils [40,46]. This anisotropic electrocarbonization might be of interest to produce "oriented electrochemical carbyne", but no systematic investigations along this way were done up to now.

Carbonization of PTFE by alkali metal amalgams was already discussed in detail in section 3.5 - 2.1. Based on the Raman spectra, this reaction yields the best-quality carbynoid materials compared to other precursors and carbonization procedures [26]. These issues are further discussed in chapter 6.1.

4.2.4. *Other perhalogenated polymers*

Due to the lability of C-Cl bond, reductive dehalogenation of poly(chloro-trifluoroethylene) (PCTFE) proceeds easier than that of PTFE [39,115] (cf. Table 3). However, carbyne yield is lower and conjugation lengths are shorter compared to products from carbonization of PTFE [26].

A direct cathodic reduction of PCTFE in γ-butyrolactone solutions of $LiBF_4$ and/or $NaBF_4$ has been interpreted in terms of a catalytic cycle including electrolyte cations (Li^+, Na^+). Chronopotentiometric measurements show that the cations not only carry the charge down to the virgin PCTFE (cf. Fig. 2) but also react reversibly with the intermediates according to a general scheme [18,19,116]:

$$-CX_i- \xrightarrow{e^-} [-CX_i-]^{\cdot -} \xrightarrow{M^+} [-CX_iM-]^{\cdot} \xrightarrow[-MX]{} [-CX_{i-1}-]^{\cdot}, \quad (0 \leq i \leq 2) \qquad (17)$$

Irradiation of PCTFE with γ-photons (^{60}Co) impedes the reductive carbonization [117] which resembles the effects of electrons and X-ray photons on PTFE, where the decreased carbonization ability was attributed to crosslinking and branching [118-120].

Relations between the crosslinking/branching and the carbonization rates of fluoropolymers were qualitatively discussed by Doblhofer *et al.* [28] who found similar carbonization blocking in a plasma-polymerized hexafluoropropene. In terms of electrochemical mechanisms (Sect. 3.5. - 2), the electron propagation along the chain should stop at the crosslinked center (**C***) where the conjugation is perturbed, and the C_4**C*** units (C-C bonds) are not reductively cracked:

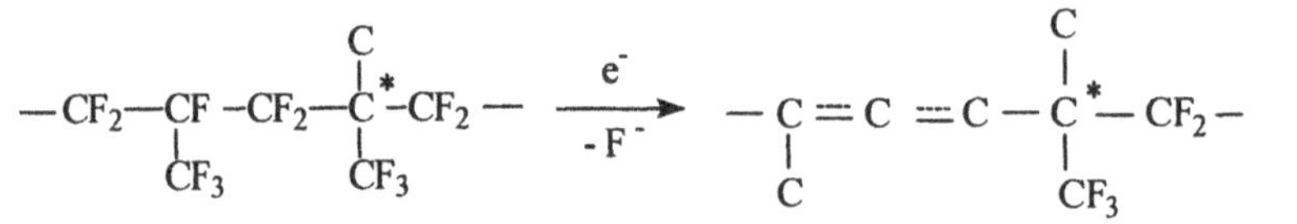

(18)

Pure, chemically-polymerized polymer (FEP) is carbonized smoothly and quantitatively to carbon by the action of alkali metal amalgams [28,30]. Carbons from FEP show IR [29] and Raman [26] active features of carbyne, but the yield of carbyne is significantly lower compared to analogous products from PTFE.

4.3. SURVEY ON MECHANISMS OF ELECTROCHEMICAL CARBONIZATION

Reductive carbonization at metal cathodes proceeds either directly or indirectly (*ec*-mechanism). Perhalogenated polymers (PTFE) show both direct reduction [14,16,40,41,46,74,107,108,110-113] and indirect reduction [12,17,102] but partly halogenated polymers (PVDF) are electroreduced only indirectly *via* suitable electron mediators [10,20,74]. A principal disadvantage of carbonization by classical "wet" electrochemistry (as well as by homogeneous reductants) is associated with the presence of solvent and protic impurities (trace water) which trigger uncontrolled irreversible formation of C-H bonds.

These complications are elegantly avoided by replacing the "wet" electrochemistry by a "dry" electrochemistry with alkali metal amalgams (section 3.5. - 2.1) [32-34,39]. Alternatively, the carbonization towards carbyne might be achieved by the action of pure alkali metals. Yamada et al. [121] have reported a successful synthesis of carbyne-like materials by the action of potassium vapor on PTFE. However, the defluorination of PTFE, e.g. with the sodium vapor, is not quantitative [95,122]. This reaction yields crystalline NaF and a complicated unsaturated and crosslinked polymer whose composition depends also on the reaction time [95,122]. Only a prolonged exposure of PTFE to alkali metals or their alloys [26,39,123] seems to lead to a pure carbon. The mechanistic details are unknown, and presumably both electrochemical and chemical pathways are to be considered (cf. section 3.5. - 2) [26,39,122].

Problems with uncontrolled kinetics (penetration of reactants through cracks in C-MF) were sometimes also mentioned in reactions of solid fluoropolymers with concentrated amalgams at high temperatures, liquid or gaseous alkali metals or liquid alloys of Na-K [5,39]. However, for diluted amalgams and lower temperatures (25-100°C), the "conductivity control" (Eq. 8) applies perfectly.

5. Electrochemical n-doping of carbyne

Reductive carbonization of perhalogenated hydrocarbons (Eq. 18) is generally accompanied by a small superstoichiometric consumption of electrons (Eq. 19), which is considered n-doping of electrochemical carbon:

$$CX_i + i\,e^- \rightarrow C + i\,X^- \qquad (19a)$$

$$C + \delta e^- \rightarrow C^{\delta-} \qquad (19b)$$

Both reactions proceed simultaneously, since the overpotential for carbonization (19a) is higher (by ca. 1 V) [40,71] than the overpotential for n-doping (19b). Cyclic voltammetry confirms the reversibility of reaction (19), although the charge recovery upon anodic oxidation is not perfect (40-94%) [71,111].

N-doping by overreduction was observed during the carbonization of PTFE both at a Pt cathode [40,110,111] and by alkali metal amalgams [33,34,39,124]. In the latter case, well reproducible doping levels (δ in the range of 0.20 - 0.21, cf. Table 4) were found, independent of the alkali metal type (Li, Na, K), reaction temperature (25-100°C) and time [39,124]. The extra negative charge (δ-) is compensated by electrolyte cations (M^+), hence the formal composition of the product is:

$$CM_\delta(MF)_2.$$

The n-doping rate of electrochemical carbon from hexachlorobuta-1,3-diene [69,70,72] increases, as expected, with decreasing size of M^+ (NBu_4^+, NEt_4^+, NMe_4^+) [71]. Table 4 summarizes the n-doping levels, δ for various systems studied.

TABLE 4. N-doping of electrochemical carbon (carbyne-like materials). Values of doping level, δ.

Precursor	Reactant/ react. medium	Carbyne formation	δ	Refs.
PTFE	Li amalgam	YES	0.21	[39,124]
PTFE	Na amalgam	YES	0.2	[39]
PTFE	K amalgam	YES	0.2	[39]
PTFE	Pt cathode/DMF+NBu_4BF_4	no data	0.15-0.45	[40]
PTFE	Pt cathode/DMF+NBu_4BF_4	no data	0.05-0.25	[41,110,111]
PCTFE	Li amalgam	YES	0.32	[39]
Nafion® 117	Li amalgam	no data	0.20	[39]
PTFE-PPVE	Li amalgam	no data	0.21	[39]
FEP	Li amalgam	YES	0.2	[28,30]
PVDF	Pt cathode/DMF+NBu_4ClO_4+BuOH	YES	0.15	[20]
C_6F_{14}	Li amalgam	YES	0.07-0.25	[27]
C_6F_6	Li amalgam	NO	0.07-0.25	[27]
IC≡CI	Pt cathode/DMF-NBu_4ClO_4	YES	[a]-	[21]
HCBD	Pt cathode/DMF-NBu_4ClO_4	YES	[b](0.02, 0.05)	[21]
HCBD	Pt cathode/acetonitrile+NBu_4BF_4	NO	[b](>0.04)	[71]

a: no δ values available but qualitative evidence for n-doping
b: estimated value(s) from literature experimental data
(For explanation of the other symbols see Tables 2 and 3)

Some authors [13,40,71,111] have regarded the interaction between the inserted cations, M^+ and the n-doped carbon skeleton, $C^{\delta-}$ as an intercalation into the graphite lattice. However, the potential at which the electrocarbonization (18) takes place (about -2 V *vs.* SHE, cf. Table 2) is too high to promote intercalation into graphite at δ ≈0.2 [40,124]. A hypothetical graphite intercalation compound, $CM_{0.2}$ would have a sufficiently negative potential to drive a subsequent carbonization [40] of the precursor, e.g. PTFE:

$$CM_{0.2} + x\,(\text{-}CF_2\text{-}) \rightarrow x\,(\text{-}C\text{-}) + 2x\,MF + CM_{0.2-2x} \qquad (20)$$

This was not observed: the doping level at the $CM_{0.2}$/PTFE junction remained unchanged even after years of contact [39]. Secondly, there is an apparent lack of graphite ordering in most electrochemical carbons. A third argument against intercalation is the absence of H_2 in the hydrolysis products of CM_δ [124]. Hydrolysis yields quantitatively alkali metal hydroxide, MOH and the corresponding "hydrocarbon", CH_δ [124,125]:

$$CM_\delta + \delta\, H_2O \rightarrow \delta\, MOH + CH_\delta \tag{21}$$

Therefore, the redox potential of CM_δ /C (cf. Eq. 19) must be even more positive than that of H_2/H^+ in alkaline medium (ca. -0.8 V *vs.* SHE) [124]. An organometallic species similar to carbides was also considered [124] but the C - M^+ bond is most probably ionic and similar to that in n-doped polyacenic materials [39]. N-doping of electrochemical carbons is, reportedly, also responsible for the observed negative thermoelectric power [21].

6. Structure of electrochemical carbyne

Reaction (11) would hardly produce a perfect polyyne. Even in low-molecular-weight capped oligoynes, such as:

$$(CH_3)_3C - (C \equiv C)_4 - C(CH_3)_3$$

the observed (- C ≡ C - C -) bond angles deviate slightly from 180°, and larger molecules may even form macrocycles (C_{18}) or helices (C_{300}?) [126]. Secondly, a linear polymeric all-carbon chain containing 10^4-10^5 carbon atoms (as in the starting PTFE macromolecule) is highly sensitive to the formation of biradical kinks [5]:

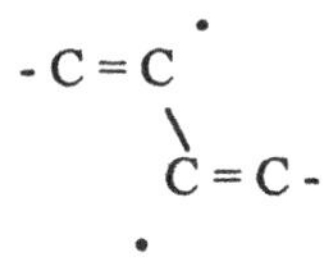

where the *sp*-conjugation is corrupted. These perturbations further promote crosslinking of carbynoid chains towards sp^2 graphene as a final product:

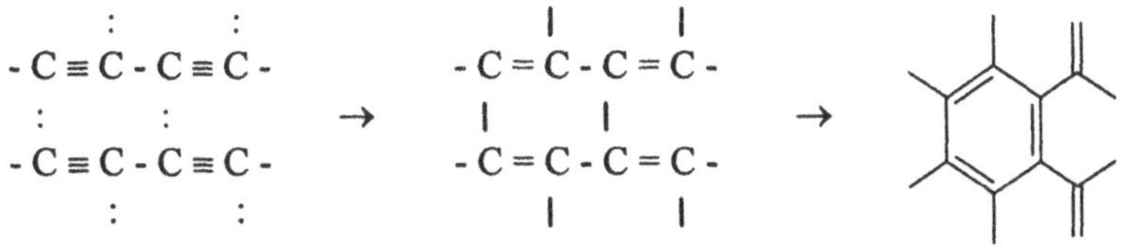

(22)

Reaction (22) is fast at sites where the neighboring carbynoid chains are at sufficiently close distances [127]. The carbyne/graphene crosslinking can be conveniently monitored by various optical and electrical properties of the product [5,6,35] (see also chapter 6.1).

The semiempirical models of carbyne crystals are mostly based on a hexagonal lattice with kinked polyyne/polycumulene chains aligned parallel to the c-axis (Fig. 3) [128-134]. The model structure in Fig. 3 is conditioned by some efficient mechanism preventing the carbyne-to-graphene crosslinking mentioned above (Eq. 22).

The fact that the carbyne structure would require stabilization, either by bulky terminal groups at the chain ends or by inserted impurities, has been addressed already in early works on carbyne [128]. The stabilization can be understood either as steric separation of individual *sp*-chains or as a promotion of sp^2 kinks with dangling bonds in the chain [128]. Capping the chain ends with $-CF_3$ or -CN groups has yielded very promising results recently [126]. Impurities, such as Fe, K, Na [130,131,134], I_2 [134], H, O and N [135] also stabilize carbyne, both synthetic [130,131,134] and natural ones [135] presumably *via* the promotion of kinks (see chapter 4.2).

The spontaneous n-doping of electrochemical carbon (Eq. 19) might favor the carbynoid structure by electrostatic charging of carbon chains and interspersing of counterions (M^+) from the electrolyte into a hypothetical carbyne crystal lattice as in Fig. 3. These effects are expressed also with chemically synthesized carbynes: Reactions of carbyne with sodium [130,134] and potassium [58,134] are interpreted as "intercalation" of alkali metal into a carbyne lattice (Fig. 3) [58,130,134], and the existence of alkali metal capped chains, $K-(C\equiv C)_n-K$; $1<n<12$ is also experimentally supported [136]. All these carbyne/metal complexes can formally be related to n-doped electrochemical carbons discussed in section 3.5. - 5.

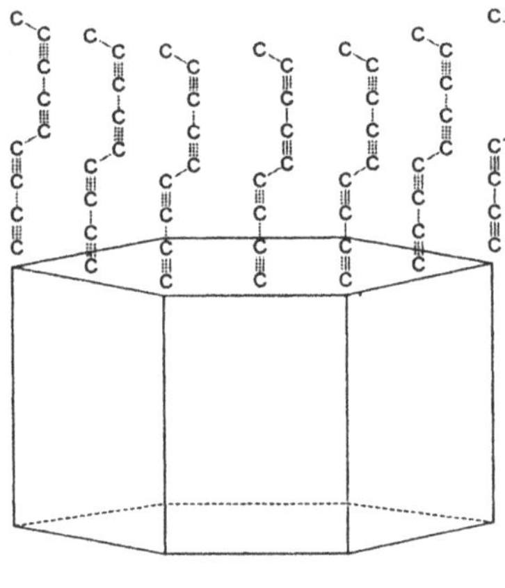

Figure 3. Hypothetical scheme of a carbyne structure. Parallel sp-carbon chains with regular kinks are packed in a hexagonal lattice. After Tanuma and Palnichenko [132-134] and Kudryavtsev et al. [131].

Strictly speaking, the occurrence of impurities or sp^2 kinks would disregard the definition of carbyne as a "carbon allotrope" or "*sp*-bonded carbon", respectively. A question is, whether an impurity-stabilized crystal can still be tolerated as a "true polymorph", a situation which is similar, e.g. to the SiO_2 phase - tridymite. If we accept this condition, the consideration of carbyne as a carbon allotrope seems to be justified. To conclude, electrochemistry offers three promising issues towards the synthesis of carbyne:

1. Under optimized conditions, the conversion of a precursor molecule into elemental carbon is stoichiometrically quantitative with no parasitic side reactions.

2. Electrosynthesis of *sp*-carbon chains from certain precursors is a well-defined, while any electrochemical cracking of C-C bonds is excluded.

3. The stabilization of carbynoid chains by n-doping is a spontaneous effect accompanying the actual electrochemical preparation procedure.

However, more work is needed to learn whether or not an "electrochemical carbyne" is eventually accessible along this way.

Acknowledgment. This work was supported by the Grant Agency of the Czech Republic and by the Ministry of Education, Youth and Sports of the Czech Republic.

7. References

1. Besenhard, J.O. and Fritz, H.P.: Elektrochemie schwarzer Kohlenstoffe, *Angew. Chem.* **95** (1983), 954-80. (Angew. Chem. Int. Ed. Eng. 1983, 22, 950).
2. Sarangapani, S., Akridge, J.R.and Schumm, B.: *Proceedings of the Workshop on the Electrochemistry of Carbon,* The Electrochemical Society, Pennington, NJ, 1983.
3. Kinoshita, K.: *Carbon. Electrochemical and Physicochemical Properties,* J.Wiley, New York, 1988.
4. McCreery, R.L.: Carbon electrodes: structural effects on electron transfer kinetics, in Bard, A.J. (ed.), *Electroanalytical Chemistry, Vol. 17,* M. Dekker, New York, (1991), p. 221-374.
5. Kavan, L.: Electrochemical carbonization of fluoropolymers, in Thrower, P.A. (ed.), *Chemistry and Physics of Carbon, Vol. 23,* Marcel Dekker, New York, (1991), p. 69-171.
6. Kavan, L.: Electrochemical carbon, *Chem. Rev.* **97** (1997), 3061-3082.
7. Hsu, W.K., Hare, J.P., Terrones, M., Kroto, H.W., Walton, D.R.M. and Harris, P.J.F.: Condensed-phase nanotubes, *Nature* **377** (1995), 687.
8. Hsu, W.K., Terrones, M., Hare, J.P., Terrones, H., Kroto, H.W. and Walton, D.R.M.: Electrochemical formation of carbon nanostructures, *Chem. Phys. Lett.* **262** (1996), 161
9. Terrones, M., Hsu, W.K., Hare, J.P., Kroto, H.W., Terrones, H. and Walton, D.R.M.: Graphitic structures: from planar to spheres, toroids and helices, *Phil. Trans. R. Soc. Lond. A* **324** (1996), 2025
10. Kijima, M.: Electrochemical synthesis of linear carbons, *Recent Res. Devel. Pure Appl. Chem.* **1** (1997), 27-33.
11. Ohmura, K., Kijima, M. and Shirakawa, H.: Synthesis of conducting polymers with conjugated carbon-carbon triple bonds by electrochemical condensation of acetylene derivatives catalyzed by coper complex, *Synth. Metals* **84** (1997), 417-8.
12. Pud, A.A. and Shapoval, G.S.: Cathodic transformation of PTFE in the presence of electron mediator, *Theor. Exp. Khim.* **32** (1996), 40-2. (Theor. Exp. Chem. 1996, 32, 38).
13. Pud, A.A., Mikulina, O.E., Shapoval, G.S. and Tomilov, A.P.: Cathodic decomposition of PTFE in galvanostatic regime, *Elektrokhimiya* **32** (1996), 1000-3. (Russ. J. Electrochem. 1996, 32, 923).
14. Pud, A.A. and Shapoval, G.S.: Electrochemistry as the way to transform polymers, *Macrom. Reports* **A32(Suppls. 5&6)** (1995), 629-38.

15. Pud, A.A. and Shapoval, G.S.: Electrochemical decomposition of solid polymers on a cathode, *Elektrokhimiya* **28** (1992), 654-68. (Russ. J. Electrochem. 1992, 28, 529).
16. Oka, T., Araki, T., Oya, A., Yamada, Y., Shiraishi, M., Miyashita, K. and Ozawa, T.: Preparation and characterization of carbyne derived from aliphatic compounds, *Proc. 4th IUMRS Conference, Makuhari, Japan, Sept. 16-18, 1997*, p. I 5,19.
17. Matsui, H., Yamaguchi, C., Yasuda, A. and Kawase, N.: Preparation of carbyne by electrochemical reduction of PTFE with reactive anode, *Proc. 4th IUMRS Conference, Makuhari, Japan, Sept. 16-18, 1997*, p. I 5,20.
18. Stezeryanskiy, E.A., Litovcenko, K.I. and Kublanovskiy, V.S.: Effect of the electrolyte cation on electrochemical reduction of poly(chlorotrifluoroethylene), *Ukr. Khim. Zhur.* **55** (1989), 826-9. (Soviet Prog. Chem. 1989, 55, 42).
19. Stezeryanskiy, E.A., Litovcenko, K.I. and Kublanovskiy, V.S.: Kinetics and mechanism of electrochemical reduction of poly(chlorotrifluoroethylene), *Ukr. Khim. Zhur.* **56** (1990), 487-92. (Soviet Prog. Chem. 1990, 56, 45).
20. Kijima, M., Toyabe, T. and Shirakawa, H.: Electrocatalytic dehydrofluorination of PVDF, *Chem. Lett.* **1995** (1995), 553-4.
21. Kijima, M., Sakai, Y. and Shirakawa, H.: A novel approach for synthesis of carbyne by electroreductive polymerization of diiodoacetylene catalyzed by Ni complex, *Chem. Lett.* (1994), 2011-4.
22. Kijima, M., Sakai, Y. and Shirakawa, H.: Electrochemical synthesis of carbyne catalyzed by nickel complex, *Synth. Metals* **71** (1995), 1837-40.
23. Kijima, M., Toyabe, T. and Shirakawa, H.: Novel approach to synthesis of a carbyne film by electrochemical reduction of hexachlorobuta-1,3-diene, *Chem. Commun.* (1996), 2273-4.
24. Kijima, M., Toyabe, T., Shirakawa, H., Kawata, S., Kyotani, H., Nakamura, Y. and Endo, M.: Synthesis of carbyne by cathodic electrolysis of hexachlorobutadiene, *Synth. Metals* **86** (1997), 2279-80.
25. Kastner, J., Kuzmany, H., Kavan, L., Dousek, F.P. and Kürti, J.: Reductive preparation of carbyne with high yield. An in situ Raman scattering study, *Macromolecules* **28** (1995), 344-53.
26. Kavan, L., Hlavatý, J., Kastner, J. and Kuzmany, H.: Electrochemical carbyne: synthesis and stability studied by Raman scattering, *Carbon* **33** (1995), 1321-9.
27. Kavan, L. and Dousek, F.P.: Reaction of hexafluorobenzene and hexadecafluorohexane with lithium amalgam, *J. Fluorine Chem.* **41** (1988), 383-91.
28. Kavan, L., Doblhofer, K. and Dousek, F.P.: Electrochemical reductive carbonization of polymers derived from hexafluoropropene, *J. Fluorine Chem.* **55** (1991), 37-45.
29. Kavan, L. and Dousek, F.P.: Carbynoid species in electrochemical polymeric carbon, *Synth. Metals* **58** (1993), 63-72.
30. Kavan, L., Micka, K. and Kastner, J.: UV-Vis absorption of thin electrochemical carbon layers on poly(tetrafluoroethylene-co-hexafluoropropene), *Synth. Metals* **63** (1994), 147-52.
31. Savéant, J.M.: Electron transfer, bond breaking and bond formation, *Acc. Chem. Res.* **26** (1993), 455-61.
32. Jansta, J. and Dousek, F.P.: Electrochemical corrosion of PTFE contacting lithium amalgam, *Electrochim. Acta* **18** (1973), 673-4.
33. Dousek, F.P. and Jansta, J.: Kinetics of electrochemical corrosion of PTFE by alkali metal amalgams, *Electrochim. Acta* **20** (1975), 1-6.
34. Jansta, J., Dousek, F.P. and Ríha, J.: Quantitative explanation of the mechanism of corrosion of PTFE caused by active alkali metals, *J. Appl. Polym. Sci.* **19** (1975), 3201-10.
35. Kavan, L. and Kastner, J.: Carbyne forms of carbon: continuation of the story, *Carbon* **32** (1994), 1533-6.
36. Kavan, L., Dousek, F.P., Weber, J. and Micka, K.: Electronic resistivity of carbon in the products of electrochemical reduction of poly(tetrafluoroethylene), *Carbon* **26** (1988), 235-43.
37. Kavan, L., Dousek, F.P., Weber, J. and Micka, K.: Electronic resistivity of carbon in the products of electrochemical reduction of poly(tetrafluoroethylene). theoretical considerations, *Carbon* **26** (1988), 245-7.
38. Kavan, L., Dousek, F.P. and Micka, K.: Time-dependent electrical resistivity of carbon, *J. Phys. Chem.* **94** (1990), 5127-34.
39. Kavan, L., Dousek, F.P. and Micka, K.: The role of ion transport in the electrochemical corrosion of fluoropolymers. preparation and properties of n-doped polymeric carbon with mixed ion/electron conductivity, *Solid State Ionics* **38** (1990), 109-18.

40. Dahm, R.H., Barker, D.J., Brewis, D.M. and Hoy, L.R.J.: Surface treatment of PTFE, in Allen, K.W. (ed.), *Adhesion 4*, Applied Science Publ. London, (1980), p. 215-32.
41. Barker, D.J., Brewis, D.M., Dahm, R.H. and Hoy, L.R.J.: The electrochemical reduction of PTFE, *Electrochim. Acta* **23** (1978), 1107-10.
42. Costello, C.A. and McCarthy, T.J.: Surface selective introduction of specific functionalities onto PTFE, *Macromolecules* **20** (1987), 2819-28.
43. Kavan, L., Dousek, F.P. and Weber, J.: Electrochemical reduction of NAFION 117, *J. Electroanal. Chem.* **208** (1986), 109-16.
44. Bening, R.C. and McCarthy, T.J.: Surface treatment of poly(tetrafluoroethylene-co-hexafluoropropene), introduction of alcohol functionality, *Macromolecules* **23** (1990), 2648-55.
45. Pud, A.A., Shapoval, G.S. and Mikulina, O.E.: Cells for the study of electrochemical properties and surface modification of polymers, *Pribory Tekh. Exp.* (1991), 213-5. (Instrum. Exp. Techn. 1991, 34, 974).
46. Barker, D.J., Brewis, D.M., Dahm, R.H. and Hoy, L.R.J.: Anisotropic electrochemical reduction of PTFE, *J. Mater. Sci.* **14** (1979), 749-51.
47. Cooney, R.P. and Mernagh, T.P.: The chemical origins of SERS effect: a review of microzone damage, *J. Electroanal. Chem.* **168** (1984), 67-90.
48. Mernagh, T.P., Cooney, R.P. and Spink, J.A.: Chemical origin of SERS by water on Ag and carbon overlayered Ag electrodes, *J. Chem. Soc. Faraday Trans. I* **80** (1984), 3469-79.
49. Matsuda, H., Nakanishi, M. and Kato, M.: Neco o oxidaci butadiynu, *J. Polym. Sci. C: Polym. Lett.* **22** (1984), 107
50. Hay, A.S.: Oxidative polymerization of diethynyl compound, *J. Polym. Sci. A1* 7 (1969), 1625-34.
51. Cataldo, F.: Synthesis of carbonaceous material rich in carbyne, a carbon allotrope by coupling reaction of copper acetylide, *Eur. J. Solid State Inorg. Chem.* **34** (1997), 53-63.
52. Hattori, T., Kijima, M. and Shirakawa, H.: Oxidative polycondensation of acetylene by iodine in the presence of a palladium-copper catalyst, *Synth. Metals* **84** (1997), 357-8.
53. Diederich, F.: Carbon scaffolding: building acetylenic all-carbon and carbon-rich compounds, *Nature* **369** (1994), 199-207.
54. Kudryavtsev, Y.P., Evsyukov, S.E., Guseva, M.B., Babaev, V.G. and Khvostov, V.V.: Carbyne- a linear chain-like carbon allotrope, in Thrower, P.A. (ed.), *Chemistry and Physics of Carbon*, M.Dekker, New York, (1997), p. 1-69.
55. Kudryavtsev, Y.P., Heimann, R.B. and Evsyukov, S.E.: Carbynes: Advances in the field of linear carbon-chain compounds, *J. Mater. Sci.* **31** (1996), 5557-71.
56. Fuchigami, T.: Electrochemical reactions of fluoro organic compounds, *Topics Curr. Chem.* **170** (1994), 1-36.
57. Wawzonek, S. and Duty, R.C.: Polarographic studies in acetonitrile and dimethylformamide VI. The formation of dihalocarbenes, *J. Electrochem. Soc.* **108** (1961), 1135-8.
58. Udod, I.A., Shchurik, V.I., Bulychev, B.M., Sirotinkin, S.P., Guseva, M.B., Babaev, V.G., Kudryavtsev, Y.P. and Evsyukov, S.E.: Formation of carbyne in the interaction of polyacetylene with potassium under high quasi-hydrostatic pressure, *J. Mater. Chem.* **3** (1993), 413-6.
59. Tammann, G.: Uber Kohlenstoff, der bei der Einwirkung von Quecksilber auf CCl_4, CBr_4 und CJ_4 entsteht, *Z. Anorg. Allg. Chem.* **115** (1921), 145-59.
60. Holcombe, C.E., Condon, J.B. and Johnson, D.H.: Metastable carbon phases from CF_4 reactions. Part I. reaction with SiC and Si, *High Temp. Sci.* **10** (1978), 183-95.
61. Holcombe, C.E., Condon, J.B. and Johnson, D.H.: Metastable carbon phases from CF_4 reactions. Part II. Reactions with CH_3SiCl_3, *High Temp. Sci.* **10** (1978), 197-210.
62. Fetkenheuer, B.: Uber die Einwirkung von Natriumamalgam auf Tetrachlorkohlenstoff, *Z. Anorg. Allg. Chem.* (1923), 281-2.
63. Herrick, R.D., Kaplan, A.S., Chinh, B.K., Shane, M.J., Sailor, M.J., Kavanagh, K.L., McCreery, R.L. and Zhao, J.: Room-temperature electrosynthesis of carbonaceous fibers, *Adv. Mater.* 7 (1995), 398-401.
64. Visscher, G.T. and Bianconi, P.A.: Synthesis and characterization of polycarbynes, a new class of carbon-based network polymers, *J. Am. Chem. Soc.* **116** (1994), 1805-11.
65. Visscher, G.T., Nesting, D.C., Badding, J.V. and Bianconi, P.A.: Poly(phenylcarbyne): a polymer precursor to diamond-like carbon, *Science* **260** (1993), 1496-9.
66. Rieke, R.D. and Chen, T.: Facile synthesis of poly[phenylcarbyne]: a precursor for diamondlike carbon, *Chem. Mater.* **6** (1994), 576-7.

67. Best, S.A., Bianconi, P.A. and Merz, K.M.: Structural analysis of carbyne network polymers, *ACS Symp. Series* **589** (1995), 304-15.
68. Best, S.A., Bianconi, P.A. and Merz, K.M.: Structural analysis of carbyne network polymers, *J. Am. Chem. Soc.* **117** (1995), 9251-8.
69. Nishihara, H., Harada, H., Kaneko, S., Tateishi, M. and Aramaki, K.: Synthesis of graphite film by electrochemical reduction of hexachlorobuta-1,3-diene, *J. Chem. Soc. Chem. Commun.* (1990), 26-7.
70. Nishihara, H., Harada, H., Tateishi, M., Ohashi, K. and Aramaki, K.: Graphite synthesis by electrochemical reduction of hexachlorobuta-1,3-diene, *J. Chem. Soc. Faraday Trans.* **87** (1991), 1187-92.
71. Nishihara, H., Harada, H., Kaneko, S., Tateishi, M., Aramaki, K. and Murray, R.W.: Electrochemical and electrical properties of poly(hexachlorobuta-1,3-diene), an electrochemically graphitized film, *J. Chem. Soc. Faraday Trans.* **87** (1991), 319-24.
72. Nishihara, H., Ohashi, K., Kaneko, S., Tanaka, F. and Aramaki, K.: Electrochemical and electrical properties of carbonaceous films formed by electroreduction of perhalogenocarbons, *Synth. Metals* **41-43** (1991), 1495-9.
73. Wall, L.A.: *Fluoropolymers*, Wiley-Interscience, New York, 1972.
74. Pud, A.A., Mikulina, O.E. and Shapoval, G.S.: Electrochemically induced functionalization of fluorocontaining polyolefins, *Macrom. Reports* **A32(Suppls. 5&6)** (1995), 621-8.
75. Dias, A.J. and McCarthy, T.J.: Dehydrofluorination of PVDF in DMF solution: Synthesis of an operationally soluble semiconducting polymer, *J. Polym. Sci. Polym. Chem. Ed.* **23** (1985), 1057-61.
76. Komatsu, T., Nomura, M., Kakudate, Y., Fujiwara, S. and Heimann, R.B.: Characterization of dehydrochlorinated poly(vinylidene chloride) and the shock-compressed material, *Macromol. Chem. Phys.* **196** (1995), 3031-40.
77. Korshak, V.V., Kudryavtsev, Y.P., Korshak, Y.V., Evsyukov, S.E., Khvostov, V.V., Babaev, V.G. and Guseva, M.B.: Formation of carbyne by dehydrohalogenation, *Makrom. Chem. Rapid Commun.* **9** (1988), 135-40.
78. Cho, J.W. and Song, H.Y.: Dehydrofluorination of a compolymer of vinylidene fluoride and tetrafluoroethylene by phase transfer catalysis reaction, *J. Polym. Sci. A: Polym. Chem.* **33** (1995), 2109-12.
79. Vointseva, I.I., Gilman, L.M., Kudryavtsev, Y.P., Evsyukov, S.E., Pesin, L.A., Gribov, I.V., Moskvina, N.A. and Khvostov, V.V.: Chemical dehydrochlorination of polytrichlorobutadienes - a new route to carbynes, *Eur. Polym. J.* **32** (1996), 61-8.
80. Shapoval, G.S., Kontsur, Y.V., Pud, A.A. and Tomilov, A.P.: Indirect electrochemical triggering of dehydrohalogentaion of polyvinylchloride in acetonitrile, *Elektrokhimiya* **32** (1996), 138-41. (Russ. J. Electrochem. 1996, 32, 130).
81. Shapoval, G.S., Kontsur, Y.V. and Pud, A.A.: Electrochemical dehydrohalogenation of polyvinylchloride, *Elektrokhimiya* **32** (1996), 142-7. (Russ. J. Electrochem. 1996, 32, 127).
82. Kontsur, Y.V., Shapoval, G.S. and Pud, A.A.: Indirect electrochemical dehydrochlorination of polyvinychloride, *J. Macrom. Sci. Pure Appl. Chem.* **A32** (1995), 687-93.
83. Shapoval, G.S. and Kontsur, Y.V.: Electrochemical initiation of surface dehydrohalogenation of polyvinylchloride films, *J. Macrom. Sci. Pure Appl. Chem.* **A32** (1995), 49-54.
84. Akagi, K., Nishiguchi, M. and Shirakawa, H.: One-dimensional conjugated carbyne - synthesis and properties, *Synth. Metals* **17** (1987), 557-62.
85. Decker, C.: Synthesis of conductive polymers by laser irradiation of chlorinated polyvinylchloride, *J. Polym. Sci. C: Polym. Lett.* **25** (1987), 5-10.
86. Decker, C., Eckhard, A. and Ehrburger, P.: Structure of laser-carbonized polymers, *Carbon* **28** (1990), 246-8.
87. Shimoyama, M., Niino, H. and Yabe, A.: A KrF excimer laser induced dehydrochlorination of chlorinated poly(vinylchloride): preparation of conjugated polyene and polyyne, *Makrom. Chem.* **193** (1992), 569-74.
88. Wittmann, J.C. and Smith, P.: Highly oriented PTFE as a substrate for oriented growth of materials, *Science* **352** (1991), 414-7.
89. Kavan, L.: Electrochemical preparation of hydrogen-free carbyne-like materials, *Carbon* in press.
90. Dietz, P., Hansma, P.K., Ihn, K.J., Motamedi, F. and Smith, P.: Molecular structure and thickness of highly oriented PTFE measured by atomic force microscopy, *J. Mater. Sci.* **28** (1993), 1372-6.
91. Fenwick, D., Ihn, K.J., Motamedi, F., Wittmann, J.C. and Smith, P.: Characterization of friction-deposited PTFE transfer films, *J. Appl. Polym. Sci.* **50** (1993), 1151-7.

92. Dannetun, P., Schott, M. and Vilar, M.R.: High-resolution electron energy loss spectroscopy of thin crystalline highly oriented films of PTFE, *Thin Solid Films* **286** (1996), 321-9.
93. Bodo, P. and Schott, M.: Highly oriented PTFE films: a force microscopy study, *Thin Solid Films* **286** (1996), 98-106.
94. Beamson, G., Clark, D.T., Deegan, D.E., Hayes, N.W., Law, D.S.L. and Rasmusson, J.: Characterization of PTFE on silicon wafer tribological transfer films by XPS, imaging XPS and AFM, *Surf. Interf. Anal.* **24** (1996), 204-10.
95. Schott, M.: Preparation and properties of highly oriented PTFE films, *Synth. Metals* **67** (1994), 55-61.
96. Siperko, L.M. and Thomas, R.R.: Chemical and physical modification of fluoropolymer surfaces for adhesion enhancement: a review, *J. Adhesion Sci. Technol.* **3** (1989), 157-73.
97. Mathieson, I., Brewis, D.M., Sutherland, I. and Cayless, R.A.: Pretreatments of fluoropolymers, *J. Adhesion* **46** (1994), 49-56.
98. Brecht, H., Mayer, F. and Binder, H.: ESCA-Untersuchungen an geätzten PTFE Folien, *Angew. Makrom. Chem.* **33** (1973), 89-100.
99. Riggs, W.M. and Dwight, D.W.: Characterization of the PTFE surface, *J. Electron Spectr. Rel. Phen.* **5** (1974), 447-60.
100. Brace, K., Combellas, C., Fritch, A., Kanoufi, F., Shanahan, M.E.R. and Thiébault, A.: A new reagent for surface treatment of PTFE (Mg), *Chem. Commun.* **1996** (1996), 403-4.
101. Chakrabarti, N. and Jacobus, J.: The chemical reduction of PTFE, *Macromolecules* **21** (1988), 3011-4.
102. Brewis, D.M., Dahm, R.H. and Konieczko, M.B.: Reaction of PTFE with electrochemically generated intermediates, *Angew. Makrom. Chem.* **43** (1975), 191-4.
103. Costello, C.A. and McCarthy, T.J.: Surface modification of PTFE with benzoin dianion, *Macromolecules* **17** (1984), 2940-2.
104. Fouletier, M.: *Preparation of Metastable Carbon and Silicon by Topochemical Reactions, Thesis,* Grenoble Polytechnics, Grenoble, 1983.
105. Fouletier, M. and Armand, M.B.: Derivatives of polymeric carbons, their preparation and chemical and electrochemical properties, Fr.Pat. 2 552 092 (1983).
106. Iqbal, Z., Ivory, D.M., Szobota, J.S., Elsenbaumer, R.L. and Baughman, R.H.: Polyacetylene formed by reduction of PTFE with benzoin dianion, *Macromolecules* **19** (1986), 2992-6.
107. Decheneaux, V.I., Drovet, F., Gabano, J.P. and Jumel, Y.: Elektrochemisches Lithium-Element, Ger.Pat. (DOS) 2210370 (1971).
108. Ge, P.: *Synthesis and Functionalization of Carbon Ex-PTFE, Thesis,* Grenoble Polytechnics, Grenoble, 1987.
109. Li, G., Xue, R. and Chen, L.: The influence of PTFE reduction on the capacity loss of the carbon anode for lithium ion battery, *Solid State Ionics* **90** (1996), 221-5.
110. Barker, D.J., Brewis, D.M., Dahm, R.H., Gribbin, J. and Hoy, L.R.J.: Surface treatment of PTFE, *J. Adhesion* **13** (1981), 67-76.
111. Barker, D.J., Brewis, D.M., Dahm, R.H. and Hoy, L.R.J.: Study of the intercalated carbon formed by electrochemical reduction of PTFE, *Polymer* **19** (1978), 856-8.
112. Shapoval, G.S., Pud, A.A. and Tomilov, A.P.: Electrochemical reductive decomposition of some polymer dielectrics, *Dokl. Akad. Nauk SSSR* **283** (1985), 434-8. (Dokl. Phys. Chem. 1985, 283, 706).
113. Shapoval, G.S., Tomilov, A.P., Pud, A.A. and Vonsyackiy, V.A.: Electrochemical reductive destruction of PTFE, *Theor. Exp. Khim.* **20** (1984), 247-9. (Theor. Exp. Chem. 1984, 20, 234).
114. Pud, A.A., Shapoval, G.S. and Mikulina, O.E.: Electroreduction of insoluble fluorocarbons in dimethylformamide in the presence of tetrabutylammonium salts, *Elektrokhimiya* **31** (1995), 555. (Russ. J. Electrochem. 31, 509 (1995)).
115. Zwier, T.A. and Burke, M.F.: Chemical modification of carbon adsorbent for liquid chromatography, *Anal. Chem.* **53** (1981), 812-6.
116. Kublanovskiy, V.S., Litovcenko, K.I., Stezeryanskiy, E.A. and Shpak, A.P.: Mechanism of electrochemical reduction of poly(chlorotrifluoroethylene), *Dokl. Akad. Nauk Ukr. SSR Ser. B Geol. Khim. Biol. Nauki* (1989), 36-9.
117. Kublanovskiy, V.S., Litovcenko, K.I., Zub, V.Y. and Stezeryanskiy, E.A.: Effect of irradiation on electrochemical reduction of poly(chlorotrifluorowethylene), *Dokl. Akad. Nauk Ukr. SSR Ser. B Geol. Khim. Biol. Nauki* (1990), 39-42.
118. Rye, R.: Spectroscopic evidence for radiation-induced crosslinking of PTFE, *J. Polym. Sci. B: Polym. Phys.* **31** (1993), 357-64.

119. Rye, R.R. and Shinn, N.D.: Synchrotron radiation studies of PTFE photochemistry, *Langmuir* **6** (1990), 142-6.
120. Rye, R.R.: Electron irradiation of PTFE: effect on adhesion and comparison with X-rays, *Langmuir* **6** (1990), 338-44.
121. Yamada, Y., Inada, Y., Shiraishi, M., Oka, T., Oya, A., Miyashita, K. and Ozawa, T.: Preparation of carbyne by chemical reduction of PTFE, *Proc. 4th IUMRS Conference, Makuhari, Japan, Sept. 16-18, 1997*, p. 13,5.
122. Tasker, S., Chambers, R.D. and Badyal, J.P.S.: Surface defluorination of PTFE by sodium atoms, *J. Phys. Chem.* **98** (1994), 12442-6.
123. Dousek, F.P. and Jansta, J.: Increase of the surface area of a Na/K alloy (in Czech), CS Pat. 184,444 (1974).
124. Jansta, J. and Dousek, F.P.: Electrochemical reduction of polymeric carbon, *Electrochim. Acta* **26** (1981), 233-7.
125. Kavan, L., Dousek, F.P. and Kubelková, L.: Chemisorption of oxygen on carbon produced by electrochemical reduction of poly(tetrafluoroethylene), *Carbon* **24** (1986), 671-6.
126. Lagow, R.J., Kampa, J.J., Wei, H.C., Battle, S.C., Genge, J.W., Laude, D.A., Harper, C.J., Bau, R., Stevens, R.C., Haw, J.F., et al.: Synthesis of linear acetylenic carbon: The sp carbon allotrope, *Science* **267** (1995), 362-7.
127. Springborg, M. and Kavan, L.: On the stability of polyyne, *Chem. Phys.* **168** (1992), 249-58.
128. Heimann, R.B., Kleiman, J. and Salansky, N.M.: Structural aspects and conformation of linear carbon polytypes, *Carbon* **22** (1984), 147-56.
129. Heimann, R.B.: Linear finite carbon chains (carbynes), *Diamond Rel. Mater.* **3** (1994), 1151-7.
130. Udod, I.A., Bulychev, B.M. and Sirotinkin, S.P.: An interaction of amorphous carbyne with sodium: the formation of carbyne intercalation compounds, *Synth. Metals* **60** (1993), 57-62.
131. Kudryavtsev, Y.P., Evsyukov, S.E., Babaev, V.G., Guseva, M.B., Khvostov, V.V. and Krechko, L.M.: Oriented carbyne layers, *Carbon* **30** (1992), 213-21.
132. Tanuma, S. and Palnichenko, A.V.: Synthesis of low density carbon crystal "carbolite" by quenching of carbon gas, *J. Mater. Res.* **10** (1995), 1120-5.
133. Tanuma, S., Palnichenko, A.V. and Satoh, N.: Synthetic of low density carbon crystals by quenching gaseous carbon and intercalation of alkali metal atoms into these crystals, *Synth. Metals* **71** (1995), 1841-4.
134. Palnichenko, A.V. and Tanuma, S.: Intercalation into carbolite, *J. Phys. Chem. Solids* **57** (1996), 1163-6.
135. Rietmeijer, F.J.M.: Are crystalline C-(H-O-N) carbons the elusive meteoritic carbyne? *Meteoritics* **28** (1993), 242-5.
136. Rohlfing, E.A., Cox, D.M. and Kaldor, A.: Production and characterization of supersonic carbon cluster beams, *J. Chem. Phys.* **81** (1984), 3322-30.

Chapter 4: Structural models of carbyne

4.1 STRUCTURAL AND ELECTRONIC PROPERTIES OF POLYYNE

M. SPRINGBORG

Department of Chemistry,
University of Konstanz
D -78457 Konstanz, Germany

Abstract. Results of theoretical studies of electronic and structural properties of carbon chains are reviewed. For an isolated, neutral C_N system the calculations predict that a chain with alternating CC bond lengths has the lowest total energy for N smaller than about 30. For larger values of N, other structures are found to have the lowest total energy. Infinite chains have a lower total energy for alternating bond lengths as compared to those with non-alternating bond lengths. When the chains are being charged (for instance through doping), structural distortions like solitons and polarons will occur, and these will lead to extra single-particle levels in the optical gap. The chains are soft against bending and if two chains are in close contact they interact. This leads to strong modifications of the structure of the individual chains. Therefore, long carbon chains can only be synthesized when assuring that the individual chains will not interact, for example through the introduction of spacers. Modifying the structures by incorporating metal atoms in the chains leads to hybridization effects between carbon π functions and metal d functions. It is predicted that such systems should be interesting within the context of non-linear optics.

1. Introduction

The rich chemistry of carbon may be interpreted as being due to carbon's capability of adjusting its electronic configuration to different bonding situations, i.e., to form bonds from sp, sp^2, and sp^3 hybrids. This arises from the relatively small energy difference of atomic $2s$ and $2p$ functions. (Within the Hartree-Fock approximation, this difference equals 7.4 eV [1], whereas density-functional calculations predict 8.2 eV). The energy difference is, however, not vanishing so that extended compounds based on sp^2-bonded carbon atoms are far more common than those based on sp-bonded atoms. Thus, whenever different isomers exist mainly differing in whether the carbon atoms form sp or sp^2 hybrids, it is likely that the sp^2-bonded isomer will be the more stable one.

Both for threefold coordinated sp^2-bonded and for twofold coordinated sp-bonded carbon atoms the interatomic bonds are partly due to strong, energetically low-lying σ

R.B. Heimann et al. (eds.), Carbyne and Carbynoid Structures, 215–233.

bonds of the sp^n hybrids and partly to weaker, higher-lying π bonds occupied by the remaining carbon valence electrons. The latter are those appearing closest to the gap separating occupied and unoccupied orbitals and are those involved in low-energy excitations. Moreover, they couple strongly to the structure, so that any modification of the distribution of the π electrons leads to a simultaneous modification of the structure and *vice versa*. This electron-phonon coupling has important consequences, as we now shall see.

One consequence of the electron-phonon coupling is the occurrence of the so-called dimerization. Peierls [2] considered an infinite one-dimensional chain of identical atoms with one electron per atom. The electronic orbitals form one band and the Fermi level cuts it in such a way that the band becomes exactly half filled. Peierls showed that due to electron-phonon couplings this system is unstable against a distortion that results in alternating shorter and longer bonds and in the simultaneous occurrence of a gap at the Fermi level.

Polyenes, $(CH)_x$, may to a certain approximation be considered a realisation of these systems when only considering the single π electron per CH unit and neglecting that the systems are not truly one-dimensional. Also polyynes may be described within this model, although one here has two π electrons per carbon atom. These electrons are, however, energetically degenerate. For polyenes, the bond-length alternation results, therefore, essentially in alternating single and double bonds, but for polyynes in alternating single and triple bonds.

As a further consequence of the electron-phonon coupling, solitons and polarons may occur. For the one-dimensional model system of Peierls, the two structures with alternating bond lengths (but differing in this alternation) will be energetically degenerate. Then, interfaces separating these two structures may occur. According to the model of Su, Schrieffer, and Heeger [3] for infinite polyenes, these are stable when the chain contains a small number of extra or less electrons. The interfaces are believed to be not sharp but extended over some tens of CH units. Moreover, they result in extra levels in the middle of the gap, and the corresponding orbitals are localized to the regions of the interfaces. The combined object of lattice distortion and localized orbital is the so-called soliton. Also for the polyynes these quasi-particles may exist (cf. Fig. 1).

Real chains are three-dimensional objects. The structures may (but need not) be such that they are no longer degenerate if they differ only in the bond-length alternation. Also in that case, extra charges (electrons or holes) may be confined to finite regions of the chain instead of entering band orbitals. Here, however, the structure is locally deformed from the low-energy structure towards the high-energy structure. Due to the electron-phonon couplings, these lattice distortions lead to localized orbitals that often have energies in the gap at the Fermi level. In this case, the quasi-particles are called polarons.

In the present contribution we shall explore whether - and how far - these qualitative considerations are recovered in theoretical studies of the properties of polyynes. We shall thereby concentrate on studies based on parameter-free methods.

These methods will be described briefly in the next Section. In Section 3 we shall review the results obtained for structural properties of single, finite or infinite, neutral polyynes. In Section 4 we shall look at properties of single chains containing solitons or polarons, whereby mainly model Hamiltonians will be employed. Section 5 is devoted to a discussion of the effects of interchain interactions that were so far mostly neglected, and in Section 6 we will discuss polyynes containing heteroatoms in the backbone. Section 7 contains the main conclusions.

2. Theoretical methods

In the present work we shall focus on results of theoretical studies of ground-state properties of carbon chains. Most of the studies, we shall refer to, are based on parameter-free methods. They are implemented either within the Hartree-Fock approximation or improvement thereof, or within the density-functional formalism. For the sake of convenience, we shall here give a brief outline of the general principles behind those approaches.

We assume that the Born-Oppenheimer approximation is valid. Consequently, the nuclei are treated as point charges at fixed positions. The total energy of the system is written as one term from the electrons moving in the field of the nuclei and of each other plus one term due to the electrostatic interactions of the nuclei. The electronic part is calculated by employing quantum theory. This can be done by solving the many-body Schrödinger equation:

$$\hat{H}\Psi = E\Psi \tag{1}$$

where $\hat{H}$ is the Hamilton operator for the (N) electrons moving in the field of the nuclei and of each other, and Ψ is the N-electron wavefunction.

Within the Hartree-Fock approximation one assumes that Ψ is written as a single Slater determinant containing N single-electron wavefunctions ψ_i. Eq. (1) is then transformed into the Hartree-Fock equations:

$$\hat{F}\psi_i = \varepsilon_i \psi_i \tag{2}$$

where $\hat{F}$ is the Fock operator. The operator $\hat{F}$ depends on all the occupied orbitals, such that solving Eq. (2) requires an iterative process. The total electronic energy, E of Eq. (1) can be calculated from the orbitals ψ_i and the total electron density is simply given by:

$$\rho(\vec{r}) = \sum_{i=1}^{N} \left|\psi_i(\vec{r})\right|^2 \tag{3}$$

Per definition, the Hartree-Fock approximation neglects correlation effects. These can be included when writing Ψ as a linear combination of more Slater determinants, whereby the calculations become more involved, but here we shall not discuss this further.

In solving Eq. (2) one expands the eigenfunctions ψ_i in some set of basis functions:

$$\psi_i(\vec{r}) = \sum_{j=1}^{K} c_{ij}\chi_j(\vec{r}) \tag{4}$$

where most often Gaussians are selected, but other choices are frequently made, too.

An alternative approach for calculating the energy E from Eq. (1) is offered by the density-functional formalism of Hohenberg and Kohn [2]. They showed that any ground-state property of a given system is a unique functional of the density. Thus, E is a unique functional of ρ, but unfortunately the functional is unknown. Within the formulation of Kohn and Sham [5], the problem of calculating E is transformed into that of solving the Kohn-Sham single-particle equations:

$$\hat{H}_{eff}\psi_i(\vec{r}) = \varepsilon_i\psi_i(\vec{r}) \tag{5}$$

These equations have the same structure as the Hartree-Fock equations (2), but in contrast to the operator $\hat{F}$, here the operator $\hat{H}_{\text{eff}}$ does not depend on the individual orbitals but only on the total density. This is given by Eq. (3). as in the Hartree-Fock case.

The density-functional approach does in principle include correlation effects, but the main problem is that the precise form of $\hat{H}_{\text{eff}}$ is not known. However, the currently applied approximations have proven to provide accurate results for a large number of systems and properties.

Also here, a self-consistent approach has to be applied in order to solve Eq. (5). As for the Hartree-Fock studies, it is most often solved by expanding the single-particle eigenfunctions in a set of basis functions as in Eq. (4).

Independent on whether the Hartree-Fock-based or the density-functional-based methods are used, infinite, periodic systems (like an infinite polyyne chain) are treated by making use of the translational periodicity. Thus, the orbitals are classified according to a continuous (one-dimensional) Bloch wavenumber (k).

3. Properties of neutral polyynes

One of the main questions is whether polyynes are stable at all. As discussed in the Introduction, compounds based on sp^2 hybrids are believed to be more stable than those based on sp hybrids, and one may thus speculate that (when comparing the two types for compounds containing only carbon atoms) the former will be the more stable ones. There have been only few theoretical studies devoted to this question, in particular following the observation that certain finite all-carbon structures, such as the C_{60} molecule, can be highly stable.

Parasuk and Almlöf [6] studied different forms of C_{20} including ring molecules resembling those of Fig. 1 (a) and 1 (b), i.e., chains of carbon atoms with either constant or alternating bond lengths. In contrast with the structures of Fig. 1, Parasuk and Almlöf assumed that the chains were closed rings. They compared these structures with a compact fullerene-like molecule. Applying the Hartree-Fock approximation they found that the ring molecules were more stable ones, but upon inclusion of some correlation effects they found that the fullerene-like molecule was the most stable one. Comparing only the ring (i.e., chain-like) structures, their results predict that the structure with non-alternating bond lengths has a lower total energy than that with alternating bond lengths. This finding was independent on whether or not correlation effects were included, and it is in contrast to the intuitive expectations for the infinite chain that were discussed in the Introduction.

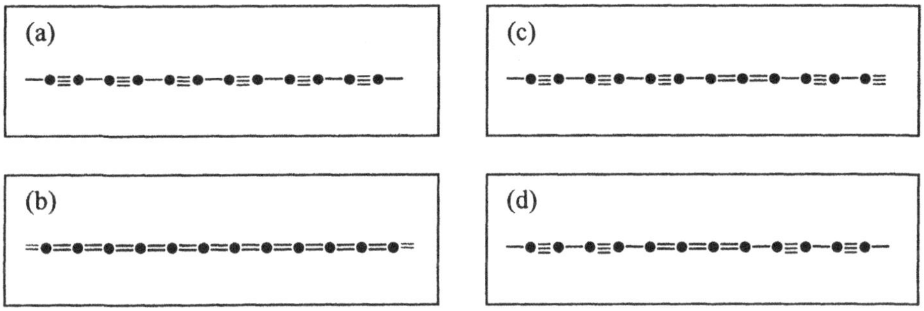

Figure 1. Structure of a linear carbon chain with (a) alternating and (b) constant bond lengths. Figure (c) shows a chain containing soliton and Figure (d) shows a chain containing polaron.

A related but more extended study was carried out by Feyereisen *et al.* [7]. They considered structures with 20 - 60 carbon atoms and found that the fullerene-like structures were stable for molecules containing more than about 30 carbon atoms, independent on whether or not the correlation effects were included. In agreement with intuition, they found - within the Hartree-Fock approximation - that the structure with alternating bond lengths possesses a lower total energy than that with constant bond lengths. It is more surprising that the inclusion of correlation caused a reversion of this finding for larger systems. Finally, their calculations predicted that the lengths of single, double, and triple bonds are 136, 128, and 120 pm, respectively.

Raghavachari *et al.* [8] studied different isomers of C_{20}. They applied both Hartree-Fock-based and density-functional-based methods and arrived at the somewhat unsatisfying conclusion that the relative stability depends on the method used. The overall result is, however, that the ring structure has a lower total energy than those of a fullerene-like structure and of a cage structure that resembles a finite part of C_{60}.

The general conclusion of these studies is that polyynes, C_N are stable over other isomers when N is not too large, i.e., about 30. The bond lengths of these structures are converged for much smaller N. Hence, the C_{30} systems can be considered good approximations to the infinite chains, and one should thus expect similar bond lengths for the finite ring molecules and for the infinite, linear chains. The question whether a bond-length alternation occurs for finite polyynes, seems, however, not to have been settled theoretically yet. The results of Feyereisen *et al.* [7] indicate that the outcome may depend on the inclusion of correlation effects, but we shall now explore the results of calculations on truly infinite chains.

The properties of infinite polyyne were first theoretically explored by Kertész *et al.* [9, 10] who used the Hartree-Fock method. By considering a finite set of structures, both with and without a bond-length alternation, they found that structures with alternating bond lengths have the lowest total energy. The bond-length alternation is accompanied by an opening up of a gap at the Fermi level in agreement with the general considerations described in Section 1. They estimated that the lowest total energy occurs for bond lengths of 141 and 115 pm for single and triple bonds, respectively.

Similar results were also found by Karpfen [11] in his Hartree-Fock study on infinite polyyne. He estimated, however, a somewhat smaller difference in the bond lengths: 137 and 119 pm for single and triple bonds, respectively, whereas the lengths of double bonds were estimated to be 127 pm.

Teramae *et al.* [12] used a slightly different approach based on the Hartree-Fock approximation but with effective potentials such that the core electrons were treated only indirectly. Their results on infinite polyyne differ slightly from those discussed above and led to bond lengths of 132 and 113 pm.

A density-functional study by the present author [13] gave larger bond lengths than those of the Hartree-Fock studies, i.e., 144, 133, and 125 pm for single, double, and triple bonds, respectively. In a more recent study with improved computational codes, values of 138, 129, and 123 pm have been obtained [14]. In both studies, the structure with alternating bond lengths is found to have a lower total energy than that with constant bond lengths.

Thus, for the infinite polyynes there is a general agreement that the lowest total energy occurs for a structure with alternating bond lengths. This result is independent on the applied computational approach and on whether or not correlation effects are included. This apparently contrasts to the case for the finite polyynes.

In a recent experimental study, Lagow *et al.* [15] determined the bond lengths of finite molecules containing segments of a linear carbon chain with up to four triple bonds. According to the discussion above, this may resemble an infinite polyyne. They reported lengths of 117-122 and 135-138 pm for the CC bonds. These values can only be interpreted as being due to triple and single bonds, respectively, i.e., a bond-length alternation is observed. The bond lengths agree well with those found in the theoretical studies on infinite polyyne, while the earlier density-functional values being slightly larger and the Hartree-Fock+core-potentials values being slightly smaller. The differences are, however, so small that we arrive at a general conclusion that the total energy is lowered for structures with alternating bond lengths. This is in agreement with the consensus of the Peierls' distortion.

A very important quantity is the variation in the total energy upon the bond-length alternation. Fig. 2 shows the total energy per atom for a linear polyyne chain with a unit-cell length kept fixed at 259 pm as a function of the difference in lengths between shorter and longer bonds. These results were obtained using a density-functional method [14], considering structures with two atoms per unit cell and gradually moving every second atom in unison away from the positions at the midpoints between the nearest neighbours. It is obvious that the total energy varies very little upon the bond-length alternation. Hence, regions, where the bond-length alternation is modified or even suppressed, may easily occur at moderate temperatures or as a consequence of other external perturbations.

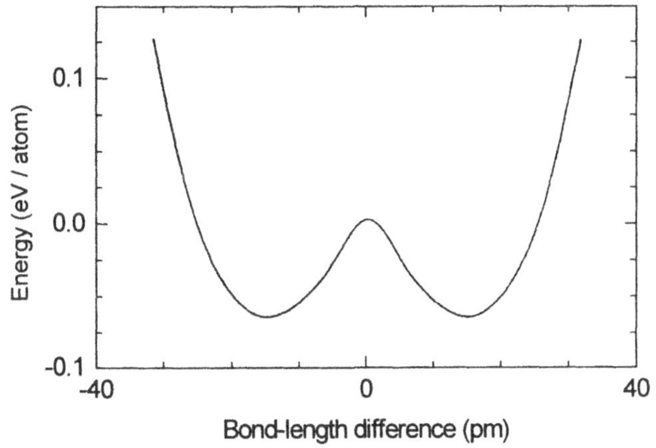

Figure 2. Relative total energy (in eV per carbon atom) as a function of bond-length alternation (expressed as a difference in lengths of shorter and longer bonds) for an infinite polyyne with a unit-cell length of 259 pm.

As discussed in Section 1, the bond-length alternation is accompanied by the occurrence of a gap at the Fermi level. This is clearly seen in Fig. 3, where we show the band structures from a density-functional study on polyyne [14] for the structures of the local total-energy maxima and minima in Fig. 2.

The band structures of Fig. 3 show also how the electronic orbitals split into energetically deep lying σ orbitals. They give rise to two bands and the π bands around the Fermi level. Although these findings are common to all electronic-structure studies of polyyne, the results of Hartree-Fock and density-functional calculations differ in two important aspects. First, none of these methods is capable of describing the gap at the Fermi level correctly. The Hartree-Fock methods lead to a large overestimate (up to several eV), whereas the density-functional methods produce too small gaps (by about 0.5-1 eV). Second, the occupied bands from Hartree-Fock studies are, in general, too broad whereas those from density-functional studies most often compare well with experimental data. This has been discussed, e.g., by Seki *et al.* [16] and Springborg and Lev [17] for polyethylene. These results are recovered here.

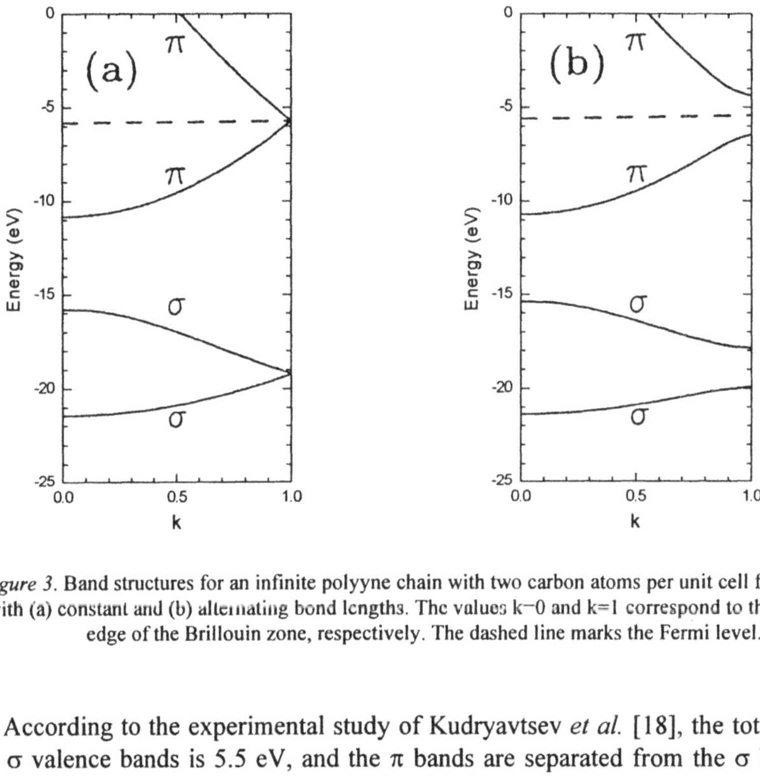

Figure 3. Band structures for an infinite polyyne chain with two carbon atoms per unit cell for a structure with (a) constant and (b) alternating bond lengths. The values k=0 and k=1 correspond to the center and edge of the Brillouin zone, respectively. The dashed line marks the Fermi level.

According to the experimental study of Kudryavtsev *et al.* [18], the total width of the σ valence bands is 5.5 eV, and the π bands are separated from the σ bands by a gap of 5 eV. These values agree perfectly with the results of Fig. 3. On the other hand, the Hartree-Fock study of Karpfen [11] gave a width of the σ valence bands over 10 eV and the gap between the σ and π valence bands of 5.5 eV.

4. Solitons and polarons

One of the main results of the previous Section is that for a long (infinite) polyyne chain, the structure with alternating bond lengths has a lower total energy than that with constant bond lengths. It is now obvious from Fig. 1 (a) and Fig. 2 that the two structures, differing only in an interchange of single and triple bonds, are energetically degenerate. Interfaces or domain walls separating the two structures may thus occur, as shown in Fig.1 (c). Their existence for polyenes was predicted by Su *et al.* [3] who also presented a simple model exploring the properties of these so-called solitons.

The model was explicitly developed for polyenes, $(CH)_x$, which differ from polyynes by the fact that the CC bonds are alternating single and double bonds or, equivalently, that there is only one π electron per carbon atom. Su *et al.* observed that, for low-energy excitations, only the π electrons were involved. Therefore, they treated only the π electrons explicitly, and applied furthermore a single-particle, tight-binding model. Compared with the simple Hückel approach the only difference is that the resonance integrals are assumed to depend (linearly) on the corresponding bond lengths.

The contribution from the π electrons to the total energy is subsequently approximated as the sum of the single-particle energies of all occupied orbitals:

$$E_\pi = \sum_{i=1}^{N} \varepsilon_i \tag{6}$$

where N is the number of electrons. The energy ε_i is determined from the single-particle equation:

$$\hat{H}_\pi \psi = \varepsilon_i \psi \tag{7}$$

ψ_i is written as a linear combination of π functions on all sites (i.e., carbon atoms):

$$\psi_i = \sum_{j=1}^{N} c_{ij} \chi_j \tag{8}$$

where χ_j is the π function of the jth site. Finally, $\hat{H}_\pi$ is defined through its matrix elements:

$$\langle \chi_k | \hat{H}_\pi | \chi_l \rangle = -\delta_{k,l\pm1}(t_0 - \alpha d_{kl}) \tag{9}$$

where d_{kl} is the bond length between carbon atom k and l.

The total energy is written as a sum of E_π and a harmonic term:

$$E_{tot} = E_\pi + E_\sigma \tag{10}$$

with:

$$E_\sigma = \frac{K}{2} \sum_{i=1}^{N-1} d_{i,i+1}^2 \tag{11}$$

This model contains three parameters, t_0, α and K. Su *et al.* determined their values, while the experimentally observed structure (with a bond-length alternation) and band gap for infinite polyenes were reproduced.

The model has subsequently been used to study other conjugated polymers as well as polarons (see, e.g., Ref. [19]). A polaron is a local distortion in the otherwise regular backbone, as shown in Fig. 1 (d). Although the model has been successful in explaining - qualitatively and in some cases also quantitatively - a number of experimentally observed features, it does rely on some approximations that may or may not be justified. Thus, both the explicit forms of $\hat{H}_\pi$ and E_σ may be questioned, as well as the values of the parameters. Further complications arise when this model together with the parameter values is applied for other systems than polyenes.

One of the predictions of the model of Su *et al.* [3] is that a soliton leads to a single-particle level exactly at the mid-gap position, and that a polaron produces two levels that are placed exactly symmetric in the gap. Each of these levels can accommodate two electrons (due to spin degeneracy), and an interesting consequence is that the infinite, neutral chain containing a soliton has a spin equal to ½, whereas the singly charged, infinite chain is spin-less.

Rice and coworkers [20, 21] noticed that polyyne is very similar to polyene but with the extra feature that the number of π electrons is doubled. Thus, all levels are not two-fold but four-fold degenerate (see Fig. 4). Solitons will accordingly lead to four degenerate levels in the gap, and polarons to eight levels that appear as two sets of four degenerate levels. For the neutral system, half of these are occupied. However, both for the neutral and the charged systems, the different ways of occupying these levels lead to a large number of possible states, as noticed by Rice *et al.* [20].

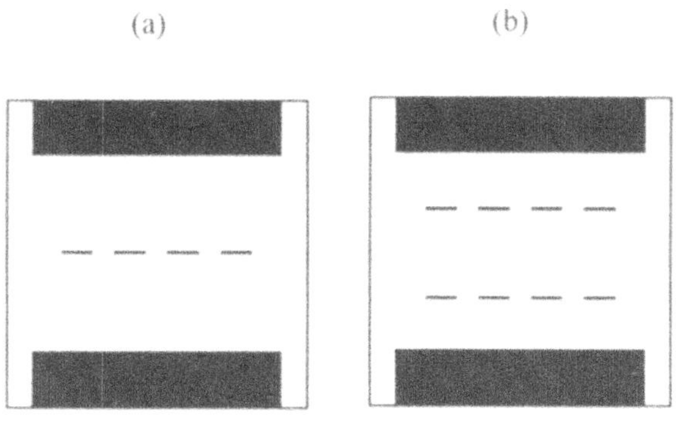

Figure 4. Energy levels for a polyyne chain containing (a) soliton and (b) polaron. The filled areas represent the valence and conduction bands separated by a gap. This gap contains four (eight) extra levels for a soliton (polaron) that, for the neutral system, contain in total two (four) electrons.

Rice and coworkers had no experimental or theoretical data from which they could determine the precise values of the parameters entering the model, as well as exploring the validity of the model. Therefore, they used the parameter values valid for polyenes. It can, however, not be excluded that a model explicitly developed for polyynes would lead to other conclusions. Therefore, we decided to study the properties of polyyne in detail [22].

We assume that Eq. (10) is still valid, and that E_π is given by Eqs. (6) and (7). However, for the matrix elements of Eq. (9) we will allow those of more distant functions to be non-vanishing. We will determine their precise values as well as the dependence on bond lengths from the results of density-functional calculations on different periodic structures of polyyne.

We fix the length of the unit cell containing two carbon atoms and move gradually every second carbon atom in unison from the position at the midpoint between neighbouring carbon atoms towards one of the two neighbours. The total energy will then vary as in Fig. 2, which is supposed to be given by Eq. (10). Simultaneously, the band structures will change as in Fig. 3. The band structures of the occupied bands are to be given by Eqs. (7) and (8) and by analyzing all the occupied π bands for the different structures, the values of $<\chi_k | \hat{H}_\pi | \chi_\pi>$ can be determined. It turned out [22] that also the second-nearest-neighbour interactions shall be included and that the parameter values are markedly different from those of polyene.

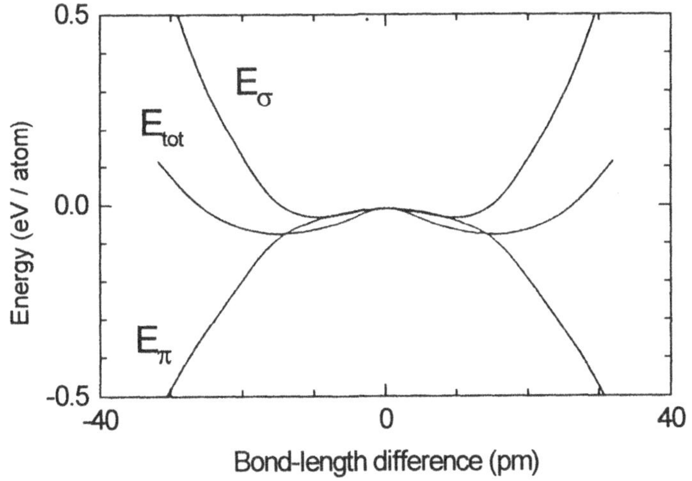

Figure 5. An example of how the total energy per carbon atom is split into a π-term and a σ-term according to Eq. (10). The results were obtained from the density-functional calculations that also gave the results of Fig. 2 and 3.

Once $\hat{H}_\pi$ is completely defined and E_{tot} known, E_σ is easily obtained as the difference from Eq. (10). This can subsequently be fitted with a simple analytical function of the nearest-neighbour bond lengths, and it turned out that it is not a harmonic function as that of Eq. (11) but anharmonicities are to be included. Fig. 5 displays how E_{tot} of Eq. (10) is split into the two contributions according to the density-functional results.

Once the parameters of the model Hamiltonian are determined we can use the model in studying solitons and polarons. Compared with the results of Rice and coworkers, the gap is significantly reduced. Furthermore, we found a very rich spectrum of stable and metastable solitons and polarons for a given number of electrons and sites mainly differing in the way the electrons are distributed among the gap levels of Fig. 4. These gap levels were still found very close to the mid-gap position for solitons and were symmetric in the gap for polarons [22]. In total, polyyne is expected to have a very rich excitation spectrum, both in the neutral and in the charged case.

5. Interchain interactions

The results reported in Section 3 show that polyyne is not the thermodynamically most stable form of carbon, in agreement with common knowledge. Thus, a linear carbon chains are expected to transform into some other form of carbon. Kavan *et al.* [23] have proposed that the chains may interact and form graphite-like (graphene) sheets [1]. From a theoretical point of view, it is difficult to address the dynamical properties of these poorly defined materials. Accordingly, there have been so far no electronic-structure studies of the transformation of polyyne. However, some relevant information has been obtained from electronic-structure calculations, which we shall now review.

The bond angles of polyyne and graphene equal 180° and 120°, respectively. Thus, forming graphene from polyyne requires both bending the chains and bringing them into contact. Each of these processes has been addressed by density-functional studies.

It was found [22] that a single, isolated, infinite polyyne it is very soft against changing it into a zigzag configuration (cf. Fig. 6). The total energy shows very small changes for bond angles varying between 180° and 150°, but it increases for smaller bond angles. However, the *sp* hybrids have partly changed into sp^2 hybrids of which one per carbon atom becomes unsaturated. If we bring the chains into contact it is likely that these orbitals will interact. They ultimately form chemical bonds between the chains, which stabilise the zigzag structure.

In another study [24] two zigzag chain were brought into contact in order to see whether they will bond. The variations in the total energy (Fig. 7) upon distance between the two chains show that this really occurs. The fact that the lowest total

[1] See also Fig. 4 in Chapter 6.1 of this book

energy is found for a structure consisting of two linear (i.e., not zigzag) chains may once again be related to unsaturated sp^2 hybrids.

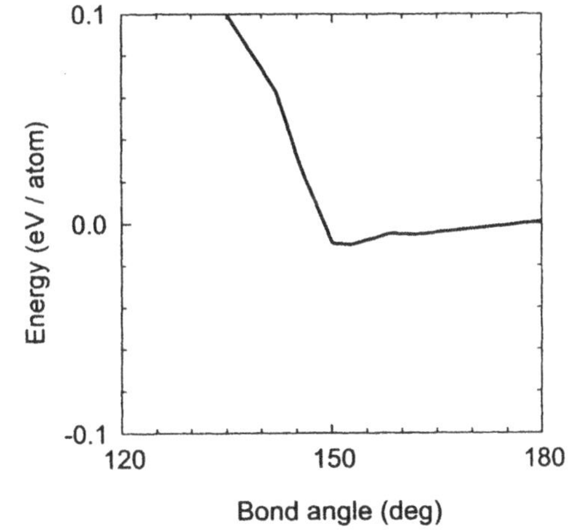

Figure 6. Variation in total energy for a single polyyne chain as a function of bond angles. The bond lengths are alternating and kept fixed at 126 and 150 pm.

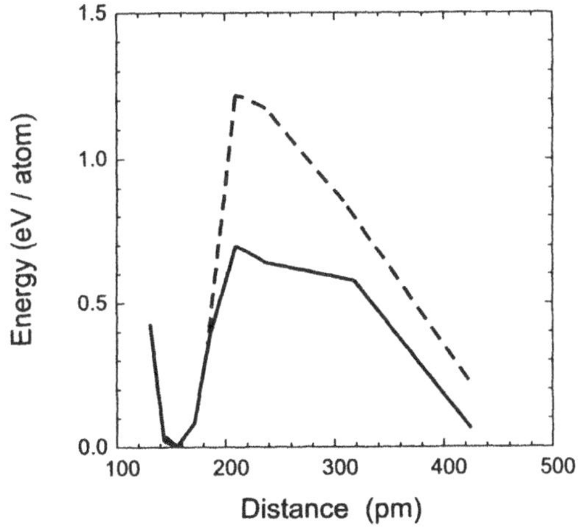

Figure 7. Variation in total energy for two interacting polyyne chains as a function of the distance between them. The dashed curve corresponds to results for all bond angles equal to 180° whereas the solid curve refers to the case where the bond angles have been varied in order to find the local minimum for fixed interchain distance. All intra-chain bond lengths were kept fixed at 153 pm.

Interchain interactions may also have consequences for solitons. Without these interactions, the two structures of Fig. 8 will be energetically degenerate. This means in turn that a soliton created on one chain will be able to move up and down it without energy costs. However, interchain interactions may lead to the preference of one of the two structures of Fig. 8 over the other. In that case, a soliton on one chain will move in that direction which leads to the largest segment of the low-energy conformation.

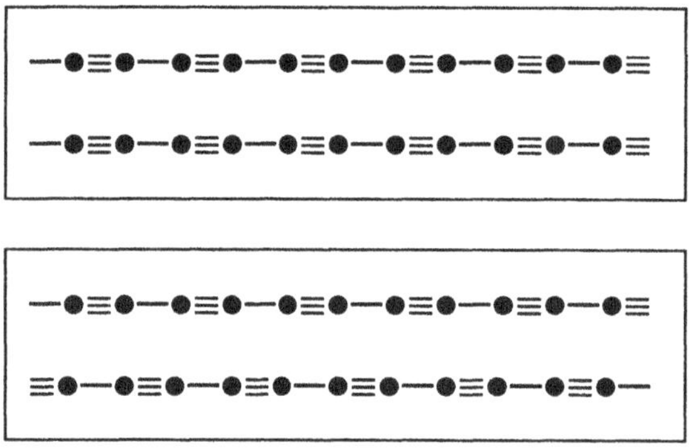

Figure 8. Two examples of the arrangement of two parallel polyyne chains.

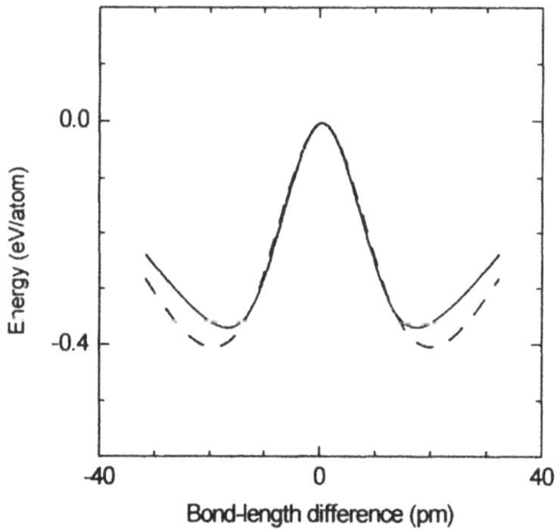

Figure 9. Relative total energy per two atoms for the two two-chain arrangements of Fig. 8 as a function of bond-length alternation. The solid and dashed curves correspond to the parallel and the antiparallel arrangement (upper and lower part of Fig. 8), respectively.

In order to address this issue, density-functional calculations were performed for different arrangements of two parallel chains [25]. Fig. 9 shows the results for the two arrangements of Fig. 8. The interchain distance was kept fixed at 420 pm, and the bond lengths of the two chains were varied in unison so that the bond-length alternation was the same on them. The length of one unit cell containing two atoms per chain was kept at 270 pm. The results shown in Fig. 9 demonstrate that there is a total-energy difference, although it is small. Therefore, the solitons may be severely influenced by the interchain interactions, and, ultimately, charged chains may contain polarons instead, or a soliton may be delocalized over more chains.

6. Metal-containing polyynes

There are several reasons for incorporating heteroatoms into the backbone of a polyyne. First, as discussed in Section 3, an infinite polyyne is not the stable form of carbon, whereas finite oligomers might well be. Thus, by inserting heteroatoms in between finite segments of the polyyne, the resulting structure may be more stable. Secondly, by modifying the polymer, one may also modify and ultimately control the physical and chemical properties of the material. Finally, from a theoretical point of view, polyynes are ideal systems for studying many of the properties of the sp^2-based conjugated polymers like polyenes. Many of those are determined by the orbitals closest to the Fermi level, i.e., by the π orbitals that are common to both types of polymers. Metal-containing polyynes serve then as model systems for the larger class of metal-containing conjugated polymers.

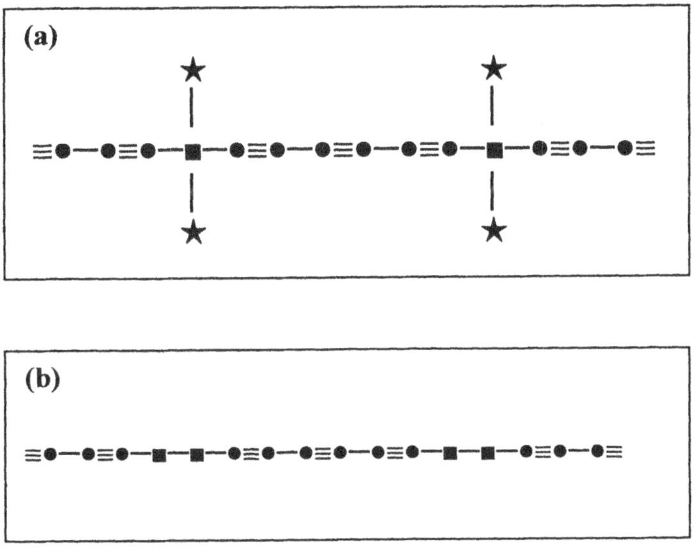

Figure 10. Schematic representation of metal-containing polyynes. The solid circles mark carbon atoms; (a) the solid squares mark Pt atoms and the stars PH_3 groups, (b) the solid squares mark Al, Na, or Pd atoms.

Fig. 10 (a) shows a polymer containing six-atom polyyne segments separated by Pt atoms. This polymer is closely related to one recently studied experimentally by Wittmann *et al.* [26], except that the middle CC pair was replaced by a phenylene ring and that the hydrogen atoms of the PH_3 groups were replaced by C_4H_9 groups. However, for a general understanding of such systems, the system of Fig. 10 (a).

provides a good approximation. This is confirmed by the semiempirical study of Lhost *et al.* [27] who compared the system of Fig. 10 (a) with that studied experimentally by Wittmann *et al.* [26]. They showed that for the bands closest to the Fermi level the two systems possess very similar properties.

One of the most interesting questions is whether the presence of the Pt atoms changes the system into one that essentially consists of non-interacting C_6 segments, or whether the electrons (at least those closest to the Fermi level) are still delocalized all over the backbone. According to a recent density-functional study [28] the frontier orbitals have large components both on the C atoms and on the Pt atoms, indicating that the electrons are, in fact, delocalized. This conclusion is supported by analyzing the electron distribution, both that of single orbitals and the total one.

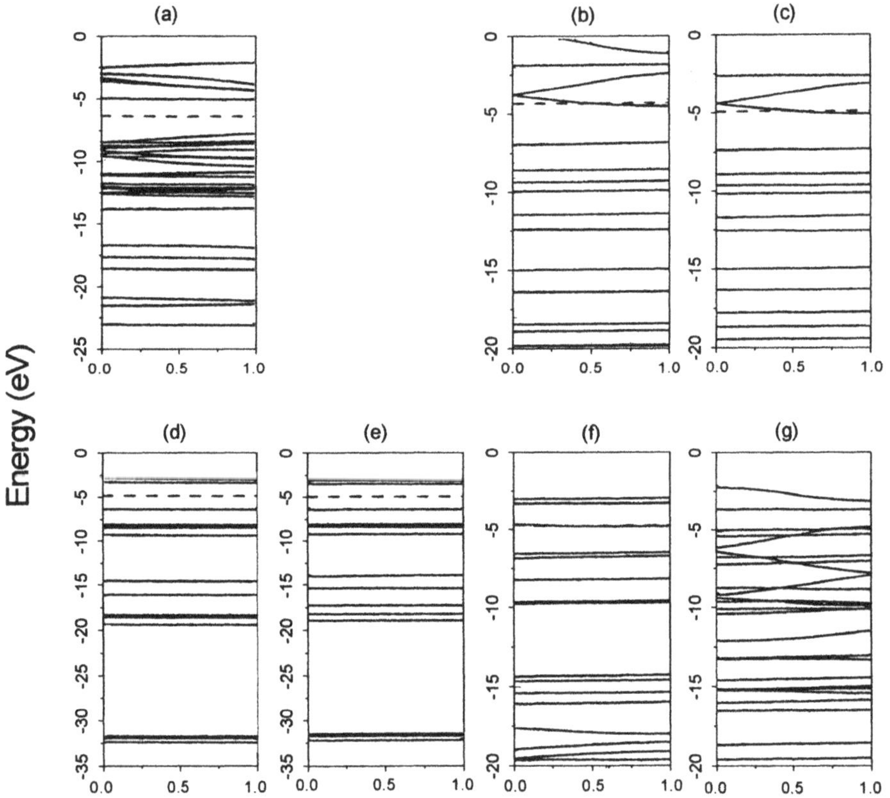

Figure 11. The band structures for different metal-containing polyynes. Plot (a) corresponds to the structure of Fig. 10 (a); the others to that of Fig. 10 (b). Plots (b) and (c) correspond to Al. Plots (d) and (e) to Na, and (f) and (g) to Pd. In (c), (e), and (g) an extra DC field has been applied, which is not the case for the other plots. The notation is otherwise as in Fig. 3. Note the different energy scales.

Nevertheless, the band structures [Fig. 11 (a)] show some effects that are due to localization of the electrons. First, compared with the band structures of pure polyyne (Fig. 3), the gap at the Fermi level is larger. Secondly, the bands are split into a number of narrower bands separated by non-negligible gaps. However, it is obvious that the incorporation of metal atoms into the backbone of a polyyne opens up new possibilities of controlling and fine-tuning of the properties of polyyne.

This possibility attracts practical interest within the field of non-linear optics. Conjugated polymers based on a backbone of sp^2 bonded carbon atoms show unusually large non-linear responses to electromagnetic fields, which often is ascribed to the π electrons. These show large responses due to their semi-free-electron-like behaviour, but not completely linear responses, since they are bound to the atomic nuclei. Polyynes should therefore also possess large non-linear responses.

As a first approach to study these effects, we have considered the polyynes of Fig. 10 (b) where different types of metal atoms (Al, Na, or Pd) were incorporated into the backbone. It should be stressed that these systems do not represent any existing systems (yet) but they were rather studied in order to arrive at general conclusions. Applying a DC field perpendicular to the chain axis leads to redistributions of the electrons. The changes in the band structures provide one way of monitoring the responses of the electrons to this field. Figures 11 (b) - 11 (g) show the band structures both without and with an applied field.

In order to study the effects of the field more easily, its strength is larger than that currently available experimentally. Fig. 11 demonstrates that the changes are the smallest for the Na-containing polyyne and the largest for the Pd-containing polyyne. This result can be understood by applying a perturbation-theoretical argument. The changes are larger if more orbitals appear close to the Fermi level. Therefore, metal atoms with a large number of occupied and unoccupied orbitals around the Fermi level (like Pd, but in contrast to Na) should be the most interesting candidates for improving the responses of polyyne to applied electromagnetic fields.

7. Conclusions

In the present contribution we have focused on the interplay between electronic properties on one hand, and the structure and composition of polyyne on the other hand. In particular, we have focused on how these aspects can be described using theoretical methods, especially by the so-called parameter-free methods.

In agreement with simple principles, an infinite carbon chain is found to prefer a structure with alternating bond lengths. This alternation is accompanied by the occurrence of a band gap at the Fermi level. The alternating structure presumably prevails also for finite, not too short chains, although the theoretical studies so far have been less conclusive in this respect. Long polyyne chains are thermodynamically unstable and structures containing sp^2-bonded carbon atoms instead of sp-bonded ones will be preferred. The calculations predict that a single, infinite polyyne chain is very soft against bending. Two parallel polyyne chains show a high tendency toward crosslinking if they are brought into a bonding distance. For finite polyynes, sp^2-bonded systems are found to have total energies comparable to those of the *sp*-bonded systems. Only for the absolutely smallest systems, polyynes appear to represent the most stable forms. Thus, in order to produce longer, stable polyynes, one need to stabilize them, e.g., by introducing spacers between different chains or within the chains (see, for example, Chapter 4.2 of this book).

The strong coupling between structural and electronic degrees of freedom leads to the possibility of the existence of solitons and polarons [1]. The calculations could be used to develop a model explicitly for polyyne that subsequently could be employed to study these structures. It turned out that polyyne possesses a very rich spectrum of (meta-)stable solitonic and polaronic structures, both for neutral and for charged chains. However, for real systems containing more than one chain, the solitons will not be free to move up and down the chain due to interchain interactions. With respect to the richness of the excitation spectrum, relaxation processes will be very complex.

One way of stabilizing the polyynes is to incorporate heteroatoms into their backbone. We considered the possibility of including metal atoms and showed that these will lead to only modest changes in the conjugation and that they, moreover, may have very positive effects on the (non-linear) response to external electromagnetic fields.

8. References

1. Huzinaga, S. and Klobukowski, M. (1993) Well-tempered Gaussian basis sets for the calculation of matrix Hartree-Fock wavefunctions, *Chem. Phys. Lett.* **212**, 260-264.
2. Peierls, R. (1955) *Quantum Theory of Solids*. Clarendon Press, Oxford, 108.
3. Su, W.P., Schrieffer, J.R. and Heeger, A.J. (1979) Solitons in polyacetylene, *Phys. Rev. Lett.* **42**, 1698-1701.
4. Hohenberg, P. and Kohn, W. (1964) Inhomogeneous electron gas, *Phys. Rev.* 136, B864-B871.
5. Kohn, W. and Sham, L.J. (1965) Self-consistent equations including exchange and correlation effects, *Phys. Rev.* **140**, A1133-A1138.
6. Parasuk, V. and Almlöf, J. (1991) C_{20}: the smallest fullerene?, *Chem. Phys. Lett.* **184**, 187-190.
7. Feyereisen, M., Gutowski, M. and Simons, J. (1992) Relative stabilities of cumulene and polyacetylene structures for C_n: n = 18-60, *J. Chem. Phys.* **96**, 2926-2932.
8. Raghavachari, K., Strout, D.L., Odom, G.K., Scuseria, G.E., Pople, J.A.,Johnson, B.G. and Gill, P.M.W. (1993) Isomers of C_{20}. Dramatic effect of gradient corrections in density functional theory, *Chem. Phys. Lett.* **214**, 357-361.
9. Kertész, M., Koller, J. and Ažman, A. (1978) Ab initio Hartree-Fock crystal orbital studies. II. Energy bands of an infinite carbon chain, *J. Chem. Phys.* **68**, 2779-2782.
10. Kertész, M., Koller, J. and Ažman, A. (1979) Different orbitals for different spins for solids: Fully variational ab initio studies on hydrogen and carbon atomic chains, polyene, and poly(sulphur nitride), *Phys. Rev. B* **19**, 2034-2040.
11. Karpfen, A. (1979) Ab initio studies on polymers. I: The linear infinite polyyne, *J. Phys. C* **12**, 3227-3237.
12. Teramae, H., Yamabe, T. and Imamura, A. (1983) Ab initio effective core potential studies on polymers, *Theor. Chim. Acta (Berlin)* **64**, 1-12.

[1] (Note by the Editor) Such solitons (topological solitons) have been found experimentally by ESR studies of carbynoid structures (see Chapter 6.2 of this book).

13. Springborg, M. (1986) Self-consistent, first principles calculations of the electronic structures of a linear, infinite carbon chain, *J. Phys. C* **19**, 4473-4482.
14. Springborg, M. (unpublished).
15. Lagow, R.J., Kampa, J.J., Wei, H.C., Battle, S.L., Genge, J.W., Laude, D.A., Harper, C.J., Bau, R., Stevens, R.C., Haw, J.F. and Munson, E. (1993) Synthesis of linear acetylenic carbon: the sp carbon allotrope, *Science* **267**, 362-367.
16. Seki, K., Ueno, N., Karlsson, U.O., Engelhardt, R., Koch, E. E. (1986) Valence bands of oriented finite linear chain molecular solids as model compounds of polyethylene studied by angle-resolved photoemission, *Chem. Phys.* **105**, 247-265.
17. Springborg, M. and Lev, M. (1989) Electronic structures of polyethylene and poly(tetrafluoroethylene), *Phys. Rev. B* **40**, 3333-3339.
18. Kudryavtsev, Yu.P., Evsyukov, S.E., Babaev, V.G., Guseva, M.B., Khvostov, V.V. and Krechko, L.M. (1992) Oriented carbyne layers, *Carbon* **30**, 213-221.
19. Heeger, A.J., Kivelson, S., Schrieffer, J.R. and Su, W. P. (1988) Solitons in conducting polymers, *Rev. Mod. Phys.* **60**, 781-850.
20. Rice, M.J., Phillpot, S.R., Bishop, A.R., Campbell, D.K. (1986) Solitons, polarons, and phonons in the infinite polyyne chain, *Phys. Rev. B* **34**, 4139-4149.
21. Phillpot, S.R., Rice, M.J., Bishop, A.R., Campbell, D.K. (1987) Dynamics, photoexcitation, and coherent anharmonicity in polyyne, *Phys. Rev. B* **36**, 1735-1744.
22. Springborg, M., Drechsler, S.L. and Málek, J. (1990) Anharmonic model for polyyne, *Phys. Rev. B* **41**, 11954-11966.
23. Kavan, L., Dousek, F.P. and Micka, K. (1990) Time-dependent electrical resistivity of carbon, *J. Phys. Chem.* **94**, 5127-5134.
24. Springborg, M. and Kavan, L. (1992) On the stability of polyyne, *Chem. Phys.* **168**, 249-258.
25. Springborg, M. (1989) Interactions between two identical polymer chains studied with first-principles calculations, *Phys. Rev. B* **40**, 5774-5779.
26. Wittmann, H.F., Friend, R.H., Khan, M.S. and Lewis, J. (1994) Optical spectroscopy of platinum and palladium containing polyynes, *J. Chem. Phys.* **101**, 2693-2698.
27. Lhost, O., Toussaint, J.M., Bredas, J.L., Wittmann, H.F., Fuhrmann, K., Friend, R.H., Khan, M.S. and Lewis, J. (1993) Electronic structure of platinum-containing polyynes, *Synth. Metals* **55-57**, 525-4530.
28. Springborg, M. and Albers, R.C. (1996) Electronic structure of Pt in polyyne, *Phys. Rev. B* **53**, 10626-10631.

4.2 KINKED CHAINS AND LAYERED STRUCTURE

R.B.HEIMANN
Chair of Technical Mineralogy, Department of Mineralogy, Freiberg University of Mining and Technology, D09596 Freiberg, Germany

Abstract - Carbyne is considered to be built up from conjugated triple-bonded (polyyne, polyethynylene) or double-bonded (cumulene, polyethenediylidene) chains of carbon atoms arranged in a parallel fashion along the c-axis of a hexagonal or rhombohedral lattice. A structural model is proposed that provides a crystallographic identity period by introducing kinks between the rectilinear parts of the chains. Several possible ways are being discussed how to form these kinks and how to stabilize the carbon chains against collapse. A macroscopic layered crystal lattice can be formed by stacking two kinds of layer parallel to the crystallographic c-axis. One layer type contains three non-equivalent close-packed carbon chains and the other one only two thus forming a 2D-lattice with $\sqrt{3} \times \sqrt{3}$ -vacancies. Very recent work suggests a relationship between carbyne polytypes and chiral poly(diacetylene)s.

1. Introduction

During extensive transmission electronmicroscopy work supporting a research project on shock wave-assisted transformation of graphite into diamond [1], performed between the years 1979 and 1982 by 3M Canada,Inc. at the Institute of Aerospace Studies, University of Toronto, a variety of carbon particles were found that obviously were neither attributable to graphite nor diamond. In fact, their electron diffraction pattern did not match any allotropic form of carbon known to the investigators at this time. A thorough literature search turned up an amazing number of papers from Russian journals contributed by a group at the A.N.Nesmeyanov Institute of Organoelement Compounds, Soviet Academy of Sciences, Moscow (for example [2-7]) as well as from international journals contributed by a group from The Aerospace Corporation, El Segundo, CA, USA (for example [8-14]).

The XRD and ED spacings reported in these papers matched our findings sufficiently well so that we did not hesitate to group our hitherto unassigned carbon forms with a high temperature, low pressure linear carbon allotrope with sp-hybridization and carbon-carbon chains running parallel to the c-axes - carbyne, as it turned out [3]. These forms appeared also to be related to a natural carbon found earlier in shock-fused graphitic gneisses of the Ries Crater impact, Nördlingen,

R.B. Heimann et al. (eds.), Carbyne and Carbynoid Structures, 235–268.

Bavaria, Germany that had been named 'chaoite' [15]. Armed with this knowledge, reexamination of our TEM plates revealed an initially confusing multitude of seemingly related but not quite identical 'carbynes' that differed in their a_0-values that were readily obtained from d_{max} values on a c-axis pattern by selected area electron diffraction (SAD). This was reminiscent of problems Arthur Whittaker faced when attempting to assign, by TEM-ED studies, crystallographic data to carbon particles found in profusion in carbonaceous meteorites (Allende and Murchison) [12]. It was subsequently observed in graphite samples exposed to a shock wave and simultaneously heated to different temperatures by flash-heating through the discharge pulse of a 160-μF capacitor (maximum energy of 32 kJ at a maximum voltage of 20 kV [1]) that some a_0-values of carbynes appeared preferentially, i.e. that at a given temperature carbyne polytypes with a more or less distinct a_0-value prevailed. This suggested a temperature-dependent 'equilibrium' of chain lengths consistent with the exceptionally narrow stability range of individual carbyne polytypes conjectured earlier by Whittaker [10] and shown in his tentative high temperature carbon phase diagram (see Fig. 4, Chapter 7.1).

This conjecture likewise suggested the idea that order may be brought into the carbyne „zoo“ by assigning to the carbyne polytypic forms different lengths of carbon atom chains, either conjugated triple bonded (polyyne, polyethynylene) or cumulated double bonded (polycumulene, polyethenediylidene) that are arranged parallel to the c-axis of a hexagonal lattice with a tentative space group symmetry P3 or $P3_{1,2}$ [4]. This intuitive, pragmatic approach appeared to work well and hence formed the foundation of a unified structural model of linear sp-hybridized carbon allotropes [16,17]. This structural model will be explored in more detail below.

2. Conjugated Carbon Chains in Carbyne

Any attempt to discuss the nature of conjugated carbon chains in carbyne must answer a variety of intimately related questions. The first pertains to the bond lengths between adjacent intrachain carbon atoms and attempts to answer the question whether there is an alternation of the bond lengths as in a conjugated triply bonded -C≡C- (polyyne) chain or whether the bond lengths are identical as in a cumulated doubly bonded model =C=C= (polycumulene). The second question is whether the structures are even- or odd-membered, a question that was answered based on theoretical considerations (for example [18]). Since both polyyne- and cumulene-bonded carbon atoms can be linked either linearly or in cyclic rings, the question of the stability of either structure arises that can be considered a function of the number of carbon atoms involved [19]. If it turns out that the carbon atoms are linearly linked like pearls on a string we ought to question whether those chains are straight, helically twisted, kinked or show other defects. Finally, if there is a twisted or kinked conformation another topological question must be considered: does the system have *trans-* or *cis-*isomery?

In the following text we will attempt to shed light on these questions but are very conscious of the fact that many of the questions posed above cannot be answered conclusively at this time. The reasons for these ambiguities are manifold but are all related to the present inability of chemists and materials scientists to synthesize crystalline 'ideal' carbyne in sufficient quantities and crystal dimensions to permit structural investigations. These should include x-ray and electron diffraction studies on three-dimensional crystals to determine the spatial relationships of the carbon chains in the crystalline lattice; Auger (AES), photoelectron (XPS), electron energy loss (EELS) and electron paramagnetic resonance (EPR) spectroscopy to determine the electronic density of states (DOS) and the extent of π-bond interactions; and infrared (IR), Raman as well as nuclear magnetic resonance (NMR) spectroscopy to ascertain the carbon bond multiplicity. In addition, a second problem relates to the observation that there appear to be several carbyne polytypes whose conjugation (chain) lengths are functions of the temperature of formation. Since it is also conjectured that a temperature-dependent quasi-equilibrium of conjugation lengths exists with very narrow stability ranges of individual carbyne forms (see Fig. 4, Chapter 7.1), a random distribution around a mean value may be expected that poses problems in identification and characterization by the various analytical methods.

Since the questions outlined above are intimately connected within the framework of structural chemistry it is neither feasible nor desirable to unravel their interconnectiveness. Hence their treatment will be integral rather than differential, a fact for which we apologize.

To relate the various carbyne forms with different crystallographic date (Table 1) by a simple unified classification scheme it was assumed the (i) ideal carbyne consist of a linear arrangement of carbon chains with either conjugated triple $(\text{-C}\equiv\text{C-})_{n/2}$ or cumulated double $(\text{=C=C=})_{n/2}$ bonds, and (ii) the lattice metric is determined by the number of carbon atoms, n, in the chains [17].

TABLE 1. Crystallographic data of carbyne polytypes (Heimann et al [16])

	ß-carbyne	C XIV	C VI	C IX	Chaoite	C VIII	α-carbyne
a_0 (pm)	824	870	923	944	894.8	909.78	892
c_0 (pm)	768	956	1224	1250	1407.8	1482	1536
a_0/c_0	1.073	0.910	0.754	0.755	0.635	0.614	0,581
V/6 (nm^3)[a]	0.451	0.626	0.903	0.965	0.976	1.062	1.058
Z	72	-	-	-	168?	-	144
ρ (g cm^{-3})	3.13	-	2.9	-	3.43	-	2.68
n	6	8	10	10	11	12	12
Space group	$P3_{1,2}$	-	P3 or $P3_{1,2}$	R3?	P6/mmm?, P3 or $P3_{1,2}$	-	$P3_{1,2}$

[a] $V = 3\sqrt{3}\cdot a_0^2\cdot c_0$

Hence carbyne is a linear macromolecule whose bond lengths between neighbouring carbon atoms alternate between 137.9 pm (≡C-C≡) [20,21] and 120.7 pm (-C≡C-) [20,22] for a polyyne chain but are identical at 128.4 pm [23] for a linear cumulene chain (butatriene). The magnitude of the bond length alternation in a polyyne chain is weakly dependent on the value of n. On the other hand, the C-C and C=C bond lengths vary systematically with the changes in bond environment, i.e. the bond length increase approximately linearly with increasing number of adjacent atoms. This effect is known as Stoicheff's relation and has first been expressed numerically in the 1962 seminal paper of this author [20]. The relation describes the change of -C-C- and =C=C= bond lengths with bond multiplicity that is predominantly caused by electron delocalization in conjugated systems by redistribution of π-electrons. As we will see below, Stoicheff's relation will play a decisive role in the development of a working, although initially heuristic chain model to explain the carbyne structure.

'Naked' carbyne chains, i.e. chains terminated by hydrogen atoms or delocalized electrons [24] and losely bound in a parallel fashion by van der Waals interactions ('pencil-in-box' model) will be stable only at low temperatures when the lack of thermally activated large scale oscillations of carbon atoms in the chain prevent cross-linking, i.e. establishing strong σ-bonds. Carbynes stable at higher temperatures in solution or in the gas phase are end-capped by bulky alkyl or aryl groups such as triethylsilyl [25], *tert*-butyl or trifluoromethyl [24] groups, but also large metal (K) ions [26]. Linear carbon chains appear to be stabilized in the solid state by metal ions such as K or Fe [27,28]. Also, pairing of surface electrons of carbon chains by adsorption of CO molecules was invoked to explain the stabilization of the carbyne structure [29]. All these terminal groups serve to provide steric stabilization by keeping the parallel carbon-carbon chains apart at a distance of about 297 pm [27] that is consistent with the van der Waals diameter of sp-hybridized carbon atoms. This characteristic carbon chain distance has been experimentally confirmed by several authors [30,31]. The terminal groups exert a substantial influence on the interchain carbon distance for $n \leq 20$. This was confirmed by quantum-chemical calculations [32] using the extended Hückel molecular orbital (EHMO) valence approximation that showed that the carbon-carbon bond length is related to the Mulliken total (σ+π) overlap population P as

$$r(\text{C-C}) = 1.860297 - 0.548695\,P + 0.109747\,P^2 \qquad (1)$$

The Mulliken bond population P [33] was determined at distances of 121, 134, 140 and 154 pm for acetylene, ethylene, benzene and ethane. Since for chains with a small number of n (allene, butatriene, diacetylene, triacetylene) good agreement with experimentally determined bond length values was obtained, the approach was extended to longer chains, C_n with $n < 18$. It was found that (i) the total energy per carbon atom increases with increasing n, (ii) for $n \leq 10$ the chain prefers a linear conformation while for $n > 10$ cyclic rings are more stable, and (iii) linear chains with $n = 2m + 1$ (odd) are more stable than those with $n = 2m$ (even) [18]. That n-odd linear C_n molecules are particularly stable due to charge-density alternation was also confirmed by single determinant Hartree-Fock calculations [34]. Furthermore, there are large contributions to bond length variations from resonance structures. Such resonance structures are consistent with the direction of overlap population variations as well as the persistence of charge alternations in n-odd species [32].

Calculations performed on model linear systems C_n consisting of an infinite number of carbon atoms by a large variety of computational methods and approximations [35] revealed that the minimum potential energy of the C_∞ system is reached when bond lengths alternations are assumed. Hence a polyyne chain ought to be more stable than a polycumulene chain. This conclusion is also in accord with the Peierls theorem of the instability of one-dimensional systems of metallic type: alternations of the bond lengths result in the appearance of a non-zero width forbidden band (HOMO-LUMO band gap) in the electron spectrum. Such a band gap has been experimentally confirmed [36,37]. There is also a strong suggestion of the existence of paramagnetic solitons in the ground state at kink sites of odd-numbered polyyne chains [38] that are thought to provide chain stabilization [39,40].

Hence there appears to be sufficient evidence to assume, on theoretical grounds, that the carbon chains in carbyne are (i) of polyyne-type, i.e. with bond length alternations and that (ii) n-odd C_n chains should be particularly stable against cross-linking, i.e. collapse into a graphene-like structure [41]. It also seems to be reasonably well established that end-capping of the chains with „spacers" (bulky organic terminal groups, metal ions, alkali fluoride) provides steric stabilization against cross-linking [42].

However, experimental results obtained by many workers in the field point to a preponderance of polycumulene-type chains [43]. In particular, chemical dehydrohalogenation reactions of poly(vinylidene halides, PVDH) appear to favour the formation of cumulene-type chains by alkali-induced abstraction of HCl or HF [44]. This idea was rejected, however, in the light of a recent ^{13}C-MAS-NMR study [45] that showed conclusively that concurrent formation of polyyne- and cumulene-type fragments occurs through a complex interaction of the target reaction with ethoxylation and hydroxylation reactions, nucleophilic addition to multiple bonds, and secondary transformations of ethoxylated and hydroxylated structures under alkaline conditons (cp. Chapter 3.2.1). Hence, cumulene structures can be expected resulting from exhaustive dehydroxyhalogenation of only those PVDH fragments that are located between ethoxylated carbon atoms, whereas polyyne structures will be formed only at carbon atoms adjacent to a carbonyl group.

A different mechanism of formation of mixed polyyne-polyene chains as a dehydroxyhalogenation product of cPVC (chlorinated poly(vinyl chloride)) and PVDC was observed using a pulsed KrF eximer laser in an ultrahigh vacuum [46,47]. Chain stabilization against cross-linking was found to occur through carbon radicals on conjugated bonds since unpaired electrons are delocalized on the polymer chain [48]. Since no oxygen atoms are present in the system, ethoxylation reactions as found to interfere with carbyne formation in the studies quoted above are excluded. Hence, laser-assisted abstraction reactions (cp. Chapter 3.2.2) appear to be much better tools towards 'ideal' carbyne.

The question whether, in carbyne, polyyne- or cumulene-type chains are preferred was theoretically tackled by Bochvar *et al* [49] using an extended Hückel approach to calculate the average total energy per carbon atom, E_{av} as well as the HOMO-LUMO band gap, $\Delta\varepsilon$ for short-membered (n = 4,6) chains. The results are shown in Table 2.

TABLE 2. Total bond energy per carbon atom (E_{av}) and HOMO-LUMO band gap ($\Delta\varepsilon$) for differently kinked polyyne and polycumulene chains (Bochvar et al, [49]).

Conformation		-Eav (eV)	Δε (eV)	Conformation		-Eav (eV)	Δε (eV)
Polyyne				Cumulene			
linear		72.01	1.53	linear		71.90	0.00
n = 4 (cis)	Fig. 3a	71.10	0.41	n = 4 (cis)	Fig. 3c	71.69	0.15
n = 6 (cis)	Fig. 3a	71.37	0.13	n = 6 (cis)	Fig. 3c	71.75	0.00
n = 4 (trans)	Fig. 3b	71.97	0.63	n = 4 (trans)	Fig. 3d	71.67	0.12
n = 6 (trans)	Fig. 3b	71.38	0.36	n = 6 (trans)	Fig. 3d	71.75	0.00

It was found that (i) the polyyne structure is (slightly) more stable than the cumulene structure for straight chain segments, (ii) the average total energy, E_{av} of a cumulene chain with defects (kinks) is higher than for the corresponding polyyne chain, (iii) increase of n in a polyyne chain results in an increase in E_{av}, (iv) the *cis*-isomer configuration is slightly more stable than the *trans*-configuration for both kinked polyyne- and cumulene-type chains with n = 4, (v) the HOMO-LUMO band gap width for polyyne structures is always non-zero and decreases with increasing n, and (vi) the HOMO-LUMO band gap width for a straight cumulene chain is zero but the introduction of defects (kinks) into the chain with n = 4 leads to non-zero values. These findings are in general accord with calculations performed by Hoffmann [32] for C_n with cumulene character. A non-zero gap in the calculations based on an extended Hückel approach (extension of Hückel π-electron approximation to include σ-electrons) has also been found for equal bond length in short chains [50] but it was concluded that with increasing chain length the gap width approaches zero, presumably owing to realization of an antiferromagnetic state.

In conclusion, the energy difference between long polyyne- and cumulene-type chains is rather small so that the probability of the occurence of either type is unlikely to stray much from 0.5, depending on the type of chain stabilization as well as the method of abstraction of side chains of the carbon-carbon backbone chain as a precursor of the carbyne structure. As the introduction of defects (kinks) into a conjugated chain of *cis*-configuration tends to increase the stability of structures with equal bond lengths [49] (Table 2), we should expect cumulene-type bonds alternating with polyyne-type bonds in 'real' carbyne (cp.Table 3). It should be stressed, however, that the use of mean values of the total energy per carbon atom to estimate the preponderance of one bond type over the other as done by Bochvar *et al* [49] results in a rather crude approximation of the real situation. The reason for this is that the total energy values will tend towards the same limit with increasing chain lengths for both conformations [32,35].

At the end of this discussion there should be given some experimental evidence of the existence of linear carbon chains in carbyne, twisted, kinked or otherwise. Much of it has been collected in several review papers on carbyne [43,51,52]. As mentioned above, no macroscopic three-dimensional crystals of carbyne are available so far for in-depth study of its electronic structure to test the contention of a parallel, partially cross-linked arrangement of carbon chains with alternating or equal bond lengths, or both.

Hence, all data generated to date were obtained from (oriented) films or crystalline microinclusions in these films [27].

Auger electron spectroscopy (AES) [53] was used to calculate the electronic structure of carbyne by self-convolution of experimentally obtained spectra of samples produced by chemical dehydrochlorination of poly(vinylidene chloride) as well as by Ar^+ ion sputtering of graphite [54,55]. It was found that (i) the width of the σ-subband of the two samples were 7.5 and 5 eV, respectively, consistent with carbon chains having ≥ 8 and 5 atoms, respectively, (ii) the density of valence states (DOVS) of carbyne corresponds to the electronic spectrum of cumulenic chains, calculated by the LCAO approximation [56], (iii) the upper part of the DOVS plot, i.e. states with $E < 17$ eV corresponds to a one-dimensional π-subband whose relatively deep position is characteristic for cumulene-type chains, (iv) by substracting the contribution of the graphite π-subband (states with $4.7 < E < 10$ eV), information on the hybridization of σ-orbitals and bond types in the chains could be obtained, and (v) the width of the gap between the top of the π-subband and the bottom of the σ-subband can be correlated with the chain length. It decreases, on annealing to 600°C, from $\Delta = 3.5$ ($n = 5$) to $\Delta = 2.0$ ($n \geq 8$) thus suggesting chain lengthening by end-to-end linkage. There should be a word of caution, however. Since the experimental AES data were interpreted mostly based on theoretical model calculations their quantitative evaluation should not be taken at face value but merely as an indication of the validity of the structural concept of carbyne as being composed of linear carbon chains.

Electron energy loss spectroscopy (EELS) data also confirmed the linear structure of carbyne macromolecules [53]. Using Kramers-Kroning's theorem, it could be shown that (i) electron excitations with an energy of ≈ 1 eV are associated with a transition in the π-band between states separated by a gap of width 1 - 2 eV, (ii) electron excitations with an energy of ≈ 8 eV coincide with a transition from the bonding σ-band to the antibonding σ*-band at $k = \pi/a$[1] , and (iii) electron excitations at $E = 9.9$ eV are caused by a transition from the bottom of the π-subband to the antibonding σ*-band at $k = 0$. All these features are consistent with the notion of a linear chain structure as well as the assumption of semiconducting properties of carbyne. For additional details the reader is referred to Chapters 6.3 and 6.4.

Lastly, measurements of the specific heat of carbyne at low temperature [57] showed a temperature dependence conforming to the equation $C = \alpha T^3 + \gamma T$. The substantial linear term strongly suggests the presence of a one-dimensional phonon spectrum consistent with the assumption of a linear structure of the carbon chains in carbyne. Moreover, neutron irradiation leads to an increase of the coeffient γ and a decrease of α, suggesting breaking of interchain cross-linking bonds.

[1] The term k is the wave vector. For the electron energy in the Brillouin zone of a linear chain with a unit cell parameter (periodicity), a, the dispersion is given by $E(k) \propto \cos(ka)$. In the case of a one-dimensional chain the DOVS has maxima at singular points corresponding to the bottom ($k = 0$) and top ($k = \pi/a$) of the valence band.

3. Twisted, zig-zag, and kinked chains

To assign a crystallographic identity period, c_0 to the different carbyne polytypic forms listed in Table 1, the conjugated or cumulated carbon chains described in the preceding paragraph have to be terminated in a more or less regular fashion. Clearly, a discrete identity period cannot be assigned in the c-direction for infinite and rectilinear C-C chains. When faced with this problem, we adopted very pragmatic and intuitive approaches of terminating the chains. These approaches were of three types.

(i) An identity period is introduced by considering n-membered C-C chains with a helical structure whose threads with a pitch $\tau = 360°/n$ generate a lattice translation analoguous to a crystallographic n-fold screw axis. Chains of this type are present in poly(acrylonitrile) [58] for which a hexagonal packing of chains with a nearest chain-to-chain distance of about 600 pm has been estimated, twice the value found for the interchain distance in carbyne (297 pm). From the measured density and the lateral hexagonal packing of the chains a c-periodicity of approximately 230 pm was calculated. Hence it was concluded that the chains cannot be straight but must be helically twisted. It should be pointed out that this analysis was performed within the framework of paracrystal theory [59] that may have important bearing on the interpretation of the carbyne structure.

A helical structure of the carbon chains has been proposed on theoretical grounds by cross-linking of chains of carbon atoms of identical or opposite chirality with multiplicities of n = 3,4 and 6 where for n = 6 the chains degenerate into a planar regular hexagon, a precursor unit of a single graphene layer [35]. Structures formed by cross-linking of chains with the same chirality are being designated $C_1(n_j)$ where n_j refers to the type of the n-fold screw axis. Structures formed by cross-linking of chains where each chain with an n-th order screw axis is surrounded by n chains with opposite chirality, are called $C_2(n)$. Finally, the infinite set of structures formed by random cross-linking of chains differing in chirality is being denoted $C^i_3(n)$ with i being a running number [35,60,61].

Figure 1 shows the arrangement of helical carbon chains with a symmetry $C_1(4_1)$ (a) and their projection onto (001) (b). This $C_1(4_1)$ carbon 'modification' was formed by cross-linking of carbon chains with 4_1 screw axes of symmetry. Under the assumption that all carbon bonds are equivalent, a space group of $I4_132$ can be assigned. The Bravais lattice is body-centred cubic (b.c.c), the unit cell with a lattice constant $a = 2\sqrt{2}\,r$ contains eight carbon atoms (Fig. 1b). It should be pointed out that this lattice is strainfree since all bond angles are exactly 120°.

Figure 2 shows a hypothetical carbon atom network with a symmetry $C_1(3_1)$ (a) and its projection onto (001) (b). This lattice has a space group of $P6_222$, a primitive hexagonal Bravais lattice with six carbon atoms per unit cell, and the unit cell constants $a = (2r_1 + 3r_2)/\sqrt{3}$ and $c = r_1\sqrt{6}$. Here r_1 is the interatomic distance in the chain (1-2 in Fig. 2b), and r_2 the interchain distance at the point of the closest approach of two neighbouring chains (1-4 in Fig. 2b).

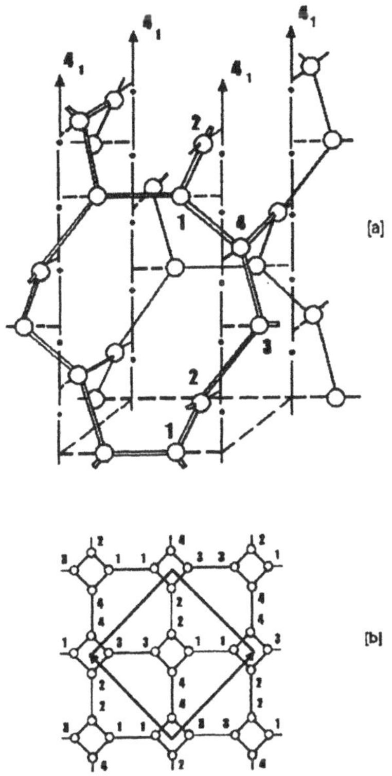

Figure 1. Fragment of crystal structur of carbon modification $C_1(4_1)$ (a), and view from above (b) [33].

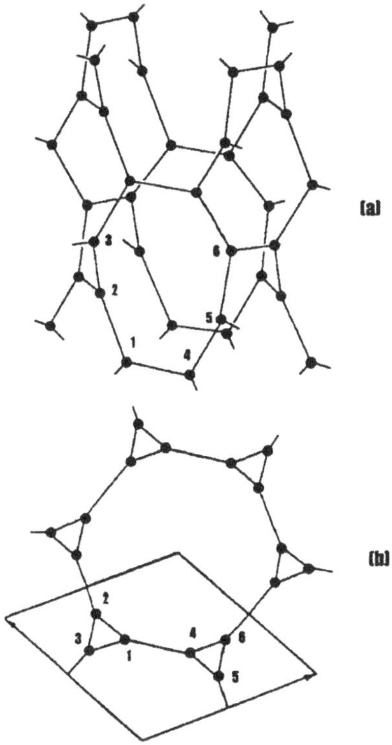

Figure 2. Fragment of crystal modification of carbon C_1 (3_1) (a), and view from above (b) [33].

Cross-linking of carbon chains with screw axes 3_1 and 3_2 of opposite chirality yields the $C_2(3)$ modification with a space group R3m, a double body-centred hexagonal Bravais lattice, and the unit cell constants $a = 2r_1 + 3r_2$ and $c = r_1 \sqrt{6}$. Again, both $C_1(3_1)$ and $C_2(3)$ structures have bond angles of 120° and hence are thought to be strainfree. For further details the papers by Bochvar *et al* [35,49,60,61] should be consulted.

In terms of stability, a 3_1 (or 3_2) arrangement of carbon atoms should be more stable than a 4_1 (or 4_3) arrangement according to calculations by Bochvar *et al* [49] using an extended HMO approximation. It was found that the mean total energy per carbon atom is little dependent on the r_1-values in the range of 138 and 145 pm but that the chain configuration with trigonal screw axes is more stable than that with tetragonal screw axes (-71.63 eV/atom vs. -70.14 eV/atom for r_1 = 145 pm) (Table 2).

(ii) A second way to subdivide a linear carbon chain can be found in zig-zag chains. This is the type of conformation that two linear parallel carbon chains may attain as revealed by parameter-free calculations of interactions [62]. The repulsion between the chains at interchain distances between 200 and 400 pm results in a transition from sp-bonded carbon atoms to sp^2-bonded atoms with dangling bonds. As these dangling bonds can be saturated by cross-linking with other neighbouring chains, a three-dimensional lattice can be built up with dimensions conforming to carbyne (for details cp. Chapter 4.1). Chains of the zig-zag type were identified in *trans*-polyalkenamers [63]. The first member of the homologous series is poly(acetylene), the last is poly(ethylene). The lengths of the planar zig-zag chains are shortened on the basis of the principle of staggered bonds extended to the double bond [64].

(iii) Lastly, a discrete identity period could be generated by kinks, constituting 2-fold screw axes, that separate straight strings of carbon atoms. These kinks can impose on the carbon chain either a *cis*- or a *trans*-isomery as shown in Figures 3a-d. In contrast to the helical structure of the chains discussed above, the kinked chains have no chirality, i.e. neighbouring kinks are situated in the same plane.

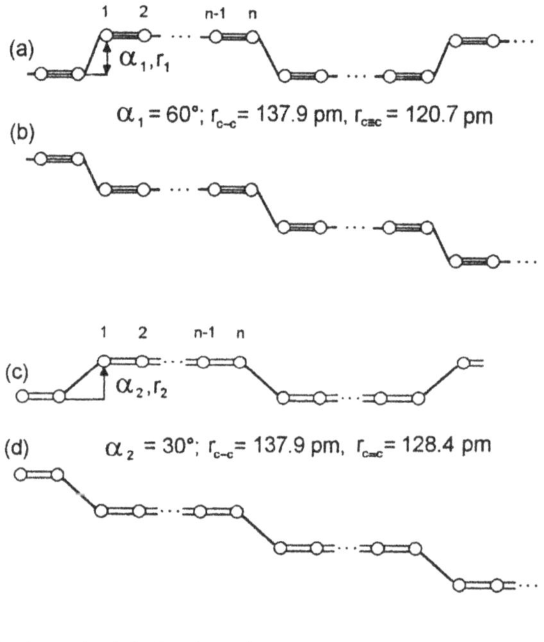

Figure 3. Polyyne (conjugated triple bonds with (a) cis-transoid- and (b) all-trans-configuration) and polycumulene chains (cumulated double bonds with (c) cis-transoid and (d) all-trans-configuration) [112].

Chains of this type have been identified in crystalline poly(ethylene) [65] and collagen [66]. The elucidation of the structure of poly(ethylene) was associated with many changes in opinion over the last fifty years. One of the earlier structural models [67] suggested, based on a diffuse meridian small angle reflection in hot stretched poly(urethene) and poly(ethylene) that parallel macromolecules (fibers) are densely packed in a paracrystalline [59] hexagonal microlattice [68]. These three-dimensionally ordered molecules are terminated by defects whose distances along the chains fluctuate according to the coordination statistics of paracrystal theory. Similar amorphous 'kinks' along the fiber axes were described in Stratton's model of the poly(ethylene) structure [69], a precursor theory of the 'shish kebab' structure model generally accepted today. Figure 4 shows the Hess-Kiessig model of stretched poly(ethylene) with 'kinks'[67].

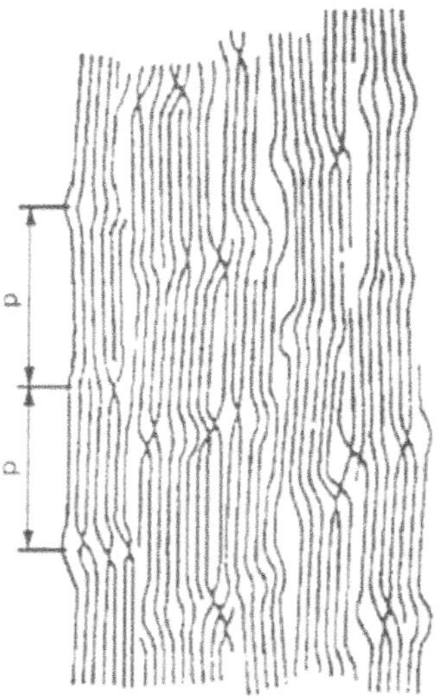

Figure 4. Hess-Kiessig model of streched poly(ethylene) displaying kinks whose spacings fluctuate according to coordination statistics of paracrystal theory [65].

In the light of these models it would be worthwhile to reconsider the kinked carbon chain model to explain the carbyne structure in terms of paracrystal theory as already suggested earlier by this author [17]. Such a paracrystalline model may be thought of as the usual 3D-ordered crystal structure that is perturbed ('defocused') by two distinct kinds of statistics: (i) the *displacement statistics* that describes the displacement of carbon atoms from their equilibrium lattice positions, and (ii) the *coordination statistics* (lattice statistics) that deals with the distortion of the actual lattice points [58]. If the distribution functions obeying these statistics approach unity, a paracrystal attains the limiting structure of an ordinary real crystal. On the other hand, if the distribution functions become very broad, the 'crystal' structure changes continuously through a diffuse liquid-like lattice to the continuous scattering characteristic of a completely disordered gas-like 'lattice'. Hence, within the framework of paracrystal theory it is possible to statistically describe structures that range from highly crystalline to highly disordered, i.e. amorphous states.

Applied to the carbyne structure, the succession of regularly spaced defects, i.e. kinks along the straight carbon chains as described in Heimann *et al* [16,17] may be replaced by the notion of kinks whose spacing obey the principle of coordination statistics in paracrystal theory [112]. If this is being done it may also turn out that inconsistencies in the assignment of n-values to individual carbyne polytypes (Table 1) can be reconciled. For example, carbon VI (a_0 = 923 pm, c_0 = 1224 pm [8]) and carbon IX (a_0 = 944 pm, c_0 = 1250 pm [70]) with n = 10 [16] as well as carbon VIII (a_0 = 909.8 pm, c_0 = 1482 pm [11] and α-carbyne (a_0 = 892 pm, c_0 = 1536 pm [3]) with n = 12 [16] may actually be identical carbyne polytypes whose c_0-distances, given by the kink spacings, fluctuate around the theoretical values observed for poly(alkenamers) with n = 10 (1240 pm) and n = 12 (1491 pm) [63]. Since the a_0-values are also a function of n as predicted by Heimann *et al* [16,17] and confirmed by AES and EELS data [53], the same argument would apply. Then paracrystal theory assumes statistically fluctuating kink angles, α and displacement distances, r (cp. Figure 1, Chapter 3.4.1.1. of this book) with *a priori* frequencies [71] $H_k(x)$, $H_k(y)$ and $H_k(z)$ in x-, y- and z-directions [72].

3.1. GRAPHITE BOND SPLITTING MODEL

Option (i) involving carbon chain termination by twisting, resulting in n-membered C-C chains with a pitch $\tau = 360°/n$ as described in detail above is unlikely to occur in carbyne. The reason is that polyyne and cumulene chains are very rigid owing to the strong π-component of the linear MO bond system and hence would resist torsional forces.

Zig-zag (option (ii)) and/or kinked (option (iii)) chains can be generated by a simple mechanism introduced by Whittaker [10]. Noting hat carbyne can be formed from graphitic carbon by laser ablation or resistive heating through a vapourization/ condensation sequence (cp. Chapter 3.4.1.1) he conjectured that polyyne-type chains will be created by breaking of single bonds in 'meta'-position in a graphene sheet and shifting of electrons towards adjacent double bonds (see Fig. 1a, Chapter 3.4.1.1.). Hence, isolated chains parallel <10.0> of graphite will be formed that are, in principle, straight. However, kinks will be generated due to one-dimensional (dislocations) or two-dimensional (stacking faults) defects in the original graphite lattice. An analogous mechanism was conjectured for the formation of cumulene-type chains [16] and is shown in Figure 1b, Chapter 3.4.1.1.. In this mechanism, bond breaking occurs on carbon bonds in 'para'-position thus generating a chain parallel <11.0>. It should be noted that in this model the graphite (graphene) sheet is thought to contain formal double bonds since the fourth electron of each carbon atom is available for a mixed role of coplanar and interplanar bonding ($\sigma^3\pi$-distribution). Hence, fluctuating non-localized double bonds are present within the coplanar graphene sheets. This is entirely consistent with Anno and Coulson's view [73] that a graphite layer can be considered as being built up from fused polyacene strips. These cumulene-type chains provide free bonds by splitting double bonds that are now available for cross-linking. Splitting of polyyne-type triple bonds leads also to free bonds adjacent to kink sites but requires a reorganization of the remainder of the bonds thus locally switching the structure to a cumulene-type double bond. This would imply that in both chain types cross-linking sites are intrinsic structural defects and that all efforts to synthesize 'ideal' carbyne have been twarted so far precisely because of this effect. From a purely geometrical point of view cumulene-type straight chains are terminated by a C-C single bond whereas polyyne-type chains can also be terminated by a double bond, or a string of double bonds. This latter statement may be at the heart of the frequently observed fact that chemical carbonization (dehydrohalogenation) products of halogenated polymers such as PVDH always consist of 'carbynoid' structures with hybrid chains, i.e. chains constituted of both polyyne, and cumulene or polyene moieties of different proportions (for example, [45,46,49,74]). An alternative explanation for these hybrid chains is based on kink stabilization by paramagnetic solitons [39] and will briefly be dealt with later in this chapter.

However, unpaired electrons are also available for bonding stabilizing bulky alkyl or aryl groups or large metal ions that are thought to prevent chain collapse by keeping sufficient distance between them. Hence, cross-linking and chain stabilization by 'spacers' compete for free bonds and unpaired electrons, respectively.

Using the bond lengths and kink angles shown in Figure 1, Chapter 3.4.1.1., tentative kink spacings can be calculated.[2] These spacings define the c_0-axes lengths as

$$c_0(\text{polyyne}) = (n/2)\, r_3 + (\{n/2\}-1)\, r_1 + r^*_p \qquad (2)$$

$$c_0(\text{cumulene}) = (n-1)\, r_2 + r^*_c \qquad (3)$$

where r_1 = r(C-C) = 137.9 pm, r_2 = r(C=C) = 128.3 pm, r_3 = r(C≡C) = 120.7 pm as obtained from Stoicheff's relation [20], and $r^*_p = r_1 \cos 60^\circ$ and $r^*_c = r_1 \cos 30^\circ$ as obtained from the geometrical relations shown in Figure 1, Chapter 3.4.1.1. This simple and purely intuitive model is corroborated by the Pariser-Parr-Pople theory [75] that predicts C-C bond distances in terms of the π-bond order [76].

The kink angles, $\alpha_p = 60^\circ$ for a polyyne chain and $\alpha_c = 30^\circ$ for a cumulene chain, were originally thought of being remnants of the graphite structure, i.e. reflecting the bonding angles in the basal plane of graphite. This assumption, however, meets intrinsic difficulties since the nature of the driving force of kink formation is not obvious if one considers a stable, strainfree graphene sheet.

3.2 TETRAHEDRON MODEL

As an alternative way to explain the formation of kinks in conjugated carbon chains, a tetrahedron model was employed based on the results by Natta *et al* [63] on linear *trans*-poly(alkenamers). Members of this homologous series $(\text{-CH=CH-(CH}_2)_q\text{-})_p$ (q = 2m+1 for odd, q = 2m for even *trans*-poly(alkenamers)) consist of planar zig-zag chains. The first member of the series is poly(acetylene) with q = 0, the last is poly(ethylene) with q = ∞. The lengths of the chains are shortened as required by the principle of staggered bonds extended to double bonds [64]. According to this principle, twisting of the n-membered chains (n > 2) around the single bond adjacent to each

[2] This is valid only if one assumes a 3D-lattice with crystallographic long range order (LRO). The paracrystal approach (see above) would yield a range of spacings.

double bond produces kinks acting a 2-fold screw axes. Figure 5a shows an example of chain conformation of an even *trans*-poly(alkenamer) (*trans*-poly (octenamer)). The chain symmetry is monoclinic and symmetry centers occur in the middle of the double bonds forming the kink and in the middle of the bond at half identity period. Note that all carbon atoms of the repeating units lie on a plane inclined towards the c-axis at an angle δ defined by

$$\tan\delta = (1.16\,\sin\sigma_1)/d \tag{4}$$

where σ_1 is the torsional angle around a single bond adjacent to a double bond (130°) and d is the projection of the fiber axis (c_0-axis) onto the plane on which the carbon atoms lie. The angle δ decreases with increasing q towards zero and hence the unit cell tends to become orthorhombic for $q \rightarrow \infty$, i.e. for poly(ethylene) [77].

Figure 5b shows the chain conformation of an odd *trans*-poly(alkenamer) (*trans*-poly(heptenamer)). Here the chain axis is parallel to the fiber axis. Two repeating chain units are contained in the identity period. There is a symmetry center in the middle of the double bonds forming the kink and also a glide plane parallel to the chain axis. In addition to these symmetry elements, there are two 2-fold axes, perpendicular to the glide plane, in the center of the C atom dividing the repeating unit in two symmetrical parts.

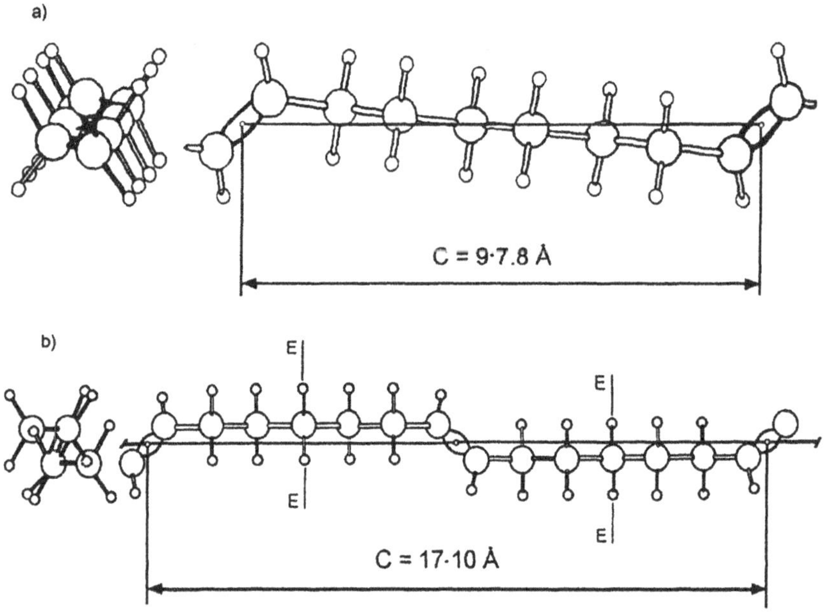

Figure 5. Conformation of the chains of (a) an even trans-polyalkenamer (trans-polyoctenamer) and (b) an odd trans-polyalkenamer (trans-polyheptenamer) whose fiber axis contains two chemical units [63].

The identity period of an even *trans*-poly(alkenamer) can be calculated [63] as

$$c_0 = (n-1)\, r_1 \cdot \sin(\Phi_1/2) + r_2 \cdot \sin(\Phi_2 - \Phi_1/2) \qquad (5)$$

where $r_1 = 154$ pm, $r_2 = 134$ pm, $\Phi_1 = 110^\circ$, $\Phi_2 = 120^\circ$, and n number of carbon atoms in the repeating unit. The term $\sin(\Phi_2-\Phi_1/2)$ represents the torsional angle of 65°.

In the light of the tetrahedron model, carbyne polytypes can be considered fully dehydrogenated poly(alkenamer)s in which the single bonds connecting the methylene groups are converted to either double or triple bonds and the repeating units are separated by kinks whose original double bonds are split, providing radical centers with sp^2 configuration [78]. During dehydrogenation the carbon chains are loosing their zig-zag conformation forming rigid straight chains owing to the π-component in the linear sp-hybridized system. Under these conditions eq. (5) is transformed into

$$c_0 = (n-1) r_2 \cdot \sin(\Phi_1^*/2) + r_1 \cdot \sin(\Phi_2 - \Phi_1/2) \qquad (6)$$

where $\Phi_1^* = 180^\circ$ as required for a strictly rectilinear macromolecular chain of carbon atoms with sp-hybridization, i.e. the generic zig-zag configuration of the poly(alkenamer) is lost by dehydrogenation. Note that eq. (6) is essentially identical to the empirically derived eq. (3).

The only variable in eqs. (2), (3), (5) and (6) is n, the length of the straight portion of the chain („conjugation length"). By varying n, one can attempt to match c_0-axes lengths obtained experimentally by electron and x-ray diffraction, respectively with the conjugation lengths obtained according to the graphite bond splitting model (eqs. (2) and (3)) and the tetrahedron model (eqs. (5) and (6)). The results are shown in Table 3.

From the calculated kink angle (column 5 of Table 3) that gives ideal fit, the ranges $23^\circ < \alpha_c < 25^\circ$ for cumulene chains and $60^\circ < \alpha_p < 65^\circ$ for polyyne chains are obtained. Table 3 also shows observed c_0-distances for *trans*-poly(octenamer), *trans*-poly(decenamer) and *trans*-poly(dodecenamer) (column 7) as well as the calculated values according to a modified equation (5) (column 8). The last column of Table 3 gives the c_0-values calculated for fully dehydrogenated poly(alkenamer)s according to eq. (6) with carbon bond lengths shortened according to Stoicheff's rule [20]. These values are identical to those experimentally obtained for ß-carbyne (c_0(obs) - c_0(calc) = Δ = 2 pm), chaoite (Δ = 1 pm) and α-carbyne (Δ = 1 pm). Larger deviations are observed for carbon XIV (n = 8, Δ = -66 pm), carbon VI (n = 10, Δ = -54 pm), carbon IX (n = 10, Δ = -28 pm) and carbon VIII (n = 12, Δ = -53 pm). For these carbyne polytypes the graphite bond splitting model appears to yield better fit: carbon XIV, Δ = -9 pm; carbon VI, Δ = 0 pm; carbon IX, Δ = 26 pm (cumulene), -23 pm (polyyne); carbon VIII, Δ = -1 pm. For both models, carbon IX yields large deviations between the observed and the calculated values suggesting that either this carbon form has a lattice different from that assumed for carbyne or that it incorporates larger contaminating or stabilizing metal ions that provide lengthening of the c_0-axis. This latter explanation would be similar to the concept of a 'stuffed' derivative structure as found, for example, in the SiO_2-phase tridymite where foreign ions (Na,K,Ca) incorporated into lattice interstices will buttress the structure and prevent its transformation or collapse to different polymorphic phases [109].

TABLE 3. Comparison of observed and calculated c_0 lengths of different carbyne polytypes [16]

Phase	n	c_0 (obs) [pm]	c_0(calc) [pm][a]	α(theo)[b]	Bond type[c]	Poly(alkenamer)[d] c_0(obs)	c_0(calc)[e]	DHPA[f] c_0(calc)[g]
ß-carbyne	6	768	760	22.93	C	-	742	766
C XIV	8	956	965	64.44	P	990	991	1022
C VI	10	1224	1224	60.00	P			
C IX	10	1250	1224/1273	46.5/45.7	P/C	1240	1240	1278
chaoite	11	1408	1401	23.97	C			1407
CV								
C VIII	12	1482	1483	60.21	P	1485	1491	
α-carbyne	12	1536	1530	24.18	C			1535

[a] according to eqs.(2) and (3), [b] kink angle required to match observed and calculated c_0 axes lengths exactly, [c] C cumulene, P polyyne, d Ref. [63], e according to eq.(5), f dehydrogenated poly(alkenamer), g according to eq.(6).

Figure 6 shows the relationship between the number of carbon atoms in the straight rectilinear chain, n and the c_0-parameter (Fig. 6a) and the a_0-parameter (Fig. 6b) for all carbyne polytypic forms shown in Tables 1 and 3. The a_0-values of carbon X - XIII were taken from d_{max} values on c-axis pattern as obtained from electron diffraction studies by Whittaker *et al* [12]. The upper line in Figure 6a relates to the polycumulene-type bond and obeys the equation

$$c_0\ [\text{pm}] = 128.5\ n - 10 \qquad (7)$$

the lower line refers to the polyyne-type bond with

$$c_0\ [\text{pm}] = 129.5\ n - 71 \qquad (8).$$

The fit of the data points to both theoretical functions is excellent as confirmed by linear regression coefficients of 0.999 and 0.997, respectively.

Figure 6b shows that the a_0-values of the carbynes investigated obey two functional relationships with very different slopes for polyyne- and cumulene-type structures. It has been conjectured [16] that the shallower slope of the cumulene trend as opposed to the polyyne trend reflects a lattice contraction in a-direction owing to increased cross-linking between kinked cumulene chains [79,80].

It must be admitted that the assigment of either cumulene- or polyyne-type chains was done only on purely geometrical grounds to obtain the best possible fit between observed and calculated lengths of the kinked chains. This was critically noted by Smith and Buseck [81] as well as by Korshak *et al* [53]. During Auger electron spectroscopy (AES) studies of carbyne films, however, the cumulene nature of chains with n = 5-6 ($a_0 \approx 820$ pm) and, after annealing at 600°C, n = 12 ($a_0 \approx 900$ pm) was established form the density of valence states (DOVS) [53]. The former may be identical with β-carbyne (a_0 = 824 pm, Table 1), the latter with α-carbyne (a_0 = 892 pm) both of which have been assigned a cumulene chain structure (Table 3). Hence, AES data appear to corroborate at least some of the rather intuitive and arbitrary conclusions drawn from the kink model. However, Smith and Buseck's allegation that 'the relationship between the a_0 parameter and the chain length n appears to be little better than random' [81], must be rejected. The linear regression coefficients as obtained from Figure 6b are R = 0.874 for the polyyne trend and R = 0.983 for the cumulene trend. Since application of the F-test shows that for both lines $F_0 > F_C$ at a 95% confidence level, we conclude that a_0 and n are highly correlated.

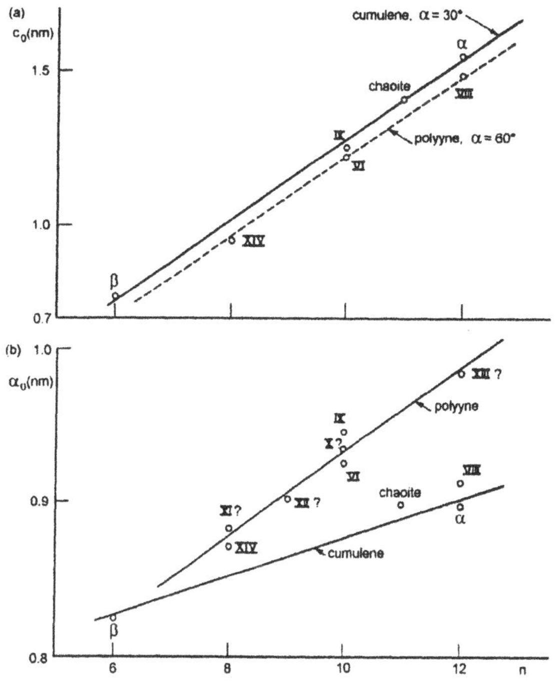

Figure 6. Relationship between the number of atoms in the chain, n and (a) the c_0 parameter for polyyne ($\alpha = 60°$) and cumulene ($\alpha = 30°$) arrangement; (b) the a_0 parameter.

As mentioned above, the kinked carbon chains interact with each other by van der Waals forces in the absence of cross-linked bonds, i.e. at a separation distance > 300 pm. The most important contributions to the total van der Waals interaction are the London dispersion forces of attraction that vary with the inverse sixth power of the interchain separation distance, d, but also with the length of the macromolecular chain, i.e. the number of carbon atoms n = 2m and the carbon-carbon intrachain distance, a. Considering two parallel chains of interchain distance d that are displaced through a kink with a kink angle γ, the following mathematical relations apply.

The London interaction energy of a pair of carbon atoms facing each other in the two chains can be written as

$$E\,(\text{dipole-dipole}) = -(3/4)\,\alpha^2\,I\,d^{-6} \qquad (9)$$

where α is the polarizability and I the ionization energy of the carbon atoms, and d is the distance of the two chains. The Hamiltonian of the interaction of the electrons of the two atoms i and j is

$$H = \Sigma(e^2/d_{ij}). \tag{10}$$

The polarizability can be expressed as

$$\alpha = [8e^2b^2(2m+1)^3]/\pi^5\beta \tag{11}$$

where b is the projection of the C-C bond lengths along the axis of the molecule (r^*_c = r(C-C) cos 30° or r^*_p = r(C-C) cos 60°) and β is the resonance integral between neighbouring atoms. From eqs.(9) and (11) under the assumption that for initially straight chains b ≡ a and that only the longitudinal polarizability is significant whereby the coefficient (3/4) in eq. (9) can be replaced by (1/8) it follows

$$E = -[8e^4a^4(2m+1)^6\ I\ ]/\pi^{10}\beta^2d^6 \tag{12}$$

and with $I = 2\beta\pi/(2m+1)$

$$E^{(2)} = -[16e^4a^4(2m+1)^5]/d^6\pi^9\beta. \tag{13}$$

The term $E^{(2)}$ denotes the second-order perturbation term of the interaction energy. Since the chain lengths are given by L = 2ma with a = 128.3 pm for polycumulene or polyene, and 137.9 pm for the single and 120.7 pm for the triple bond in a polyyne chain, for small d/L $E^{(2)}$ is proportional to 2m, and for large d/L it is proportional to $(2m)^5$.

Considering the offset of two parallel carbon chains introduced by a kink with a kink angle γ = arctg (r*/d), eq. (13) is transformed into the following asymptotic analytical expression

$$E^{(2)} = -[16e^4a^4(2m+1)^5 \cos^2\gamma(3\sin^2\gamma-1)^2]/d^6\pi^9\beta. \tag{14}$$

With γ = 0, i.e. for strictly parallel chains the expression (14) reduces to eq. (13). According to Coulson and Davies [82], the direction of the minimum of $E^{(2)}$ is γ = 66° for a linear hexatriene macromolecule (2m = 6) with an equilibrium chain distance d = 300 pm. On the other hand, for γ = 90° is $E^{(2)}$ = 0, i.e. total chain collapse occurs. The direction γ of the minimum in the London dispersion energy for 2m = 6 is remarkably similar to the kink angle calculated based on purely geometrical reasoning as demonstrated above.

3.3 SOLITON STABILIZATION OF KINKS

The graphite bond splitting model attempted to explain the origin of kinks in the macromolecular carbon chains as being the remainder of the original structure of a graphene layer. It was pointed out that it is rather unlikely that the bonding angles will be exactly retained after splitting, and that alternative driving forces for kink formation must be considered. Consequently, the tetrahedron model was developed that discusses kink formation based on the conformation of fully dehydrogenated even *trans*-poly(alkenamer)s. But this model also failed to explain why kinks should be formed at all in an otherwise rectilinear macromolecular carbon chain whose sp-hybridization state and its associated strong π-bond component provide a degree of rigidity that would effectively resist kink formation.

Another approach is based on the displacement of carbon atoms in the conjugated chains by the movement of charge-density solitons [39]. According to Rice *et al* [38], strong electron-phonon coupling occurs in linear carbon chains at the half-filled band level. This in turn leads to kink soliton and polaron states that appear as possible solutions of the static, semiclassical N = 4 Gross-Neveu Model relativistic field theory. In effect, a polyyne chain can be described as a Peierls insulator whose π-electron degeneracy is N = 4 as opposed to the normal one-band Peierls insulator with N = 2. The reason for this can be found in the fact that the π-electrons in the polyyne chain half-fill two degenerate bond levels that are created by the two carbon p-orbitals, p_x and p_y, available in the linear sp σ-bonding system. We conjecture that in the polyyne chain charge-density (CDW) soliton waves exist that move along the one-dimensional carbon chain but do not spread between neighbouring chains. This is because in such one-dimensional structures (1D-Peierls insulator) the electrons are confined to a linear chain of atoms which results in a Fermi surface consisting of two parallel planes. When a periodic lattice modulation such as a CDW couples electronic states across the Fermi surface through electron-ion interaction, the added periodic potential will truncate the Fermi surface and thus create an energy gap, confining the CDW.

The modulation of the electron density of the carbon atoms by passing of a CDW causes a displacement of the atoms to new equilibrium positions. Strong electron-phonon coupling in polyyne chains as postulated by Rice *et al* [38] can drive a commensurate-incommensurate transition whose elastic term leads to a (kink) displacement. These generic lattice kinks are in effect the domain walls of a transition [83], and hence the soliton separates two energetically equivalent but conformationally distinct segments of the polyyne chain [84]. This is entirely analogous to the description of the electrical conductivity of one-dimensional conductors. In this case, the linear chain can lower its energy by a periodic displacement of the lattice positions, depending on the electron-phonon coupling constant. The electrons lock to the displaced ions forming a CDW, the configuration of the one stabilizing the other. Hence the system forms a so-called Fröhlich-Peierls condensate with some of the properties being that of a superfluid [85]. Since this condensate in practice is pinned by either impurity centers [86], interchain coupling or lattice commensurability with a pinning potential of period $2\pi/a$, the kinks and antikinks have the properties of a sine-Gordon soliton that carries charges proportional to $\pm 2\pi/a$. At thermal energies below the pinning potential these kinks or antikinks carry the current.

A soliton in a linear chain of carbon atoms can, in principle, be observed by the occurence of new diffraction satellites from which location the period of the CDW can be calculated [87]. There is some circumstantial evidence that such satellites have been observed during mapping of the crystal potential surface U(r) of carbyne, separated by a distance of 70 pm from the carbon chains (Fig. 10b, [27]).

The moving CDW soliton shortens the distance between carbon atoms in the chains owing to the elastic transition described above. This in turn affects the stiffness of the chains and results in strained, i.e. compressed chains in which the soliton is pinned between adjacent kinks. Referring back to the tetrahedron model, since solitons must be localized they can only exist in *trans*-, not in *cis*-poly(alkenamer)s [88] because free rotation along the chains is prevented by the kinks. This is analogous to the formation and movement of kinks in dislocations that were described in term of the sine-Gordon soliton already by Frenkel in 1939 [85] (see also [89]).

The analogy can be understood by defining a kink in a moving dislocation. A dislocation line crossing over from one potential trough of the energy surface for dislocation movement in a glide plane (Peierls valley) to the neighbouring one has to overcome a substantial energy barrier. The kinks occur in pairs and their driving force is provided by thermal fluctuations (at higher temperature) or quantum effects (at low temperatures). Dislocation movement in the presence of an applied force results in the collaps of a kink/antikink pair on one side of the dislocation line, and the spreading of the kink pair on the preferred side of the dislocation line. This spreading causes eventually the movement of the entire dislocation line into the next Peierls valley [90]. Similarly, the sine-Gordon soliton originates in one global vacuum state and crosses over an energy barrier separating two vacuum states into another zero-energy vacuum state [91].

The structural defects, i.e. the kinks in the carbon chain are localized as required by Stoicheff's primary enviroment effect [92] and appear as a soliton/antisoliton pair in n-even linear C_n macromolecules. In such n-even finite (protonated) polyyne chains the ground state,

$$\text{H-C≡C-C≡C-C≡C-C≡C-H,} \tag{15a}$$

transforms by excitation of four valence electrons into

$$\text{H-C≡C-C≡C-C:-C:-C≡C-H} \tag{15b}$$

that separate into soliton/antisoliton pairs [38],

$$\text{H-C≡C-C:-C≡C-C:-C≡C-H.} \tag{15c}$$

In n-odd chains, a soliton state can occur already in the ground state without any excitation. The only carbyne with an odd number of carbon atoms in the chain found so far is chaoite (n = 11, Table 3) that appears to be exceptionally stable: it occurs naturally and has presumably formed $6 \cdot 10^7$ years ago [15]. It is also the carbyne polytypic form with the lowest possible „magic number„ (n = 4m+7) [93]. Its stability is consistent with results obtained by single determinant Hartree-Fock calculations that showed that n-odd linear C_n molecules are particularly stable because of charge-density alternations [34]. The abundance and stability of carbon clusters with n = 11,15,19 and 23 (m = 1,2,3,4) has been repeatedly confirmed by TOF-MS of the products of laser-ablated carbon [94,95]. Recent results of ablation of graphite with a pulsed Nd:YAG laser revealed that in the presence of a magnetic field the growth of larger carbon chain clusters ions with C_{11}^+ dominance was promoted (see Fig. 2, Chapter 3.4.1.1. of this book), presumably through an enhancement of ion-neutral reactions [96]. These reactions occur because of an increase of ionic species, resulting from collisional ionization of neutral species through a confinement of electrons in the magnetic field. This process could be at the heart of formation of carbyne observed during condensation of carbon vapour from laser vapourization of pyrolytic graphite [97,98].

In the following, we will attempt, in the light of the soliton approach, to explain the fact that in 'real' carbynes, i.e. carbynoid structures with short conjugated sp-hybridized chains separated by sp^2-hybridized structural units, polyyne and polyene moieties appear together in products obtained by CW-laser ablation of graphite and pulsed eximer laser-induced dehydrohalogenation of PVDH (see, for example [46,47]). Photoexcitation of odd-numbered chains with n = 2m+1 (m odd) in the ground state can lead to positive-sign solitons that introduce strings of double (polyene, cumulene) bonds into a polyyne chain according to -C≡C -[$C^\bullet$=C=$C^\bullet$]-C≡C-. Hence, finite hybrid polyyne/cumulene (polyene) chains will be generated as the result of strong electron-phonon coupling leading to kink solitons as described above. It should be noted that this explanation is an alternative to the possibility of introducing polyyne strings in polycumulene or polyene chains produced by chemical dehydrohalogenation as a result of structural defects caused either by *in situ* defects in the PVDH precursor or by cross-linking or ethoxylation reactions [30,44,45,99].

Similar mechanisms appear to operate in fullerene-doped polymers [100] as well as in iodine-doped *trans-cis*oid polyene chains [101]. Also, a dynamic stabilization mechanism of linear polyyne- and cumulene-type carbon chains through interaction of mobile paramagnetic solitons with the chains was proposed recently [102]. Solitons were identified by ESR analysis of uniaxially oriented carbynoid films (cp. Chapter 6.2).

3.4 PHONON SPECTRA OF KINKED CHAINS

A few additional supporting arguments in favour of the kink model will be given below. To aid in the interpretation of IR spectra obtained from carbyne, produced by chemical dehydrohalogenation of PVDC with a saturated solution of KOH in ethyl alcohol and THF at -15°C for 24 hours, phonon spectral simulation was done [103] using a linear chain of carbon atoms interacting with each other according to a harmonic oscillation law in a simple ball-and-spring model. This is a gross oversimplification. A more detailed treatment is being discussed in Chapter 6.1.The equation of motion along the chain axis (x-coordinate) can be expressed by

$$m\,(\partial^2 X_n/\partial t^2) = k_{n+1}\,(X_{n+1}-X_n) + k_n\,(X_{n-1}-X_n) \qquad (16)$$

where X_n is the displacement from the equilibrium position, i.e. from the energy potential minimum of a Morse function of the nth carbon atom, k_n is the stiffness of the nth bond, and m is the mass of a carbon atom. A cumulene-type chain containing N = 40 carbon atoms with a length of the repeating unit n = 4 separated by kinks was used to calculate the oscillation frequency ω. The stiffness of the bonds, k (spring constant) in the cumulene repeating unit was assumed to equal that of the C=C double bond with an oscillation frequency $\omega = \sqrt{2}k/m$ (1600 cm^{-1}). The stiffness of the bond forming the kink, k_0, was varied in the range $0 < k_0/k < 1$. The results of the simulation are shown in Figure 7 for $k_0/k = 1$, 0.5 and 0. One of the limiting cases, $k_0/k = 1$ (Fig. 7a) relates to a straight cumulene chain without any kinks. In this case the oscillation frequencies are expressed as one acoustic branch extending from $\omega = 0$ to a maximum that corresponds to the absorption wave vector of 2260 cm^{-1}.

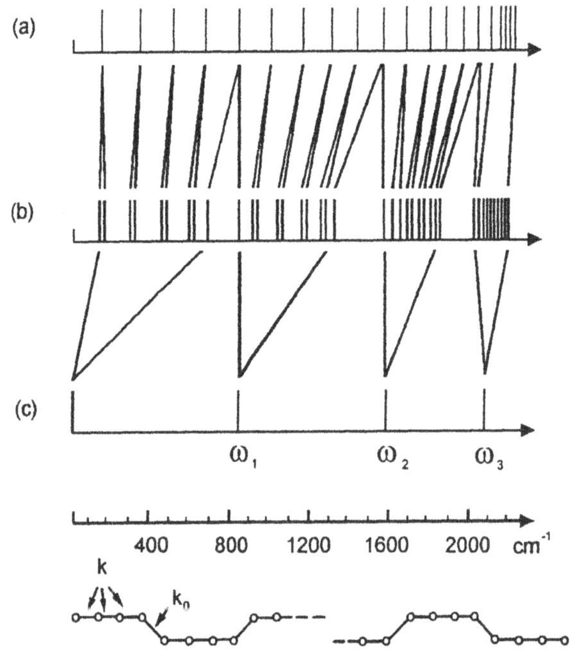

Figure 7. Characteristic oscillation frequencies of a kinked carbon chain with 40 atoms and kink spacing n = 4 [103]. (a) $k_0/k = 1$, (b) $k_0/k = 0.5$, (c)$k_0/k = 0$.

The second limiting case, $k_0/k = 0$ (Fig. 7c) corresponds to a model of unconnected cumulene chain fragments that form the carbon backbone of the butatriene molecule. In this case there are only three isolated frequencies at $\omega_1 = 865$, $\omega_2 = 1600$ and $\omega_3 = 2100$ cm^{-1} that can be directly calculated from

$$\omega_1 = (2-\sqrt{2})^{1/2}\,(k/m)^{1/2} \tag{17a}$$
$$\omega_2 = (2 \pm 0)^{1/2}\,(k/m)^{1/2} \tag{17b}$$
$$\omega_3 = (2+\sqrt{2})^{1/2}\,(k/m)^{1/2} \tag{17c}$$

The model spectrum corresponds well to the basic oscillation frequencies of the butatriene molecule [104].

When $k_0/k = 0.5$ (Fig. 7b), the cumulene chain fragments are connected via elastic bonding with an oscillation frequency corresponding to the absorption wave vector of 1000 cm^{-1}. Comparing Figure 7b to Figures 7a and c, it can be concluded that the progressive weakening of the stiffness of the periodically repeated kink in the chain leads to splitting of the oscillation frequencies, i.e. to removal of the spectral degradation and hence to the appearance of „forbidden zones" in the frequency spectrum. This further leads to a gradual narrowing of the frequency zones and to their conversion, at the limit $k_0/k = 0$, into the oscillation frequencies of butatriene.

Figure 8 shows a comparison between an experimental IR spectrum of β-carbyne and a phonon spectrum simulation calculated for a linear chain of poly(butatriene) at $k_0/k = 0.5$. The absorption band in the frequency range of 2100 - 2200 cm^{-1} corresponds to the band of oscillation frequencies of poly(butatriene) that at the limit $k_0/k = 0$ transforms into fully symmetrical oscillations of the butatriene molecule. The broad band in the range 1000 - 1700 cm^{-1} relates to two oscillation zones that converge at the limit $k_0/k = 0$ into a symmetric ($\omega_1 = 865$ cm^{-1}) and an antisymmetric ($\omega_2 = 1600$ cm^{-1}) phonon of the butatriene molecule. This broad band appears to be typical of carbyne formed by dehydrohalogenation of PVDH (see, for example [45,99,105]; also Fig. 1 in Chapter 7.1). Another typical feature is a „dip" in the broad band centred around 1500 cm^{-1} that may be an expression of the presence of the two separated oscillation zones. In conclusion, the phonon spectrum of carbyne is characterized by a narrow band in the 2100-2200 cm^{-1} region [106] and a very broad band in the 1000-1700 cm^{-1} region with a gap centred around 1500 cm^{-1}. For more details on IR spectroscopy of carbyne the reader is referred to Chapter 6.1.

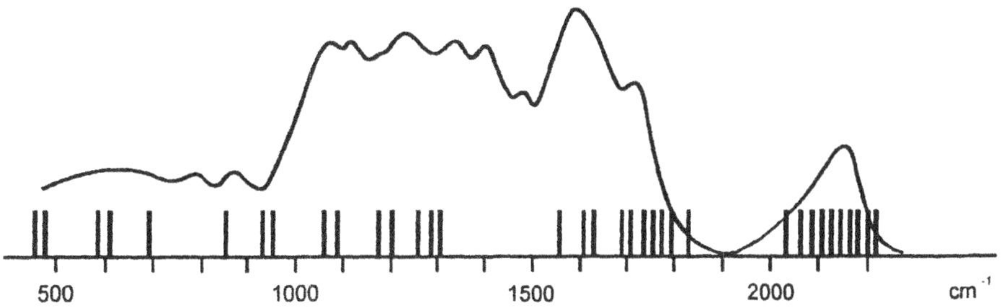

Figure 8. Comparsion of an experimental IR spectrum of β-carbyne (top) and a calculated phonon spectrum of carbon chain shown in Fig. 7 at $k_0/k = 0.5$ [103].

4. Layered structure of carbyne

The kink model has drawn considerable criticism from researchers who alleged that the always observed lamellar morphology of carbyne microcrystals with large {00.1} faces as revealed under the electron beam of a transmission electronmicrocope is incompatible with the notion of extended carbon chains parallel to the c-axis [107,108] (cp. also Chapter 1.2). This inconsistency can be resolved by the layered carbyne lattice model [27] that is entirely based on the kink model [16,17].

Extensive electron diffraction work performed on carbyne [27] showed that the diffraction pattern of an oriented carbyne single crystal film produced by ion sputtering of graphite in conjunction with a bombardment of the growing film by Ar+ ions, is characterized by a high intensity of the first lattice spacing (11.0) at 435 pm. Accordingly, the carbyne polytype that can be assigned is C XIV with a_0 = 870 pm (Table 1). This carbyne polytype was originally discovered surrounding tetragonally disordered diamond grains (p-diamond [1], Fig. 9) in shock-quenched graphite samples. It is thought that after pressure decay the prevailing residual high temperatures in so-called ‘hot spots’ lead to partial reversal of the diamond structure to low pressure carbon forms including carbyne („regraphitization“). For details concerning the high temperature-low pressure phase diagram of carbon see Chapter 7.1, Figure 4.

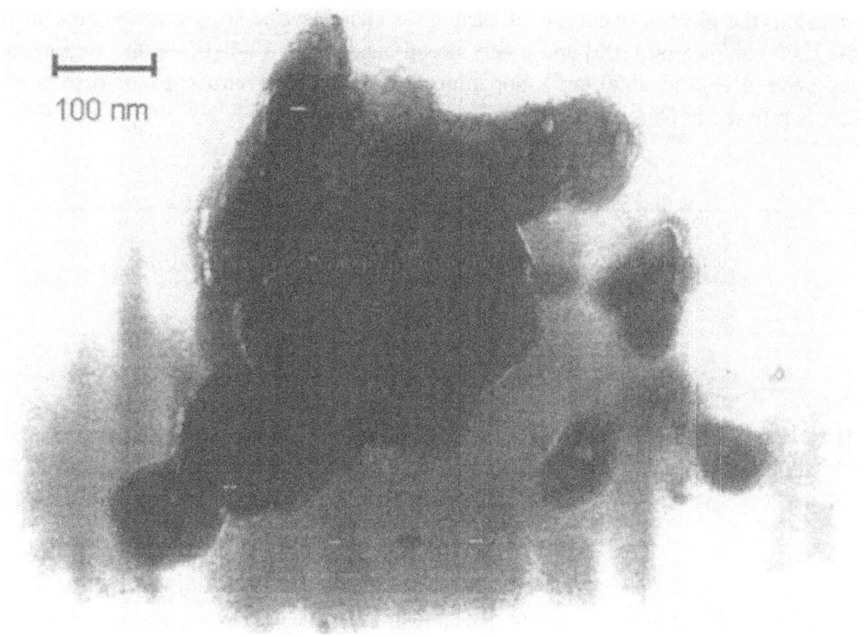

Figure 9 Transmission electronmicrograph of a shocked graphite grain [1]. Halos around p-diamond cores consist of carbon XIV with a_0 = 870 pm, c_0 = 956 pm.

Microcrystalline inclusions showed a very high degree of order with more than 140 hki0 reflexes present. These inclusions could be assigned to an α-carbyne lattice with a_0 = 892 pm. The large intensity of the (11.0) reflexes of the film, more than 10 times that of the other reflexes, was explained by the authors as due to a random deviation of some of the carbyne layers from their ideal positions in a perfect hexagonal lattice. From the standard deviation σ of the position of the layers, the diffraction amplitudes (intensities) can be estimated as

$$A_{hk0} \propto |S_{hk0}|^{-1} \cdot [(-S^2_{hk0} \cdot \sigma^2)/2] \quad (18)$$

where

$$|S_{hk0}| = (4\pi/\sqrt{3}a_0)\,[h^2+hk+k^2] \quad (19)$$

is the diffraction vector of the hexagonal lattice. With the experimentally obtained values of the intensities A(11.0)/A(30.0) = 10, the standard deviation is σ = 90 pm suggesting that this rather large value cannot be accounted for by thermal vibrations but instead supporting the notion of a layered structure of the carbyne lattice.

With the 140 individual reflexes of the highly crystalline α-carbyne microinclusions the Patterson function was calculated that in the case of electron diffraction is a self-convolution of the crystal potential U(xyz). Its Fourier transform is the electron scattering function from the potentials of the crystal array. As input values to calculate the Patterson function the squares of the structural amplitudes $|F_{hkl}|^2$ obtained from 72 independent reflexes were used. From the maxima of the Patterson function a map of the crystal potential was obtained as shown in Figure 10a. The local potential maxima correspond to the position of the carbon chains (1) and (2) at the corners of the hexagonal cell, and (3) in the center. Two other chains (4) and (5) are shifted by 149 pm along a translation vector relative to the chains (1) and (2). Based on this model the phases of the electron scattering amplitudes were calculated, and with this information and the experimentally obtained structural amplitudes $|F_{hk0}|$ a 2D-Fourier synthesis was performed of the crystal potential U(r) of the carbyne lattice (Fig. 10b). The solid lines represent the positive equipotentials around the carbon atoms in the chains, the dashed lines indicate the potential of the electron clouds of the π-orbitals. The additional satellite maxima at a distance of 70 pm from the chains were attributed to (stabilizing) impurity atoms by the authors [27] but could also be interpreted as manifestations of a soliton as invoked above during discussion of the soliton stabilization of kinks.

Since the short distance of 149 pm between, respectively chains (1) and (4), and (2) and (5) would result in σ-bond interaction and thus collapse, a layered structure was assumed, i.e. the chains (1) to (3), and (4) and (5), respectively are located in adjacent layers. Figure 11 shows a spatial model with a lower layer consisting of a close-packed arrangement of three non-equivalent chains of five carbon atoms each and an upper layer containing two non-equivalent chains with vacancies of a period of 515 pm (= $a_0/\sqrt{3}$) hence forming an hexagonal superlattice of the type $\sqrt{3} \times \sqrt{3}$. Indicated are also large impurity atoms (K,Fe) intercalated between the two layers at a kink site that may provide saturation of dangling bonds. In Figure 5 of Chapter 4.3. three possible types of

2D-layered carbyne lattices are shown: (a) without vacancies, i.e. with close-packed carbon chains perpendicular to (00.1) and an interchain distance a_0 of 297 pm, (b) a √3 x √3 vacancy superlattice with a chain distance of 515 pm (=a_0/√3), and (c) a superlattice 2 x 2 with a chain distance of 594 pm (= 2 a_0).

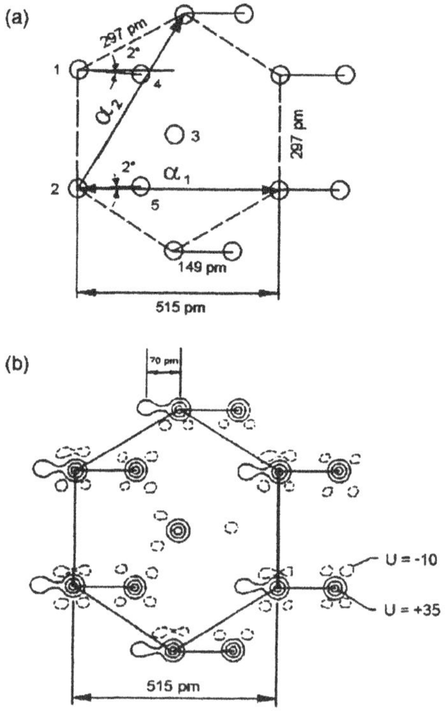

Figure 10. Crystal potential map of the layered structure of carbyne (a) and Fourier synthesis obtained from the phases of the electron scattering amplitudes and the experimentally obtained structural amplitudes (b).

Recently, Tanuma and Pal'nichenko [105] discovered a new carbon allotrope ('carbolite') by quenching, on a copper plate to ambient temperature, a carbon gas obtained by Joule heating of graphite to temperatures exceeding 2600 K. This carbon form manifests itself in two crystallographically distinct structures consisting of an essentially layered arrangement of respectively, two and three kinked carbon chains parallel to the c-axes containing four carbon atoms each. The unit cell metric of the two forms were determined from XRD measurements to be hexagonal with a_0 = 1193 pm, c_0 = 1062 pm (two layers), and a_0 = 1166 pm, c_0 = 1568 pm (three layers), giving an average layer thickness of 527 pm. The latter form obeys the rhombohedral extinction rule and can thus be transformed into a rhombohedral lattice with a_0/√3 = 673 pm, c_0 = 1568 pm with the rhombohedral edge length of 852 pm and a rhombohedral angle α = 86.34°. The interchain distance r appears to be much larger than that for the known carbyne forms, being on average r = 340 pm as compared to 297 pm for carbyne. This value is in the range of the interlayer distance of turbostratic graphite and accounts for the extraordinary low density of 'carbolite' of 1.46 g/cm³. An interchain distance of 340 pm was also found for carbyne intercalated with sodium (CIC-Na [110]). With the van der Waals diameter of the sp-hybridized carbon atom (300 pm) the potential minimum of intermolecular interaction of polyyne chains in the case of sodium intercalation requires a tetragonally-coordinated sodium atom. Hence, in CIC-Na the central Na atom is surrounded by four carbon chains whose distance of each other is 340 pm. The question arises how the chains in Tanuma's 'carbolite' keep their distance of 340 pm without the aid of a hetero-atom acting as 'spacer'.

The hexagonally indexed 'carbolite' forms contain 36 chains in their unit cells. Reduction of the unit cell size by a factor of four leads to a subcell containing 9 chains with a lattice constant a_0 = 596 or 583 pm corresponding to $r\cdot\sqrt{3}$. This is close to the chain arrangement of a 2 x 2 superlattice with a_0 = 594 pm (Fig. 5, Chapter 4.3). With this in mind, 'carbolite' can be tentatively incorporated into the carbyne family of carbon allotropes as a somewhat 'bloated' structure with larger interchain distance, similar to the form obtained by Heimann *et al* recently [30]. For more details the reader is referred to Chapter 3.4.1.2.

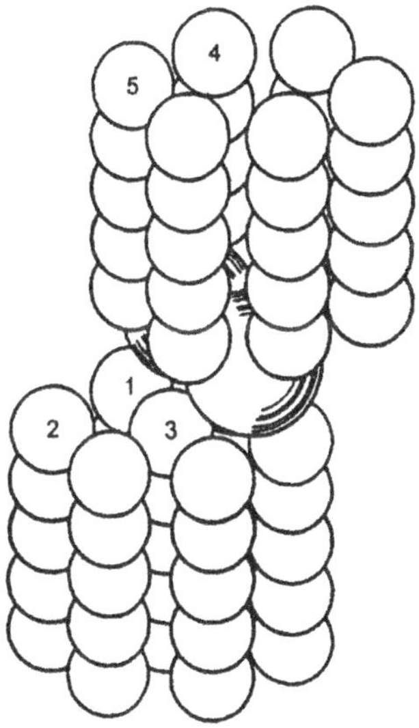

Figure 11 Spatial model of the layered structure of carbyne [27].

In conclusion, the always observed lamellar morphology of carbyne can be explained by the stacking of two kinked layers consisting of short carbon chains, one layer having a close-packed arrangement of three non-equivalent carbon chains, the other having two chains with vacancies of the type $\sqrt{3}$ x $\sqrt{3}$. The minium thickness of a lamella corresponds to the length of the rectilinear carbon chains. In the new allotropic carbon form 'carbolite', either ABA stacking of two or ABCA stacking of three kinked layers occurs with carbon chains forming a 2 x 2 superlattice whose mode of stabilization is hitherto unknown.

5. Carbyne as chiral poly(diacetylene)?

Baughman *et al* [111] recently predicted the existence of two different types of crystalline poly(diacetylene) carbon networks, 'hinged poly(diacetylene)' and 'chiral poly(diacetylene)'. They are characterized by a spatial arrangement of parallel diacetylene chains. Figure 12a shows the chiral phase with the vertical $C(sp^2)$-$C(sp^2)$ bonds connecting the polyacetylene chains $(-C{\equiv}C-)_n$ situated in three different spatial directions. Figure 12b shows a projection down the c-axis (polyacetylene chain axis) of the chiral phase displaying layers of parallel poly(diacetylene) chains oriented in three equivalent directions of the hexagonal lattice. The chiral poly(diacetylene) crystal obeys a hexagonal space group $P6_122$ for a unit cell containing 24 carbon atoms, and with unit cell parameters $a_0 = 491$ pm and $c_0 = 1444$ pm.

Based on this predicted all-carbon phase. Baughman *et al* [111] developed an intriguing notion that carbyne may be intimately related to this hexagonal chiral (*cis-cisoid*) poly(diacetylene). The (rhombohedrally indexed) a_0-values of carbyne polytypic phases known from the literature [16] (see Table 1) are all within 10%, and the majority within 5%, of the predicted a_0-value of chiral poly(diacetylene). The largest deviations of about 10% are observed for carbon VI and carbon IX consistent with the suggestion by Whittaker that these carbyne forms may be 'bloated' chaoite obtained by intercalation of impurities [8]. This explanation is in line with the assumption that a_0 of chiral poly(diacetylene) may increase by up to 10% if the chains are highly strained owing to impurity intercalation. The c_0-values of all known carbyne forms appear to be within <6% of the values of integer multiples (3,4,5 and 6 times) of the $c_0/6$-distance (=241 pm) of the chiral poly(diacetylene). The sublayer distance of 1/6 of the unit cell length in c-direction would be consistent with the hexagonal screw axis 6_1.

Since according to the authors the packing sequence of the layered array of $(-C{\equiv}C-)_n$ chains can be interrupted by interspersed sequences of a so-called *trans*-hinged phase, the notion of coordination statistics in paracrystal theory would apply as discussed in section 2 of this chapter.

Furthermore, Baughman *et al* [111] point out that the accuracy of the unit cell dimensions reported in the literature for the various carbyne polytypes may be affected by one or all of the following : mixtures of phases, dynamic diffraction effects for the electron diffraction data, and unknown amount of intercalation of impurities. The unusual structure of the chiral poly(diacetylene) provides large electron density variations with long periodicities that produces a strong 100 electron diffraction intensity. This reflection is typically observed in carbyne to be strong, and is, in fact, the only interplanar spacing found in (hk0) patterns of disordered, i.e. intercalated carbyne [110]. While the authors also compare Raman data of poly(diacetylene) and carbyne, and find satisfactory correlation, they emphasize the need for caution when drawing far-reaching conclusions from this agreement since the spectroscopic techniques lack the high resolution required to obtain unequivocal spectra from the carbyne microcrystals that are always embedded in a matrix of amorphous carbon.

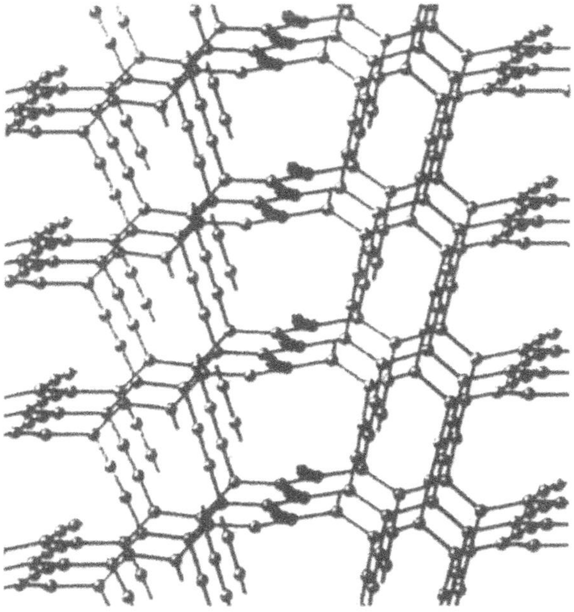

Figure 12a. Structure of chiral poly(diacetylene) carbon crystal. Parallel poly(diacetylene) chains in one layer are bonded to chains in equivalent neighboring layers [111]

Figure 12b. Projection along the poly(diacetylene) chains axis (c-axis) of a chiral poly(diacetylene) crystal [111] (for explanation see text).

6. References

1. Kleiman, J., Heimann, R.B., Hawken, D. and Salansky, N.M. (1984) Shock compression and flash heating of graphite/metal mixtures at temperatures up to 3200 K and pressures up to 25 GPa, *J.Appl.Phys.* **56** (5), 1440-1454.
2. Korshak, V.V., Kasatochkin, V.I., Sladkov, A.M., Kudryavtsev, Yu.P. and Usenbaev, K (1961) Synthesis and properties of polyacetylene, *Dokl.Akad,Nauk SSSR* **136(6)**, 1342-1344 (in Russian)
3. Sladkov, A.M., Kasatochkin, V.I., Kudryavtsev, Yu.P. and Korshak, V.V (1968) Synthesis and properties of linear carbons. *Izv.Akad.Nauk SSSR, Ser.Khim.* **12**, 2697-2704 (in Russian)
4. Sladkov, A.M.and Kudryavtsev, Yu.P. (1969) Diamond, graphite, carbyne - allotropic forms of carbon, *Priroda* **5**, 37-44 (in Russian)
5. Kasatochkin, V.I., Savranski, V.V., Smirnov, B.N. and Mel'nichenko, V.M. (1974) Investigation of carbyne condensed from carbon gas, *Dokl.Akad.Nauk SSSR* **217**(4), 796-799 (in Russian)
6. Sergushin, I.P., Kudryavtsev, Yu.P., Elizen, V.M.,Sadovskii, A.P., Sladkov, A.M., Nefedov, V.I. and Korshak, V.V. (1977) X-ray electron and X-ray spectroscopic investigation of carbyne, *Zhurn.Strukt.Khim.* **18**(4), 698-700 (in Russian)
7. Sladkov, A.M (1981) Carbyne-a new allotropic form of carbon, *Sov.Sci.Rev.* **B3**, 75-110.
8. Whittaker, A.G. and Wolten, G.M.(1972) Carbon: a suggested new hexagonal crystal form, *Science* **178**, 54-56.
9. Whittaker, A.G. and Evangelides, J.S. (1977) Influence of carbon polymorphs on ablation of graphitic materials, Abstr. 13th Biennial Conf. on Carbon, p.410.
10. Whittaker, A.G. (1978) Carbon: a new view on its high-temperature behavior, *Science* **200**, 763-764.
11. Whittaker, A.G. (1979) Carbon: occurence of carbyne forms of carbon in natural graphite, *Carbon* **17**, 21-24.
12. Whittaker, A.G., Watts, E.J., Lewis, R.S. and Anders, E, (1980) Carbynes: carriers of primordial noble gases in meteorites, *Science* **209**, 1512-1513.
13. Whittaker, A.G. (1981) Chaoite coating process, U.S.Patent 4,248,909.
14. Whittaker, A.G., and Kintner, P.L. (1981) Carbon Vapor Pressure in the Range 3450 to 4500 K and Evidence for Melting at 3800 K. Rpt.Aerospace Corp., Materials Science Lab., El Segundo, CA.
15. El Goresy, A. and Donnay, G., (1968) A new allotropic form of carbon from the Ries crater, *Science* **161**, 363-364.
16. Heimann, R.B., Kleiman, J. and Salansky, N.M. (1983) A unified structural approach to linear carbon polytypes, *Nature* **306**, 164-167.
17. Heimann, R.B., Kleiman, J. and Salansky, N.M. (1984) Structural aspects and conformation of linear carbon polytypes, *Carbon* **22**, 147-156.
18. Pitzer, K.S. and Clementi, E. (1959) Large molecules in carbon vapor, *J.Am.Chem.Soc.* **81**, 4477-4485.
19. Brown, W.L., Freeman, R.R., Raghavachari, K. and Schlüter, M. (1987) Covalent group IV atomic clusters. *Science* **235**, 860-865
20. Stoicheff, B.P. (1962) The variation of carbon-carbon bond lengths with environment as determined by spectroscopic studies of simple polyatomic molecules, *Tetrahedron* **17**, 135-145.
21. Bastiansen, O. and Traetteberg, M. (1962) The nature of bonds between carbon atoms. How they vary with environment, *Tetrahedron* **17**, 147-154.
22. Christensen, M.T., Eaton, D.R., Green, B.A. and Thompson, H.W. (1956) Vibration-rotation bands of acetylene and the molecular dimensions, *Proc.Roy. Soc.* **A237**, 15-30.
23. Stoicheff, B.P. (1957) High-resolution Raman spectroscopy of gases. X.Rotational spectrum of butatriene, *Canad.J.Phys.* **35**, 837-841.
24. Lagow, R.J., Kampa, J.J., Wei, H-C., Battle, S.L., Genge, J.W., Laude, D.A., Harper, C.J., Bau, R., Stevens, R.C., Haw, J.F. and Munson, E. (1995) Synthesis of linear acetylenic carbon: the „sp" carbon allotrope, *Science* **267**, 362-367.
25. Eastmond, R., Johnston, T.R. and Walton, D.R.M. (1972) Silylation as a protective method for terminal alkynes in oxidative couplings, *Tetrahedron* **28**, 4601-4616.
26. Rohlfing, E.A., Cox, D.M. and Kaldor, A. (1984) Production and characterization of supersonic carbon cluster beams, *J.Chem.Phys.* **81**, 3322-3330.

27. Kudryavtsev, Yu.P., Evsyukov, S.E., Babaev, V.G., Guseva, M.B., Khvostov, V.V. and Krechko, L.M.(1992) Oriented carbyne layers, *Carbon* **30** (2), 213-221.
28. Udod, I.A., Shchurik, V.I., Bulychev, B.M., Sirotinkin, S.P., Guseva, M.B., Babaev, V.G., Kudryavtsev, Yu.P. and Evsyukov, S.E. (1993) Formation of carbyne in the interaction of polyacetylene with potassium under high quasi-hydrostatic pressure, *J.Mater.Chem.* **3**(4), 413-416.
29. van Thiel, M. and Ree, F.H. (1978) Properties of carbon clusters in TNT detonation products: graphite-diamind transition, *J.Appl.Phys.* **62**, 1761-1767.
30. Heimann, R.B., Fujiwara, S., Kakudate, Y., Koga, Y., Komatsu, T. and Nomura, M. (1995) A new carbon form obtained by weak shock compression of carbyne, *Carbon* **33**(6), 859-863.
31. Yamada, K., Burkhard, G., Tanabe, T. and Sawaoka, A.B. (1996) An in-situ observation of amorphization of carbyne particles under electron-beam irradiation, *Carbon* **34**, 1601-1602.
32. Hoffmann, R. (1966) Extended Hückel theory. V. Cumulenes, polyenes, polyacetylenes and C_n, *Tetrahedron* **22**, 521-538.
33. Mulliken, R.S. (1955) Electronic population analysis on LCAO-MO molecular wave functions I. *J.Chem.Phys.* **23**, 1833-1841; Mulliken, R.S. (1955) Overlap populations, bond orders and covalent bond energies, *J.Chem.Phys.* **23**, 1841-1846.
34. Ewing, D.W. and Pfeiffer, G.V. (1982) Structure and properties of linear C_n molecules, *Chem.Phys.Lett.* **86**(4), 365-368.
35. Bochvar, D.A., Nikerov, M.V. and Stankevich, I.V. (1987) Problems in the structural chemistry of crystalline carbon, *Sov.Sci.Rev.B.Chem.* **9**, 483-527.
36. Popov, E.M. and Kogan, G.A. (1968) The structrue of conjugated open-chain hydrocarbons, *Russ.Chem.Rev.*, **37**(2), 119-141.
37. Krauch, H.(1958) Spectral comparison of polyenes, cumulenes and poly-ynes *J.Chem.Phys.* **28**, 898-899.
38. Rice, M.J., Philpot, S.R., Bishop, A.R. and Campbell, D.K. (1986) Solitons, polarons, and phonons in the infinite polyyne chain, *Phys.Rev.B* **34**(6), 4139-4149.
39. Heimann, R.B. (1994) Linear finite carbon chains (carbyne): their role during dynamic transformation of graphite to diamond, and their geometrical and electronic structure, *Diamond Rel.Mater.* **3**, 1151-1157.
40. Ertchak, D., personal communication.
41. Kavan, L. and Kastner, J. (1994) Carbyne forms of carbon: continuation of the story, *Carbon* **32**(8), 1533-1536.
42. Kijima, M. (1997) Recent approaches for chemical and electrochemical synthesis of linear carbon, in: Recent Res.Dev.in Pure & Appl.Chem., Transworld Research Network, in the press.
43. Kudryavtsev, Yu.P., Evsyukov, S.E., Guseva, M.B., Babaev, V.G. and Khvostov, V.V. (1993) Carbyne-the third allotropic form of carbon, *Russ.Chem.Bull.* **42**(3), 399-413.
44. Korshak, V.V., Kudryavtsev, Yu.P., Korshak, Yu.V., Evsyukov, S.E., Khvostov, V.V., Babaev, V.G. and Guseva, M.B. (1988) Formation of ß-carbyne by dehydohalogenation, *Makromol.Chem., Rapid Commun.* **9**, 135-140.
45. Evsyukov, S.E., Paasch, S., Thomas, B. and Heimann, R.B. (1997) Formation of carbynoid structures by chemical dehydrohalogenation of poly(vinylidene chloride). A ^{13}C solid-state NMR study, *Ber.Bunsenges. Phys.Chem.* **101**(5), 837-841.
46. Shimoyama, M., Niino, H. and Yabe, A. (1992) A KrF excimer laser induced dehydrochlorination of a chlorinated poly(vinyl chloride): preparation of conjugated polyene and polyyne, *Makromol.Chem.* **193**, 569-574.
47. Niino, H. and Yabe, A., Formation of conjugated polyene and polyyne structure by KrF eximer laser-induced dehydrochlorination of polyvinylidene chloride, *Macromol.Chem*, submitted for publication.
48. Tanaka, K., Nishio, S., Matsuura, Y. and Yamabe, T. (1994) Electron spin resonance studies of organic semiconductive thin films by plasma polymerozation of 3,4,9,10-perylene tetracarboxylic dianhydride, *Synth.Met.* **64**, 209-215.
49. Bochvar, D.A., Stankevich, I.V., Gal'pern, E.G. and Bakuradse, R.Sh. (1987) Investigation of the influence of defects in carbon chains on the electronic structure and the related stability of carbyne macromolecules, *Zhurn. Strukt. Khim.* **28** (4), 23-27 (in Russian).
50. Misurkin, I.A. and Ovchinnikov, A.A. (1977) The electronic structures and properties of polymeric molecules with conjugated bonds, *Russ.Chem.Rev.* **46**(10), 967-987.

51. Mel'nichenko, V.M., Sladkov, A.M. and Nikulin, Yu.N. (1982) Structure of polymeric carbon, *Russ.Chem.Rev.* **51**(5), 421-438.
52. Kudryavtsev, Yu.P., Heimann, R.B. and Evsyukov, S.E. (1996) Carbynes: advances in the field of linear carbon chain compounds, *J.Mater.Sci.* **31**, 5557-5571.
53. Korshak, V.V., Kudryavtsev, Yu.P., Khvostov, V.V., Guseva, M.B., Babaev, V.G. and Rylova, O.Yu. (1987) Electronic structure of carbynes studied by Auger and electron energy loss spectroscopy, *Carbon* **25**(6), 735-738.
54. Khvostov, V.V., Guseva, M.B., Babaev, V.G. and Rylova, O.Yu. (1986) Auger-spectroscopy studies of the electronic structure of amorphous carbon films, *Surf.Science* **169**, L253-L258.
55. Martinez, V. (1979) Global methods in the inversion of a self-convolution, *J.Electron Spectr.* **17**, 33-43.
56. Leleyter, M. and Joyes, P. (1975) Experimental and theoretical study of secondary emission of molecular Group IVB elements, *J.de Physique* **36**, 343-355 (in French).
57. Vagh, A.S., Carton, B. and Mrozowski, S. (1974) Specific heat of glassy carbon between 0.4 and 4.5 K I, *Carbon* **12**(6), 645-650.
58. Lindenmeyer, P.H. and Hosemann, R. (1963) Application of the theory of paracrystals to the crystal structure analysis of polyacrylonitrile, *J.Appl.Phys.* **34**(1), 42-45.
59. R.Hosemann and S.N.Bagchi, Direct Analysis of Diffraction by Matter. North Holland: Amsterdam, 1962.
60. Nikerov, M.V., Bochvar, D.A. and Stankevich, I.V. (1981) On new crystalline modifications of carbon and boron nitride, *Izv.Akad.Nauk SSSR, Ser Khim.* **1981**(5), 1177-1178 (in Russian).
61. Nikerov, M.V., Bochvar, D.A. and Stankevich, I.V, (1982) On allotropic modifications of carbon, *Zhurn.Strukt.Khim.* **23**(1), 177-179 (in Russian).
62. Springborg. M. and Kavan, L. (1992) On the stability of polyyne, *Chem.Phys.* **168**, 249-258.
63. Natta, G., Bassi, I.W. and Fagherazzi, G. (1969) Crystal structure of odd and even *trans* polyalkenamers, *Eur.Polym.J.* **5**, 239-260.
64. Perego, G. and Bassi, I.W. (1963) The crystal structure of isotactic trans-1,4 poly-1-ethyl-butadiene, *Makromolek. Chem.* **61**, 198-206.
65. Reneker, D.H. (1962) Point dislocations in crystals of high polymer molecules, *J.Polym.Sci.* **59**, S39-S42.
66. Pechhold, W. and Blasenbrey, S. (1967) Cooperative rotational isomerism in polymers. I. Melting theory and lump concentrations, *Kolloidz.u.Z.Polym.* **216**, 235-244.
Rich, A. and Crick, R.H.C. (1961) The molecular structure of collagen, *J.Mol. Biol.* **3**, 483-506.
67. Hess, K. and Kiessig, H. (1943) Die Feinstruktur von Polyamidfasern, *Naturwiss.* **31**, 171 (in German).
68. Hosemann, R. (1963) Crystalline and paracrystalline order in high polymers, *J.Appl.Phys.* **34**(1), 25-41.
69. Stratton, W.O. (1959) *J.Polymer Sci.* **41**, 143.
70. Whittaker, A.G., Neudorffer, M.E. and Watts, E.J. (1983) Carbon: a rhombo-hedral carbyne form, *Carbon* **21**, 597-599.
71. Debye, P. (1930) Isomeric molecules studied by x-ray diffraction, *Phys.Z.* **31**, 142.
72. Hosemann, R. (1973) Paracrystals in biopolymers and synthetic polymers, *Endeavour* **117**, 99-105.
73. Anno, T. and Coulson, C.A. (1961) The structure of graphite, *Proc.A.Soc.* **A264**, 165-175.
74. Kijima, M., Toyabe, T. and Shirakawa, H. (1995) Electrocatalytic dehydrofluori-nation of poly(vinylidene fluoride), *Chem.Letters* **1995**, 553-554.
75. Julg, A. (1968) Origin of the relation between the bond index and the interatomic distance in conjugated molecules, *J.Chem.Phys.* **65**, 541-548.
76. Coulson, C.A.(1952) *Valence*, Clarendon: Oxford.
77. Bunn, C.W. (1939) Crystal structure of long-chain normal paraffin hydrocarbons „shape" of the methylene group, *Trans.Faraday Soc.* **35**, 482-491.
Seto, T., Hara, T. and Tanaka, K. (1968) Phase transformation and deformation processes in oriented polyethylene, *Jpn.J.Appl.Phys.* **7**, 31-42.
78. Jansta, J. and Dousek, F.P (1980) Some aspects of existence of elementary carbon with sp-hybridized bonds, *Carbon* **18**, 443-437.
79. Kasatochkin, V.I., Sladkov, A.M., Aseev, Yu.G., Kudryavtsev, Yu.P., Egorova, O.I. and Korshak, V.V. (1963) Infrared spectra of polyynes, *Dokl.Akad.Nauk SSSR* **153**(2), 346-349 (in Russian).

80. Hay, A.S. (1969) Oxidative polymerization of diethynyl compounds *J.Polymer.Sci.* **A1,V7**, 1625-1634.
81. Smith, P.P.K. and Buseck, P.R. (1985) Carbyne forms of carbon: evidence for their existence (reply), *Science* **229**, 486-487.
82. Coulson, C.A. and Davies, P.L. (1951) Long range forces between large chain molecules, *Trans.Faraday Soc.* **48**, 777-789.
83. Bak, P. (1978) Solitons in incommensurate systems, in: Solitons and Condensed Matter Physics, A.R.Bishop and T.Schneider (eds.), Springer:Berlin, p.216.
84. Kürti, J. and Surján, P.R. (1992) Embedded units in conjugated polymers, *J.Math.Chem.* **10**, 313-327.
85. Bullough, R.K.(1978) Solitons, *Phys.Bull.* February **1978**, 78-82.
86. Fogel, M.B., Trullinger, S.E., Bishop, A.R. and Krumhansl, J.A. (1976) Classical particlelike behavior of sine-Gordon solitons in scattering potentials and applied fields, *Phys.Rev.Lett.* **36**, 1411-1414.
87. Di Salvo, F.J. and Rice, T.M. (1979) Charge-density waves in transition-metal compounds, *Phys.Today* April **1979**, 32-38.
88. Su, W.P., Schrieffer, J.R. and Heeger, A.J. (1979) Solitons in polyacetylene, *Phys.Rev.Lett.* **42**, 1698-1710.
89. Seeger, A. and Schiller, P. (1966) in: Physical Acoustics, Vol.III, W.P.Mason (ed.), Academic Press:New York, p.361.
90. Bohm, J. (1995) *Realstruktur von Kristallen*, E.Schweizerbart'sche Verlagsbuchhandlung (Nägele und Obermiller):Stuttgart, pp.213-215.
91. Rebbi, C. (1979) Solitons, *Sci.Am.* **240**(2), 92-116.
92. Costain, G.C. and Stoicheff, B.P. (1959) Microwave spectrum, molecular structure of vinyl cyanide, and a summary of carbon-carbon, carbon-hydrogen bond lengths in simple molecules, *J.Phys.Chem.* **30**, 777-782.
93. Heath, J.R., Zhang, Q., O'Brian, S.C., Curl, R.F., Kroto, H.W. and Smalley, R.E. (1987) The formation of long carbon chain molecules during laser vaporization of graphite, *J.Am.Chem.Soc.* **109**, 359-363.
94. McElvany, S.M., Ross, M.M. and Callahan, J.H. (1992) *Acc.Chem.Res.* **25** 162.
95. Smalley, R.E. (1992) Self-assembly of the fullerenes, *Acc.Chem.Res.* **25**, 98-105.
96. Kokai, F., Koga, Y. and Heimann, R.B. (1996) Magnetic field enhanced growth of carbon cluster ions in the laser ablation plume of graphite, *Appl.Surf.Science* **96/98**, 261-266.
97. Kasatochkin, V.I., Kazakov, M.E., Savranskij, V.V., Nabatnikov, A.P. and Radimov, N.P. (1971) Synthesis of new allotropic forms of carbon from graphite, *Dokl.Akad.Nauk SSSR* **201**(5), 1104-1105. (in Russian).
98. Nelson, L.S., Whittaker, A.G. and Tooper, B. (1972) The formation of new polymorphs of carbon and fluid flow patterns by irradiating carbon with a CO_2 laser, *High Temp.Sci.* **4**. 445-477.
99. Komatsu, T., Nomura, M., Kakudate, Y., Fujiwara, S. and Heimann, R.B. (1995) Characterization of dehydrochlorinated poly(vinylidene chloride) and the shock-compressed material, *Macromol.Chem.Phys.* **196**, 3031-3040.
100. Wang, Y., (1992) Photoconductivity of fullerene-doped polymers, *Nature* **356**, 585-587.
101. Schopov, I. and Mladenova, L. (1992) Synthesis and properties of methylene-bridged polyenes, *Synth.Met.* **48**, 249-258.
102. Ertschak, D.P. (1996), personal communication.
103. Korshak, V.V., Kudryavtsev, Yu.P., Evsyukov, S.E., Korshak, Yu.V., Guseva, M.B., Babaev, V.G. and Kostishko, B.M. (1988) Infrared spectra of carbyne, *Dokl.Akad.Nauk SSSR* **298**(6), 1421-1424 (in Russian).
104. Bower, D.I. and Maddams, W.F. (1989) The Vibrational Spectroscopy of Polymers, Cambridge University Press:Cambridge.
105. Tanuma, S. and Pal'nichenko, A.V. (1995) Synthesis of low density carbon crystal „carbolite" by quenching of carbon gas, *J.Mater.Res.* **10**(5), 1120-1125.
106. Kudryavtsev, Yu.P., Evsyukov, S.E., Guseva, M.B., Babaev, V.G. and Khvostov, V.V. (1993) Carbyne-the third allotropic form of carbon, *Russ.Chem.Bull.* **42**(3), 399-413.
107. Smith, P.P.K. and Buseck, P.R. (1982) Carbyne forms of carbon: do they exist?, *Science* **216**, 984-986.

108. Rietmeijer, F.J.M. (1993) Are crystalline C-(H-O-N) carbons the elusive meteoritic carbyne?, *Meteoritics* **28**, 242-245..
109. Heimann, R.B. (1977) High temperature and high pressure modifications of silica, *Minerals.Sci.Engng.* **9**(2), 57-67.
110. Udod, I.A., Bulychev, B.M., Sirotinkin, S.P., Babaev, V.G and Kudryavtsev, Yu.P. (1993) An interaction of amorphous carbyne with sodium: the formation of carbyne intercalation compounds, *Synth.Metal* **60**, 57-62
111. Baughman, R.H., Galvao, D.S., Cui, C. and Dantas, S.O. (1997) Hinged and chiral polydiacetylene carbon crystals, *Chem.Phys.Letters* **269**, 356-364.
112. Heimann, R.B. (1997) Should carbyne be described by paracrystal theory?, Carbon **35**, 1669-1971.

4.3. CARBYNE INTERCALATION COMPOUNDS

I.A. UDOD
Department of Chemistry,
M.V.Lomonosov Moscow State University,
Moscow 119899, Russia

Abstract - Structural data about linear carbon modifications within the span of more than 30 years of rather 'mysterious' carbyne history were analyzed. A two-dimensional structure with a hexagonal parameter a = 505-525 pm is considered the basis for all carbyne modifications. The large value of this parameter, larger than a 300 pm-van der Waals diameter of *sp*-carbon atoms, is accounted for by the existence of regular vacancies. The three-dimensional packing of zigzag-like carbon chains with a $3^{1/2}\times3^{1/2}$ vacancy superlattice is an ideal matrix for doping with metals. The capture of metal atoms during crystallization stabilizes additionally linear carbon chains in the solid state. Finally, a polycrystalline mosaic structure of carbyne macrocrystals is proposed that can explain the textured (spotted) diffraction pattern of hexagonal type observed experimentally.

1. Structural Investigations of Linear Carbon Matrix

Undoubtedly, the basis for recognition of any allotropic form is the individuality of its structure. Therefore, structural investigations play a particularly important role in studies of the elusive, to some extent 'mysterious' form of carbon - carbyne. Strictly speaking, it was the diffraction pattern, which does not belong to any other known carbon structure, that has initiated numerous investigations and controversies lasting already more than 30 years. Unfortunately, only small particles of several microns in size were reported to be available in the vast majority of studies as impurities in well crystallized (diamond and graphite) and amorphous carbon phases. Evidently, the mechanical isolation of the new substance in its pure state was very difficult in those cases. Furthermore, the insolubility and nonvolatility of the substance did not allow its isolation by chemical methods in amounts that would be sufficient to study all its properties. Thus, it is just this 'adverse' chemical nature of carbyne that is responsible for the unprecedented case, that a substance observed in dozens of independent scientific studies and synthesized by several totally different methods still has not been worldwide recognized by the scientific community. Consequently, owing to the lack of rigorous and reliable structural data a number of hypothetical models often based on the subjective opinion of the authors were published. These facts suggest that the 'mystery' of carbyne requires one additional careful, objective, and detailed review of all what has been done to date.

R.B. Heimann et al. (eds.), Carbyne and Carbynoid Structures, 269–294.

The term *"carbyne"* was introduced into the literature in 1961 by Russian researchers [1-3]. The simple but quite attractive idea concerning the existence of a carbon form constituted of *sp*-hybridized carbon atoms (along with sp^3 and sp^2 carbon allotropes, *i.e.* diamond and graphite) provided a cornerstone for their research. The suffix *"yne"* used in organic chemistry to designate acetylene derivatives, had to reflect, according to the authors' intention, the *sp*-nature of the substance. Unfortunately, later the IUPAC adopted this term to refer to carbon-centered triradicals, used in particular for synthesizing *polycarbynes* - crimped sp^3 carbon networks stabilized by phenyl groups [4,5]. The use of the term *carbyne* thus requires some caution. For example, it is necessary to emphasize that *polycarbyne* has nothing to do with one-dimensional carbon since it is a totally different substance.

In the first attempts, polycondensation reactions widely used to synthesize organic polymers were chosen to obtain carbyne. It was supposed that the sequence of chemical transformations would determine the nature of the linear structure of carbon expected to form. However, the polymeric nature of this material along with the extreme reactivity of unsaturated carbon fragments does not preclude the participation of other atoms and functional groups, present in the chemical reactor, in the formation of interchain cross-links and generally branched structures. The low-pressure thermolysis at 1000 °C could lead not only to removal of solvent molecules, but also could cause profound structural transformations in the entire sample. Thus, the mechanism and, therefore, the linear structure might not necessarily follow from the sequense of chemical reactions considered in those studies [6,7]. In particular, this remark concerns two possible electronic structures of polymeric molecules, *viz.*, polyyne and polycumulene types of the *sp*-bond symmetry. Nevertheless, the authors did manage to isolate pure (according to chemical analysis data) phase of gray carbon, using selective oxidation of black amorphous particles. It is just these structural data reported by Kasatochkin *et al.* [8,9] that should establish a priority in the extending number of discoveries in this field. According to electron diffraction data, two types of crystals were observed with rhombohedral[1] parameters a = 508 pm, c = 780 pm and a = 476 pm, c = 258 pm, called α- and β-carbyne, respectively. It was supposed, that the former phase was less dense and might transform into the latter phase under thermobaric treatment [10]. The use of electron microscopy instead of usual x-ray diffraction method allowed to separate these phases, thus showing for the first time that the new carbon form can exist in at least two modifications.

This fact was not taken into account in studies of natural carbyne (according to the authors' term - Ries crater carbon [11]). Some 22 lines of a new substance in addition to 12 strong lines of graphite were present in the x-ray diffraction pattern. Simple mathematical treatment of the interplanar distance values on the assumption that all reflections belonged to one single phase gave the hexagonal parameters a = 897 pm and c = 1240 pm. Nevertheless, even in that case some lines, including a very strong line (d = 412 pm), probably corresponding to the *(100)* reflection of β-carbyne [12], were outside the experimental error.

Whittaker *et al.* [13-16] observed carbyne in the carbon soot produced by sublimating graphite in vacuum. The variability of parameter c and relatively

---

[1] To convert this to hexagonal parameters the value of a has to be multiplied by $3^{1/2}$.

inconspicuous changes in parameter *a* (I consider here the value of parameter *a* as a minimum, following from *hk0* reflections given in Ref. 13) were observed for carbyne polytypes. Indeed, the hexagonal parameters of Carbon VI and the phase identical to the Ries crater carbon are a = 533 pm, c = 1224 pm and a = 516 pm, c =1461 pm, respectively. Moreover, it was discovered, that carbyne particles often possess a polycrystalline structure, and the size of single crystals is essentially smaller than the width of the electron beam [13]. The textured (spotted) *hk0* diffraction pattern observed in that case testified that all crystallites were mutually oriented.

From my point of view a very important progress in carbyne crystallochemistry was made by Heimann *et al.* [17,18]. The authors conjectured that all modifications of carbon possessing hexagonal symmetry of two-dimensional sublattice with parameter *a* being larger than that in graphite belong to the carbyne family. The large value of parameter *c* (substantially larger than that one could expect for certain links of polyyne or polycumulene structure) might be a consequence of the presence of regular kinks in polymer molecules. Unfortunately, Heimann [17,18] does not answer the main question: how are linear carbon forms being built spatially?

The scarcity and discrepancy of structural data on carbyne and the complexity of long-sought quantitative syntheses of the pure phase resulted in publications with a somewhat provocative tone. It should be emphasized here that some authors of critical papers [*e.g.*, 19,20] used to invoke some isolated facts without considering the whole body of experimantal data known to date. For example, the strange property of capturing atoms of other elements (including silicon) into the crystal lattice observed in carbynes was used to argue that all diffraction patterns reported hitherto as belonging to the carbyne family were those from some types of layered silicates (clays). The electron diffraction techniques used at that time were not capable of making local chemical analyses precisely from the site of the electron diffraction survey. The worldwide predominance of silicates in nature coupled with the use of silicon-based materials in vacuum pumping system of electron microscopes might have cast some doubt on the correctness of the experiments performed. However, we have to remember, that the authors did not describe only one single cases of diffraction in their papers, but reported on typical diffraction patterns from crystalline parts of the samples. The x-ray diffraction data, being quite ineffective for precise determination of the lattice parameters of linear carbon, testify to the quantitative presence of carbyne in the samples [21-24]. It is a remarkable enough fact that the problem of silicate contaminations in electron diffraction studies was discussed only in connection with carbyne research and nowhere else. Unfortunately, the results reported by El Goresy and Donnay [11] were later not reproduced, and we still do not know much about carbyne properties. Actually, if a carbyne crystal can be entirely destroyed under irradiation with an electron beam of low intensity, then the possibility exists of similar processes upon expose to light for long times. To keep the truth it is pertinent to note that all silicates as typically ionic compounds do not 'burn' during electron microscopic measurements. Thus, we are talking here about the substance, which has been observed in dozens of fundamentally different chemical investigations. The starting materials in those experiments were carbon-containing substances (sometimes graphite or diamond), but never silicon-containing ones. The chemical analyses showed that the basic element in the samples studied was carbon, while other elements were encountered as random impurities. Probably, the atoms of these impurities play a certain role in the crystallization of carbyne but do not affect its molecular structure. These considerations bring up the questions: what does carbyne represent by itself and where are the proofs of its linear structure? If we really deal with a polymer, we should expect globular or fibrillar structures rather than layered or plate-like ones. Finally, which structural fragment does pool all the carbyne forms together?

Actually, the layered lamellar structure is characteristic primarily of the substances with predominant two-dimensional type of strong chemical bonds (for example graphite or hexagonal BN). In this case a *00l* family of reflections should be present in the diffraction pattern, whose intensity should stand out sharply as a result of a high density of layers in the *c*-axis direction. The intensity of these reflections should increase when surveying by x-ray powder analysis method as a consequence of preferable orientation of lamellar crystals in the sample. However, currently available x-ray analysis data [11,21-24] testify persuasively to the absence of layer anisotropy. Quite opposite, the same diffraction pattern comprising *hk0* reflections of a two-dimensional hexagonal lattice was observed in almost all instances reported so far[2] [2,8,9,12,18,24-38].

Strictly speaking, it is this diffraction pattern that is traditionally used to prove the presence of carbyne in carbon samples (Fig. 1). Although the parameter *a* is varied through a range of 508-545 pm, it should be noted that the absence of an internal standard in some studies could bring about the non-systematic error when determining

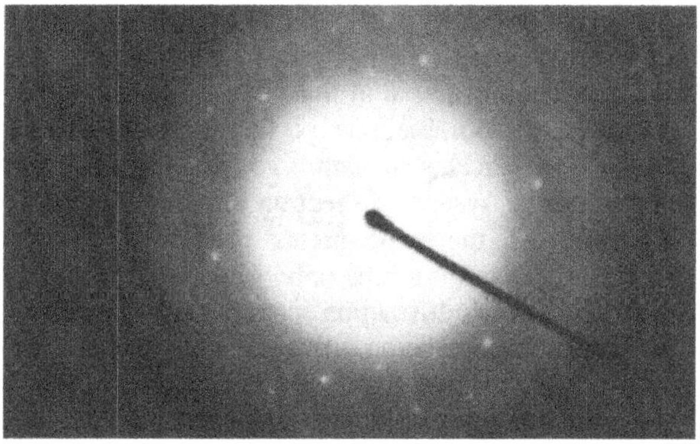

Figure 1. Typical electron diffraction pattern from carbyne crystal (*hk0* section of inverse lattice). It is a two-dimensional structure corresponding to this diffraction pattern, that pools all linear carbon forms together.

the value of this parameter. Moreover, diffraction patterns comprising only six first-order reflections of the hexagonal symmetry (Fig. 2) were observed in recent studies (parameter *a* was reported to be 508-512 pm) [35,36,38,41]. The reason for the absence of the rest reflections of higher orders will be discussed below when considering recent models. Thus, the common fragment of all carbyne modifications is

---

[2] I do not consider here β-carbyne [12] and the carbon phase observed recently by Heimann *et al.* [39,40]. They have been synthesized under high-pressure conditions and may therefore have a denser structure than usual linear carbon matrices.

a two-dimensional hexagonal lattice with parameter *a* [11,17,18]. The existence of a two-dimensional lattice as a structural basis for the whole three-dimensional crystal proves, probably indirectly, the prevailing one-dimensional structure of carbon molecules in carbyne. Indeed, such a typical polymeric material as polyacetylene also possesses two-dimensional packing of molecules resulting from van der Waals interactions between them [42]. The parameter *c* in polyacetylene is determined by the length of intermolecular covalent interaction between carbon atoms. The presence of hydrogen atoms impairs the symmetry of polyacetylene molecules and, as a consequence, their packing in the polymer lattice as a whole. This is why polyacetylene has two isomers (*cis* and *trans*) with orthorhombic symmetry [42]. In turn, we might expect hexagonal symmetry of the carbyne structure to result from the possible cylindrical symmetry of linear carbon molecules. However, it is still not clear why the parameter *a* is substantially larger than that we could expect from the value of the van der Waals diameter of *sp*-hybridized carbon atoms.

Figure 2. Typical electron diffraction pattern from oriented carbyne layers. It is evident that the spatial character of packing of carbon chains in the carbyne crystal and its oriented film is very similar.

An attempt to build a spatial model of carbyne was made by Kudryavtsev *et al.* [35]. In order to calculate a Patterson function, the authors have taken into account only the *hk0* diffraction, and only dispersion from the electron density of carbon atoms was taken into consideration for calculations. Moreover, that model[3] was created on the assumption that the spotted *hk0* electron diffraction pattern corresponded to a three-dimensional single-crystalline structure. However, from the viewpoint of this assumption we could maintain only the single-crystal structure of a two-dimensional projection, but the *hk0* diffraction could be compounded (combined). In the latter case the dispersion from impurity atoms possessing electron densities comparable with that from certain fragment of the carbon chain should also be considered. Furthermore, the proposed spatial model of the two-layered carbyne lattice implies regular alternation of the densest carbyne layers and the layers with vacancies. The formation of vacancies is possible only on the condition that the carbon chains have regular breaks, that is the

[3] See Fig. 11 in Chapter 4.2 of this book.

whole structure should be built up of carbon molecules with finite and equal length. Unfortunately, to the best of my knowledge, there is no means to control the length uniformity of all carbon chains during the processes of both their formation and crystallization.

Despite all the odds, exploration into the structure of oriented carbyne layers [35] yielded a number of interesting results and new ideas:

(i) the shortest distance between carbon chains was reported to be *ca.* 300 pm that satisfactorily correlates with previous data [2,8,18];

(ii) the large value of parameter *a* of the two-dimensional hexagonal carbyne lattice could result from the existence of a regular vacancy sublattice;

(iii) impurity atoms located in the vacancies were surmised to promote the crystallization of linear carbon.

Unfortunately, the task of calculating the structure of carbyne on the basis of the *hk0* intensities is very difficult. It is almost intractable for the following reason alone: the intensity distribution in the *hk0* section of the inverse carbyne lattice varies from one sample to another and, strictly speaking, did not always conform to the hexagonal symmetry [2,8,9,12,18,24-38]. Moreover, the manner in which intensities differ from the expected hexagonal symmetry can be explained neither by a small inclination of the sample relative to the incident beam nor by thermal dynamic vibrations of the lattice. Perhaps, the reason for such unusual behavior has the same origin as does the lamellar and layered structure of carbyne crystals being unexpected in the polymeric substance.

Thus, the two-dimensional sublattice (or its projection on a {*ab*} plane) is a common fragment for the substances which we are calling *carbynes* here. The parameter *c*, in turn, can vary significantly. Moreover, as discussed previously [41,43], this fact could be connected with local concentration of the impurity atoms in the system during its crystallization. The precise determination of parameter *c* by electron diffraction is associated with at least two difficulties. First, the rotation of carbyne crystal through a certain angle to obtain good inclined diffraction requires some time that is quite sufficient to destroy the sample under the electron beam. Secondly, the appearance of a 'twinning' effect in the inclined diffraction pattern instead of the spotted *hk0* inverse lattice has been observed when rotating many carbyne crystals [44] (Fig. 3). Observations of the diffraction from several types of crystals in this case suggests that the real carbyne crystal possesses a polycrystalline character. It is important at this point to recall the data by Whittaker *et al.* [12]. The authors described the carbyne macrocrystal as a polycrystal with textured (spotted) *hk0* diffraction. The crystallites constituting the macrocrystal were mutually oriented and hence the spotted *hk0* diffraction observed was a combination of the pattern of the individual subcrystals. Unfortunately, we can not be sure that the parameter *c* is the same for all crystallites constituting the macrocrystal. The accuracy of the determination of parameter *c*, especially in the absence of the internal standard, is less than that of parameter *a*. Therefore, the direct rigorous calculation of interatomic distances in the molecules seems still uncertain. However, to estimate the number of carbon atoms in a rectilinear fragment we can divide the value of parameter *c* by 120-130 pm.

Thus, the structural data accumulated to date suffice to recognize the reality of some particular phase consisting basically of carbon. The analysis of these data points out the preferred one-dimensional (polymeric) structure of the carbon molecules. Unfortunately, the complex nature of this substance still does not allow to accomplish the synthesis of carbyne single crystals that would be suitable for direct x-ray diffraction measurements. In this situation the creation of structural model satisfying all experimental facts mentioned above and complying with known chemical properties of carbon atoms is much needed. The design of such a spatial model can be conditionally subdivided into three parts: first, *molecular structure of linear carbon*; secondly, *the packing of carbon chains in the microcrystal*, and finally, *the oriented structure of microcrystals in the carbyne macrocrystal*.

2. Molecular Structure of Linear Carbon

Two types of molecular structure can be envisaged for polymers possessing a crystal lattice with parameter *c* being essentially larger than shortest covalent interatomic distances within their molecules. According to Heimann *et al.* [17,18], the zigzag-chain model has indisputable advantages over a chiral (helical) one [45]. Actually, the value of parameter *c* in the helical structure of polymeric molecules should depend on the van der Waals diameter of *sp*-carbon atoms to a greater extent than parameter *a* does. Moreover, the realization of linear C-C-bond conjugation (polyyne or polycumulene) within the fragment of a limited length is unlikely for well-known values of parameter *a*. In turn, the formation of van der Waals-type bonds between parallel-oriented rectilinear fragments of carbon chains is, probably, an additional factor stabilizing linear carbon in the condensed state.

Compounds containing short linear fragments with both polyyne and polycumulene structure are well-known in chemistry. The polyyne type of the C-C bonds is characteristic of short chains with an even number of carbon atoms [45-48]. Such compounds can be quite stable in the condensed state in the presence of large end-capping groups. To date relatively long linear chains (more than 10 carbon atoms) have been synthesized. The polymeric character of these compounds was proven by direct x-ray diffraction measurements [45]. Based on these data, however, we can conclude that the stability of short polyyne carbon chains in the condensed state is due to their separation from each other by distances exceeding their van der Waals diameter [45].

The polyyne-type structures can be stabilized by metal atoms as in copper(I) polyacetylides [2] and short-chain intercalates of alkali metals [36,41,44] (see this book, Chapter 3.2.3). The stability of such compounds in the latter case can be easily accounted for in terms of organometallic compounds.

The cumulene type of chemical C-C bonds is known to occur, for instance, in carbon suboxides being stable with odd numbers and unstable with even numbers of carbon atoms [49-52], and also in a number of organometallic compounds, *e.g.* $R_2C{=}C{=}CMC{=}C{=}CR_2$ [53,54]. Probably, the type of the C-C bond symmetry in the short carbon chains depends on the order of combination and the chemical character of stabilizing functional groups.

Based on ozonolysis results [8], the discoverers of α- and β-carbyne suggested for them polyyne and polycumulene structures, respectively. Unfortunately, these forms

a

b

c

Figure 3. Manifestation of the twinning effect on rotating the carbyne crystal; (*a*) transmission electron micrograph of the crystal; (*b*) spotted *hk0* diffraction pattern from this crystal; (*c*) its typical inclined diffraction.

were not isolated as pure phases. Therefore, we can rigorously accept only the presence of triple C≡C bonds in the samples described by Kasatochkin *et al.* [8], possibly not associated with typical diffraction pattern discussed in their study.

To date we could not find any structural evidence for the existence of zigzag-like carbon chains with polyyne type of chemical bonds in the condensed state. From the viewpoint of classical chemistry such zigzag chains look very strange indeed and, strictly speaking, unusually. On the other hand, the polycumulene zigzag carbon chain complies satisfactorily with commonly known general concepts of chemistry (the carbon valency and the parameters of chemical bonds). Moreover, there is some theoretical basis to recognize, that the zigzag polycumulene carbon chain is the most stable among simple polymeric carbon chains [55,56], i.e. the formation of regular kinks is energetically prefered.

In order to obtain a linear carbon allotrope through a direct phase transition, one has to use thermobaric conditions presently inaccessible in laboratories [57]. Fast non-equilibrium processes occurring upon vaporization of graphite by various high-energy techniques lead usually to the formation of amorphous carbon soot. Nevertheless, in spite of some fortuitous character of recombination reactions, some opportunities exist to control these processes and increase the yield of ordered carbon molecules and phases, for example fullerenes [56], nanotubes [57], and oriented carbyne films [35].

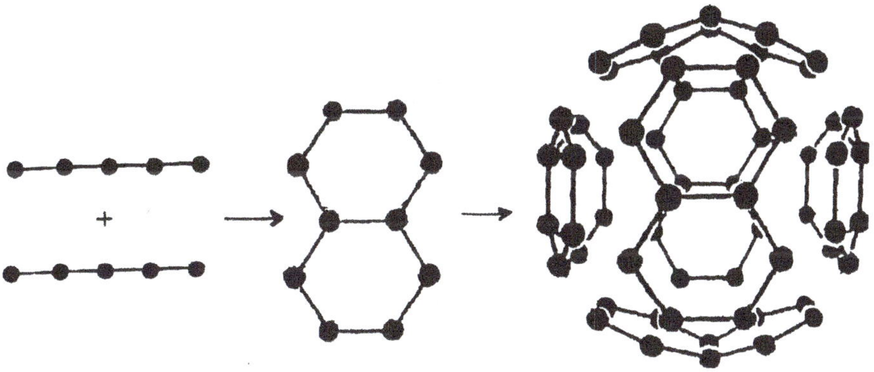

Figure 4. Possible mechanism of C_{60} assembling in the quasi-condensed state of a noble gas environment. The recombination of carbon clusters proceeds only through the formation of new C-C bonds without any breaks of existing ones [43].

Actually, we would expect it to be more likely that a new carbon particle formed by recombination of two carbon radicals would have a number of carbon atoms that is the sum of the original numbers of carbon atoms in the parent particles, and no breaks of existing covalent σ-bonds would take place during the formation of new carbon bonds, i.e. the aggregation would proceed only through addition reactions to the unsaturated bonds. Considering the very high strength of ordinary C-C bond and the high rate of cooling after vaporization of the carbon source (usually graphite), the above-mentioned conjecture seems in my view to be more likely than those made earlier for self-assembling of the fullerene-type molecules [60-62]. In fact, the inert gas environment used in syntheses of fullerenes is needed to create a quasi-condensed state very close to the high-temperature zone. In that case the short carbon chains can recombine by the Baeyer's mechanism [63] to form a system of aromatic rings. According to the mechanism proposed by Bulychev and Udod [43], the formation of the C_{60} molecule can be illustrated, as shown in Fig. 4. Thus, the unstable and very reactive short carbon chains C_5 (odd number of carbon atoms) can be a real source for the synthesis of spherical molecules. Partially, this mechanism has been already confirmed by Taylor *et al.* [64].

The situation can be changed dramatically during recombination of carbon chains in high vacuum. In this case we can consider the stepwise recombination of short carbon chains through a lengthening mechanism [43]. The zigzag-like polycumulene carbon chain of our interest can be formed as a result of a simple polyrecombination reaction of short-chain clusters with an odd number of carbon atoms (*Scheme 1*).

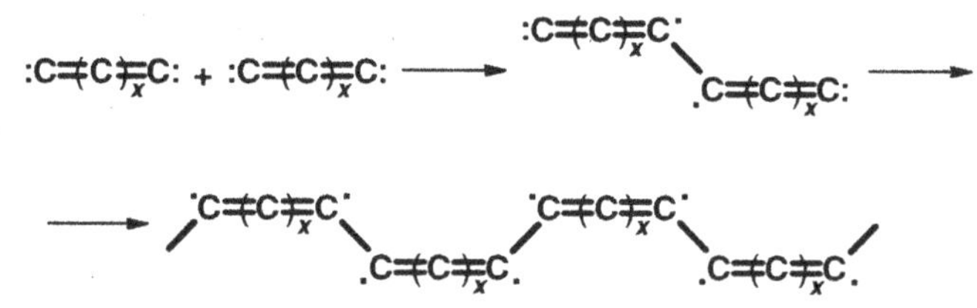

Scheme 1

As obviuous from *Scheme 1*, the zigzag polycumulene chain is a polyradical, which might be reasonably stabilized by delocalization of non-paired electrons due to π-π-conjugation of multiple C-C-bonds of the rectilinear fragments [44]. In fact, some intermediate state different from the pure cumulene-type of the C-C bonds could result from such stabilization. There is no clear distinction between the two types of conjugation (polyyne and cumulene) even in compounds containing diacetylenic fragments, -C≡C-C≡C-. Indeed, the central single bond (132-140 pm) is in some cases even shorter than the double C=C-bond in ethylene (134 pm) and is much shorter than the typical single bond in ethane (154 pm). One can assume that this bond shortens as the conjugation length increases. The emergence of some polyyne character in the zigzag polycumulene chain might lead to a *flat structure* of the whole polymer molecule at any number of carbon atoms in the rectilinear fragments, whereas hydrogen atoms in allene, the most simple cumulene $H_2C=C=CH_2$, belong to planes perpendicular to one another.

The additional stabilization affording a stable substance in the condensed state can be probably attained through chemical bonding of non-paired electrons by the impurity (or admixture) atoms. This can be realized in two ways: (i) through formation of covalent bonds with non-metal atoms, or (ii) through coordination bonds with metal atoms. In the former case amorphous substances named *carbynoids* are mainly formed [65-74]. The presence of large (usually oxygen-containing) groups [69], able to stabilize polymeric carbon chains, does not allow to build up a three-dimensionally ordered structure. Nevertheless, the principally linear structure with presumed zigzag-like polycumulene carbon chains as a structural basis accounts satisfactorily for the spectroscopic data [75-77].

The influence of hydrogen atoms on the stability of carbon chains should be particularly emphasized [78-80]. Indeed, the presence of hydrogen as an admixture to the inert gas during the synthesis of fullerenes increased the yield of linear clusters. Moreover, long-chain carbon molecules of approximately the same length (*ca.* 70 carbon atoms) soluble in some organic solvents were isolated [78]. The oriented carbyne films revealing a typical six-spot diffraction pattern, corresponding to the hexagonal lattice with parameter a = 508 pm, were formed after recrystallization [78]. Evidently, the small size of hydrogen atoms did not prevent carbon chains to align to form an ordered structure.

According to the proposed structure of the zigzag polycumulene carbon chain, the carbon atoms at kink sites should have a formal valency close to 3. Therefore, the polymeric carbon molecules could be more stable, if the carbon atoms at the kinks would be replaced by nitrogen atoms. By this means one could pass from polyradicals to quite stable polymeric molecules with no dangling bonds. Probably, a similar expedient was performed during ion-assisted deposition of carbon films in a nitrogen gas environment [38]. The formation of well-crystallized carbyne containing no metal atoms in its lattice can be explained by the insertion of nitrogen atoms into the carbon skeleton. Presumably, the reaction occurring in the course of depositing carbon-nitrogen films on hot substrates [38] can be envisioned as polymerization of cyanopolyynes NC-$(C{\equiv}C)_x$-CN (*Scheme 2*). The possibility of the formation of similar dicyanopolyynes during vaporization of carbon in a nitrogen gas environment was discussed earlier [81].

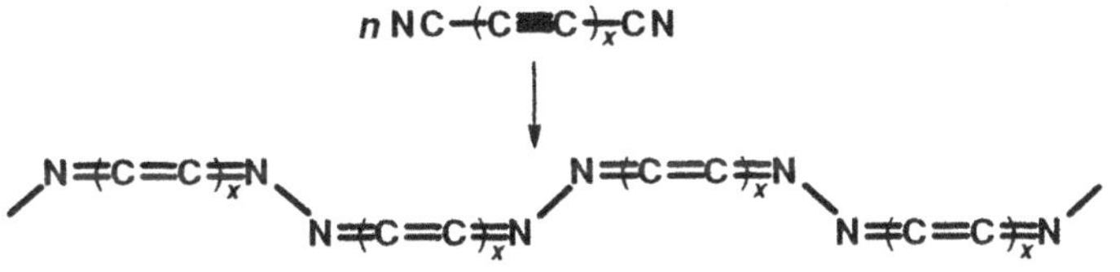

Scheme 2

3. Spatial Structure of the Linear Carbon Matrix

The presence of many unsaturated bonds and non-paired electrons in the zigzag carbon molecules makes them a unique object, able to enter into chemical interactions of different nature with dopant atoms. In my opinion, the possible compounds of carbyne with metals are best to be considered intercalation compounds of a linear carbon matrix. The term 'intercalation compound' combines a large number of substances, generally different in their chemical nature. The particularity of any intercalation system lies in the possibility of subdividing (sometimes quite conditionally) such a chemical system into two components, *viz.*, the original matrix (host) and the intercalant (guest). Numerous experiments show that parameters of chemical interactions between host atoms change insignificantly upon intercalation. This fact allows to summarize the common property of intercalation compounds consisting in general retention of the skeleton structure of original matrix and also of the type of chemical bonding in it. Therefore, we can expect that the basic contribution to the stability of an intercalation system should be made by the energy of chemical interaction between the host atoms themselves. At the same time the interactions of intercalated atoms with the host atoms and between themselves have usually additional minor character. Hence we have to consider not just coupled atom-atom interactions, but a collective influence of the matrix on the substance intercalated. The chemical system can be stable (i.e. intercalation compound exists) even in the absence of significant chemical interaction of the host-guest type. A similar situation can be observed, for example, in *clathrate* compounds of water with various gases. The melting points of such compounds are very close to those of ordinary ice. From the viewpoint of quantum chemistry such a system can be easily conceived as a particle inside the potential well. Thus, the consideration of the collective influence of matrix atoms on the substance intercalated (for example, metal atoms) makes an understanding of the chemical stability of such compounds essentially easier. The chemical nature of the intercalation system, its structural and other properties depend on the chemical nature of the original matrix and its spatial structure that determines the spatial arrangement of the intercalant.

So we can expect the formation of an isolated arrangement of the intercalant in three-dimensional frame matrices, as in the case of doped fullerenes [82]. In turn, the layered matrices (for example graphite) promote the formation of their own intercalated layer, in which the interaction between intercalated atoms becomes possible. Finally, the formation of one-dimensional channels of intercalated metal atoms is characteristic of polymer matrices of the polyacetylene type. It is interesting that the coordination number of polyacetylene chains around the metal channel depends on the size of the metal atoms. Indeed, this value is 3 in lithium-doped polyacetylene [83], whereas it is 4 in similar compounds with heavier alkali metals [84]. The van der Waals size of *sp*-carbon atoms, being still lesser, might determine the coordination number to be 6 for the majority of metals [44].

Although the binding energy between the matrix and intercalant is essentially less than the bond energy of host atoms, the former is a driving force for the intercalation reaction. The nature of this interaction may vary over a wide range from quite strong interactions with preferred ionic character (layered aluminosilicates) to very weak van der Waals interactions (clathrate compounds of water with gases). Evidently, the type of the host-guest interaction will depend on the type of the original matrix, i.e. it will be determined by its chemical nature. Therefore, all attempts to create a unified theory of chemical bonding that would adequately describe all known intercalation systems are, probably, doomed to fail. Thus, in order to consider a specified intercalation system we should first of all give much attention to chemical particularities of the original matrix structure.

Carbon matrices containing a system of delocalized π-electrons and having weak van der Waals bonds in their crystal structures are capable of forming intercalation compounds with metals. The driving force for the intercalation reaction in that case is the formation of the matrix-intercalant interaction, whose chemical nature is probably similar to that in organic charge-transfer complexes. The observed ability of linear carbon (carbyne) also possessing a delocalized π–electron system to incorporate impurity atoms into its lattice may be due to the formation of the intercalation compound. To propose a tentative structure of carbyne intercalation compounds, it is necessary to consider first the structure of the original linear carbon matrix.

Unfortunately, rigorous diffraction data concerning impurity-free carbyne structures are lacking so far. In my view there are several reasons for that. First, we deal with a polymer. Hence it is reasonable to expect a diffraction pattern corresponding to a one-dimensional structure, i.e. few reflections, usually of *hk0* orientation with low indices. Secondly, the existence of pure stable linear carbon in the condensed state, particularly under x-ray or electron irradiation, is still questionable. Thirdly, the synthesis of carbyne phases containing no other crystalline or amorphous phases is still associated with severe difficulties.

Nevertheless, based on reported data [35,36,38,41,44], one can conclude with a fair degree of confidence that the two-dimensional lattice of carbon chains in the absence of the metal dopant atoms also possesses a hexagonal symmetry with parameter a = 508-512 pm (Fig. 2). In other words, the intercalation of metals into the carbyne matrix presumably stabilizes the crystal structure as a whole, but does not dramatically change the packing mode of the carbon chains. Thus, the capability of the carbon matrix to incorporate metal atoms without profound structural transformations should be characteristic of the original structure. Indeed, according to data reported by Kudryavtsev *et al.* [35], a sublattice of regular vacancies exists in the two-dimensional carbyne layer.

The theoretical estimation of the van der Waals diameter for *sp*-carbon atoms correlates satisfactorily with experimental data [35] and is close to 300 pm. This fact points to the existence of a van der Waals interaction between *sp*-carbon atoms, that is stronger than that between sp^2-carbon atoms (for instance, in graphite or polyacetylene). Perhaps, it is this stronger van der Waals interaction that might be realized in the event of the parallel orientation of rectilinear fragments of the zigzag-like polymer molecules, and hence constitutes an additional factor stabilizing carbon chains in the condensed state. Possible types of two-dimensional packing of carbon chains containing different vacancy sublattices are shown in Fig. 5. As can be seen from Fig. 5, the only packing which possesses a hexagonal parameter a corresponding to the experimentally observed value [35,36,38,41,44] is that including the $3^{1/2}\times3^{1/2}$ vacancy sublattice.

The realization of the densest packing shown in Fig. 5(a) seems to be possible only for short straight polyyne chains with metal atoms at both ends [41]. At the same time a packing can be conceived, in which van der Waals interactions between flat zigzag-like chains can exist even if the length of rectilinear fragments is variable (Fig. 6) [43,44]. As evident from Fig. 6, this structure is just the one including the sought vacancy sublattice in the two-dimensional layer (section A-A). Thus, the capability of linear carbon to incorporate impurity atoms into existing vacancies can be easily understood.

The insertion of the impurity or admixture atoms into the carbyne lattice, captured during crystallization, is not just a mechanical act. It is a chemical reaction (intercalation) instead leading to a leveling of all rectilinear fragments at least in a certain two-dimensional layer (Fig. 7). This may give rise to a structural periodicity along its c-axis. The main driving force for such intramolecular transformation of carbon chains is the partial interaction of their non-paired electrons at kink sites with doping metal atoms through the formation of chemical bonds. A substance with a structure shown in Fig. 7 would be expected to have very interesting electronic properties. Indeed, the defects (breaks) in conjugation at the kink sites of the conductive chain can be repaired through electron transfer from metal atoms to the delocalized π-system of carbon molecules. In this case we would deal with a true possibility of one-dimensional conductivity through the whole length of the polymeric molecules. Thus, we are approaching a new promising direction in the chemistry of intercalation compounds.

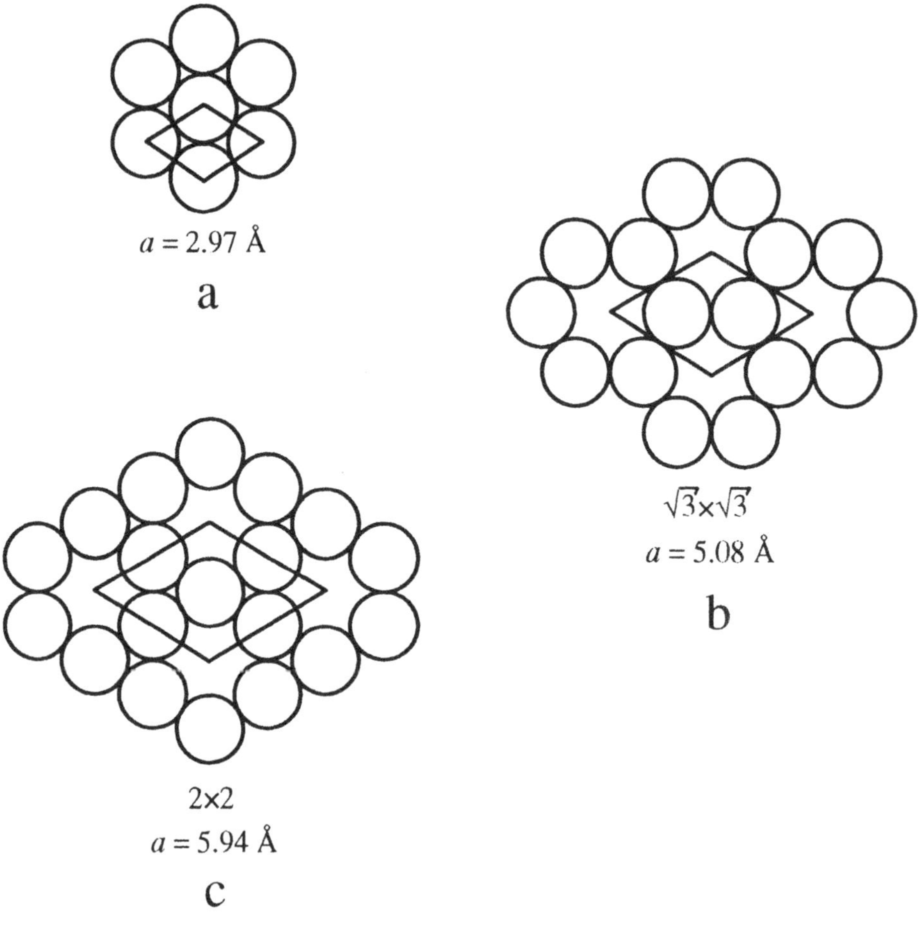

Figure 5. Possible variations of two-dimensional packing of carbon chains without vacancies (a) and those including $3^{1/2}\times3^{1/2}$ (b) and 2×2 (c) vacancy sublattices. According to Kudryavtsev *et al.* [35], the van der Waals diameter is considered to be ~300 pm.

Indeed, according to Udod *et al.* [36,41,43,85], the destruction of short-chain organometallic compounds under severe conditions (hydrolysis with strong mineral acids) takes place through another mechanism compared to treatment with water (see Chapter 3.2.3 of this book). This reaction is supposed to bring about the oriented lengthening of the chains with simultaneous accommodation of metal atoms into the existing vacancies of the linear carbon matrix. The formation of intercalation compounds should affect the C-C bond lengths and kink angles, thus leading to a slight increase in parameter *a* of the hexagonal carbyne lattice. The experimental values of parameter *a* in the carbyne intercalation compounds vary within the 515-525 pm range, whereas those in original oriented carbyne layers range from 508 to 512 pm. However, this increase does not change the structural basis of the original matrix wherein, according to the model shown in Fig. 7, the parameter *a* is determined by the van der Waals size of *sp*-atoms and depends only slightly on the size of atoms intercalated. On the other hand, the smallest possible length of rectilinear fragments in the zigzag-like carbon chain (parameter *c*) should depend on the size of metal atoms intercalated into the carbyne matrix. The experimental value of the smallest length in sodium-intercalated carbyne is 690 pm (five carbon atoms) [36], whereas it is 800 pm (six carbon atoms) in carbyne intercalated with potassium [85].

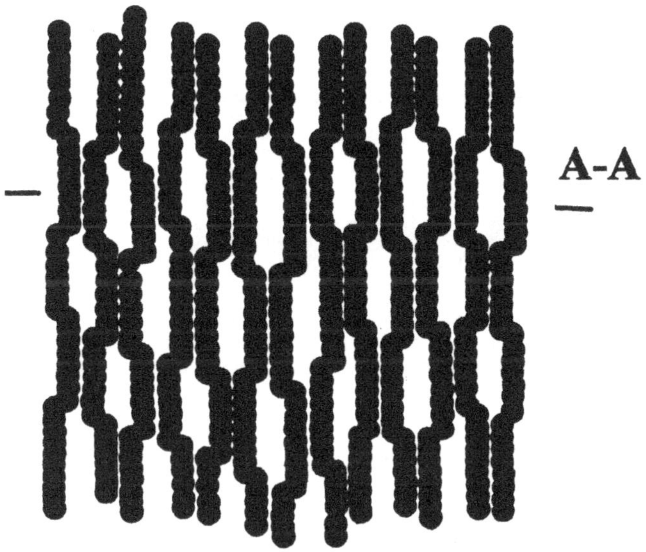

Figure 6. Proposed spatial model of the original linear carbon matrix. Polymeric carbon chains with variable length of rectilinear fragments can form such a structure.

The model of the carbyne structure shown in Fig. 7 quite adequately reflects chemical nature of metal inclusions, the constancy of parameter *a*, and the variability of parameter *c* for carbyne modifications synthesized by different methods. Such a packing can be assembled from the carbon chains of different lengths. Indeed, no length differences between polymer molecules should lead to structural disorder in the single crystallite as a whole. Thus, the growth of such a crystallite, according to our model, will take place preferably on an $\{ab\}$ plane as a result of both van der Waals interactions between carbon chains and the formation of coordination bonds with metal atoms. At the same time, the dimension along the c-axis can be evidently limited by the average length of the carbon chains.

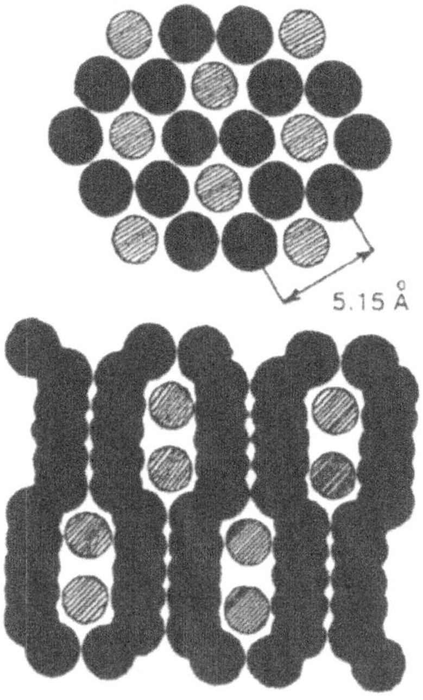

Figure 7. Proposed spatial model of the carbyne intercalation compounds (real carbyne single crystal). The lengths of all rectilinear fragments became the same as a result of chemical interaction of carbon atoms at kink sites with intercalated metal atoms.

4. Structure of Carbyne Macrocrystal

The analysis of experimental facts presently available shows that the ideal single crystal of carbyne suitable for direct x-ray diffraction measurements cannot be synthesized easily at this time. To come to such a pessimistic conclusion it suffices to consider the tendency towards variability of dimensions in single crystals of naturally occurring carbon allotropes: from diamond to graphite and further to chaoite (natural carbyne). The greatest possible size of single crystals in this series decreases markedly with decreasing contribution of covalent chemical bonds and increasing contribution of van der Waals interactions to the total energy of the crystal lattice. While natural graphite single crystals of several millimeters in size are available, we cannot expect the same from a polymer. Moreover, the processes of carbyne crystallization described in literature [35,43,77,85,86] employ drastic non-equilibrium conditions wherein the formation of large amounts of crystal nuclei is preferred. Thus, one might expect a polycrystalline structure in carbyne macrocrystals. Actually, it was noted already in early studies [12] that the carbyne sample examined with an electron microscope was a kind of mosaic of crystallites grown together and oriented so that the overall *hk0* diffraction was still the spotted one. The compound character of the *hk0* diffraction can be discovered when rotating the carbyne crystal. In this case one can observe the changing of the *hk0* family of spot reflections into inclined diffraction from at least two phases, i.e. the twinning effect appears (Fig. 3) [44]. Thus, the real carbyne crystal with spotted *hk0* diffraction pattern is probably not a single crystal.

A similar situation exists with another allotropic form of carbon, *viz.*, graphite. The macrocrystals of graphite with disorientation angle of crystallites along the *c*-axis of less than 1° are called quasi-single crystals [87]. The orienting factor leading to the quasi-single-crystal structure is the flat lamellar shape of graphite crystallites. In the case that oriented structures have been built from one-dimensional molecules the common hexagonal habit of crystallites can be such a factor. Indeed, carbyne crystallites of approximately equal size can be closely packed in a mosaic crystal (Fig. 8). Comparable dimensions of crystallites can in turn be determined by similar growth conditions of the crystal nuclei. An important factor affecting the crystallite growth is probably the local concentration of impurity (admixture) atoms in the crystallization zone. This concentration can have a pronounced impact on the length of rectilinear fragments (parameter *c*) and hence on the thickness of carbyne crystallites along the *c*-axis. Moreover, the concentration of metal atoms along with the concentration of carbon chains can influence the growth kinetics and thus the size of the carbyne crystallites.

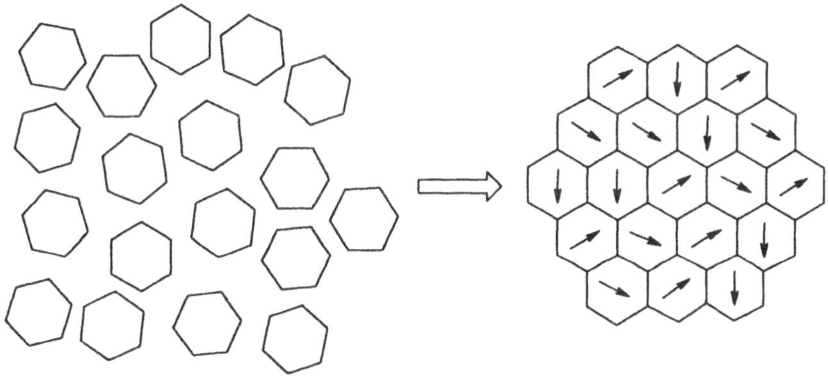

Figure 8. Formation of mosaic carbyne macrocrystal from crystallites possessing a common hexagonal habit and close sizes.

In the case of significant differences in the crystallite sizes the oriented combination does not occur and polycrystals with the typical diffraction pattern shown in Fig. 9 can be observed. As can be seen from Fig. 9(*a*), the carbyne polycrystal obtained by thermobaric treatment of a polyacetylene-sodium system consists of several types of crystal lattices. This diffraction pattern can be arbitrarily subdivided into two parts corresponding to a spotted inclined *hkl* section from the basis crystal (Fig. 9(*b*)) and the diffraction from macrocrystals included in the basis crystal (Fig. 9(*c*)). As evident from Fig. 9(*b*), the inclined diffraction is the section of the inverse lattice whose plane is rotated around the *-1-10* and *110* spots (Table 1). This allows to identify all other

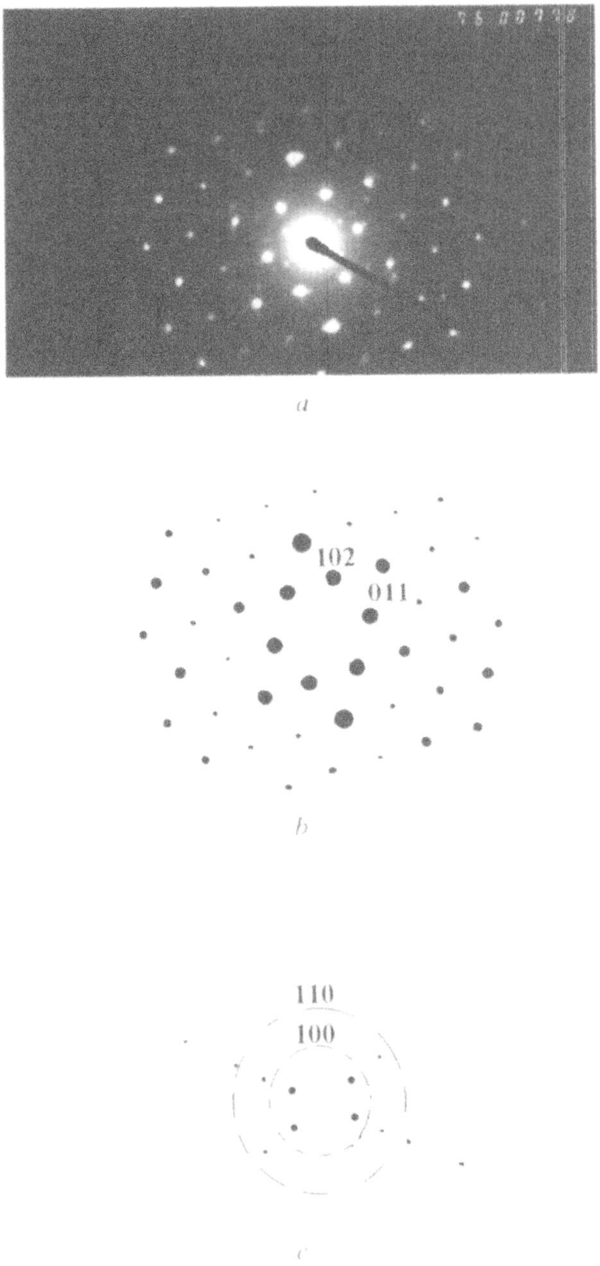

Figure 9. Typical electron diffraction pattern from a carbyne polycrystal obtained by thermobaric treatment of a polyacetylene/sodium system (*a*) along with schematics of inclined diffraction from the basis crystal (*b*) and diffraction from microcrystals included in the basis crystal (*c*).

reflections and determine the hexagonal parameters a = 516 pm and c = 2653 pm. The value of parameter c corresponds to 22 carbon atoms. According to Heimann *et al.* [17], the length of rectilinear fragments is in the range from 10 to 15 carbon atoms even with rather fortuitous character of admixture inclusions. Moreover, there are strong grounds to believe that longer linear carbon chains are less stable and may form cyclo[N]carbons, for example C_{18} [88]. Thus, the structure of linear carbon is indeed two-layered (22 : 2 = 11) that is in satisfactory agreement with the model shown in Fig. 7.

TABLE 1. Electron diffraction data of the carbyne macrocrystal shown in Fig. 9(*b*)

d_{exp}, pm	d_{calc}*, pm	Δd, pm	*hk.l*
440.3	440.5	0.2	*10.1*
422.6	423.1	-0.5	*10.2*
257.5	258.0	-0.5	*11.0*
247.1	247.5	-0.4	*11.3*
221.1	220.3	0.8	*20.2*
211.7	211.6	0.1	*20.4*
169.3	168.5	0.8	*21.1*
164.0	163.6	0.4	*21.4*
160.6	160.8	-0.2	*21.5*
147.2	146.8	0.4	*30.3*
140.4	141.0	-0.6	*30.6*
129.2	129.0	0.2	*22.0*
123.5	123.8	-0.3	*22.6*
123.0	123.4	-0.4	*31.2*
121.6	120.6	1.0	*31.5*
118.1	117.7	0.4	*31.7*
110.8	110.1	0.7	*40.4*
103.0	102.4	0.6	*32.1*
96.5	96.9	-0.4	*41.3*

* The calculation was performed for a hexagonal lattice with parameters a = 516 and c = 2653 pm.

The microcrystalline inclusions exhibit a diffraction pattern with two diffuse rings (*100*) and (*110*) (Fig. 9(*c*)) that allows to identify them as carbyne microcrystals with a hexagonal parameter a of 515 pm. In view of the obvious structural anisotropy the spot reflections observed are attributable to the *00l*-family. This suggests a different orientation of these carbyne particles in relation to the basis crystal.

The value of periodicity spacing along the c-axis is 695 pm corresponding to the minimum length of rectilinear fragment in carbyne doped with sodium. From the viewpoint of the two-layered model shown in Fig. 7 the real value of parameter c in this case should be twice as large since the presence of a 2_1-screw axis leads to the extinction of all *00l* reflections for which l is odd. Thus, a large amount of microcrystals is formed under non-equilibrium conditions of the carbyne crystallization, e.g., during a burst-like reaction (accompanied by local burning) of short-chain organometallic compounds with mineral acids [36,41]. These microcrystals agglomerate into larger macrocrystals. The possibility exists of both oriented (if the sizes of crystallites are similar) and non-oriented (if their sizes are different) intergrowth. Thus, the picture shown in Fig. 9(*b*) is the superposition of *hkl* diffraction from carbyne microcrystals possessing the same orientation and the same parameter c.

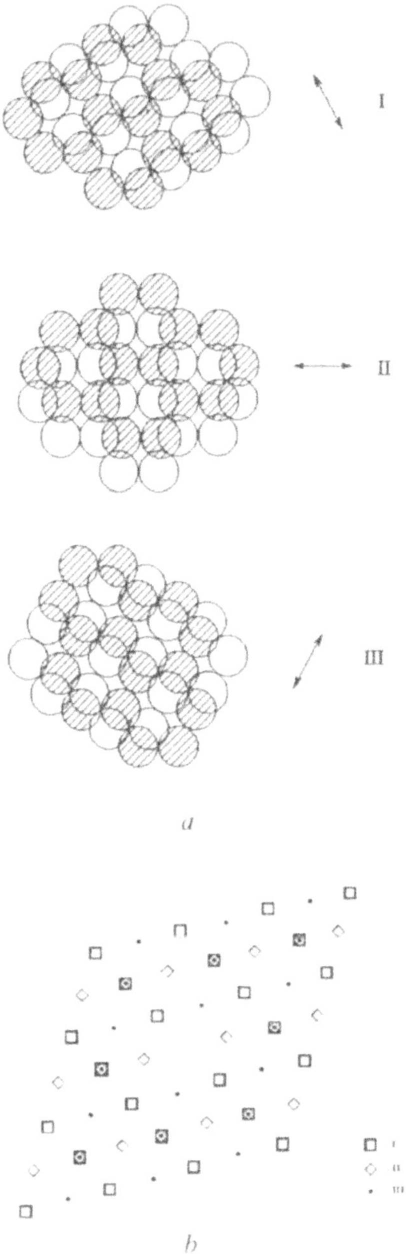

Figure 10. Three possible types of carbyne crystallites in the mosaic crystal (*a*) and their overall *hk0* diffraction pattern (*b*).

Summarizing the aforesaid, the growth of the carbyne macrocrystal can be conveniently subdivided into two stages. The first stage includes aggregation of free carbon chains and metal atoms into crystal nuclei and their preferred two-dimensional growth within the {*ab*} plane. Microcrystals with well-ordered structure and a thickness equal to the average length of carbon chains are formed in the first stage. The combination of these microcrystals into larger macrocrystals takes place in the second stage. Microcrystals with similar dimensions may be united through oriented intergrowth to form a mosaic crystal (Fig. 8). Thus, according to the above-described model, the carbyne macrocrystal presents itself as a quasi-single crystal with a spotted *hk0* diffraction pattern, and consisting of microcrystals resembling real single crystals.

Actually, there are three types of carbyne microcrystals distinguishable by the direction of the shift vector of the 2_1-screw axes (Fig. 10(*a*)). In fact, Fig. 10(*a*) presents a structure identical to that shown in Fig. 7 but having the macrocrystal oriented in three different ways. The *hk0* section of diffraction from each individual microcrystal possesses a tetragonal 2D-point group. However, the overall *hk0* diffraction from the entire mosaic crystal will be hexagonal because of superposition of the *hk0* diffraction from individual microcrystals mutually oriented in three different directions (Fig. 10(*b*)). This model thus allows to explain for the first time why the location of *hk0* spots corresponds to the hexagonal symmetry, whereas the distribution of *hk0* intensities for the majority of experimentally observed carbyne crystals contains only the twofold rotational symmetry axis (2) as a consequence of the presence of a tetragonal superlattice (Fig. 11).

Figure 11. Typical electron diffraction pattern of a carbyne crystal with non-hexagonal distribution of *hk0* intensities.

5. Conclusions

The present review of currently available structural data on linear carbon allotropes was performed first of all in an attempt to find common criteria for detecting carbyne in carbon materials. Researchers working in the field of carbon chemistry, particularly those using non-equilibrium, high-velocity recombination reactions (including explosive shock compression) have to be prepared to encounter this non-trivial subject. Otherwise they should find out another, probably even less understandable explanation for their results. For example, Krätschmer *et al.* [89] have accounted for the occurence of several absorption bands, forbidden for spherical C_{60} molecule in IR spectrum, by the presence of CO and CO_2 molecules. More recently, Spitsina *et al.* [78] have shown that linear carbon chains comprising dozens of carbon atoms can be observed in carbon soot produced in an electrical arc during fullerene synthesis in the presence of hydrogen. Therefore, one can speculate that the additional lines observed in the IR spectrum of fullerene [89] might belong to inclusions of linear carbon.

Another example concerns the recent discovery of an already familiar, but new in the authors' view, carbon chain structure called "*carbolite*" [79,80] (cp. Chapter 3.4.1.2). Meagre experimental details given by the authors allow nonetheless to expect the formation of carbyne during their experiments (especially in the part concerning the role of a hydrogen admixture). The application of the x-ray diffraction method, as shown above, is absolutely useless to determine lattice parameters of a new chain-like carbon structure. Actually, making a reflection survey of carbyne particles with lamellar, layered morphology, we should expect the presence of the primarily *00l* and *hkl* lines in diffraction patterns. The appearance of the *hk0*-family of reflections as a result of oriented arrangement of particles in this case might be unlikely. According to our data, it is just the *hk0* reflections that characterize the carbyne family of carbon allotropes. The parameter *c* can vary not only from one macrocrystal to another, but also within the boundaries of a certain polycrystal. The x-ray diffraction pattern observed by Tanuma *et al.* [79,80] might be due to the diffraction from a mixture of chain-like structures possessing a common two-dimensional sublattice, but different values of parameter *c*. In such a situation, the use of a mathematical method for indexing reflections on the assumption that all of them belong to a single phase is, mildly speaking, incorrect.

The design of the structural model of crystalline carbyne [36,41,43,45,91] was necessary first of all in order to proceed from mysterious speculations and arrive at a real understanding, at least in our view, of this quite widespread but still elusive carbon material. In my opinion such an approach with no possibilities for direct diffraction measurements is the only way out of an impasse where the science of carbyne had been over 35 years of its history. Indeed, the variability of the *hk0* intensity distribution from one crystal to another even within the same experiment attests to a very complex structure of this object. Recently, it was clearly demonstrated once again [90] that changes in *hk0* intensities observed as a result of the carbyne destruction under electron beam irradiation reduce to almost zero the chance to obtain rigorous and precise diffraction data.

Nevertheless, one can distinguish some general features of the diffraction from carbyne crystals. First, carbyne crystals with a textured (spotted) *hk0* diffraction pattern are not single crystals. For example, in some cases we could observe the appearance of a 'twinning' effect when rotating the crystal (Fig. 3). The mosaic crystal discussed above (Fig. 8), composed from oriented carbyne microcrystals, should possess the same property.

Secondly, departures of the *hk0* intensity distribution from the hexagonal symmetry have been noted in many carbyne crystals (Fig. 11). This can possibly be explained by the tetragonal symmetry of an {*ab*} projection of the two-layered structure in carbyne microcrystals (Fig. 10).

Lastly, the presence of the third-order reflections (*300*, *030*, *etc.*), strong enough for this region of interplanar distances, is characteristic of the majority of carbyne crystals stable against irradiation with an electron beam (Fig. 1). This fact can be accounted for, as simple calculation shows, by the existence of a $3^{1/2}\times3^{1/2}$ vacancy superlattice in the two-dimensional layer (Fig. 7).

Thus, while the idea of the carbyne structure presented above is based on real experimental results, it nevertheless does not pretend to be the only and perfect structural model. Undoubtedly, these ideas can be developed further, and should be corrected as we go along. However, with no other more or less chemically justified spatial model available this model was needed for rational planning of further experimental studies in the field of carbyne. Evidently, the main challenge in the science of linear carbon is to accomplish the synthesis and isolation of crystalline carbyne in high yield that would suffice for comprehensive investigation of all its properties. The mosaic structural model of the carbyne intercalation compounds shows clearly the way of resolving the issue through a deliberate introduction of metal atoms into the carbyne matrix during its crystallization.

6. References

1. Sladkov, A.M. and Kudryavtsev, Yu.P.: Diamond, graphite, carbyne - allotropic form of carbon, *Priroda* (5) (1969), 37-44 (in Russian).
2. Sladkov, A.M.: Carbyne - a new allotropic form of carbon, *Sov.Sci.Rev.* **B3** (1981) 75-110.
3. Sladkov, A.M., Kasatochkin, V.I., Korshak, V.V., and Kudryavtsev, Yu.P.: The phenomenon of existence of a new crystalline form of carbon - carbyne, Inventor's certificate No 107, December 7, 1971. Priority date November 4, 1960, Publ. *Bull. Izobr.* (No. 6) (1972), 3 (in Russian).
4. Visscher, G.T. and Bianconi, P.A.: Synthesis and characterization of polycarbynes, a new class of carbon-based network polymers, *J. Am. Chem. Soc.* **116** (1994), 1805-1811.
5. Visscher, G.T., Nesting, D.C., Badding, J.V., and Bianconi, P.A.: Poly(phenylcarbyne) - a polymer precursor to diamond-like carbon, *Science* **260** (1993), 1496-1499.
6. Korshak, V.V., Kasatochkin, V.I., Sladkov, A.M., Kudryavtsev, Yu.P., and Usenbaev, K.: Concerning the synthesis and properties of polyacetylene, *Dokl. Akad. Nauk SSSR* **136** (1961), 1342-1344 (in Russian).
7. Sladkov, A.M., Korshak, V.V., and Nepochatykh, V.P.: Polymers with cumulated double bonds in the chain, *Izv. Akad. Nauk SSSR, Ser. Khim.* (1) (1967), 196-197 (in Russian).
8. Kasatochkin, V.I., Sladkov, A.M., Kudryavtsev, Yu.P., Popov, N.M., and Korshak, V.V.: Crystalline forms of the linear modification of carbon, *Dokl. Akad. Nauk SSSR* **177** (1967), 358-360 (in Russian).
9. Kasatochkin, V.I., Babchinitser, T.M., Kudryavtsev, Yu.P., Sladkov, A.M., and Korshak, V.V.: The chain length of the linear carbon polymer - carbyne, *Dokl. Akad. Nauk SSSR* **184** (1969), 353-354 (in Russian).
10. Kasatochkin, V.I., Shterenberg, L.E., Kazakov, M.E., Slesarev, V.N., and Belousova, L.V.: The thermal transformation of carbyne under pressure, *Dokl. Akad. Nauk SSSR* **209** (1973), 388-391 (in Russian).
11. El Goresy, A. and Donnay, G.: A new allotropic form of carbon from the Ries crater, *Science* **161** (1968), 363-364.
12. Kasatochkin, V.I., Korshak, V.V., Kudryavtsev, Yu.P., Sladkov, A.M., and Shterenberg, L.E.: On crystalline structure of carbyne, *Carbon* **11** (1973), 70-72.
13. Whittaker, A.G. and Kintner, P.L.: Carbon: observations on the new allotropic form, *Science* **165** (1969), 589-591.
14. Whittaker, A.G. and Wolten, G.M.: Carbon: a suggested new hexagonal crystal form, *Science* **178** (1972), 54-56.
15. Whittaker, A.G., Neudorffer, M.E., and Watts, E.J.: Carbon: a rhombohedral carbyne form, *Carbon* **21** (1983), 597-599.
16. Whittaker, A.G., Watts, E.J., Lewis, R.S., and Anders, E.: Carbynes: carriers of primordial noble gases in meteorites, *Science* **209** (1980), 1512-1514.
17. Heimann, R.B., Kleiman, J., and Salansky, N.M.: A unified structural approach to linear carbon polytypes, *Nature* **306** (1983), 164-167.
18. Heimann, R.B., Kleiman, J., and Salansky, N.M.: Structural aspects and conformation of linear carbon polytypes (carbynes), *Carbon* **22** (1984), 147-156.

19. Smith, P.P.K. and Buseck, P.R.: Carbyne forms of carbon: do they exist?, *Science* **216** (1982), 984-986.
20. Man, L.I., Malinovski, Yu.A., and Semiletov, S.A.: Crystal phases of carbon, *Sov. Phys. Crystallogr.* **35** (1990), 608-616.
21. Kasatochkin, V.I., Kazakov, M.E., Savranskii, V.V., Nabatnikov, A.P., and Radimov, N.P.: Synthesis of a new allotropic form of carbon from graphite, *Dokl. Akad. Nauk SSSR* **201** (1971), 1104-1105 (in Russian).
22. Nikulin, Yu.N., Mel'nichenko, V.M., Kudryavtsev, Yu.P., and Sladkov, A.M.: Plasmochemical synthesis of carbyne, *Izv. Akad. Nauk SSSR, Ser. Khim.*, (9) (1981), 1943-1948 (in Russian).
23. Nikulin, Yu.N., Mel'nichenko, V.M., Kudryavtsev, Yu.P., and Sladkov, A.M.: Synthesis of carbyne from organic compounds in a low-temperature plasma, *Dokl. Akad. Nauk SSSR* **256** (1981), 1162-1165 (in Russian).
24. Sladkov, A.M., Kasatochkin,V.I., Kudryavtsev, Yu.P., and Korshak, V.V.: Synthesis and properties of chain-like polymers of carbon, *Izv. Akad. Nauk SSSR, Ser. Khim.*, (12) (1968), 2697-2704 (in Russian).
25. Nelson, L.S., Whittaker, A.G., and Tooper B.: The formation of new polymorphs of carbon and fluid flow patterns by irradiating solid carbons with a CO_2 laser, *High Temp. Sci.* **4** (1972), 445-477.
26. Setaka, N. and Sekikawa, Y.: Chaoite, a new allotropic form on carbon, produced by shock compression, *J. Amer. Ceram. Soc.* **63** (1980), 238-239.
27. Fedoseev, D.V., Deryagin, B.V., Varshavskaya, I.G., Lavrent'ev, A.V., and Matveev, V.V.: Homogeneous formation of metastable carbon phases at high degrees of supersaturation, *Sov. Phys. JEPT* **53** (1981), 210-213.
28. Mel'nichenko, V.M., Sladkov, A.M., and Nikulin, Yu.N.: Structure of polymeric carbon, *Russ.Chem. Rev.* **51** (1982), 421-438.
29. Fitzgerald, A.G., Simpson, M., Dederski, G.A., Moir, P.A., Matthews, A., and Tither, D.: Triode technology in the sputer deposition of carbon films, *Carbon* **26** (1988), 229-234.
30. Pompe, W., Scheibe, H.-J., Richter, A., Baure, H.-D., Bewilogua, K., and Weissmantel, C.: On the influence of residual stresses and density fluctuations on the crystallization of amorphous carbon films, *Thin Solid Films* **144** (1986), 77-92.
31. Fitzgerald, A.G., Simpson, M., Dederski, G.A., Storey, B.E. , Moir, P.A., and Tither, D.: Carbon films produced by fast atom bombardment, *Carbon* **26** (1988), 547-552.
32. Degen, M.G., Drozdova, I.A., and Frantsev, A.I.: Electron microscopy and electron diffraction study of the crystallization of carbon film on heated substrates, *Zh. Prikl. Khim.* **62** (1989) 1698-1703 (in Russian).
33. Tither, D., Matthews, A., Fitzgerald, A.G., Storey, B.E., Henderson, A.E., Moir, P.A., Dines, T.J., Bower, D.I., Lewis, E.L.V., Doughty, G., and Foster, W.: Characterization of carbon films containing boron and nitrogen, *Carbon* **27** (1989), 899-907.
34. Kimstach, G.M., Urtaev, A.A., and Molodtsova, T.D.: Concerning the existence of carbyne in the structure of austenite cast iron, *Metallovedenie i Termicheskaya Obrabotka Metallov* (2) (1991), 17-18 (in Russian).
35. Kudryavtsev, Yu.P., Evsyukov, S.E., Babaev, V.G., Guseva, M.B., Khvostov, V.V., and Krechko, L.M.: Oriented carbyne layers, *Carbon* **30** (1992), 213-221.
36. Udod, I.A., Bulychev, B.M., Sirotinkin, S.P., Babaev, V.G., and Kudryavtsev, Yu.P.: An interaction of amorphous carbyne with sodium: the formation of carbyne intercalation compounds, *Synth. Met.* **60** (1993), 57-62.
37. Ando, Yo.: A carbyne observed in a marketing microgrid, *Carbon* **33** (1995), 171-175.
38. Guseva, M.B., Babaev, V.G., Babina, V.M., and Khvostov, V.V.: Formation of oriented nitrogen doped carbon films, *JEPT Lett.* **62** (1995), 715-718.
39. Heimann, R.B., Fujiwara, S., Kakudate, Y., Koga, Y., Komatsu, T., and Nomura, M.: A new carbon form obtained by weak shock compression of carbyne, *Carbon* **33** (1995), 859-861.
40. Komatsu, T., Nomura, M., Kakudate, Y., Fujiwara, S., and Heimann, R.B.: Characterization of dehydrochlorinated poly(vinylidene chloride) and the shock-compressed material, *Macromol. Chem. Phys.* **196** (1995), 3031-3040.
41. Udod, I.A., Bulychev, B.M., Babaev, V.G., and Kudryavtsev, Yu.P.: Formation of carbyne intercalation compounds with alkali metals, *Zh. Neorg. Khim.* **41** (1996), 382-387 (in Russian).
42. Chien, J.C.W.: *Polyacetylene. Chemistry, Physics and Material Science*, Academic Press, Orlando, 1984.
43. Bulychev, B.M. and Udod, I.A.: Linear carbon (carbyne): approaches to synthesis, identification of structure and intercalation, *Ross. Khim. Zh.* **39** (1995), 9-18 (in Russian).
44. Udod, I.A.: Synthesis and structure of carbyne and graphite intercalation compounds with sodium and potassium, *Ph.D. Thesis*, M.V.Lomonosov Moscow State University, Moscow, 1994 (in Russian).

45. Lagow, R.J., Kampa, J.J., Wei, H.-Ch., Battle, S.L. Gende, J.W., Laude, D.A., Harper, C.J., Bau, R., Stevens, R.C., Haw, J.F., and Munson, E.: Synthesis of linear acetylenic carbon: the "sp" carbon allotrope, *Science* **267** (1995), 362-367.
46. Eastmond, R., Johnson, T.R., and Walton, D.R.M.: Silylation as a protective method for terminal alkynes in oxidative couplings. A general synthesis of the parent polyynes $H(C{\equiv}C)_nH$ (n = 4-10,12), *Tetrahedron* **28** (1972), 4601-4616.
47. Eastmond, R. and Walton, D.R.M.: Silylation as a protective method in Cadiot-Chodkiewicz couplings. Synthesis of aryl-butadiynes and -hexatriynes, *Tetrahedron* **28** (1972), 4591-4599.
48. Johnson, T.R. and Walton, D.R.M.: Silylation as a protective method in acetylene chemistry. Polyyne chain extensions using the reagents, $Et_3Si(C{\equiv}C)_mH$ (m = 1,2,4) in mixed oxidative couplings, *Tetrahedron* **28** (1972), 5221-5236.
49. Livingston, R.L. and Rao, C.N.R.: The molecular structure of carbon suboxide, *J. Am. Chem. Soc.* **81** (1959), 285-287.
50. Maier, G., Reisenauer, H.P., Schafer, U., and Bali, H: C_5O_2 (1,2,3,4-pentatetraen-1,5-dion), ein neues Oxid des Kohlenstoffs, *Angew Chem.* **100** (1988), 590-592.
51. Maier, G., Reisenauer, H.P., Bali, H., Brandt, W., and Janoschek, R.: C_4O_2 (1,2,3-butatriene-1,4-dione), the first dioxide of carbon with an even number of C atoms, *Angew Chem. Int. Ed. Engl.* **102** (1990), 905-908.
52. Sutzle, D. and Schwarz, H.: Identification of butatatrienedione, its radical anion, and its radical cation in the gas phase, *Angew Chem. Int. Ed. Engl.* **102** (1990), 908-909.
53. Pirio, N., Touchard, D., Dixneuf, P.H., Fettouhi, M., and Ouahab, L.: Metallacumulenes: activation of diynes and access to a novel bis(alkenylallenylidene)-ruthenium complex with the structural unit $R_2C{=}C{=}C{=}Ru{=}C{=}C{=}CR_2$, *Angew. Chem. Int. Ed. Engl.* **31** (1992), 651-653.
54. Zhou, Y., Seyler, J.W., Weng, W., Arif, A.M., and Gladysz, J.A.: New families of coordinated carbon: oxidative coupling of an ethynyl complex to isolable and crystallographically characterized $MC{=}CC{=}CM$ and $M^+{=}C{=}C{=}C{=}C{=}M^+$, *J. Am. Chem. Soc.* **115** (1993), 8509-8510.
55. Rice, M.J., Phillpot, S.R., Bishop, A.R. and Campbell, D.K.: Solitons, polarons, and phonons in the infinite polyyne chain, *Phys. Rev. B* **34** (1986), 4139-4149.
56. Heimann R.B.: Linear finite carbon chains (carbynes): their role during dynamic transformation of graphite to diamond, and their geometric and electronic structure, *Diamond Relat. Mater.* **3** (1994), 1151-1157.
57. Whittaker, A.G.: Carbon a new view of its high-temperature behavior, *Science* **200** (1978), 763-764.
58. Kroto, H.W., Heath, J.R., O'Brien, S.C., Curl, R.F., and Smalley, R.E.: C_{60}: Buckminsterfullerene, *Nature* **318** (1985), 162-163.
59. Iijima, S.: Helical microtubules of graphitic carbon, *Nature* **354** (1991), 56-58.
60. Curl, R.F. and Smalley, R.E.: Probing C_{60}, *Science* **242** (1988), 1017-1022.
61. Kroto, H.: Space, stars, C_{60}, and soot, *Science* **242** (1988), 1139-1145.
62. Goeres, A. and Sedlmayr, E.: On the nucleation mechanism of effective fullerite condensation, *Chem. Phys. Lett.* **184** (1991), 310-317.
63. Baeyer, A., Ueber Polyacetylenverbindungen, *Ber.Deuts. Chem. Ges.* **18** (1885), 2269-2281.
64. Taylor, R., Langley, G.J., Kroto, H.W., and Walton, D.R.M.: Formation of C_{60} by pyrolysis of naphthalene, *Nature* **366** (1993), 728-731.
65. Akagi, K., Nishiguchi, M., Shirakawa, H., Furukawa, Y., and Harada, I.: One-dimensional conjugated carbyne - synthesis and properties, *Synth. Met.* **17** (1987), 557-562.
66. Korshak, V.V., Kudryavtsev, Yu.P., Korshak, Yu.V., Evsyukov S.E., Khvostov, V.V., Babaev, V.G., and Guseva, M.B.: Formation of β-carbyne by dehydrohalogenation, *Macromol.Chem., Rapid Commun.* **9** (1988), 135-140.
67. Kavan, L., Dousek, F.P., and Micka, K.: The role of ion transport in the electrochemical corrossion of fluoropolymers. Preparation and properties of *n*-doped polymeric carbon with mixed electron/ion conductivity, *Solid State Ionics* **38** (1990), 109-118.
68. Shimoyama, M., Niino, H., and Yabe, A.: A KrF excimer laser induced dehydrochlorination of a chlorinated poly(vinyl chloride): preparation of conjugated polyene and polyyne, *Macromol. Chem.* **193** (1992), 569-574.
69. Kavan, L. and Dousek, F.P.: Carbynoid species in electrochemical polymeric carbon, *Synth.Met.* **58** (1993), 63-72.
70. Vointseva, I.I., Gil'man, L.M., Kudryavtsev, Yu.P., Evsyukov, S.E., and Valetskii, P.M.: Synthesis of polymorphic modification of carbyne by dehydrochlorination of poly(trichlorobutadiene) isomers, *Polym. Sci., Ser A* **38** (1996), 707-711.

71. Evsyukov S.E., Paasch, S., Thomas, B., and Heimann, R.B.: Formation of carbynoid structures by chemical dehydrohalogenation of poly(vinylidene chloride). A ^{13}C solid-state NMR study, *Ber. Bunsen-Ges. Phys. Chem.* **101** (1997), 837-841.
72. Kijima, M., Sakai, Yu., and Shirakawa, H.: A novel approach for synthesis of carbyne by electroreductive polymerization of diidoacetylene catalyzed by Ni-complex, *Chem. Lett.* (1994), 2011-2014.
73. Kavan, L. and Kastner, J.: Carbyne forms of carbon: continuation of the story, *Carbon* **32** (1994), 1533-1536.
74. Kavan, L., Hlavaty, J., Kastner J., and Kuzmany, H: Electrochemical carbyne from perfluorinated hydrocarbons: synthesis and stability studied by Raman scattering, *Carbon* **33** (1995), 1321-1329.
75. Berdyugin, V.V., Kudryavtsev, Yu.P., Evsyukov, S.E., Korshak, Yu.V., Shorygin, P.P., and Korshak, V.V.: Resonance Raman spectra of β-carbyne, *Dokl. Akad. Nauk SSSR* **305** (1989), 362-365 (in Russian).
76. Korshak, V.V., Kudryavtsev, Yu.P., Khvostov, V.V., Babaev, V.G., Guseva, M.B., and Rylova, O.Yu.: Electronic structure of carbynes studied by Auger and electron energy loss spectroscopy, *Carbon* **25** (1987), 735-738.
77. Kudryavtsev, Yu.P., Evsyukov S.E., Guseva, M.B., Babaev, V.G., and Khvostov, V.V.:Carbyne - the third allotropic form of carbon, *Russ. Chem. Bull.* **42** (1993), 399-413.
78. Spitsina N.G., Boiko G.N., Kudryavtsev, Yu.P., Babaev, V.G., Guseva, M.B., and Evsyukov, S.E.: Concerning the existence of linear carbon molecules in soot obtaned in an electric arc, *Russ. Chem. Bull.* **44** (1995), 1339-1341.
79. Tanuma, S. and Palnichenko, A.: Synthesis of low density carbon crystal "carbolite" by quenching of carbon gas, *J. Mater. Res.* **10** (1995), 1120-1125.
80. Tanuma, S., Palnichenko, A.V., and Satoh, N.: Synthesis of low density carbon crystals by quenching gaseous carbon and intercalation of alkali metal atoms into these crystals, *Synth. Met.* **71** (1995), 1841-1844.
81. Heath, J.R., Zhang, Q., O'Brien, S.C., Curl, R.F., Kroto, H.W., and Smalley, R.E.: The formation of long carbon chain molecules during laser vaporization of graphite, *J. Am. Chem. Soc.* **109** (1987), 359-363.
82. Zhou, O., Fischer, J.E., Coustel, N., Kycla, S., Zhu, Q., McGhie, A.R., Romanow, W.J., McCauley, J.P., Smith, A.B., and Cox, D.E.: Structure and bonding in alkali-metal-doped C_{60}, *Nature* **351** (1991), 462-464.
83. Murthy, N.S., Shacklette, L.W., and Baughman, R.H.: Structure of lithium-doped polyacetylene, *Phys. Rev. B* **40** (1989), 12550-12553.
84. Billaud, D., Ghanbaja, J., and Goulon, C.:Heavy alkali metal doping of $(CH)_x$: new data, *Synth. Met.* **17** (1987), 497-502.
85. Udod, I.A., Shchurik, V.I., Bulychev, B.M., Sirotinkin, S.P., Guseva, M.B., Babaev, V.G., Kudryavtsev, Yu.P., and Evsyukov, S.E.: Formation of 'carbyne' in the interaction of polyacetylene with potassium under high quasi-hydrostatic pressure, *J. Mater. Chem.* **3** (1993), 413-416.
86. Kudryavtsev, Yu.P., Heimann, R.B., and Evsyukov, S.E.: Carbynes: advances in the field of linear carbon chain compounds, *J. Mater. Sci.* **31** (1996), 5557-5571.
87. Herold, A.: Crystallochemistry of graphite intercalation compounds, in: *Physics and Chemistry of Materials with Layered Structures. Intercalated Materials*, ed. by Levy, F., Vol. 6, Riedel, Dordrecht, 1979, pp. 323-421.
88. Diederich, F., Rubin, Y., Knobler, C.B., Whetten R.L., Schriver, K.E., Houk, K.N., and Li, Y.: All-carbon molecules: evidence for the generation of cyclo[18]carbon from stable organic precursor, *Science* **245** (1989) 1088-1090.
89. Kraetschmer, W., Lamb, L.D., Fostiropoulos, K., and Huffman, D.R.: Solid C_{60}: a new form of carbon, *Nature* **347** (1990), 354-358.
90. Yamada, K., Burkhard G., Tanabe T., and Sawaoka A.B.: An in-situ observation of amorphization of carbyne particle under electron-beam irradiation, *Carbon* **34** (1996), 1601-1602.
91. Udod, I.A., Bulychev, B.M., and Babaev V.G.: Structural models of carbyne (a linear modification of carbon) and its intercalation compounds, *Zhurn. Neorg. Khim.* **42** (1997), 521-528 (in Russian).

4.4. ELECTRON DIFFRACTION AND MICROSCOPY

A. G. FITZGERALD
University of Dundee, Carnegie Laboratory of Physics
Dept. of Applied Physics and Electronic & Mechanical Engineering,
University of Dundee
Dundee DD1 4HN, Scotland UK

Abstract. Electron microscopy and electron diffraction have proved to be the most appropriate techniques for investigating the crystal structure and microstructure of carbyne crystals because of the crystal size range. Carbyne was first identified in meteorites by electron diffraction. With the synthesis of carbynes, electron microscopy and electron diffraction have proved to be important techniques in identifying these materials and comparing carbynes prepared by different methods. These techniques have also proved useful in investigating the effects of annealing, electron irradiation and ion irradiation. On the basis of electron diffraction data a number of models for the crystal structure of carbynes have been developed. In one model it is proposed that kinks occur in the carbon chains. The existence of these kinks can explain the lamellar nature of carbyne crystals observed in electron micrographs. A number of suggestions have been advanced to explain the occurrence of these kinks including the application of paracrystalline theory and the existence of charge-density soliton waves in the one dimensional carbon chain.

1. Introduction

Electron diffraction has proved to be the most effective technique so far, in analysing the crystal structure of the carbyne allotropes of carbon. In general carbynes detected in naturally occurring minerals, particularly meteoritic material, occur in polycrystalline form [1]. This rules out the application of the single crystal x-ray diffraction techniques which could provide an unequivocal determination of the carbyne crystal structures. Crystals of carbynes which have been formed synthetically rarely exceed a micrometre in size [2] and therefore single crystal x-ray diffraction methods are also unsuitable for comprehensive crystal structure determinations from this material.

The size range of carbyne crystals has made transmission electron microscopy and electron microdiffraction the most appropriate techniques for investigation of the structure, microstructure and structural changes in carbyne. High resolution electron microscopy also has the added advantage that structural and microstructural changes that occur at high temperature and under electron irradiation can be followed *in situ* in the instrument.

R.B. Heimann et al. (eds.), Carbyne and Carbynoid Structures, 295–308.

In addition to the application of electron microscopy and electron diffraction to high resolution structural studies of carbyne, the associated techniques of energy dispersive x-ray microanalysis (EDX) and electron energy loss spectroscopy (EELS) have been particularly important in providing supporting evidence for structural investigations. These techniques have enabled the analysis of submicrometre areas of carbyne crystals during electron microscopic studies, particularly x-ray microanalysis [3].

Transmission electron microscopy has also proved to be an important means of investigating the morphology of carbyne crystals and the relation of the morphology to the atomic arrangement, in this case, the polymeric carbon chain proposed for the arrangement of carbon atoms.

2. Electron Microscopy and Diffraction of Meteoritic Carbynes[1]

Controversy has surrounded the existence of carbyne since the early 1960s when a material variously called "white carbon", "chaoite" and "Ries crater carbon" was first detected [1,4,5]. This new allotropic form of carbon had a hexagonal unit cell with a_o = 894.8 pm and c_o = 1407.8 pm.

Whittaker [7] extended these studies of this new naturally occurring allotrope of carbon to natural graphitic formations from geographical locations ranging from Sri Lanka to California. His investigations, which involved scanning electron microscopy, high energy electron diffraction, and x-ray diffraction confirmed the existence of a form of carbon which was designated as carbon VII (a_o = 909.8 pm, c_o = 1482 pm). These studies indicated that carbyne forms of carbon can be found to occur naturally all over the world and are at least 60 million years old.

In further investigations of naturally occurring carbonaceous chondrites, the Allende and Murchison chondrites, Whittaker *et al.* [8] used electron diffraction to detect five carbynes which they labelled carbon VI, VIII, X, XI and tentatively XII. A number of experiments were carried out on these materials and it was found that three of the four Allende carbynes transformed to graphite on heating. Some of the carbyne crystals were also sensitive to the electron beam in the electron microscope forming graphite as an end product. Each of the carbynes detected gave a hexagonal array of electron diffraction spots in single crystal patterns. The maximum interatomic planar spacing d_{max} for each structure detected was compared with *d*-values for α-carbyne or chaoite, β-carbyne, graphite, londsdaleite and diamond.

The existence of carbynes was disputed in the early eighties by Smith [9] and Smith and Busek [10]. As a result of electron diffraction combined with x-ray microanalysis of a carbon separate from the Allende material they claimed that their investigations implied that the diffraction patterns obtained by Whittaker *et al.* [8] were in fact due to carbon contaminated with sheet silicates. Smith and Busek's conclusion was that the four published examples of naturally occurring carbynes did not stand close scrutiny as a new naturally occurring phase of carbon.

---

[1] For further discussion of meteoritic carbynes see also Chapter 2.1 of this book

The suggestion by Smith and Busek [10] that carbynes had not been detected in nature was investigated further by Whittaker [11]. He concluded that Smith and Busek had not made use of all of the diffraction data available in analysing the earlier electron diffraction work [8]. Whittaker's contention was, that if all of the diffraction maxima available had been used in analysing the structure, other minerals would not have been confused with the chaoite form of carbon. Whittaker also drew attention to the large body of data on the preparation and properties of carbynes by Russian researchers [12]. He also produced electron diffraction results from synthetic carbyne films prepared by quenching carbon vapour on a polished metal substrate as evidence in support of the existence of carbyne, since this material was prepared, in this instance, from pure carbon with no possibility of silicate contamination. Although Smith and Busek [13] in a reply to the Whittaker's suggestions, [11] continued to refute these arguments that supported the identification of carbonaceous material in meteorites as carbyne, they did not discuss the growing electron diffraction evidence for the formation of carbyne by synthetic routes.

Recently Rietmeijer [14] re-interpreted the carbyne electron diffraction data of Whittaker and Wolten [15], Whittaker *et al.* [16] and Kasatochkin *et al.* [17]. He arrived at a conclusion which was quite different from that currently accepted. Rietmeijer suggested that the carbynes that exist in meteorites could be crystalline C-(H-O-N) carbons which are the metastable relics of kinetically-inhibited incomplete pyrolysis rather than carbon allotropes. His conclusions supported the proposition that carbynes should be crystalline randomly interstratified (or mixed-layered) carbons with a variable C/(C+H+O+N) ratio rather than thermodynamically stable carbon allotropes. Rietmeijer believed that this proposal reconciled the apparent lack of equilibrium phase relations between graphite, diamond and carbynes and the incompatibility of the probable sheet structure with a structure of chains running parallel to a six-fold symmetry carbyne axis.

3. Electron Microscopy and Electron Diffraction of Synthetic Carbynes

Over the past twenty years carbyne has been synthesised by a variety of widely differing methods. Among the first reports of the synthesis of carbyne were those of Katsatochkin *et al.* [18] who synthesised material by passing acetylene through a heated solution of copper chloride in a mixture of pyridine and methyl alcohol. The resulting black precipitate of polyenes and copper acetylides was oxidised with an aqueous solution of ferric chloride and then treated successively with hot hydrochloric acid (1:1), ammonia and water. The product was dried in the absence of air, and was then heated under vacuum with a gradual increase in temperature to 1000°C.

Electron microdiffraction from 0.05μm areas of this material gave single crystal and amorphous ring patterns. The hexagonal cell edges of the crystals were found to be a_0 = 508 pm and c_0 = 780 pm. The cell contains three hexatomic carbon chains disposed parallel to the *c*-axis. Fig. 1 shows schematically the disposition of the linear carbon chains parallel to the *c*-axis as projection on the basal plane of the carbyne crystal as proposed by Kasatochkin *et al.* [18]. The shortest distance is R = 295 pm between neighbouring chains.

Another type of electron diffraction pattern accompanied the carbyne spot diffraction patterns. This spot pattern corresponded to the basal (*hk0*) plane of the reciprocal lattice of a hexagonal second crystalline phase of carbon. This phase consisted of crystalline cumulene $(=C=C=C=)_n$. Ring patterns from this phase were also observed. The dimensions of the unit cell of this phase were a_0 = 476 pm and c_0 = 258 pm. The cumulene cell includes six atoms from two of the three cumulene carbon chains (=C=C=) passing through the cell parallel to the *c*-axis of the crystal. The results of this investigation showed that the carbon samples studied constituted a three-phase system in which carbyne and cumulene crystallites are combined with an amorphous phase of carbon.

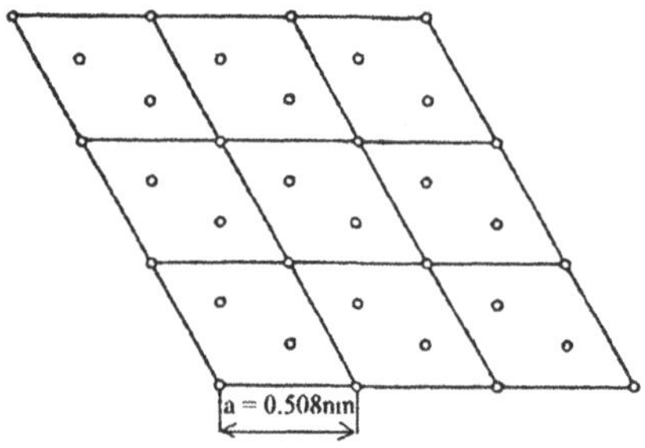

Figure 1. The structure of carbyne proposed by Kasatochkin et al. [18].

The cumulene modification of carbyne was synthesised for the first time in 1968. A two step synthetic method was developed and two types of single crystals corresponding to the α and β modifications of carbyne were identified by electron diffraction [19].

Kasatochkin and co-workers [20] have also prepared carbyne by laser-induced sublimation from pyrographite. The following values of the unit cell parameter a_0 were determined for different single crystal particles 824 pm, 880 pm and 894 pm.

Sladkov and co-workers [21] have also identified carbyne crystals by electron microscopy and diffraction. The carbyne was synthesised by the oxidative polydehydrocondensation of acetylene in the presence of salts of divalent copper. The products were then oxidised with an aqueous solution of iron perchlorate. Copper was then removed from the material. Two structures were identified in this material by electron diffraction. The first was hexagonal with a_0 = 509 pm and c_0 = 780 pm. The unit cell was believed to include 18 atoms, from three carbon chains, passing through it parallel to the c-axis. The second structure was also hexagonal with a_0 = 476 pm and c_0 = 258 pm. This cell included six carbon atoms, in pairs, from three cumulene carbon chains =C=C= , passing through the cell parallel to the *c*-axis of the crystal. The first more complex cell was assigned to crystals of polyyne on the basis of general considerations on alternation of the bonds in the carbon chain. The simpler cell was assigned to crystals of polycumulenes with a more uniform distribution of π-electrons along the carbon chain.

Using an entirely different preparation technique, deposition from an rf plasma discharge in acetylene, Zorin *et al.* [22] detected two modifications of carbyne, the α and β-forms. Graphite was also present in the films. With increase in substrate temperature a transition from carbyne to diamondlike carbon was observed. The authors concluded that the carbyne forms as a natural result of the deposition of -C≡C- radicals, which are structural parts of acetylene, followed by the subsequent polymerisation of these groups with the α-carbyne form consisting of -C≡C- chains, and the more stable β-carbyne form consisting of =C=C=.

Heimann *et al.* [23] have investigated the effects of a shock wave on graphite-copper mixtures. These experiments yielded several polymorphic forms of carbon which have been identified by electron microscopy and electron diffraction. These included chaoite, a new form of carbon with a_o = 870 pm and c_o = 956 pm and crystals they described as belonging to the carbyne family.

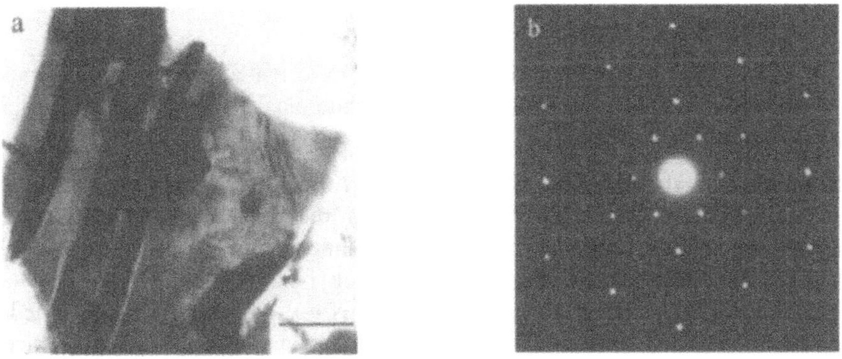

Figure 2. (a) Electron micrograph of an α-carbyne crystal (scale mark 0.3μm), (b) Electron diffraction pattern from a crystallite of carbyne in a film prepared by a fast atom beam source.

During the characterisation of thin carbon films prepared by sputter deposition from a dc triode sputtering gun, Fitzgerald and co-workers [24] identified crystallites of α-carbyne (Fig. 2a) distributed across the amorphous carbon film surface. Similar crystallites were also identified by electron diffraction (Fig. 2b) in carbon films produced by fast atom bombardment [25] using butane and butane-argon gas sources for the fast atom beam.

The first preparation of oriented single crystal carbyne layers using chemical (low temperature carbonisation of a poly(vinylidene fluoride) (PVDF) film using a chemical dehydrofluorination reaction) has been described by Kudryavtsev *et al.* [26]. The dehydrofluorinated films were amorphous with the first diffraction ring corresponding to a value of $d = 435$ pm. Annealing the films at 400°C in vacuum on the (100) surface of the KBr substrate crystals resulted in partial crystallisation with the diffraction pattern corresponding to a hexagonal lattice with $a_0 = 870$ pm. The single crystal regions occupied about 30% of the film.

In this investigation films were also prepared by ion sputtering of graphite and by electrical arc discharge from carbon. The electron diffraction patterns from the ion sputtered films had an overall oriented structure containing some separated single-crystal regions (~1-2μm). The diffraction pattern corresponded to a hexagonal lattice with $a_0 = 870$ pm and included (*hki0*) reflections. This indicated that the carbon chains were oriented perpendicular to the surface. A specific feature of the electron diffraction patterns from the single crystal films of carbyne was the intensity of the (11.0) reflections which were more that 10 times higher than that of any other (*hki0*) reflection. The diffraction patterns from the single crystal microinclusions were extremely sharp. The *d*-values from these patterns gave a lattice parameter $a_0 = 892$ pm indicating the crystals were carbyne.

In the films produced by electrical arc discharge, oriented areas were observed as well as amorphous regions corresponding to graphite clusters. The oriented film areas had a hexagonal lattice with $a_0 = 870$ pm.

The single crystal films with high intensity (11.0) reflections were explained as a random deviation of some carbyne layers from the ideal position in the perfect hexagonal lattice. The single crystal microcrystalline inclusions in the ion sputtered films were believed to be ideal single crystals of carbyne in which the strong orientation between the layers was due to impurity atoms. X-ray microanalysis confirmed that these carbyne crystals contained iron and potassium impurities with small traces of silicon due to contamination from the vacuum pumping system.

The location of the carbon atoms in the carbyne single crystals was determined by calculating Patterson functions. The diagram for the carbyne structure (Fig. 3) deduced by Kudryavtsev *et al.* [26] shows a carbyne spatial model in which the carbyne lattice is composed of double-layers. The lower layer consists of closely packed carbon chains. The columns of the chains in the upper layer have vacancies that are located with a period of 515 pm. These atoms can provide saturation for the dangling bonds at the chain ends. In the model the chains are denoted by A, B, C, D and E. The distance between the chains in the carbyne crystal is close to 300 pm. This value is much smaller than expected in a carbon polytype consisting of polyene-type chains. The calculations of Kudryavtsev *et al.* showed that a polyyne chain is favoured for an approximately 300 pm interchain spacing.

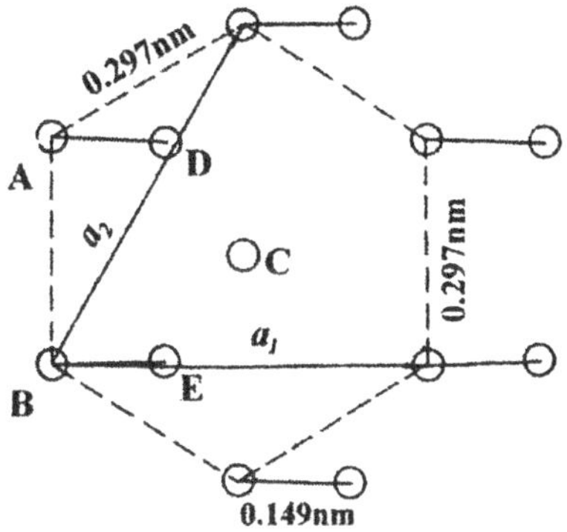

Figure 3. Position of carbon chains in a unit cell of a carbyne crystal in a (0001) projection after Kudryavtsev et al. [26].

Kudryavtsev and co-workers [26] concluded that carbyne has a layered structure. Each layer is formed by short carbon chains and the chemical bonds among the carbon atoms in these chains are characterised by sp hybridisation of carbon. Metal (Fe, K) atoms can be intercalated between the layers. It should be noted that in some cases the atoms of other impurities (e.g. Si, Al, Mo etc.) can also be intercalated (see Chapter 4.3 of this book). This is why in some cases, the presence of Al, Si and Mo has led to the incorrect interpretation of the carbyne structure as a silicate structure [9,10].

A recent study has shown that it is also possible to synthesise carbyne by dehydrogenation of polyacetylene (see Chapter 3.2.3 of this book). Electron diffraction patterns obtained from crystals prepared by this method indicated that the crystals had the hexagonal crystal lattice of carbyne with the unit cell parameters a_0 = 886 pm and c_0 = 1600 pm.

4. Electron Beam and Electron Bombardment Effects in Carbynes

Electron beam effects in carbynes were reported almost immediately after the first discovery of the material. Radiation damage was observed by Whittaker and Wolten [27] in chaoite and carbon VI carbyne at electron beam energies in the electron microscope above 75 keV. Radiation damage was also observed in the electron microscope in the rhombohedral carbyne (carbon IX). This damage was observed for accelerating voltages of greater than 60 kV. Amorphisation of carbyne under electron irradiation in the electron microscope has also been observed[1] [28]. The sharp electron

[1] (Note by the Editor) Amorphisation of carbyne under electron beam irradiation has been recently studied also by Yamada et al., *Carbon* **34**, 1601 (1996) and others (cf. Chapter 2.1 of this book).

diffraction patterns, both ring and spot patterns, normally obtained from carbyne, have been observed to gradually fade with the formation of diffuse amorphous rings. This observation was attributed to weak bonding between the molecules which form the basic units in the crystal structure. As amorphisation proceeded, features described as lamel pits or ulcers were observed in transmission electron microscope images from the crystalline lamellae.

Fitzgerald *et al.* [25] observed similar electron beam effects in the carbyne crystals which accompany amorphous carbon films prepared by means of a fast atom source using butane and butane-argon mixtures. The sensitivity of carbyne to the electron beam in the transmission electron microscope was followed with time. Areas of carbyne crystals showed enhanced electron transmission (Fig. 4a) which suggested sublimation of some of the crystalline material.

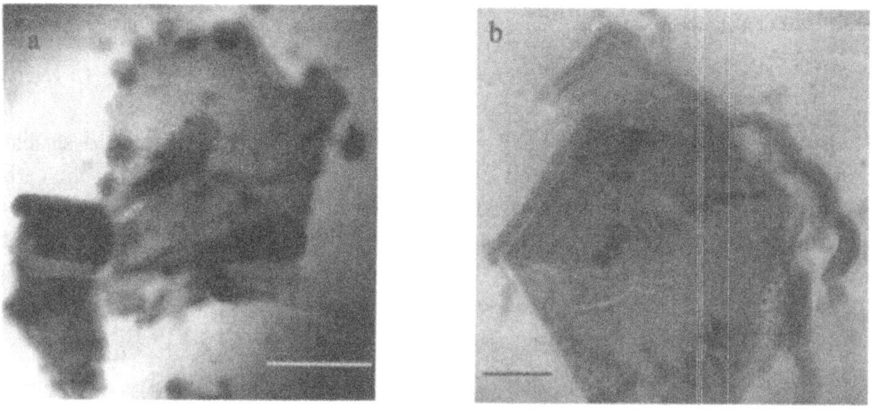

Figure 4. (a) Micrograph showing the effects of electron irradiation on a carbyne crystal (scale mark 0.3μm), (b) α-carbyne crystal exhibiting larger lattice parameters (scale mark 0.2μm).

5. Experimental Studies of Carbyne Crystals

A number of electron diffraction studies of carbyne have been made after subjecting the material to various physical conditions. For example, the effect of shock loading amorphous 10μm carbyne films on the microstructure has been explored. Transmission electron microscopy and electron diffraction studies of carbyne films recovered after loading up between 18 and 26 GPa showed that the material transformed to a mixture of crystalline and amorphous carbyne. A small amount of graphite and diamond was also observed in several samples.

The electron microdiffraction examination of 10μm carbyne films recovered after loading up to 36GPa contained crystalline phases with diffraction patterns of two types. The diffraction pattern of the first phase exhibited a narrow ring with a *d*-spacing of about 460 pm which has also been detected by means of x-ray diffraction. The second phase had interplanar distances which are similar to those of the hexagonal carbyne structure with a_0 = 509 pm [2, 29]. From transmission electron microscopy observations the typical size of these carbyne crystals was of the order of 1μm.

Recovered samples from 150μm carbyne films loaded up to peak pressures of between 26 and 36 GPa contained a mixture of amorphous carbon and a small but detectable amount of cubic diamond with crystal size less than 0.1μm.

The effects of ion bombardment and annealing have also been studied by electron diffraction. One method of preparation of carbyne films is ion-assisted condensation of carbyne [30]. Films prepared by alternate condensation and ion irradiation were found to be generally amorphous, but their structure and properties depended greatly on the irradiation conditions. Microcrystalline inclusions between 10 and 100 nm appeared in these films at ion energies greater than 30 eV, with electron diffraction patterns corresponding to carbyne [31]. The concentration of carbyne crystallites depended on the number of ions impinging on the substrate during film growth and increased as the ion flux increased. Variation of the ion energy within the range studied did not have an appreciable effect on the density of the crystalline inclusions. However, an increase in the energy of the ions caused a very significant change in the carbyne crystal lattice parameter from a_0 = 103 pm at an argon energy of 30 eV to 831 pm at an argon ion energy of 150eV.

Fitzgerald *et al.* detected an increase in lattice parameter to 941.7 pm in carbyne crystals of the type shown in Fig. 4b that were prepared using a dual ion beam system [32]. Similar changes in the parameter a_0 were obtained after a high temperature anneal of the carbyne single crystals, which appeared in an amorphous matrix during the crystallisation process at 600°C [33]. At first, the crystals formed were flat and acute-angled exhibiting low contrast. When the annealing time and the temperature were increased, they grew rapidly and reached a size of 1μm at 650°C. The electron microdiffraction pattern from these crystals contained diffraction spots corresponding to a hexagonal structure with a parameter a_0 = 901 pm. Successive increases in the annealing temperature resulted in alteration of the carbyne unit cell parameter a_0. At first, the value of a_0 decreased with increasing temperature from 901 pm at 700 °C to 840 pm at 750 °C, and then it increased back up to 901 pm at 800 °C. Subsequent annealing at 850 °C resulted in the graphitisation of carbyne.

6. Crystal Structure Models for Carbynes from Electron Diffraction Studies

Since the first reports on carbyne compounds (α- and β-carbynes) [17] their polymorphism has been explained on the basis of electron diffraction data by assuming these various carbyne materials contained carbon chains of different lengths parallel to the hexagonal c-axis. The α-carbyne form is believed to contain a polymeric carbon chain in which a 2-atomic unit of 256 pm length is repeated six times, yielding a polyyne-type chain of twelve atoms. Likewise, β-carbyne is assumed to possess a cumulene-type chain of six atoms length.

Heimann *et al.* [23] have attempted to describe the host of carbyne forms as linear carbon polytypes with varying numbers of carbon atoms in the conjugated triple-bonded or cumulated double-bonded chains parallel to the hexagonal *c*-axes. The bonding of the chains to each other is achieved by intermolecular linkage. The interchain distance of 293 pm [34,35] is sufficiently close to permit overlap of unoccupied π-orbitals in a bonding mode [36].

However, as pointed out by Heimann *et al.*, [23] this mechanism constitutes an apparent ambiguity insofar as the resulting weak van der Waals bonds in two dimensions by no means account for the high thermal stability and the frequently observed hardness of some carbyne forms [37]. Another problem with the linear form for carbon polytypes is the assignment of a definite identity period in the *c*-direction which defines the elementary unit cell of the various carbyne compounds. Infinite, straight and evenly spaced C-C chains with either conjugated triple or cumulated double bonds do not permit the assignment of discrete c_0 values. There are however, two ways that this can occur either by *n*-membered carbon-carbon chains with a helical structure having a pitch of $2\pi/n$ or a straight portion of chain separated by kinks.

Both polyyne and cumulene chains possess sufficient rigidity owing to the π-component in the linear molecular orbital bonding system and are unlikely to form a helix. On the other hand, introduction of kinks in the chains will terminate a unit cell. This requires that adjacent unit cells are laterally displaced with respect to each other by a distance which is an integer part of the inter-chain distance $a^* = 293$ pm. Such a displacement is equal to $2a^*/5$ for a cumulene chain (kink angle 25°) and $a^*/5$ for a polyyne chain (kink angle 60°), respectively [38]. This latter conversion of the graphite structure to a carbyne structure based on polyyne has been first proposed by Whittaker [39,40] and involves the transformation of the graphite basal plane sheet of atoms into $(C\equiv C)_n$ chains. A unified structural approach containing both polyyne and polycumulene has been suggested by Heimann et al. [38]. The occurrence of kinks can be thought of as a deformed remainder of the original bonding angle in the basal plane of graphite (cf. Fig. 1 in Chapter 3.4.1.1). It was assumed by Heimann et al. [23] that kink portions are formed by single bonds with the σ-configuration. The kinked cumulene chains then provide free bonds by splitting of double bonds which can be used for either bonding of impurities such as silicon and stabilising bulky alkyl or aryl groups, or extensive cross-linking of carbon chains. This model could explain the exceptional hardness and thermal stability of chaoite [15,37] and α- and β-carbyne [12] which are all thought to possess a cumulene structure, contrary to the conclusions of Zorin *et al.* [22]. It could also explain the fact that chaoite is the only carbyne found in nature as a stable compound, having survived for at least sixty million years.

Kudryavtsev et al. [26, 41] have calculated a Patterson function on the basis of data from electron diffraction. From this work, a picture of the crystal potential as well as a model of the location of the atoms in the crystal unit cell have been developed. This model shows that carbyne has a layered structure. Fig. 3 shows the position of the carbon chains in a unit cell of carbyne in the (0001) projection obtained from their calculations. This model suggests that impurity atoms could be located at a distance of 70 pm at the position of the additional maxima in the crystal potential map of the carbyne lattice. These impurity atoms are located near the chains. The distance of 149 pm between the chains (A) and (D) and also (B) and (E) (Fig. 3) in the projection on the (000*l*) plane is too small and, therefore, it can be presumed that the structure of the carbyne crystal is layered. In this case the chains will be located in adjacent layers i.e. chains (A), (B) and (C) are located in the lower layer, while the chains (D) and (E) are located in the upper one. These results suggest that the carbyne lattice is double-layered. The lower layer consists of closely packed carbon chains. The columns of the chains in the upper layer have vacancies that are located with a period of 515 pm. Impurity atoms such as iron or potassium which are intercalated between the layers are presumed to play a rôle. These impurities have been detected by x-ray microanalysis. The chains (A) and (D), as well as (B) and (E), located in the adjacent layers are mutually shifted in a projection on the basal plane by a distance of 149 pm (Fig. 3). If the distance between adjacent layers is small enough, the existence of a σ-link between the chains (A) and (D), and (B) and (E) (Fig. 3) respectively can be assumed. In this case, the chains (A) and (D), as well as (B) and (E) (Fig. 3) would be fragments of the same larger carbon chains having regular kinks. Such a structure is similar to the structural model proposed by Heimann et al. [23] for different carbyne polytypes.

In this model it was suggested that the kinks formed within the linear carbon chains are subject to statistically fluctuating kink angles and kink offset distances. These relationships can be described in terms of the concept of paracrystal theory [42]. The kinks in the carbon chains terminate the crystallographic identity period and thus define discrete *c*-axis lengths. This layered structure of carbyne explains its lamellar morphology [28]. It is evident that the minimum thickness of a lamella corresponds to the length of the carbon chains oriented perpendicularly to the lamella surface.

An alternative model, a tetrahedron model, has also been employed by Heimann *et al.* [38] and is based on the conformation of linear polyalkenamers. In this model, the kinks act as two-fold screw axes, and carbynes can then be considered fully dehydrogenated polyalkynamers.

A quite different mechanism has also been proposed by Heimann [43] in which it is suggested for polyyne chains, that charge-density (CDW) soliton waves exist that move from point to point along the one-dimensional carbon chain but do not spread between neighbouring chains.

The modulation of the electron density of the carbon atoms by the transmission of a charge-density wave results in a displacement of the atoms to a new equilibrium position. Strong electron-phonon coupling can then drive a commensurate-incommensurate transition whose elastic term leads to a kink displacement. These lattice kinks are the domain walls of the transition [44], and thus the soliton separates two energetically equivalent but conformationally distinct segments of a polyyne chain. This can, in principle, be observed by the occurrence of new diffraction satellites, from which location of the period of the charge-density wave can be calculated [45]. Such satellites have been observed from carbynes by Kudryavtsev *et al.* [26], separated by a distance of 70 pm from the carbon chains in a map of the crystal potential and were tentatively attributed by the authors to impurity atoms.

7. Conclusions

Although some electron diffraction studies of carbyne detected in meteoritic material suggest that this material is not an allotrope of carbon, subsequent electron diffraction studies of synthetically prepared materials, where it is possible to control the starting material, have confirmed that carbyne is a new structural form of carbon.

The crystal structure of carbyne is still, however, not completely resolved. Structural evidence obtained so far by electron diffraction indicates that carbyne is a layered structure which contains kinked chains of carbon atoms in a hexagonal unit cell. A number of theories have been advanced to explain the formation of these kinks, these include a paracrystalline phenomenon and a tetrahedron model based on the conformation of linear *trans*-poly(alkenamers). An alternate driving force for kink formation has also been proposed which assumes a strong electron-phonon coupling within the linear carbyne chain. This mechanism may lead to a charge-density soliton wave resulting in a kink displacement driven by the elastic term of a commensurate-incommensurate transition.

Carbynes have also been observed to exhibit unusual properties. They are sensitive to electron beam energies above 75keV eventually transforming to amorphous material. The technique used to prepare carbyne films has also been observed to influence the carbyne lattice parameter. The concentration of carbyne crystallites and the carbyne lattice parameter have been found to depend on the ion energy in ion-assisted deposition. A range of single crystal carbyne lattice parameters have also been observed when carbyne films were annealed in the range 650°C to 800°C. Carbynes have also been demonstrated to form intercalation compounds.

Electron microscopy and electron diffraction have played a major rôle in characterising the carbyne allotropes of carbon. Carbynes are believed to be one dimensional semiconductor materials which are a missing link in the series of carbon allotropes. They open up the possibility for creation of a microelectronics based on carbon only [41]. It is also believed that defectless filamentary crystals of carbyne should in theory be the strongest of all known materials. With the prospects of a class of new materials based on carbynes and carbyne compounds it is likely that electron microscopy, electron diffraction and electron microbeam analytical techniques will continue to play a major part in the characterisation of these materials.

8. References

1. El Goresy, A., and Donnay, G.: New allotropic form of carbon from the Ries crater *Science* **161** (1968), 363-364.
2. Kasatochkin, V.I., Sladkov, A.M., Kudryavtsev, Yu.P., Popov, N.M. and Korshak, V.V.: Crystalline forms of the linear modification of carbon, *Dokl. Akad. Nauk. SSSR* **177** (1967), 358-360.
3. Heimann, R.B., Kleiman, J., and Salansky, N.M.: Structural aspects and conformation of linear carbon polytypes (carbynes), *Carbon* **22** (1984), 147-155.
4. Whittaker, A.G., and Kintner, P.L.: Carbon: Observations on the new allotropic form, *Science* **165** (1969), 589-591.
5. Hayatsu, R., Scott, R.G., Studier, M.H., Lewis, R.S., and Anders, E.: Carbynes in meteorites: detection, low-temperature origin and implications for interstellar molecules, *Science* **209** (1980), 1515-1517.
6. Kasatochkin, V.I., Shterenberg, L.E., Kazakov, M.E., Slesarev, V.N., and Belousova, L.V.: The thermic transformation of carbyne under pressure, *Dokl. Akad. Nauk. SSSR* **209** (1973), 388-391.
7. Whittaker, A.G.: Carbon: occurrence of carbyne forms of carbon in natural graphite, *Carbon* **17** (1979), 21-24.
8. Whittaker, A.G., Watts E.J., Lewis, R.S., and Anders, E: Carbynes: Carriers of primordial noble gases in meteorites, *Science* **209** (1980), 1512-1514.
9. Smith , P.P.K.: Carbon in carbonaceous chondrites: carbyne or graphite? *Nature* **291** (1981), 15-16.
10. Smith, P.P.K., and Busek, P.R.: Carbyne forms of carbon: do they exist? *Science* **216** (1982), 984-986.
11. Whittaker, A.G.: Carbyne forms of carbon: evidence for their existence, *Science* **229** (1985), 485-496.
12. Sladkov, A.M.: Carbyne - a new allotropic form of carbon, *Sov. Sci. Rev. Sect.* **B3** (1981), 75-110.
13. Smith, P.P.R. and Busek, P.R.: Reply to: Carbyne forms of carbon: evidence for their existence, *Science* **229** (1985), 486-487.
14. Rietmeijer, F.J.M.: Are crystalline C-(H-O-N) carbons the elusive meteoritic carbynes? *Meteriotics* **28** (1993), 242-245.
15. Whittaker, A.G. and Wolten, G.M.: Carbon: A suggested new hexagonal crystal form, *Science* **178** (1972), 54-56.
16. Whittaker, A.G., Neudorffer, M.E., and Watts, E.J.: Carbon a rhombohedral carbyne form, *Carbon* **21** (1983), 597-599.
17. Kasatochkin, V.I., Korshak, V.V., Kudryavtsev, Yu. P., Sladkov, A.M., and Sterenberg, I.E.: On crystalline structure of carbyne, *Carbon* **11** (1973), 70-72.
18. Kasatochkin, V.I., Sladkov, A.M., Kudryavtsev, Yu.P., Popov, N.M., and Korshak, V.V.: Crystalline forms of the linear modification of carbon, *Dokl. Akad. Nauk SSSR* **177** (1967), 358-360.
19. Sladkov, A.M., Korshak, V.V, and Nepochatykh, V.P.: Polymers with cumulated double bonds in the chain , *Bull. Acad. Sci. USSR, Div Chem. Sci.*, **1** January (1968), 191-193.
20. Kasatochkin, V.I., Savaranskiy, V.V., Smirnov, B.N., and Melnitchenko, V.M.: Carbon condensed from carbon vapours, *Dokl. Akad. Nauk. SSSR* **217**(4) (1974), 796-799.
21. Sladkov, A.M., Kasatochkin, V.I., Kudryavtsev, Yu.P., andKorshak, V.V.: Synthesis and properties of valuable polymers of carbon, *Isv. Akad. Nauk SSSR, Ser. Khim.* **12** (1968), 2697-2704.
22. Zorin, E.I., Sukhorukov, V.V. and Tetel'baum D.I.: Deposition of carbine and diamond-like films in the plasma of an rf gas discharge, *Sov. Phys. Tech. Phys.* **25**(1) (1980), 103-105.
23. Heimann, R.B., Kleiman, J. and Salansky, N.M.: Structural aspects and conformation of linear carbon polytypes (carbynes), *Carbon* **22** (1984), 147-155.
24. Fitzgerald, A.G., Simpson, M, Dederski, G.A., Moir, P.A., Matthews, A., and Tither, D.: Triode technology in the sputter deposition of carbon films, *Carbon* **26** (1988), 229-234.
25. Fitzgerald, A.G., Simpson, M, Dederski, G.A., Storey, B.E., Moir, P.A. and Tither, D.: Carbon films produced by fast atom bombardment, *Carbon* **26** (1988), 547-552.
26. Kudryavtsev, Yu. P., Evsyukov, S.E., Babaev, V.G., Guseva, M.B., Khvostov V.V. and Krechko L.M.: Oriented carbyne layers, *Carbon* **30** (1992), 213-221.
27. Whittaker, A.G. and Wolten, G.M.: Carbon: A suggested new hexagonal crystal form, *Science* **178** (1972), 54-56.
28. Melnitchenko, V.M., Smirnov, B.N., Varlakov, V.P., Nikulin, Yu.P., and Sladkov, A.M.: Lamellar morphology and inner structure of crystalline carbyne fragments, *Carbon* **21** (1983), 131-133.
29. Kasatochkin, V.I. and Babchinitser, T.M.: Chain length of carbine, a linear carbon polymer, *Dokl.*

Acad. Nauk SSSR **184** (2) (1969), 353-354.
30. Guseva, M.B.: Ion stimulation processes on the NaCl crystal-surface by ion irradiation *Izv. Akad. Nauk SSSR* **50** (1986), 459-464.
31. Guseva, M.B., Babaev, V.G., Nikiforova, N.N., and Savchenko, N.F.: Variation of the structure and properties of carbon films under slowed ion irradiation, *Vopr. At. Nauk. Tekhn., Ser. Fiz. Radiats. Poverzhdenii. Radiats. Material.* **2** (1983), 92-95.
32. Fitzgerald, A.G. and Henderson A.E.: Microbeam analysis of a-C:H films, in J.J. Pouch and S.A. Alterovitz (eds.), *Properties and Characterisation of Amorphous Carbon Films, Materials Science Forum* **52 & 53**, Trans Tech Publications, Aedermannsdorf, Switzerland, 1990, pp. 495-514.
33. Guseva, M.B., Savchenko, N.F., and Babaev, V.G.: New data on the structure of carbine, *Sov. Phys. Dokl.* **30**(8) (1986), 686-687.
34. Kasatochkin, V.I., and Sladkov, A.M.: Crystalline forms of linear modification of carbon, *Dokl. Acad. Nauk SSSR* **177**(2) (1967), 358-360.
35. Kasatochkin, V.I., Kazakov, M.E., Kudryavtsev, Yu. P., Sladkov, A.M., and Korshak, V.V.: Enthalpy of carbine - a chain polymer of carbon, *Dokl. Acad. Nauk SSSR* **183**(1) (1968), 109-111.
36. Patai, S. (Ed.): *The chemistry of the carbon-carbon triple bond*, Wiley, New Jersey, 1978.
37. Whittaker, A.G.: Chaoite coating process, US Patent 4248,909 (3 Feb. 1981).
38. Heimann, R.B., Kleiman, J., and Salansky, N.M.: A unified structural approach to linear carbon polytypes, *Nature* **306** (1983), 164-167.
39. Whittaker, A.G.: The controversial carbon solid-liquid-vapour triple point, *Nature* **276** (1978), 695-696.
40. Whittaker, A.G.: Carbon: A new view of its high-temperature behaviour, *Science* **200** (1978) 763 -764.
41. Kudryavtsev, Yu. P., Evsyukov, S.E., Guseva, M.B., Babaev, V.G., and Khvostov, V.V.: Carbyne - the third allotropic form of carbon, *Russian Chemical Bulletin* **42** (1993), 399- 413.
42. Hosemann, R: Paracrystals in biopolymers and synthetic polymers, *Endeavour* **32**(117) (1973) 99-105.
43. Heimann, R.B.: Linear finite carbon chains (carbynes): their role during dynamic transformation of graphite to diamond, and their geometric and electronic structure, *Diamond and Related Materials* **3** (1994), 1151-1157.
44. Bak, P.: Solitons in incommensurable systems, in A.R. Bishop and T. Schneider (eds.), *Solitons and Condensed Matter Physics*, Springer Series in Solid-State Sciences **8**, Berlin, Springer Verlag, 1978, pp. 216-223.
45. Di Salvo, F.J., and Rice, T.N.: Charge-density waves in transition-metal compounds, *Phys. Today* **32**(4) April (1979), 32-38.

Chapter 5: Properties of carbyne and carbynoid structures

5.1. CHEMICAL PROPERTIES

S.E. EVSYUKOV
A.N. Nesmeyanov Institute of Organoelement Compounds, Russian Academy of Sciences, 117813 Moscow, Russia

Abstract –. The information accumulated to date on chemical properties of carbyne is as controversial as that on its structure. Dramatic distinctions in chemical properties of carbyne as well as variety of properties of the latters suggest that they have different structure. Notwithstanding the fact that linear carbon chain fragments are common structural elements of both carbyne and carbynoid materials, their chemical properties are determined by differences in their molecular structure (the length of linear *sp*-carbon fragments and, correspondingly, the presence and amount of defect links such as various side groups and cross-links) along with supramolecular structure (closest environment of linear fragments and crystallinity.

1. Introduction

The information available in scientific literature concerned with chemical properties of carbynes and carbynoid structures is rather scanty. Probably, the studies of structure and a search for reliable preparative methods were considered by researchers to be much more important in order to prove the mere fact of existence of carbyne as the third allotropic form of carbon. Furthermore, a remarkable chemical inertness of carbyne reported in early studies [1–3] might dampen any interest in this area. On the other hand, the inertness of carbyne as well as the fact that it survived during the high-temperature (1000°C) annealing is in drastic contrast with all what is generally known in organic chemistry about highly reactive short-chain polyynes and cumulenes. It should be noted, however, that carbyne as a form of carbon is a subject of rather materials science than that of pure organic chemistry. In other words, it should not be regarded as a set of non-interacting polyyne or cumulene chains. Therefore, the properties of crystalline carbynes and any amorphous carbynoid material can be completely different. Indeed, the carbynoid materials produced by reductive defluorination of poly(tetrafluoroethylene) (PTFE)[1] were reported to be highly reactive.

[1] For details *cf.* chapter 3.5 of this book.

R.B. Heimann et al. (eds.), Carbyne and Carbynoid Structures, 309–316.

2. Oxidation

Crystalline carbyne prepared by oxidative dehydropolycondensation of acetylene followed by high-temperature annealing is stable with respect to boiling in a 1:1 HNO_3/H_2SO_4 mixture. Based on this property, a method for purification of crystalline carbyne and for isolation of carbynes from other (graphite-like) carbon materials was worked out [1–3]. Prolonged boiling in concentrated nitric acid resulted in the complete oxidation of carbyne to carbon dioxide [1,2].

Ozonolysis of carbyne in aqueous suspension for 30 days also resulted in the complete oxidation, and oxalic acid was isolated afterwards from the solution [1,3] thus attesting to the presence of conjugated triple bonds in the dehydropolycondensation product (*Scheme 1*).

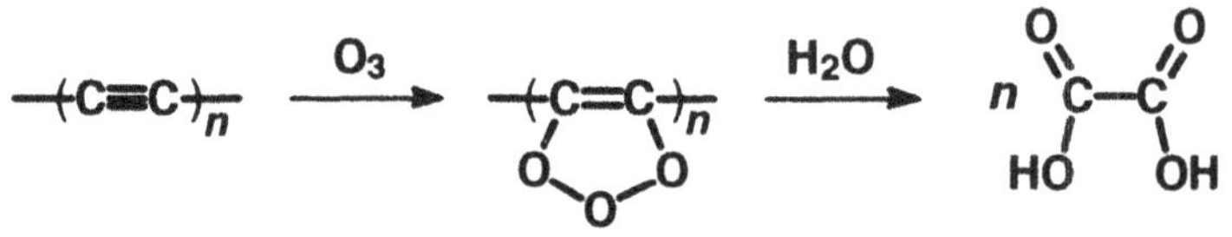

Scheme 1

On the other hand, the evolution of some amount of carbon dioxide was interpreted as an evidence of the presence of cumulenic structures that were oxidized according to *Scheme 2*.

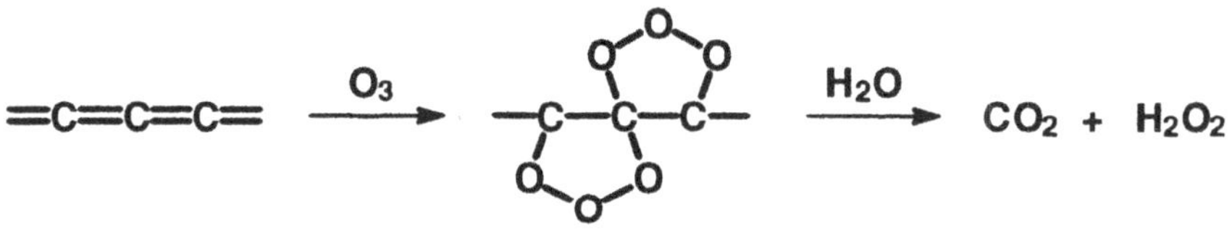

Scheme 2

The latter conclusion, however, seems to be rather ambiguous since oxidation of any organic substance under severe conditions may produce CO_2.

In contrast, carbynoid material[2] produced by an essentially equivalent method (oxidative Glaser coupling of dicopper acetylide without subsequent annealing) was recently reported to be readily oxidized during its reactions with chlorine [4] and dinitrogen tetroxide [5]. Partial destruction was also observed. Furthermore, electrochemical carbynoid materials produced by reductive defluorination of PTFE were reported to undergo rapid oxidation in air to form a layer of surface oxides containing mainly carboxylic groups [6,7].

3. Reduction

Attempts of hydrogenation of carbyne obtained by dehydropolycondensation of acetylene failed. No hydrogenation at 200°C and 20 MPa in alcoholic suspension was

[2] The author called his product carbyne [5], although it was mainly amorphous and contained *ca.* 2% of copper.

observed over Raney nickel or using homogeneous hydrogenation catalysts [1–3]. The interaction of carbyne with sodium borohydride and aluminum hydride at elevated temperatures was reported to bring about the formation of cross-linked structures [1].

The interaction of carbynoid material prepared by dehydrohalogenation of poly(vinylidene chloride) with alkali metals gives rise to chain-scission reactions (probably, at the oxygen-containing defect links [8]) followed by the formation of short-chain intercalation compounds of carbyne[3] [9–11].

Cathodic over-reduction (*n*-doping) of carbynoid structures was reported to occur during electrochemical defluorination of PTFE by treating with alkali metal amalgams [6]. An ionic interaction between metal cations and negatively charged carbon chains was proposed in that case.

4. Halogenation

Bromine does not react with carbyne prepared by dehydropolycondensation of acetylene. No addition reaction was observed even during prolonged boiling in liquid bromine [2]. Crystalline carbyne does not react with chlorine at room temperature. The interaction with chlorine leading to the degradation of carbyne under formation of undefined polychlorides was observed only at temperatures above 400°C [1–3]. By contrast, 'carbyne' prepared by the Glaser coupling of dicopper acetylide (see footnote 1) was reported to add one chlorine atom per *ca.* 20 carbon atoms[4] [4]. Moreover, electrochemical carbynoids were reported to add readily bromine and chlorine [6,12].

Kutsenok *et al.* [13] reported on fluorination of thermally pre-treated (500–600°C) carbyne by interaction with fluorine at 200–300°C and 0.3–0.5 MPa. Carbyne, which was not thermally pre-treated (or treated below 500°C), was reported to be unstable in fluorine (the formation of volatile burning products was observed). The material also burned out at pressures lower than 0.3 MPa. The resulting fluorination product was reported to be a chemically and thermally stable, insoluble white powder consisting of carbon and fluorine in a ratio of C:F=1:1.6 [13]. The fluorination conditions, however, are in contrast with observations by Vlasov *et al.* [14], who reported on fluorination of carbyne at normal pressure and 40–120°C. The fluorination was performed using fluorine diluted with an inert gas because the starting material was reported to burn in pure fluorine at ambient pressure and normal temperature. Moreover, carbyne was reported to burn even in diluted fluorine at the temperatures above 120°C. In a 80–120°C range the amount of fluorine in the reaction products was maximum. The ratio of carbon and fluorine in fluorinated carbyne was, respectively, 1:0.26 (40°C) to 1:1.16 (120°C) [14]. The product was reported to be soluble in organic solvents. The activation energy of the fluorination reaction was found to be $1.7 \cdot 10^{-4}$ J/g-at. [15]. The contradictions between the observations by Kutsenok *et al.* [13] and Vlasov *et al.* [14] suggest that the former research group used partially cross-linked carbyne whose

[3] For details *cf.* Chapter 4.3 of this book.

[4] In a subsequent paper [5], the author gave the Cl:C ratio to be 1:41 that is about twice as large as the value given in a previous study [4].

degree of cross-linking increased during its thermal pre-treatment. It should be noted that no structural evidence was given in both cases.

5. Some Other Reactions

Electrochemical carbynoids were reported to react with almost all chemical reagents [6,16]. Noteworthy, however, are some difficulties in distinguishing true chemical reactions from various types of adsorption or chemisorption. Halogenated and oxidized carbynoid materials can be further modified by reacting with various reagents like chlorosilanes or Grignard reagents [6].

The over-reduced (*n*-doped) electrochemical carbynoid materials based on PTFE and related polymers can be easily hydrolysed [6,16]. Normally, the hydrolysis reactions proceed exothermically and are accompanied by side reactions of carbon skeleton with water or OH^- ions and thermal decomposition. In contrast to graphite intercalation compounds, electrochemical carbon doped with lithium, CLi_x, does not produce hydrogen during its hydrolysis, but yields solid 'hydrocarbon', CH_x [6,17] (*Scheme 3*).

$$CLi_x + xH_2O \rightarrow CH_x + xLiOH, \quad x \approx 0.21$$

Scheme 3

The addition of water and alcohols to *sp*-hybridized carbon-carbon bonds in carbynoid materials is also possible [6,18].

The reaction of electrochemical carbynoid materials with sulphur gives rise to the formation of unidentified gaseous, liquid, and solid products [19,20]. The latter were supposed to contain fused disulphide heterocycles formed via addition of sulphur to polyyne sequences. Similarly, electrochemical carbynoids were reported to add ammonia and react with nitrogen monoxide [19]. Probably, the reaction with NO is accompanied by its dissociative chemisorption with preferential bonding of oxygen.

'Carbyne' produced by the Glaser coupling of dicopper acetylide was reported to react with dinitrogen tetroxide yielding a product containing one NO_2 group per each 15 carbon atoms as calculated from the weight increase [5].

The polyyne sequences in electrochemical carbynoid materials were recently reported to add *in situ* generated carbenes after removal of the LiF salt protecting carbon chains and preventing them from cross-linking (*Scheme 4*) [21]. The cyclopropene or cyclopropanediylidene groups formed are strained and hence reactive species, that are supposed to bring about further cross-linking.

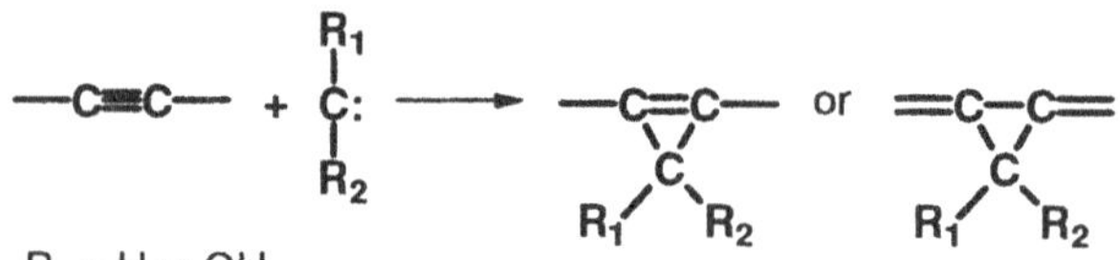

R_1 = H or CH_3

R_2 = H, CF_3, or phenyl

Scheme 4

The cross-linking behaviour of carbynes and carbynoid materials is a special case. Electrochemical carbynoids are known to undergo slow cross-linking during prolonged storage even at room temperature, yielding graphene-like material [22]. Once the fluoride salt protecting carbon chains was removed from a C-MF composite material (M = alkali metal), the cross-linking became very fast. The carbynoid material prepared by dehydrohalogenation of poly(vinylidene chloride)[5] was recently reported to transform into diamond due to extensive cross-linking in the presence of ultra-dispersed diamond [23]. In sharp contrast to this, carbyne prepared by oxidative dehydropolycondensation of acetylene was claimed to survive high-temperature annealing at 1000°C [1–3], and carbyne occurring as a natural mineral (chaoite) [24,25] has survived a meteoritic impact for at least 60 million years.

In 1973 Kasatochkin *et al.* [26,27] reported that carbyne synthesized by oxidative dehydropolycondensation of acetylene did not transform into diamond under conditions sufficient to achieve diamond synthesis from graphite (9 GPa, 1800°C, 5 min). On the other hand, in 1989 Sobolev *et al.* [28] reported on the transformation of carbyne (*viz.*, chaoite obtained in dynamic loading experiments) into lonsdaleite and cubic diamond at 1330 °C and static pressure of 5.5 GPa within a short period of time (≤1 min). The formation of graphite was observed at higher temperatures (≥1530 °C). Noteworthy is the disappearance of diffraction lines corresponding to the α-carbyne phase from the x-ray diffraction pattern and appearance of lines corresponding to β-carbyne reported to occur during the high pressure/high temperature treatment in both cases [26-28]. In this regard it is interesting to speculate whether the β-phase is really carbyne or merely is a partially (but regularly) cross-linked carbynoid structure (similar to 'layer-chain carbon' reported by Melnitchenko *et al.* [29]) being in effect a missing intermediate in the transformation sequence of carbyne to diamond. Recently, Cataldo reported on a surprising transformation of 'β-carbyne' produced by the Glaser coupling of dicopper acetylide (see footnote 2) into 'α-carbyne'[6] [4,5].

6. Catalytic Properties

Paushkin *et al.* studied catalytic properties of some conjugated polymers, including carbyne, in dehydration reactions of *iso*-propanol [30], diacetone alcohol (2-hydroxy-2-methylpentan-4-one) [31], and dimethylvinylcarbinol [32]. Carbyne prepared by dehydropolycondensation of acetylene was reported to possess higher catalytic activity compared to that of activated carbon [30,32]. The dehydrogenation reaction in the case of *iso*-propanol is preferred compared to dehydration (*Scheme 5*) [30].

On the other hand, the dehydration reaction of diacetone alcohol [31] and dimethylvinylcarbinol [32] strongly predominated over their decomposition (*Scheme 6*). Specific catalytic activities of carbyne and graphite, *i.e.*, yields of the dehydration

---

[5] For details *cf.* Chapter 3.2.1 of this book.

[6] The author has drawn his conclusion solely on the basis of changes in IR spectra [4,5]. However, the frequency of 2223 cm^{-1} is too high for polyynes, and the band at 1620 cm^{-1} is not characteristic of only cumulenes, inasmuch as polyenes can also contribute to this band. Moreover, taking into account the existence of a variety of the carbyne crystalline polytypes, one should note that the use of crystallographically derived terms, *viz.*, α- and β-carbynes, for describing the type of chemical bonds in carbynes may turn out to be incorrect.

product (*iso*-propylideneacetone [31]) calculated per 1 m^2 of the catalyst's surface, were found to be of the same order of magnitude. However, the absolute activity of carbyne was much higher because of its higher porosity and, correspondingly, specific surface area.

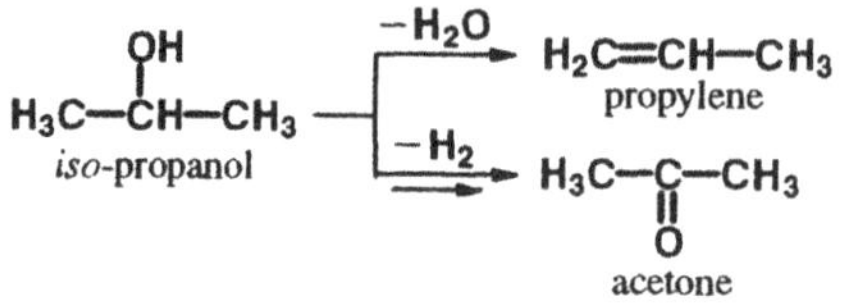

Scheme 5

The authors did not propose any mechanism to explain the catalytic action of carbyne in the above-mentioned reactions. However, they reported a clear correlation between catalytic activity and concentration of unpaired electrons (paramagnetic centres) measured by the ESR method [30,31]. All conjugated polymers were claimed to possess catalytic activity in dehydration and dehydrogenation reactions of alcohols, provided that the concentration of paramagnetic particles on their surfaces exceeds 10^{17} spin/g.

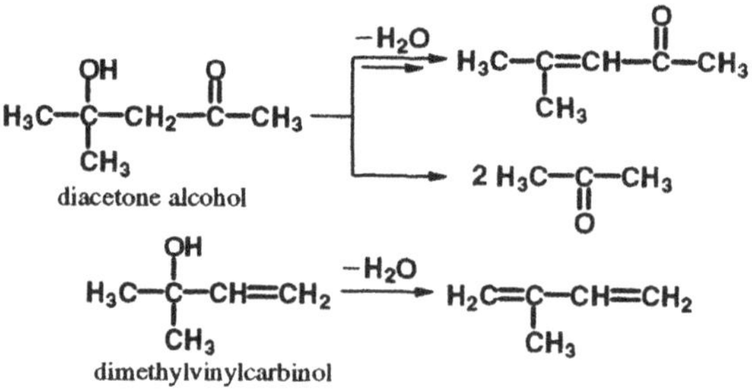

Scheme 6

Annealing of carbyne at various temperatures was reported to affect significantly the concentration of paramagnetic centres along with its catalytic activity [31]. The concentration slightly increases with increasing annealing temperature up to 300°C, and then drops drastically during annealing at 400 and 500°C. Similarly, the catalytic activity increased with increasing annealing temperature up to 300°C, then decreased dramatically at 400°C and remained almost constant beyond 600°C, approaching the catalytic activity of activated carbon, graphite, and diamond [31].

7. Conclusions

Despite the paucity of data on chemical properties of carbynes and carbynoid materials available in the literature, one can draw some general conclusions based on the brief survey given above. First of all it should be noted that the data accumulated so far are even more contradictory than those concerned with crystal and molecular structure of carbyne. Dramatic differences in chemical properties of carbyne and carbynoid materials as well as the variety of properties of the latters suggest that they have different structure. It is known that crystalline polymers are less reactive than amorphous ones [33]. However, this is a very weak explanation for the exceptional stability and chemical inertness of crystalline carbyne reported in early studies [1-3]. Notwithstanding the fact that linear carbon chain fragments are common structural elements of both carbyne and carbynoid materials, their chemical properties are determined by differences in their molecular structure (the length of linear *sp*-carbon fragments and, correspondingly, the presence and amount of defect links such as various side groups and cross-links) along with their supramolecular structure (closest environment of linear fragments) and crystallinity.

Since some authors call sometimes their products 'carbyne' just speculatively on the basis of a conjectured (or idealized) chemical structure, one should treat such studies with caution. The most reliable interpretation of chemical properties is possible when molecular and crystal structures are completely elucidated. On the other hand, the studies of chemical properties can furnish additional insights into the nature of the matter under discussion. Therefore, one can only emphasize that comprehensive studies of chemical properties must be performed each time when dealing with carbyne or carbynoid structures[7].

Acknowledgement – I gratefully acknowledge the generous support by the Alexander-von-Humboldt Foundation under the auspices of its follow-up program.

8. References

1. Kasatochkin, V.I., Sladkov, A.M., Kudryavtsev, Yu.P., Popov, N.M., and Korshak, V.V.: Crystalline forms of a linear modification of carbon, *Dokl. Akad. Nauk SSSR* **177** (1967), 358–360 (in Russian).
2. Sladkov, A.M., Kasatochkin, V.I., Kudryavtsev, Yu.P., and Korshak, V.V.: Synthesis and properties of chain-like polymers of carbon, *Izv. Akad. Nauk SSSR, Ser. Khim.* (12) (1968), 2697–2704 (in Russian).
3. Korshak, V.V., Kudryavtsev, Yu.P., and Sladkov, A.M.: Carbyne - a new allotropic form of carbon, *Vestnik Akad. Nauk SSSR* (1) (1978), 70–78 (in Russian).
4. Cataldo, F.: Synthesis of carbonaceous material rich in carbyne, a carbon allotrope by coupling reaction of copper acetylide, *Eur. J. Solid State Inorg. Chem.* **34** (1997), 53–63.
5. Cataldo, F.: A study on the structure and electrical properties of the fourth carbon allotrope: carbyne, *Polym. Int.* **44** (1997), 191–200.
6. Kavan, L.: Electrochemical carbonization of fluoropolymers, in: *Chemistry and Physics of Carbon*, Vol. 23, Thrower, P.A., Ed., Marcel Dekker, New York, 1991, pp. 69–171, and references therein.
7. Kavan, L., Dousek, F.P., and Kubelková, L.: Chemisorption of oxygen on carbon produced by electrochemical reduction of poly(tetrafluoroethylene), *Carbon* **24** (1986), 671–676.
8. Evsyukov, S.E., Paasch, S., Thomas, B., and Heimann, R.B.: Formation of carbynoid structures by chemical dehydrohalogenation of poly(vinylidene chloride). A ^{13}C solid-state NMR study, *Ber. Bunsen-Ges. Phys. Chem.* **101** (1997), 837–841.
9. Udod, I.A., Bulychev, B.M., Sirotinkin, S.P., Babaev, V.G., and Kudryavtsev, Yu.P.: An interaction of amorphous carbyne with sodium: the formation of carbyne intercalation compounds, *Synth. Met.* **60** (1993), 57–62.
10. Udod, I.A., Bulychev, B.M., Babaev, V.G., and Kudryavtsev, Yu.P.: The formation of carbyne intercalation compounds with alkali metals, *Zh. Neorg. Khim.* **41** (1996), 382–387 (in Russian).
11. Udod, I.A., Bulychev, B.M., and Babaev, V.G.: Structural models of carbyne (a linear modification of carbon) and its intercalation compounds, *Zh. Neorg. Khim.* **42** (1997), 521–528 (in Russian).

---

[7] Of course, provided that the quantity of the material is sufficient to do such research. Sometimes the presence of carbyne is revealed only in microscopic traces, and, evidently, no chemical studies are possible in this case.

12. Dousek, F.P. and Kavan, L.: Fixation of bromine to carbon in products of electrochemical reduction of poly(tetrafluoroethylene), *Carbon* **23** (1985), 405–409.
13. Kutsenok, Yu.B., Serushkin, I.L., Seryshev, G.A., and Kirillov, A.K.: Carbon fluoride, *Pat. USSR* 577175 (1977); *Chem. Abstr.* **88** (1978), 74020v.
14. Vlasov, S.V., Kuznetsov, A.N., and Semenova, L.G.: Synthesis of non-stoichiometric fluorinated carbyne, *Izv. Akad. Nauk SSSR, Neorg. Mater.* **16** (1980), 1299–130 (in Russian).
15. Vlasov, S.V.: Study of the kinetics of the fluorination of carbyne, *Izv. Akad. Nauk SSSR, Neorg. Mater.* **17** (1981), 1789–1791 (in Russian).
16. Kavan, L.: Electrochemical carbon, *Chem. Rev.* **97** (1997), 3061–3082.
17. Jansta, J. and Dousek, F.P.: Electrochemical reduction of polymeric carbon, *Electrochim. Acta* **26** (1981), 233–237.
18. Bohlmann, F. and Viehe, H.-G.: Polyacetylenverbindungen, IX. Mitteil.: Alkalische und saure Addition an Polyine, *Chem. Ber.* **88** (1955), 1017–1027.
19. Kavan, L. and Dousek, F.P.: Study of chemical reactivity of carbon prepared by electrochemical reduction of poly(tetrafluoroethylene) against NH_3, NO and S in the gas phase, *Carbon* **24** (1986), 61–65.
20. Kavan, L., Novák, P., and Dousek, F.P.: Electrochemistry of sulphur adsorbed on carbon, *Electrochim. Acta* **33** (1988), 1605–1612.
21. Hlavatý, J. and Kavan, L.: Modification of electrochemical carbon by *in situ* generated carbenes, *Carbon* **35** (1997), 127–131.
22. Kavan, L., Hlavatý, J., Kastner, J., and Kuzmany, H.: Electrochemical carbyne from perfluorinated hydrocarbons: Synthesis and stability studied by Raman scattering, *Carbon* **33** (1995), 1321–1329.
23. Kudryavtsev, Yu.P. and Evsyukov, S.E.: The formation of diamond from carbynoid material at ambient pressure, *Diamond Relat. Mater.* **6** (1997), 1743–1746.
24. El Goresy, A. and Donnay, G.: A new allotropic form of carbon from the Ries crater, *Science* **161** (1968), 363–364.
25. El Goresy, A.: Eine neue Kohlenstoff-Modifikation aus dem Nördlinger Ries, *Naturwiss.* **56** (1969), 493–494.
26. Kasatochkin, V.I., Sterenberg, L.E., Kazakov, M.E., Slesarev, V.N., and Belousova, L.V.: Thermal transformation of carbyne under pressure, *Dokl. Akad. Nauk SSSR* **209** (1973), 388–391 (in Russian).
27. Kasatochkin, V.I., Korshak, V.V., Kudryavtsev, Yu.P., Sladkov, A.M., and Sterenberg, L.E.: On crystalline structure of carbyne, *Carbon* **11** (1973), 70–72.
28. Sobolev, V.V., Slobodskoi, V.Ya., and Kozlov, I.T.: Phase transitions of chaoite under high pressures and temperatures, *Mineralog. Sbornik (L'vov)* **43** (1989), 66–68 (in Russian).
29. Melnitchenko, V.M., Nikulin, Yu.N., and Sladkov, A.M.: Layer-chain carbons, *Carbon* **23** (1985), 3–7.
30. Paushkin, Ya.M., Burova, L.M., Sladkov, A.M., Nepochatykh, V.P., Kudryavtsev, Yu.P., and Korshak, V.V.: Catalytic properties of polymers with ethynyl and cumulated groups in the chain, *Dokl. Akad. Nauk SSSR* **180** (1968), 622–625 (in Russian).
31. Paushkin, Ya.M., Panidi, I.S., Ryabtseva, N.V., Sladkov, A.M., and Kudryavtsev, Yu.P.: Catalytic properties of carbyne in the dehydration reaction of diacetone alcohol, *Dokl. Akad. Nauk SSSR* **195** (1970), 101–103 (in Russian).
32. Paushkin, Ya.M., Lunin, A.F., and Losev, Yu.P.: Concerning catalysis on high-temperature superconductors and polymeric semiconductors, *Dokl. Akad. Nauk SSSR* **307** (1989), 1415–1419 (in Russian).
33. Evsyukov, S.E., Kudryavtsev, Yu.P., and Korshak, Yu.V.: Chemical dehydrohalogenation of halogen-containing polymers, *Usp. Khim.*, **60** (1991), 764–798 (in Russian) [(*Russ. Chem. Rev.*, **60** (1991), 373–390 [Engl.Transl.]).

5.2. THERMOPHYSICAL PROPERTIES

B.V. LEBEDEV
Scientific Research Institute of Chemistry,
Nizhnii Novgorod State University,
603600 Nizhnii Novgorod, Russia

Abstract – A review is presented of all thermophysical properties of carbyne and carbynoid structures published up to the present. The data on heat capacity of carbyne in the range of (80–300) K were found to possess a high degree of probability. As for other chain polymers, for carbyne $C°_p$ is proportional to T^1. The numerical values of its heat capacity at the same temperature are several times those of graphite and diamond, consistent with a change of rigidity of the oscillating atomic skeleton in the order carbyne < graphite < diamond. From the $C°_p$ values, the functions H°(T)–H°(0), S°(T)–S°(0) and G°(T)–H°(0) for carbyne are calculated between 0 and 300 K. The heat capacities of allotropic carbon modifications in the range of (0–400) K and standard entropies of their formation $\Delta_f S°$ are compared. According to the values of $\Delta_f S°$ obtained from the temperature dependences of the heat capacities in the range of 0 to (300–400) K, the allotropic carbon modifications can be arranged in the order carbyne > C_{60} > C_{70} > graphite > diamond.

1. Introduction

The synthesis of α- and β-carbyne single crystals[1] [1,2] in amounts sufficient to conduct calorimetric measurements is difficult because of the complicated methods of their preparation, low yields and difficulties connected with their separation [2,3]. Carbyne formation via an oxidative dehydropolycondensation reaction of acetylene [3–5] and dehydrochlorination of poly(vinylidene chloride) [3,5,6] with a subsequent prolonged heating under vacuum at 1300 K give products containing up to 99.9 % of carbon. The resulting three-phase system consists of small amounts of α- and β-carbyne crystals embedded in amorphous (glassy) carbon [4,7]. An increase of a crystalline part in the system can be achieved by applying high pressure at higher temperature [1,2]. Besides, some cross-links are formed between carbon chains [7,8]. In some cases, the number of these cross-links is small, and from such carbyne samples reliable values of some thermophysical characteristics approaching those of a perfect

[1] A chain carbon macromolecules with a polyyne structure ~–C≡C–C≡C–~ was called α-carbyne and with a cumulene one ~=C=C=C=C=~ β-carbyne.

R.B. Heimann et al. (eds.), Carbyne and Carbynoid Structures, 317–331.

carbyne can be obtained [9]. For example, amorphous or partially crystalline samples can be used to study the temperature dependence of the heat capacity of carbyne, since it is known that the heat capacity of inorganic and organic polymers with a chain structure does not depend on their crystallinity [10]. The differences in their C^o_p values do not exceed the experimental errors of measurement (about 0.2% in the range (40–350) K). This seems to be valid for carbyne too.

It is necessary, however, to ensure that the number of intermolecular cross-links in carbyne is small. This can be judged rather reliably by the data on the character of the temperature dependence of the heat capacity. A specific feature of almost all organic and inorganic linear polymers investigated hitherto, independent of the composition and structure of repeating monomer units in their macromolecules, is a wide temperature interval where the heat capacity C^o_p is a linear function of the temperature. This dependence usually manifests itself in the range from (50–60) K up to the glass transition temperature for glassy and partly crystalline polymers, and up to the melting point for crystalline polymers [10,11]. The appearance of the cross-links leads to a deviation of the heat capacity of chain polymers from the relation C^o_p vs. T^1 towards an increase of a temperature exponent. In the case of complete cross-linking and transition from a chain structure of a polymer to a spatial one, C^o_p becomes proportional to T^3. Tarasov has developed an approximative quantum theory of the heat capacity of chain and layered structures [12] and mixed layer-chain structures [13]. From this it follows that for the chain structures the limiting law of heat capacity is expressed as C^o_p vs. T^1, for the layered and layer-chain structures as C^o_p vs. T^2 and C^o_p vs. T^n, respectively, where $1<n<2$. For the spatial structures the T^3-Debye law of heat capacity holds true [14]. It is obvious that for the carbyne samples whose temperature dependence is described by the relation C^o_p vs. T^1, the number of cross-links between macromolecules of the chain carbon is small and their thermodynamic characteristics are likely to be close to those of pure carbyne as postulated in Ref. [9]. In Ref. [2] it is suggested that α- and β-forms of carbyne are energetically equivalent.

Below are considered thermodynamic properties of carbyne and carbynoid structures, recent results of calorimetric measurements of heat capacities of polymers with segments of carbynoid structure, and values of thermodynamic parameters of the main reactions of carbyne synthesis. Besides, the thermodynamic properties of carbyne will be compared to those of other allotropic carbon modifications (fullerenes C_{60} and C_{70}, graphite and diamond).

2. Heat Capacity and Thermodynamic Functions

Rabinovich *et al.* [15] measured the temperature dependence of the heat capacity C^o_p of nine differently prepared carbyne samples[2] [16] between 80 K and 300 K using an

---

[2] Despite identical (or nearly identical) methods of sample preparation their thermodynamic properties differ considerably.

adiabatic vacuum calorimeter whose design, operation procedure as well as calibration and test results were described previously [17]. The calorimeter yielded the C^{o}_{p} of solid and liquid substances with an uncertainty of 0.2–0.3 %. The masses of the carbyne samples placed in a calorimetric ampoule were 5–10 g. In most cases, their heat capacities were 25–40% of the total heat capacity of the calorimetric ampoule and the substance. Using a diamond sample for comparison, the heat capacity was 1.4% at 20 K and more than 20% of the total heat capacity at a temperature above 200 K. Prior to the onset of measurements, the carbyne samples were heated *in vacuo* directly in the calorimetric ampoule, then helium was introduced into the ampoule as a heat-exchange gas, and the ampoule was sealed. Repeat measurements of C^{o}_{p} were performed approximately every 7–8 degrees. 30–35 experimental points of heat capacity were obtained for each sample and fitted graphically. The scatter of C^{o}_{p} points around the corresponding fitted lines $C^{o}_{p} = f(T)$ was within 0.5% and as the authors estimated, a measurement error of the heat capacity for carbyne was the same. All experimental points of C^{o}_{p} and the averaging curves C^{o}_{p} vs. T for four carbyne samples are shown in Fig. 1. To compare with carbyne, the heat capacities of acetylene black (the commercial product Saratovskaya™) and spectral-pure C-3 graphite were measured in the same temperature range. Between 80 K and 130 K, the heat capacity of the acetylene black is 7% larger than that of C-3 graphite; with rising temperature the difference in their heat capacities decreases to 3% at 300 K. This difference can be easily explained by the fact that the black, being a variety of graphite [18], is much more polydisperse.

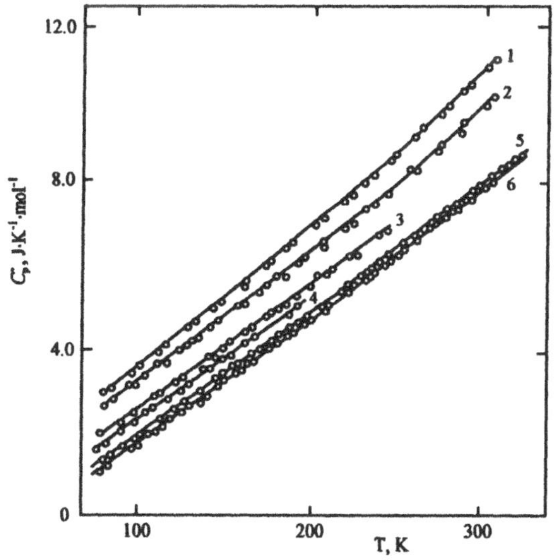

Figure 1. Temperature dependence of heat capacity: 1 – carbyne sample K-7, 2 – K-6, 3 – K-5, 4 – K-4, 5 – sample of "Saratovskaya" acetylene black, 6 – sample of C-3 graphite (the data from Ref. [15]).

It was found that for all carbyne samples studied the heat capacities were substantially greater than those of graphite. However, the heat capacities among the carbyne samples differ also. At 80 K, C^o_p of sample K-7 is 2.5 times that of C-3 graphite, at 300 K it is higher by 40%. At 80 K, the C^o_p value of the carbyne sample K-3 with the smallest heat capacity of all carbyne samples studied is higher by 45% than that of graphite and at 200 K by 10%. The heat capacities of the remaining four samples, not presented in Figure 1, are close to that of K-3.

The reason for the differences in the heat capacities of carbyne may be seen in the degree of cross-linking between the chain-like macromolecules. It is obvious that an increase in the cross-linking density leads to an enhanced rigidity of the oscillating skeleton that in turn results in a decrease of the heat capacity in the order carbyne > graphite > diamond.

As noted above, concurrent with the reliable evidence for a chain structures the relation C^o_p vs. T is linear. Indeed, for the carbyne samples K-7 and K-6 with the largest heat capacity the relation C^o_p vs. T^1 holds for a temperature interval between 80 and 240 K. In particular, for the sample K-7 the heat capacity can be described by the equation

$$C^o_p = -0.0402 + 3.674 \cdot 10^{-2}\,T, \qquad (1)$$

where C^o_p is expressed in $J \cdot K^{-1} \cdot mol^{-1}$ with an error of ±0.9%. The mean-square deviation of the C^o_p points from the corresponding values of the heat capacity calculated by eq. (1) is ±0.2%.

For the other carbyne samples, when plots of log $C^o_p = f(\log T)$ are divided into stepwise linear sections, the exponent of T varies from 1.5 to 1.2. Likewise, in the case of C-3 graphite the exponent n in the expression $C^o_p = A \cdot T^n$ changes from 1.8 to 1.3 over the temperature range (80–240) K.

Data of the heat capacity of K-7 carbyne, graphite and diamond determined by the calorimetric method in the interval (50–300) K were shown by Usenbaev [19]. He drew attention to the fact that for all compounds under study the experimental values of C^o_p coincide with those cited in Ref. [20] with an error not exceeding 0.3%. However, there are no data for carbyne in Ref. [20] and Usenbaev's reference to carbyne K-7 relates, perhaps, to the data by Rabinovich *et al.* [15] also quoted by him in the beginning of the paper.

Thus, the experimental data on the heat capacity of carbyne K-7, showing the highest C^o_p value of all carbynes investigated, and the strict linear dependence of the heat capacity on temperature, allow to conclude that only a few cross-links exist. Hence, a satisfactory agreement of the heat capacity data with that of an expected ideal carbyne in an amorphous (glassy) state can be stated. It is well known that the C^o_p values of chain polymers in the glassy, crystalline or partially crystalline states coincide within experimental measurement errors [10].

Tarasov [12] and Wunderlich [20] showed that at low temperatures ($T<0.1\theta_1$, where θ_1 is a characteristic temperature introduced by Tarasov) the heat capacity of chain polymers is the sum of contributions of the skeleton macromolecule oscillations and is well reproduced by Tarasov's two-parameter function of heat capacity for chain structures [12]

$$Cp_{1,3} = D_1\left(\frac{\theta_1}{T}\right) - \frac{\theta_3}{\theta_1}\left[D_1\left(\frac{\theta_3}{T}\right) - D_3\left(\frac{\theta_1}{T}\right)\right], \qquad (2)$$

where D_1 and D_2, respectively are symbols of Tarasov and Debye functions of the heat capacity [14], and θ_1 and θ_3 are the corresponding characteristic temperatures. From the experimental C^o_p values of carbyne K-7 in the range (80–100) K, the values θ_1=512 K and θ_3=102 K were obtained using Chernoplekov's tables of heat capacity of chain structures [21].With these values of θ_1 and θ_3, eq. (2) reproduces the experimental C^o_p values in the range (80–100) K within an error of 0.2–2%. By assuming that at T<80 K it reproduces C^o_p of carbyne with the same accuracy, the numerical magnitudes of C^o_p for carbyne between 0 K and 80 K were obtained. By taking into account these values, the heat capacity of carbyne is illustrated over the range (0–300) K in Figure 2. From 40 K to 240 K C^o_p is a linear function of the temperature and follows eq. (1). In the interval (240–300) K, the heat capacity of carbyne is a nonlinear function of the temperature

$$C^o_p = 9.667 - 4.331\cdot10^{-2}\,T + 1.659\cdot10^{-4}\cdot T^2. \qquad (3)$$

The mean-square deviation of the experimental C^o_p values from those estimated by eq. (3) does nowhere exceed 0.2%. The appearance in eq. (3) of the component of C^o_p, depending on T^2, is probably caused by the contribution to the heat capacity of atomic oscillations of –C≡C–, =C=C= and ≡C–C≡ bonds, which at T<240 K do not seem to be excited yet and at T>240 K are partly excited. This gives an increasing contribution to C^o_p with rising temperature. At T<30 K the heat capacity is well described by the polynomial function

$$C^o_p = 2.587\cdot10^{-5} + 0.0029\cdot T - 4.754\cdot10^{-5}\cdot T^2 + 1.063\cdot10^{-5}\cdot T^3 - \\ - 1.952\cdot10^{-7}\cdot T^4 + 1.209\cdot10^{-9}\cdot T^5 - 2.31\cdot10^{-12}\cdot T^6. \qquad (4)$$

In this temperature range, the deviation from the linearity is likely caused by enhanced interchain interaction with decreasing temperature. According to Tarasov [12], at $T \ll \theta_1$ the contribution of chain structures to the heat capacity becomes dominant, resulting from the oscillations of macromolecules in a force field of the interchain interaction, and at the lowest temperatures C^o_p is proportional to T^3. At intermediate temperatures the interchain interaction is characterized by the ratio θ_3/θ_1. In accordance with the theory of heat capacity of chain structures [12], the ratio θ_3/θ_1 changes from 0 to 1. θ_3/θ_1=0 describes the hypothetical case of the absence of any interchain interaction and θ_3/θ_1=1 is realized for isotropic solids. For carbyne, θ_3/θ_1=0.2 indicates a pronounced interchain interaction and agrees with its estimate obtained by other methods [4].

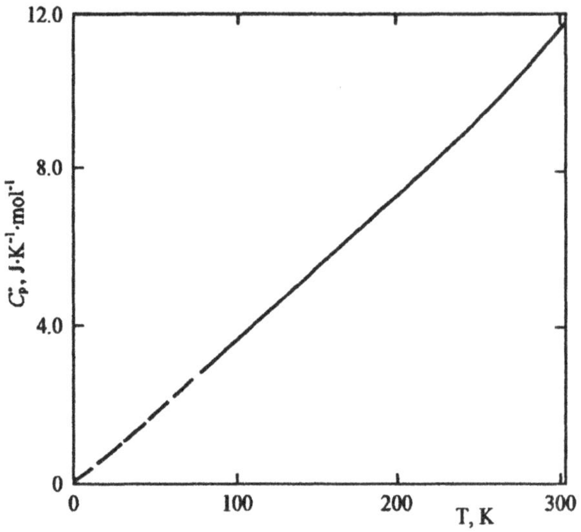

Figure 2. Heat capacity of carbyne (K-7 sample) [15].

The calculation of the thermodynamic functions of carbyne (Table 1) was made by using the well known equations

$$[H^\circ(T) - H^\circ(0)] = \int_0^T C_p^\circ(T)dT, \tag{5}$$

$$S^\circ(T) - S^\circ(0) = \int_0^T C_p^\circ(T)d\ln T, \tag{6}$$

$$G^\circ(T) - H^\circ(0) = [H^\circ(T) - H^\circ(0)] - T\cdot[S^\circ(T) - S^\circ(0)], \tag{7}$$

where $C^\circ_p(T)$ is temperature-dependent heat capacity that in the range (0–40) K as described by eq. (4), between 40 K and 240 K by eq. (1) and from 240 K to 300 K by eq. (3); $H^\circ(T) - H^\circ(0)$ is the enthalpy, $S^\circ(T) - S^\circ(0)$ denotes entropy, $S^\circ(0)$ is the residual (zero) entropy; $G^\circ(T) - H^\circ(0)$ is the Gibbs function.

3. Thermochemical Characteristics

Energies of combustion of three samples of carbyne were measured calorimetrically by Kasatochkin *et al.* [7]. The carbyne samples were synthesized in the same manner as for the study of the temperature dependence of heat capacity (see above) [16]. According to x-ray analysis data, they consisted mainly of an amorphous mixture of macromolecules of polyyne and cumulene structures with a certain admixture of crystals formed by polyyne and cumulene chains. It was also demonstrated that the carbyne samples had different ratios of polyyne and cumulene chains. Kasatochkin *et al.* [7] did not rule out that a small portion of carbon atoms in the macromolecules formed intermolecular cross-links transforming a portion of the carbyne into a substance of coal type with an aromatic structure.

TABLE 1. Thermodynamic functions of carbyne per mole of carbon atoms, M.m. = 12.011 g·mol^{-1}; p=101.325 kPa

T, K	C^o_p, J·K^{-1}·mol^{-1}	H°(T) – H°(0), kJ·mol^{-1}	S°(T) – S°(0)], J·K^{-1}·mol^{-1}	– [G°(T) – H°(0)], kJ·mol^{-1}
5	0.145(1)	0.00036	0.130(5)	0.00029
10	0.294(2)	0.00146	0.276(4)	0.00131
15	0.451(8)	0.00332	0.425(1)	0.00306
20	0.619(9)	0.00600	0.578(0)	0.00556
25	0.799(3)	0.00953	0.735(1)	0.00885
30	1.00(7)	0.0140(5)	0.899(2)	0.0129(3)
40	1.40(6)	0.0259(9)	1.24(0)	0.0236(1)
50	1.82(7)	0.0423(3)	1.60(3)	0.0378(2)
100	3.623	0.1796	3.436	0.1641
150	5.439	0.4063	5.250	0.3812
200	7.322	0.7240	7.065	0.6890
250	9.205	1.137	8.902	1.088
298	11.46	1.631	10.70	1.558

To measure energies of combustion, a KU-2 calorimetric apparatus with a self-packing calorimetric bomb was used [7,22]. The calorimeter was calibrated by burning of reference benzoic acid (the certificate value of combustion enthalpy was $\Delta_{comb}H^o(cr)$=–26443 J·g^{-1}). Complete combustion of the substance was controlled by the analysis of gaseous products. Unfortunately, Kasatochkin *et al.* [7] nowhere reported data on the calibration of the calorimetric apparatus they used and the results of actual experiments on the burning of carbyne as well as the analysis of the gaseous combustion products. The absence of these data does not allow to draw well-founded conclusions on the complete combustion of carbyne and the reliability of the results.

For the three carbyne samples denoted in [4] as 'a', 'b' and 'c', $\Delta_{comb}U$ was found to be –353.7, –356.5 and –371.5 kJ·mol^{-1}, respectively. A portion of one of the carbyne samples was heated before burning to about 3000 K. As was shown by XRD, carbyne converted into nongraphitizing carbon [7]. Its energy of combustion proved to be equal to –394.3 kJ·mol^{-1}, i.e. close to the value of graphite measured with the same apparatus

(-393.5 kJ·mol^{-1}). As the experimental data showed, an increase of the coal fraction in the composition of the carbyne sample leads to an increase of the absolute magnitude of $\Delta_{comb}U$. Judging by the experimental data, it is possible to conclude that sample 'a' had the least number of cross–links between the macromolecules. Using its $\Delta_{comb}U$ yields the enthalpy of formation of carbyne per mole of carbon atoms, $\Delta_f H^\circ = -37.03$ kJ·mol^{-1}. In turn, the calculation of the Gibbs function of carbyne from the above value of $\Delta_f H^\circ$ by the equation

$$\Delta_f G^\circ = \Delta_f H^\circ - 298.15 \cdot \Delta_f S^\circ \tag{8}$$

leads to the value of $\Delta_f G^\circ = -38.7$ kJ·mol^{-1}. This implies that carbyne is thermodynamically much more stable than graphite, a contention that on the whole contradicts the real state of affairs since the most ubiquitous allotrope of carbon in nature is still graphite. It follows that the value of $\Delta_{comb}U$ of carbyne obtained by Kasatochkin *et al.* [7] is highly erroneous. In eq. (8) $\Delta_f S^\circ$ is the entropy of formation of carbyne. It was estimated from the absolute magnitudes of the entropy S°(298.15) of carbyne and graphite [23]. To calculate S°(298.15) of the amorphous carbyne the values of S°(298.15) – S°(0) (Table 1) and the zero entropy S°(0) in the amorphous state were used. Zero entropies of linear polymers in the glassy state were shown to be about 10% of their entropies at 298.15 K [24]. For carbyne this means that its residual entropy is S°(0) ≈1 J·K^{-1}·mol^{-1}. With regard for this, $\Delta_f S^\circ$ was found to be approximately 5.6 J·K^{-1}·mol^{-1}.

In conclusion it should be noted that the enthalpy of formation of carbyne in the gaseous state at 298.15 K and standard pressure was estimated theoretically [25,26] by methods similar to those used by Benson *et al.* [27]. It was found that $\Delta_f H^\circ = 113.4$ kJ·mol^{-1} [25] and 106.3 and 110.6 kJ·mol^{-1} for carbyne of the polyyne and cumulene structure, respectively [26]. Since data on the entropy of formation of carbyne in the identical physical state and under similar physical conditions are unavailable, at this time it is impossible to evaluate the Gibbs function of formation $\Delta_f G^\circ$ of carbyne.

4. Comparison of Thermodynamic Properties of Carbyne with those of other Allotropic Carbon Modifications

According to Sladkov [3], the three valence states of a carbon atom determine three types of a spatial structure: sp^3-state – a three-dimensional structure, sp^2 – a planar structure and sp – a linear one. Diamond, graphite and carbyne correspond to the above structures. Fullerenes [28], however, can be considered a new allotropic modification although the valence state of carbon atoms in them is close to that of graphite. Moreover, the combination of the physicochemical characteristics of fullerenes is so distinct and specific just like those of diamond, graphite and carbyne that with regard to this and the chronology of discoveries of carbon allotropes, fullerenes should be, without doubt, classified as the fourth allotropic modification of carbon. Furthermore, before proceeding to a comparison of the thermodynamic properties of the allotropic modifications, let us briefly dwell upon a description of the calorimetric data, published hitherto, of diamond, graphite, and the fullerenes C_{60} and C_{70}. The proper data of carbyne were discussed above. All results cited below were obtained from data accrued from adiabatic vacuum calorimetry that presently is the most precise method of measurements of heat capacities, temperatures and enthalpies of physical transitions of substances [29].

Wunderlich and Baur [20] generalized the results of calorimetric studies of temperature-dependent heat capacities of a series of samples of natural diamonds and collected a table of numerical values of averaged heat capacities. In their opinion, the most probable accuracy of the listed values of C°_{p} is within ±20% at 13 K, ±6% at 20 K, ±8% at 100 K, ±0.2% in the range (200–280) K and ±0.5% from 280 K to 300 K. DeSorbo [30] made measurements of C°_{p} for natural diamond in the interval (17–300) K. He estimated the measurement error to be within ±30% at $T<70$ K and about ±5% for $T>70$ K. Lebedev *et al.* [31] studied the temperature dependence of heat capacity of a commercially synthesized diamond sample over the range of (9–340) K. According to the certificate data, the sample contained about 1% of impurities. The mean-square deviation of the experimental C°_{p} values from the fitting curve $C^{\circ}_{p} = f(T)$ was found to be approximately ±1% between 9 K and 40 K, ±0.8% in the range (80–160) K and ±0.5% from 160 K to 340 K. The absolute error of measurements appears to be the same as for the natural diamond [20]. The comparison of divergences of the heat capacity ΔC°_{p} of diamonds investigated by DeSorbo [30] and Lebedev *et al.* [31] from the averaged value for diamond obtained by Wunderlich [20] shows that the maximum scatter of DeSorbo's data is observed in the range (200–300) K and it amounts to 0.05 $J{\cdot}K^{-1}{\cdot}mol^{-1}$ at 280 K; at all other temperatures it does not exceed 0.01 $J{\cdot}K^{-1}{\cdot}mol^{-1}$. For the synthetic diamond the largest deviation is seen in the interval (280–300) K, ΔC°_{p} = 0.18 $J{\cdot}K^{-1}{\cdot}mol^{-1}$ at 280 K. At other temperatures ΔC°_{p} is substantially smaller: in the range (50–90) K it is +0.04 $J{\cdot}K^{-1}{\cdot}mol^{-1}$ and from 90 K to 130 K $\Delta C^{\circ}_{p} = -0.04$ $J{\cdot}K^{-1}{\cdot}mol^{-1}$. Despite such small absolute differences in heat capacities of the diamond specimens, their relative differences, especially at low temperatures, can amount to some tens of percent. With rising temperature, they diminish to a few percent or even fractions of a percent. So, for the synthetic diamond the highest deviation between 280 K and 300 K is about 3% and for the natural one about 1% in the range (200–300) K. Large relative deviations ΔC°_{p} are chiefly connected with the fact that absolute values of C°_{p} for diamond are small, in particular at lower temperatures. A decrease of the precision of C°_{p} measurements of the diamonds is also caused by the highly unfavourable (too small) ratio of the heat capacity of the diamond samples studied, and the sum of the heat capacities of the calorimetric ampoule and the substance. However, all values of heat capacity known until now for diamond at the corresponding temperatures are close to each other. A maximum deviation of the heat capacity at this temperature is observed only for the synthetic diamond, and was found to be 0.18 $J\cdot K^{-1}\cdot mol^{-1}$.

The values of C^o_p for graphites of various types (black, coke and scaly natural) were reported in Ref. [20]. It was established that C^o_p changes much with the type of graphite as determined by the perfection of the crystalline structure. As demonstrated in Ref. [20], the perfection of the graphite structure is enhanced in the order from black graphite to coke to natural graphite. In black graphite, only 1/3 of the lattice planes are located correctly relative to their neighbours. The natural graphite is more close to a perfect single crystals and consists of plates with a thickness of 10–40 nm [20]. All these factors lead to the fact that at 2 K, for example, C^o_p of black graphite exceeds by more than 200 % the heat capacity of the natural one. The uncertainty of the experimental C^o_p values for the graphites is estimated to be 5–2% at T<20 K and 0.5–0.1% in the range (20–340) K. To determine the average values of C^o_p of graphite Markelov *et al.* [32] used the experimental data of the heat capacity of natural graphites (Madagaskar [33], Canada [34], Sri Lanka [35], Acheson [36] and Taiga in Siberia [32]) and synthetic pyrolitic annealed graphite [32]. Their data were compared to the C^o_p values of glassy carbon [37] and acetylene black [15], which are variants of graphite and represent its extreme form of nonperfect crystals, and to the values of C^o_p of natural Acheson graphite as one of the natural graphites having the most perfect crystals [36]. For glassy carbon and acetylene black C^o_p is throughout larger than the averaged C^o_p values of the graphites (ΔC^o_p lies between 0.03 $J\cdot K^{-1}\cdot mol^{-1}$ at 30 K and 0.50 $J\cdot K^{-1}\cdot mol^{-1}$ at 298.15 K). It is of interest that in the range (80–210) K C^o_p of acetylene black is greater than that of glassy carbon and from 210 K to 300 K the heat capacity is smaller, the differences being within ±0.24 $J\cdot K^{-1}\cdot mol^{-1}$. In the range (10–145) K Acheson graphite has C^o_p approaching the averaged value for graphite and over the interval (145–300) K the heat capacity is always lower. Thus, the data on the heat capacity of diverse graphites differ by not more than 0.50 $J\cdot K^{-1}\cdot mol^{-1}$ that is about 6% at 298.15 K.

The numerical values of the heat capacities of the fullerenes C_{60} and C_{70} obtained in an adiabatic vacuum calorimeter over the interval (5–390) K were published elsewhere [38,39]. Measurement errors are about 2% at 5 K, 0.4% at 40 K and 0.2% in the range (40–340) K. Calculated per mole of carbon atoms, their heat capacities are close. Comparison with the average values of C^o_p calculated by the expression $\{C^o_p(C_{60}) + C^o_p(C_{70})\}/2$ shows that for fullerene C_{60} the heat capacity is throughout larger and for C_{70} it is smaller. In the range of (5–50) K the differences lie within ±(8–5)%, between 50 K and 210 K they are within ±(4–3)%, and from 210 K to 340 K ±(2–0.2)%.

Figure 3 shows the temperature dependences of the heat capacities C^o_p per mole of carbyne, fullerene C_{60} and C_{70}, graphite and diamond. The heat capacities of diamond, graphite and carbyne increase gradually with rising temperature; for the fullerenes there are temperature intervals in which anomalous effects occur associated with physical transformations on heating and cooling. Thermodynamic characteristics and interpretations of these transitions were described in detail earlier [38,39]. Over the range of (80–160) K the C^o_p values of the fullerenes are similar and approach those of graphite (the differences do not exceed 8%). It is interesting to note that C^o_p of fullerene C_{70} is slightly smaller than that of C_{60} and closer to the heat capacity of graphite. Apparently, this is not mere chance but is connected with a lower surface curvature of the fullerene C_{70} molecules compared to that of fullerene C_{60} molecules.

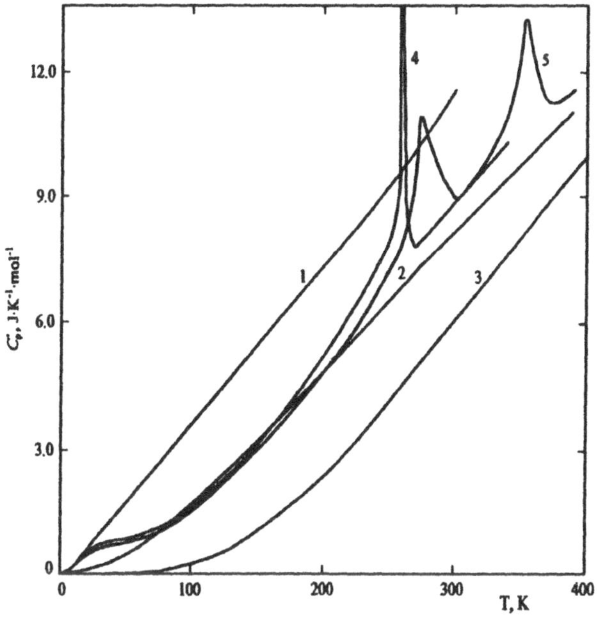

Figure 3. Temperature dependences of C^o_p of allotropic modifications of carbon: 1 – carbyne [15], 2 – graphite [32], 3 – diamond [20], 4 – fullerene C_{60} [38], 5 – fullerene C_{70} [39].

At 80 K the heat capacity of carbyne is 16 times that of the diamond and almost 9 times that of the graphite and fullerenes. At 300 K it is higher by a factor of 1.7 and 1.3, respectively. This can be explained by the fact that the rigidity of an oscillating skeleton of the allotropes decreases in the same order. Whereas a reduction of the rigidity of graphite relative to diamond is associated with the transition from a three-dimensional structure to a two-dimensional one, a subsequent decrease of the rigidity of the carbyne skeleton relative to the graphite conforms with the transition from a layer to a chain structure.

An analysis of the temperature dependence of the heat capacity of the allotropes showed that C^o_p of the diamond in the range (25–50) K is well-described by Debye's T^3-law [14] that is typical of spatial structures; the heat capacity of the graphite between 10 K and 100 K is described by a T^2-dependence according to Tarasov's theory of heat capacity for layer structures [12]; and the heat capacity of carbyne over the range of (40–240) K is described by a T^1-dependence that corresponds also to Tarasov's theory of heat capacity for chain structures with a weak intermolecular interaction [12,13].

Table 2 lists the thermodynamic functions of allotropic carbon modifications calculated from their temperature-dependent heat capacities. The calculation procedures are analogous to those used in Ref. [10]. The numerical values of C^o_p, enthalpy $H^o(T)-H^o(0)$ and entropy $S^o(T)$ decrease from carbyne to diamond, and the Gibbs function $G^o(T)-H^o(0)$, on the contrary, increases.

The changes of the thermodymamic functions on passing from carbyne to diamond result from the temperature dependence of the heat capacities of the corresponding allotropes. The enthalpy and entropy which are integrals of C^o_p over T and lnT, respectively, decrease in the range of (0–298.15) K as n increases in the relation C^o_p vs. T^n. For the enthalpies, for example, this is the areas under the curves $C^o_p=f(T)$ for the corresponding allotropes in the temperature interval studied, Figure 3.

TABLE 2. Thermodynamic functions of allotropic carbon modifications per mole of carbon atoms at T=298.15 K and standard pressure

Allotrope	C^o_p, $J\cdot K^{-1}\cdot mol^{-1}$	$H^o(T)-H^o(0)$, $kJ\cdot mol^{-1}$	$S^o(T)-S^o(0)]$, $J\cdot K^{-1}\cdot mol^{-1}$	$-[G^o(T)-H^o(0)]$, $kJ\cdot mol^{-1}$
Carbyne[a]	11.46	1.631	10.70[b]	1.558
Fullerene C_{60}	8.747	1.207	7.108	0.912
Fullerene C_{70}	8.911	1.102	6.467	0.826
Graphite	8.109	1.008	5.508	0.634
Diamond	6.117	0.5236	2.348	0.176

[a] In fact, the material studied was amorphous carbon with predominantly sp-bonded carbon atoms;
[b] Calculated without regard for zero entropy $S^o(0)$.

Table 3 shows the thermochemical characteristics of the allotropic carbon modifications obtained from the calorimetric data. The procedures of their calculation are similar to those used in Ref. [40]. The data for the fullerenes were taken from Refs. [38,39], for graphite from [23], for diamond the quantities $\Delta_{comb}H^o$ and $\Delta_f H^o$ were adopted from Refs. [41] and [28], respectively; $\Delta_f S^o$ was estimated from the value of $S^o(298.15)$ which in turn was calculated from the data of the relation $C^o_p=f(T)$ reported in Ref. [20].

According to the value and sign of the standard Gibbs function (Table 3) one can infer the order of thermodynamic stability of the allotropic carbon modifications as: graphite > diamond > fullerene C_{70} > fullerene C_{60}.

TABLE 3. Enthalpies of combustion ($\Delta_{comb}H^o$), enthalpies ($\Delta_f H^o$), entropies ($\Delta_f S^o$) and Gibbs functions ($\Delta_f G^o$) of formation of allotropic carbon modifications at T=298.15 K and standard pressure

Allotrope	$-\Delta_{comb}H^o$, kJ·mol^{-1}	$\Delta_f H^o$, kJ·mol^{-1}	$\Delta_f S^o$, J·K^{-1}·mol^{-1}	$\Delta_f G^o$ kJ·mol^{-1}
Carbyne[a]	(356.5)[b]	(–37.03)	(5.6)	(–38.7)
Fullerene C_{60}	432.8	39.25	1.54	38.79
Fullerene C_{70}	430.0	36.50	0.959	36.21
Graphite	393.51	0	0	0
Diamond	395.39	1.828	–3.16	2.770

[a] The material studied was amorphous carbon with predominantly *sp*-bonded carbon atoms.
[b] The enthalpy of combustion of carbyne reported by Kasatochkin *et al.* [7] appears to be erroneous. Therefore, the enthalpy $\Delta_f H^o$ and Gibbs function of formation $\Delta_f G^o$ of carbyne derived from the $\Delta_{comb}H^o$ value are probably invalid (*cf.* Refs. [25,26]).

Thus, the thermodynamically most stable allotropic modification of carbon among those discussed above is graphite and the least stable ones are fullerenes although their thermodynamic stabilities are similar – the differences seem to approach the errors of experimental measurements and calculations of the values of $\Delta_f H^o$ and $\Delta_f S^o$ used to estimate $\Delta_f G^o$. It should be emphasised that the order is valid only for those physical conditions (T and p) and the physical states of the allotropes at which the $\Delta_f G^o$ values were obtained. Under differing conditions, a change in the order of thermodynamic stability seems possible. In accordance with the values of standard entropies of formation, the allotropic modifications can be arranged as carbyne > fullerene C_{60} > fullerene C_{70} > graphite > diamond. This sequence corresponds to the position of the curves of the temperature-dependent heat capacities of the allotropes which were used in the calculation of their absolute entropies by the third law of thermodynamics. The absolute entropies in turn were used to evaluate the standard entropies of formation.

5. References

1. Kasatochkin, V.I., Korshak, V.V., Kudryavtsev, Yu.P., Sladkov, A.M., and Elizen, V. M.: On polymorphism of carbyne, *Dokl. Akad. Nauk SSSR* **214** (1974), 587–589 (in Russian).
2. Korshak, V.V., Kudryavtsev, Yu.P., and Sladkov, A.M.: Carbyne – a new allotropic form of carbon, *Vestn. Akad. Nauk SSSR* (1) (1978), 70–73 (in Russian).
3. Sladkov, A.M.: *Polyconjugated Polymers*, Nauka, Moscow, 1979, p. 19 (in Russian).
4. Kasatochkin, V.I., Sladkov, A.M., Kudryavtsev, Yu.P., and Korshak, V.V.: Concerning a chain polymer of carbon – carbyne, in: *Structural Chemistry of Carbon and Coals*, Nauka, Moscow, 1969, pp. 17–26 (in Russian).
5. Kudryavtsev, Yu.P., Evsyukov, S.E., Guseva, M.B., Babaev, V.G., and Khvostov, V.V.: Carbyne – the third allotropic form of carbon, *Izv. Akad. Nauk, Ser. Khim.* (3) (1993), 450–463 (in Russian) [*Russ. Chem. Bull.* **42** (1993), 399–413 (Engl. Transl.)].

6. Kudryavtsev, Yu.P., Varfolomeeva, O.B., Litovchenko, G.D., and Sladkov, A.M.: Study of dehydrochlorination reaction of polyvinylidene chloride, *Izv. Akad. Nauk SSSR, Ser. Khim.* (1) (1983), 195–197 (in Russian).
7. Kasatochkin, V.I., Kazakov, M.E., Kudryavtsev, Yu.P., Sladkov, A.M., and Korshak, V.V.: Enthalpy of a chain polymer of carbon, *Dokl. Akad. Nauk SSSR* **183** (1968), 109–111 (in Russian).
8. Melnitchenko, V.M., Sladkov, A.M., and Nikulin, Yu.N.: Structure of polymeric carbon, *Usp. Khim.* **51** (1982), 736–763 (in Russian).
9. Sladkov, A.M., Korshak, V.V., Kudryavtsev, Yu.P., and Kasatochkin, V.I.: Discovery Diploma No. 107 of 6.XII.1971, priority of 4.XI.1960, Discoveries of USSR in 1971, *Bull. Izobr.* (6) (1972), 3 (in Russian).
10. Lebedev, B.V.: *Thermodynamics of Polymers*, Gorky State University, Gorky, 1989 (in Russian).
11. Godovskii, Yu.K.: *Thermophysics of Polymers*, Khimiya, Moscow, 1982 (in Russian).
12. Tarasov, V.V.: Theory of heat capacity of chain and layer structures, *Zh. Fiz. Khim.* **24** (1950), 111–128 (in Russian).
13. Tarasov, V.V. and Yunitskii, G.A.: Theory of heat capacity of chain-layer structures, *Zh. Fiz. Khim.* **39** (1965), 2077–2079 (in Russian).
14. Mendelsson, K.: *Cryophysics*, Interscience Publishers, Inc., New York, 1960.
15. Rabinovich, I.B., Lebedev, B.V., Sladkov, A.M., Kudryavtsev, Yu.P., Martynenko, L.Ya., and Korshak, V.V.: Polymeric carbon with a higher heat capacity, *Dokl. Akad. Nauk SSSR* **168** (1966), 599–602 (in Russian).
16. Kudryavtsev, Yu.P.: Synthesis and study of properties of polymers containing conjugated triple bonds in the chain, *Ph.D. Thesis*, Institute of Organoelement Compounds, USSR Akademy of Sciences, Moscow, 1965 (in Russian).
17. Nikolaev, P.N. and Lebedev, B.V.: Calorimetric apparatus for measurements of heat capacity in range 60–300 K, *Tr. Khim. i Khim. Tekhnol. (Intercollegiate Collection)* 2 (1966), 322–343 (in Russian).
18. Cotton, F.A. and Wilkinson, G.: *Basic Inorganic Chemistry*, Wiley, New York, London, Sydney, Toronto, 1976.
19. Usenbaev, K.: Comparative investigation of low-temperature heat capacity of allotropic forms of carbon and cryptocarbons, *Izv. Akad. Nauk Kirgizskoi SSR, Khim.-Tekhnol. Biol. Nauki* (4) (1990), 17–22 (in Russian).
20. Wunderlich, B. and Baur, H.: *Heat Capacities of Linear High Polymers*, Springer-Verlag, Berlin, New-York, 1970.
21. Chernoplekov, A.G.: Tables of heat capacity functions of heterodynamic structures, *Zh. Fiz. Khim.* **25** (1951), 878–883 (in Russian).
22. Kasatochkin, V.I., Galiakbarov, M.F., Kazakov, M.E., and Smutkina, Z.S.: Enthalpy of transition forms of carbon, in: *Structural Chemistry of Carbon and Coals*, Nauka, Moscow, 1969, pp. 27–34 (in Russian).
23. *Thermal Constants of Substances*, Part IV, Ed. Glushko, V.P., VINITI, Moscow, 1970 (in Russian).
24. Lebedev, B.V. and Rabinovich, I.B.: Determination of zero entropy of some glassy polymers from calorimetric data, *Dokl. Akad. Nauk SSSR* **237** (1977), 641–644 (in Russian).
25. Vitvitskii, A.I.: Calculation of heats of formation for carbocyclic compounds, *Teor. Eksper. Khim.* **3** (1967), 82–87 (in Russian) [*Theor. Exper. Chem.* **3** (1967), 44–47 (Engl. Transl.)].
26. Baughman, R.H., Eckhardt, H., and Kertesz, M.: Structure-property predictions for new planar forms of carbon: Layered phases containing sp^2 and sp atoms, *J. Chem. Phys.* **87** (1987), 6687–6699.
27. Benson, S.W., Cruickshank, F.R., Golden, D.M., Haugen, G.R., O'Neal, H.E., Rodgers, A.S., Shaw, R., and Walsh, R.: Additivity rules for the estimation of thermochemical properties, *Chem. Rev.* **69** (1969), 279–324.
28. Kroto, H.W., Heath, J.R., O'Brien, S.C., Curl, R.F., and Smalley, R.E.: C_{60}: Buckminsterfullerene, *Nature* **318** (1985), 162–163.
29. *Experimental Thermodynamics*, Vol. 1, Ed. McCullough, J.P. and Scott, D.W., Plenum Press, New York, London, 1971.
30. DeSorbo, W.: Specific heat of diamond at low temperatures. *J. Chem. Phys.* **21** (1953), 876–880.
31. Lebedev, B.V., Tsvetkova, L.Ya., and Zhogova, K.B.: Thermodynamics of allotropic modifications of carbon (of synthetic diamond, graphite, fullerene C_{60} and carbyne), *Thermochemica Acta* (in press).
32. Markelov, N.V., Volga, V.I., and Buchnev, L.M.: Thermodynamic functions of single-crystalline graphite in the 0–3000 K temperature range, *Zh. Fiz. Khim.* **47** (1973), 1824–1827 (in Russian).

33. Hoeven, B.J. and Keesom, P.H.: Specific heat of various graphites between 0.4 and 2.0 K, *Phys. Rev.* **130** (1963), 1318–1321.
34. DeSorbo, W. and Nichols, G.E.: A calorimeter for the temperature region 1–20 K. The specific heat of some graphite specimens, *Phys. Chem. Solids* **6** (1958), 352–366.
35. DeSorbo, W.: Low temperature heat capacity of ceylon graphite, *J. Am. Chem. Soc.* **77** (1955), 4713–4715.
36. DeSorbo, W. and Tyler, W.W.: The specific heat of graphite from 13 to 300 K, *J. Chem. Phys.* **21** (1953), 1660–1663.
37. Takanashi, Y. and Westrum, E.F.: Glassy carbon low-temperature thermodynamic properties, *J. Chem. Thermodyn.* **2** (1970), 847–854.
38. Lebedev, B.V., Zhogova, K.B., Bykova, T.A., Kaverin, B.S., Karnatsevich, V.L., and Lopatin, M.A.: Thermodynamics of fullerene C_{60} in the range of 0–340 K, *Izv. Akad. Nauk, Ser. Khim.* (9) (1996), 2229–2233 (in Russian).
39. Zhogova, K.B. and Lebedev, B.V.: Thermodynamics of fullerene C_{70} in the range of 0–390 K, *Izv. Akad. Nauk, Ser. Khim.* (in press) (in Russian).
40. Lebedev, B.V. and Kiparisova, E.G.: Thermochemical characteristics of some hydrocarbon vinyl polymers at T = 298.15 K and P = 101.325 kPa, *Zh. Fiz. Khim.* **70** (1996), 1351–1358 (in Russian).
41. *Thermodynamic Properties of Individual Substances*, ed. by Glushko, V.P., Gurvich, L.V., Bergman, G.A., Veits, I.V., Medvedev, V.A., Khachkuruzov, G.A., and Yungman, V.S., USSR Acad. Sci., Moscow, 1962 (in Russian).

5.3. ELECTRICAL AND OPTICAL PROPERTIES

E.M. BAITINGER
Department of Physics,
Chelyabinsk State Pedagogical University
454080 Chelyabinsk, Russia

Abstract – Experimental data on optical and electrical parameters of carbyne and carbynoid materials are discussed, taking into consideration their band structure and electronic correlations. Doping with electronodonors and acceptors changes these properties significantly. The qualitative interpretation of this effect based on the modern ideas about electronic structure of carbyne and carbynoids is given.

1. Introduction

The electronic structure of a chain of carbon atoms is under thorough consideration by experts for a sufficiently long period of time. This situation has been stipulated by much interest in the electrical (and, in a more general sense, electronic) and optical properties of these structures. However, before discussing common regularities of the electronic structure of a carbon chain, it is necessary to specify problems facing the researchers when attempting theoretical and experimental investigations.

First, the conditions of synthesis influence greatly physical and chemical properties of carbynoid structures, including their electronic parameters. Even considering a sufficient degree of carbyne energy band rigidity, the peculiarities of the atomic and molecular structure of a chain have a considerable effect on the value of the electrical conductivity and other kinetic coefficients. Chain structure distortions generate discontinuities in the electron density distribution along it. For example, sp^2-defects at chain cross-linking sites or foreign atom attachment cause fluctuations of the energy gap width and, as a consequence, of the dynamics of charge carriers. These defects can constitute a barrier for quasi-particles associated with hopping conductivity at cross-linking sites. Foreign atoms deform energy bands even more. The integral characteristics of samples (specific electrical conductivity, optical absorption and others) are physically averaged over a large number of random as well as intrinsic factors. For this reason, the reproducibility and accuracy of the experimental data leave something to be desired.

Secondly, quasi one-dimensional systems are sensitive to multiple electronic effects to a larger degree than structures of higher dimensionality. This is because these electronic effects considerably transform the electronic spectrum and thus modify the physical and chemical properties. This also concerns the soliton and polaron systems

R.B. Heimann et al. (eds.), Carbyne and Carbynoid Structures, 333–341.

whose formation causes a group of energy levels to appear inside a gap [1,2]. In addition, the well-known Peierls' instability of a linear chain with respect to dimerization, *i.e.* alternation of bond lengths, transforms the system from a conductive state to a semiconducting phase, called a Peierls' dielectric.

Thirdly, the problem of the degree of one-dimensionality of the carbon system is important. It is known, that at ambient temperature and pressure conditions graphite is thermodynamically stable. Therefore, during synthesis, and in the presence of external effects (temperature, light, mechanical loading, *etc.*), spontaneous creation of a graphite-like phase inside the carbynoid systems is possible.

2. Brief Survey on the Electronic Structure of Carbyne

The four valence electrons of carbon do not equally participate in its electronic properties. Ideally, their role can be represented as follows. Two valence electrons are strongly localized, forming hybrid *sp*-orbitals and thus bonds with large binding energies along the chain. For example, at present it is known that the binding energy value for *sp*-electrons ranges between 6 and 20 eV [3,4]. The energies are counted from the upper edge of the valence band. The two other electrons (π-electrons), which in an ideal case are independent, form two degenerate energy bands with a binding energy between 0–7 eV. The information on the state of these electrons is most important for the description of the electrophysical and optical properties of carbon chains. Exact knowledge of the energy levels localized inside a gap and for conducting states near the gap are also necessary.

The presence of a gap in the energy spectrum of carbynoid structures is a well established fact. Its value is 1–2 eV (see, for example, [5]). Any value in this energy interval is reasonable since the colour of these materials range from black to transparent, including some colours from the visible range of radiation.

There are several reasons for the energy gap formation in the π-electrons spectrum of carbyne (Fig. 1). One reason relies on the participation of one-electron excitations to generate the Peierls' instability in relation to doubling of the period. On the other hand, there are interelectronic correlations. Previously it was considered that interelectronic interactions (both repulsion and attraction) suppress the Peierls' gap by equalizing the bond lengths along the chain [6]. Later for the spectrum of quasi one-dimensional systems it has been considered possible that both gap formation mechanisms coexist if the repulsion (or weak attraction) of π-electrons are not very significant [7].

As a rule, electronically and structurally exited states appear in the one-dimensional system whose energy spectra correspond to prohibited energies. These are solitons and, more specifically, polarons [8] (see also Chapter 6.2). In carbyne, the situation becomes more complicated owing to simultaneous soliton generation by two π-electronic systems, whose degeneration is almost not removed. Thus the number of levels inside a forbidden region becomes rather great (see Fig. 1). These levels are partly derived from the ground state, and partly originate from structural kinks or spin discontinuities. Generation of a polaron is connected with an attraction of electrons via phonons. Confirmation of the existence of such quasi-particle systems can be obtained by the indirect fact of carbon chain superconductivity [9] since the formation of Cooper pairs is stipulated by that very mechanism.

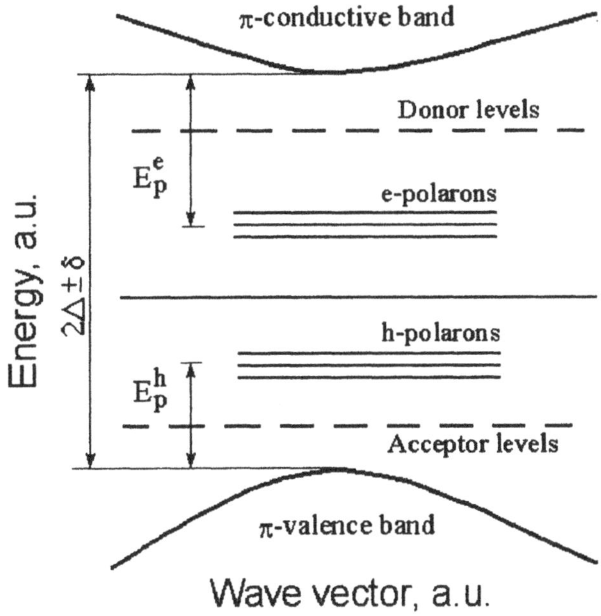

Figure 1. Schematic representation of the carbyne band structure. Δ is the Peierls' parameter, δ is a correlation parameter, E^e and E^h are polaron energies.

Estimated data of soliton energy levels inside a gap are given in Ref. [1, 2]. In view of the four-fold degeneration of bands (two spin orientations and two π-electron systems) there are six different solitons. Besides, in Ref. [1] the existence of a polarexciton was assumed arising from a chain distortion site with simultaneous creation of an electron-hole pair.

Lastly, doping is an important mechanism to alter the electronic states inside a gap and close to it. Doping can occur either in a controlled way or by random impurities during synthesis. Impurity atoms, molecules or radicals of donor or acceptor type create a system of various donor-acceptor levels inside a gap or "dissolve" among allowed states close to it.

3. Optical Studies of Carbyne

To elucidate the detailed electronic structure of carbynoid materials, optical experiments in a wide range of electromagnetic radiation frequencies from near IR region up to deep UV are important. Unfortunately, to date such information is still unavailable.

The data interpretation given below should be considered qualitative and preliminary. However, in my opinion such a description is necessary to stimulate research by others along this way.

Let us start from the results by Akagi *et al.* [10] who obtained carbynoid chains by dehydrochlorination of particularly doped polyacetylene. Optical absorption spectra were obtained for the pristine polymer, for intermediate products obtained by doping with chlorine, and also for samples of carbyne, being the final product of the reaction (Fig. 2). For polyacetylene the fundamental optical absorption edge is located in the UV region. Therefore excitonic (polaronic) effects are readily observed in the visible radiation range. In Fig. 2 they are evident as sharp maxima of light absorption at the energies 2.1, 2.2 and 2.6 eV, related to states inside a gap. One can describe these values as characteristic features of carbon chains that contain one π-electron per atom. In the paper by Mylnikov *et al.* [11], an excitonic peak in spectral dependence of the photoresistance of the polymer is detected just at a photon energy equal to 2.6 eV.

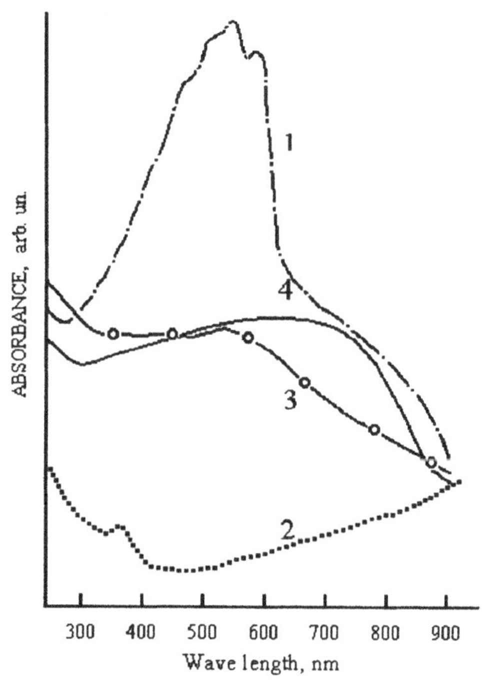

Figure 2. UV-VIS absorption spectra of *cis*-$(CH)_x$ (1), $(CHCl)_x$ (2) and carbynes (3, 4) [10].

Chlorination of the polymeric precursor localizes presumably the single π-electron, thereby changing its symmetry. Due to this effect the gap increases and the sample becomes completely transparent in the visible region (See curve 2 in Fig. 2). Only at a photon energy of 3.2 eV a small maximum is noticeable which is probably connected with the optical activity of fragments of a chain that remained undoped with chlorine.

Dehydrochlorination returns the two π-electrons to a band. This is evident from the displacement of the very shallow local absorption maximum to the region of lower energy values. This shift is the more pronounced the deeper the electron delocalization process is as seen by comparing spectra 3 and 4 in Fig. 2. The last spectrum in Fig. 2 was obtained from a sample, in which the carbyne synthesis reaction has occurred in the most complete way.

Let us now describe optical results that were obtained by Kavan and Dousek [12] by studying of carbynoid structures synthesized by an electrochemical method in the presence of alkaline metals amalgams. The general resemblance of the absorption spectrum in the photon energy interval 1.2–5 eV with that of a polyacetylene spectrum allows to conclude that as a result of the synthesis mainly one π-electron is delocalized. A very wide absorption maximum with a half-width of 2 eV is centred at an energy of 3.1 eV. This region corresponds to excitonic (polaronic) states inside a gap. In due course the number of π-electrons decreases. Consequently, an exponential decrease of the absorbance with time after synthesis was revealed.

4. Electrical Conductivity of Carbynoid Structures

The specific resistance of an ideal dielectric carbon chain most likely is large and of the order of value typical for classical dielectrics. Real samples obtained by chemical synthesis have many impurities which can account for the observed experimental values.

4.1. CONDUCTIVITY OF CARBYNE DOPED WITH DONORS

Samples obtained during earlier work [13] as well as more recent ones contain a significant amount of impurities.

The specific resistance (ρ) of specimens obtained by oxidation in the presence of copper-containing species, is ~1 Ω·cm [13]. Using an electrochemical method in the presence of alkaline metals (lithium, sodium, potassium) yields a material whose specific resistance is also of the same order [14]. In both cases the concentration of paramagnetic centres is estimated to be 10^{19} cm^{-3} (see Chapter 6.2). One can use a standard equation for the evaluation of the charge carrier mobility value:

$$\rho = (en\mu)^{-1} \tag{1}$$

where e is a charge of a carrier, n and μ are carrier concentration and mobility, respectively.

Estimation according to eq. (1) with $n = 10^{19}$ cm^{-3} gives a mobility value of ~1 cm^2 V^{-1} s^{-1}. This magnitude, according to Mott's evaluations, is intermediate between hopping and band mobility of the charge carriers. An ionic component contribution because of its small value cannot significantly alter the result of this estimation (see Table 1), but plays a distinct role in the process of carbynoid synthesis [15]. The character of the temperature coefficient of resistance can become an additional criterion, clarifying the conductivity mechanism. The experiment shows that the temperature dependence of the resistance of carbynoid samples doped with alkaline metals corresponds to hopping conductivity:

$$\rho = \rho_0 \cdot \exp(T_0/T)^{1/4} \quad (2)$$

where T is the sample temperature (below 300 K), and T_0 is a parameter dependent on the properties of the investigated material.

It has been shown by Kavan *et al.* [16] that electrons (but not more massive particles) hop between adjacent areas of metal-type conductivity, *i.e.* between microcrystals with graphite-like structure. Such interpretation requires more detailed description of the role doping impurities play in electrical conductivity both for copper and carbynoid structures doped with alkaline metals. In my opinion, there should be a small amount of metal impurities required in the structure of the material (see Chapter 4.2). As described in Ref. [13], free copper was detected by X-ray diffraction. In other carbynoids it is necessary to supplement the structural models by the presence of alkaline metals atoms in addition to fluorides of these metals[1]. The interchain environment stabilizes them. However, they are ionized, ensuring n-conductivity of a layer (see, for example, structural formula (18) in Ref. [16]). The atomic concentration of alkaline metal, intercalated between carbon chains, is less than about 10 % and consequently it is rather difficult to detect them by chemical methods. Nevertheless, the intercalated metal atoms provide an essential contribution to the electrical conductivity value. However, it is not possible to observe the dependence of band activation of the resistance on temperature, stipulated by partial electron transition from the donor levels to the carbynoid conduction band. The top of the valence band of the source polymer is located at a binding energy range I = 8-9 eV relative to the vacuum level (I is the ionization potential). During the synthesis of the carbyne π-electrons are liberated and the top of the valence band promptly expands to a value that equals the work function of a carbynoid materials of f = 4.9 eV [5]. Hence the energy levels of copper, lithium or sodium are forced to overlap inside the valence band (see the last line of Table 1), raising the density of states. The character of this "hybrid" electronic density (one-dimensional or two-dimensional) should still be studied. One can estimate the depth of the location of the donor levels inside the valence band using the difference between I and f. According to this estimation only potassium levels are inside a gap.

TABLE 1. Ionic components of the specific resistance of carbynoids [15] and ionization potentials I of some doping impurities

Dopant	Cu	Li	Na	K
Specific resist., Ohm·cm	–	10^6	10^7	10^9
I, eV	7.73	5.39	5.2	4.34
$I - f$, eV	+2.83	+0.49	+0.3	-0.54

The conductivity along a chain in case of the dissolution of donor levels in the valence band can have a hopping nature with involvement of both electrons and

[1] Metals atoms have been considered stabilizing features of the kinked sp-hybridized chains in a layered arrangement of hexagonal symmetry thought to be existing in crystalline carbyne (see Chapters 4.2 and 4.3, and references therein) (*Edt.*)

polaronic states. However, since polarons are rather heavy their influence on the measured resistance value is hardly noticeable when a direct current is used for the studies.

The change of the density of states strongly influences the values of T_0 (see eq. (4) in Ref. [17]) and thus determines the sign of the temperature dependence of the resistance.

The different efficiencies of alkaline metals are stipulated also by the ionic radius of atoms. Copper in this relation occupies a special position due to its capacity to contribute to a band up to two electrons.

Measurement of the Seebeck coefficient of samples doped with copper [13] and lithium [18] confirms a donor-like character of doping as the sign of thermoelectricity has proven to be negative.

A special type of electroresistance experiments can be used to study the time dependence of the resistance value during synthesis and doping [9,16]. Thus a precise tool has been obtained to control the state of the carbon system in the course of reaction.

For example, Palnichenko and Tanuma [9] have shown that doping for many hours leads to a peculiar time dependence of the electroresistance. At first, the resistance is almost independent of the time of doping. However after a definite time interval that is specific for each impurity, the resistance strongly decreases. For example, upon doping with potassium this change amounts to a factor of 10^6.

The magnitude of the resistance varied from a value characteristic of a dielectric up to values comparable to those of other doped carbynoid materials. The presence of a plateau on the resistance vs. time-curve can be explained by the significant time required for diffusion of the dopant into interchain positions. Due to this process a gradual filling of the entire volume of the sample takes place. The following result is also of interest. During implantation with Na the specific resistance of the carbynoid material decreased only by 2.5–3 times, considerably less than for potassium. Probably, the charge of potassium atom delocalization is more effective at higher temperature.

The measurement of the resistance has proven fruitful to control carbon film growth [16]. The interpretation of the decrease of the resistance of the sample during its growth in width as suggested in Ref. [16] describes satisfactorily the observed features in view of the influence of kinetic factors. Indeed, the best match with the experiment is reached for large periods of time after the termination of the synthesis. In freshly prepared layers the resistance is essentially higher than predicted by theory. The same is valid for the description of the electroconductivity in terms described above. During aging of the films many interchain cross-links are being formed with intermediate type of π-electron hybridization. As an accessory product of the reaction, MF crystals form (M = Li, Na, K) of large enough sizes to keep the chains apart [17].

4.2. CONDUCTIVITY OF CARBYNE DOPED WITH ACCEPTORS

The essential mechanism of the conductivity of carbynoid materials doped with acceptors (I, Br, O) does not differ from that for donor described above. There are experimental data indicating an increase of electroresistance by 10^6 times due to implantation of I_2 for many hours [10]. The initial structure was that of a dielectric (specific resistance ~10^7 Ω cm), and the finite product is characterized by a resistance typical for carbynoids with impurities, *i.e.* ~10 Ω cm.

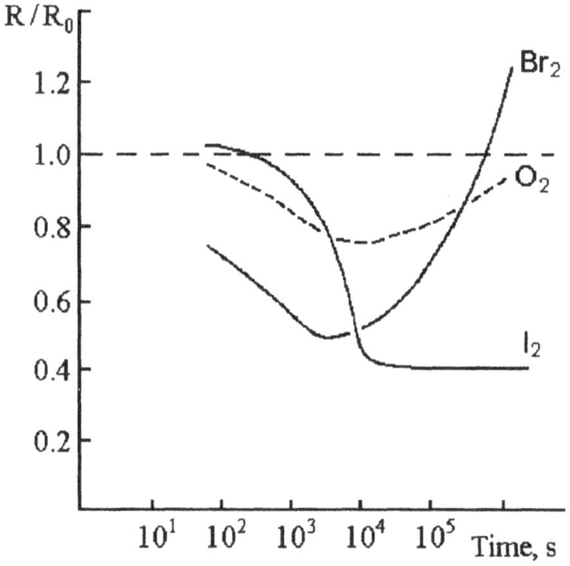

Figure 3. Resistivity changes of a C-LiF layer under exposure to Br_2, I_2 and O_2 in the gas phase [12].

Different results are obtained with implantation of acceptors by an electrochemical technique in presence of lithium [12]. In this case, the doping process is more complicated and includes partial compensation of existing electrons by holes.

In Fig.3 the experimental plots of the resistance of carbynoids electrochemically doped with lithium versus time of I_2, Br_2, O_2 implantation are displayed [12]. The behaviour of the resistance with increasing acceptor concentration is similar for all dopants. At first, during 2.5-3 hours of the reaction time all acceptors studied promote a decrease of the carbynoid's resistance, creating a conductivity of mixed type. The compensation effect which commonly stabilizes the resistance of a sample is in this stage insignificant because of the small concentration of the acceptor and the low probability of electron-hole recombination. At the end of this 'soft' period of reaction the effects of generation of 'covalent bridges' between chains due to chemical binding of an acceptor become important. This phenomenon is the more pronounced the longer the reaction takes place and changes the mode of the resistance. In case of iodine impurities the resistance is stabilized at the expense of exception of π-electrons from the system, including the excess ones obtained at donor ionization. In case of oxygen and bromine implantation, the dynamics of the electroresistance is somewhat different. It starts to increase most likely for the same reason. It is impossible to exclude also influences of the decrease of the charge carrier mobility due to partitions of a chain by 'covalent bridges'.

5. References

1. Rice, M.J., Phillpot, S.R., Bishop, A.R., and Campbell, D.K.: Solitons, polarons, and phonons in the infinite polyyne chain, *Phys. Rev., B* **34** (1986), 4139–4149.
2. Rice, M.J. and Phillpot, S.R.: Infinite and finite polyynes: strongly coupled Peierls insulators with interesting soliton, polaron and photoexcitation properties, *Synth. Met.* **17** (1987), 93–98.
3. Kudryavtsev, Yu.P., Baytinger, E.M., Kugeev, F.F., Korshak, Yu.V., and Evsyukov, S.E.: Electronic structure of carbyne studied by x-ray photoelectron spectroscopy and x-ray emission spectroscopy, *J. Electron Spectrosc. Relat. Phenom.*, **50** (1990), 295–307.
4. Baytinger, E.M. and Gagarin S.G.: Quantum-chemical calculation of the valence band of carbyne, *Izv. Vys. Uchebn. Zaved., Fizika* (7) (1989), 65–68 (in Russian).
5. Kudryavtsev, Yu.P., Evsyukov, S.E., Guseva, M.B., Babaev, V.G., and Khvostov, V.V.: Carbyne - a linear chainlike carbon allotrope, in: *Chemistry and Physics of Carbon*, ed. by Thrower, P.A., Marcel Dekker, New York, 1996, Vol. 25, pp. 1–69.
6. Ovchinnikov, A.A., Ukrainskii, I.I., and Kventsel', G.F.: Theory of one-dimensional Mott semiconductors and the electronic structure of long molecules having conjugated bonds, *Usp. Fiz. Nauk* **108** (1972), 81–111 (in Russian) [*Sov. Phys. – Usp.* **15** (1973), 575–591 (Engl. Transl.)].
7. Krivnov, V.Ya. and Ovchinnikov, A.A.: Peierls instability in weakly nonideal one-dimensional systems, *Zh. Eksp. Teor. Fiz.* **90** (1986), 709–723 (in Russian) [*Sov. Phys. JETP* **63** (1986, 414-421 (Engl. Transl.)].
8. Krive, I.V., Rozhavskii, A.S., and Kulik, I.O.: Mechanisms of nonlinear conductivity and electrodynamics of quasi-one-dimensional conductors in the Peierls insulator state, *Fiz. Nizk. Temp.* **12** (1986), 1123–1164 (in Russian).
9. Palnichenko, A.V. and Tanuma, S.-I.: Effect of intercalation of alkali and halogen species into the low density carbon crystal 'carbolite', *J. Phys. Chem. Solids* **57** (1996), 1163–1166.
10. Akagi, K., Nishiguchi, M., Shirakawa, H., Furukawa, Y., and Harada, I.: One-dimensional conjugated carbyne: Synthesis and properties, *Synth. Met.* **17** (1987), 557–562.
11. Myl'nikov, V.S., Sladkov, A.M., Kudryavtsev, Yu.P., Luneva, L.K., Korshak, V.V., and Terenin, A.N.: Photosemiconducting properties of acetylene polymers, *Dokl. Akad Nauk SSSR* **144** (1962), 840–843 (in Russian).
12. Kavan, L. and Dousek, F.P.: Carbynoid species in electrochemical polymeric carbon, *Synth. Met.* **58** (1993), 63–72.
13. Korshak, V.V., Kasatochkin, V.I., Sladkov, A.M., Kudryavtsev, Yu.P., and Usenbaev, K.: Concerning the synthesis and properties of polyacetylene, *Dokl. Akad. Nauk SSSR* **136** (1961), 1342–1344 (in Russian).
14. Kavan, L.: Electronic properties of electrochemical polymeric carbon, *Springer Ser. Solid-State Sci.* **107** (1992), 424–428.
15. Kavan, L., Dousek, F.P., and Micka, K.: The role of ion transport in the electrochemical corrosion of fluoropolymers. Preparation and properties of *n*-doped polymeric carbon with mixed ion/electron conductivity, *Solid State Ionics* **38** (1990), 109–118.
16. Kavan, L., Dousek, F.P., and Micka, K.: Time-dependent electrical resistivity of carbon, *J. Phys. Chem.* **94** (1990), 5127–5134.
17. Kavan, L., Dousek, F.P., Micka, K., and Weber, J.: Electronic resistivity of carbon in the product of electrochemical reduction of poly(tetrafluoroethylene), I, *Carbon* **26** (1988), 235–243.

Chapter 6: Molecular and electron spectroscopy of carbyne structures

6.1. RAMAN AND INFRARED SPECTROSCOPY

L. KAVAN

*J. Heyrovský Institute of Physical Chemistry,
Dolejškova 3, CZ-182 23 Prague 8, Czech Republic*

J. KASTNER

*PROFACTOR Produktionsforschung GmbH,
Wehrgrabengasse 1, A-54400 Steyr, Austria*

Abstract. Raman and IR spectra of carbyne-like materials show a characteristic band at 2000-2300 cm^{-1} which was assigned to the CC stretching mode of *sp*-bonded carbon chain. From the Raman and IR spectra, the purity of carbynoid materials can be deduced. The length of an unperturbed *sp*-carbon chain (conjugation length) can be evaluated from the band position and dispersion effects. The most detailed spectroscopic data exist for carbyne-like materials prepared by chemical or electrochemical methods from perhalo-n-alkanes, either polymers or low-molecular weight precursors. Raman and IR spectra of carbynoid products are sensitive towards elevated temperatures and/or reactive environment (air oxygen, humidity). Crosslinking of carbynoid chains can be monitored by time-dependencies of both the intensity and position of the CC stretching mode. By resonance Raman scattering the first allowed electronic transition can be identified. Anisotropy, introduced by preferential orientation of carbynoid chains, can be followed by polarized Raman measurements.

1. Introduction

Vibration (Raman and infrared) spectroscopy provides an easy and powerful tool to investigate the structure of carbon materials [1-7]. Carbyne is expected to show a characteristic vibration spectrum both in polycumulene and polyyne configurations [8-11]. Besides the usual analysis of band positions, intensities and shapes, experiments with polarized radiation provide information on the carbyne anisotropy, i.e. on the preferred orientation of *sp*-chains. Resonance Raman spectroscopy supplies data about the electronic structure, which might be also interpreted in terms of the molecular and crystallographic structure of carbonaceous materials. Hence, vibration spectroscopy is regarded as a very convincing method to detect carbyne, and this opinion is shared even among the authors of various critical reports on this topic [12,13].

R.B. Heimann et al. (eds.), Carbyne and Carbynoid Structures, 343–356.

Raman and infrared (IR) spectroscopy detect vibration modes of molecules/solids or excitations of quasiparticles like phonons or magnons, while some vibration modes are Raman- or IR-active [5,7]. The Raman/IR activity can be evaluated by using selection rules of the group theory. For instance, the face centered cubic lattice of diamond belongs to the O_h factor group [2] which gives one triply degenerated phonon with Raman activity and no IR active phonon:

$$G_{diamond} = T_{2g}\ (\text{Raman}) \tag{1}$$

The first order Raman band of diamond appears at 1332 cm^{-1}. Diamond has no first order infrared absorption, both in cubic and hexagonal (lonsdaleite) modifications.

Graphite belongs to the D^6_{4h} space-group symmetry [2,6], which gives the following irreducible representation for the optical phonons:

$$G_{graphite} = 2\ E_{2g}\ (\text{Raman}) + E_{1u}\ (\text{IR}) + 2\ B_{2g} + 2\ A_{2u}\ (\text{IR}) \tag{2}$$

The E_{2g} modes are Raman active, one is a shear-type rigid layer mode at 42 cm^{-1} and the other an in-plane stretching mode at 1582 cm^{-1}. The infrared optical vibrations are the out-of-plane A_{2u} mode and the in-plane E_{1u}.

Polycrystalline graphite shows a second, disorder-induced Raman band at 1350 cm^{-1}. The in-plane graphite crystallite size (L) can be deduced by comparing the intensity of the E_{2g}-mode at 1582 cm^{-1} (I_G) relative to the intensity of the disorder induced band at 1350 cm^{-1} (I_D) [2,6]:

$$L = 4.4\frac{I_G}{I_D} \quad [\text{in nm}] \tag{3}$$

When the disorder in a graphitic material increases and approaches that of amorphous carbon (a-C), the G and D lines get broader until they overlap. For a-C carbon this gives a broad and asymmetric band between 1000 and 1700 cm^{-1}.

As the crystal structure of carbyne is not known in detail yet, theoretical studies of its vibrational spectrum concentrate on an individual infinite *sp*-carbon chain [8,14]. Useful information follows also from theoretical and experimental studies on model linear carbon molecules, such as C_n ($2<n<10$) [15,16]:

$$\bullet C{\equiv}C - (C \equiv C)_{n/2-4} - C \equiv C \bullet \quad \leftrightarrow \quad :C = C = (C = C)_{n/2-4} = C = C:$$

and their hydrogen-capped derivatives, HC_nH [10,11,17,18] and $H_2C_nH_2$ [10,11] as well as model conjugated polymers, such as poly(butatriene) [9], poly(acetylene) and poly(diacetylene) [3,5]. These works point at characteristic carbon-carbon stretching at ca. 2000 cm^{-1} which is especially pronounced in the Raman spectrum. (For linear oligoacetylenes, HC_nH, the CC stretch (Σ_g for a $D_{\infty h}$ molecule) should be IR inactive [17], but experiments do show a strong IR band in this region [18]).

Table 1 summarizes the vibration properties of the most intense Raman band, together with the hybridization state and bond lengths in the common carbon polymorphs, i.e. cubic diamond, fullerene (C_{60}) and graphite. In addition, two model conjugated polymers, i.e. poly(acetylene) and poly(diacetylene) are quoted. Apparently, the frequency of CC stretch shifts to higher wavenumbers with increasing bond strength, i.e. with decreasing p-character in the hybridization.

By employing the Longuet-Higgins-Salem model to calculate the electronic and vibrational structure of a linear oligoyne, the normal coordinate analysis leads to (N/2) Σ_g modes for the stretching vibration of a N-membered oligoyne (N even). For N $\rightarrow \infty$, only the "k = 0" mode is Raman active, while its frequency decreases with increasing chain length [14] in analogy to the C=C stretching mode in conjugated polymers [3].

Table 1: Summary of the most intense Raman modes of diamond, fullerene (C_{60}), graphite, *trans*-poly(acetylene) and poly(diacetylene) in comparison with the hybridization and the bond lengths.

Material	Vibration mode	Frequency [cm^{-1}]	Hybridization	Bond length [pm]	Ref.
diamond	T_{2g} stretching	1332	*sp*^{3}	154	[2]
fullerene C_{60}*	A_g bond alternation	1469	≈ *sp*^{2}	140/145	[4,7]
graphite	E_{2g} stretching	1582	*sp*^{2}	142	[2,6]
trans-poly(acetylene)	C-C stretching	1100	*sp*^{2}	145	[3,5]
	C=C stretching	1500		136	
poly(diacetylene)	C=C stretching	1500	*sp*	142	[3,5]
	C ≡C stretching	2100		132	

* C_{60} has two different bonds and the hybridization is slightly higher than *sp*^{2} [4,7].

Fig. 1 displays the calculated frequencies for the Σ_g "k=0" modes of linear oligoynes with various chain lengths, N [14]. For this special mode, neighboring carbon atoms vibrate in opposite phase. The calculated frequencies, $\nu(N)$ can be fitted as follows:

$$\nu(N) = 1750 + \frac{3980}{N} \quad [\text{in cm}^{-1}] \tag{4}$$

Eq. (4) presents an easy tool to evaluate the conjugation length of actual carbynoid materials.

The vibration frequency of polyyne should also be influenced by n-doping, a characteristic property of carbynoid materials (cf. chapters 3.5 and 4.3). Model studies confirmed that various phases of alkali-doped C_{60} exhibit a characteristic shift of the

A_g bond alternation mode [4,7], while the frequency softening was about 6 cm^{-1} per charge on a C_{60} molecule. A similar behavior was observed for the C=C stretching mode of doped poly(acetylene) and poly(alkylthiophene) [3-5]. A shift to lower frequencies with doping is also observed for some IR active vibrations. For instance, the lower-frequency IR modes of C_{60} can also be used as an indicator for the charge transfer between dopant C_{60} [7]. However, the doping effects in carbyne-like materials were not yet systematically explored by vibration spectroscopy.

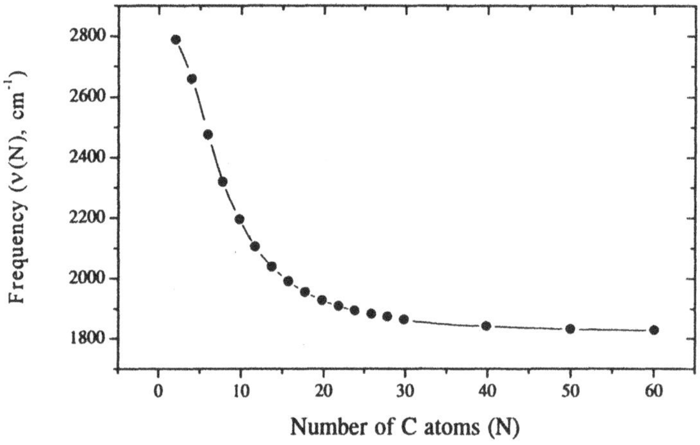

Figure 1: Calculated vibration frequency of the Σ_g "k=0" mode of linear oligoynes after Kastner *et al.* [14]. The frequency was determined from the Longuett-Higgins-Salem model, and is plotted as a function of the number of carbon atoms in the chain (N). For $N \rightarrow \infty$, this mode corresponds to optical mode of polyyne.

The existing experimental IR and Raman studies of carbynoid materials concentrate mostly on the identification of the "carbyne band" around 2000 cm^{-1}. This band shows a pronounced decrease in intensity and increase in frequency (decrease in conjugation length, Eq. 4) if a sample is exposed to elevated temperatures [19-22] and/or reactive environment [21]. Consequently, it is recommended to make the measurements *in-situ* avoiding any contact of the sample with air. This is technically more easy with Raman spectroscopy, where both the sample preparation and spectral measurement can be carried out simply in a sealed silica glass apparatus under vacuum [14]. On the other hand, many IR spectroscopic studies on carbyne should be considered with care, if the sample was handled in air and/or prepared by a drastic KBr-pellet technique [23-27].

Fig. 2 shows that the characteristic CC stretch "carbyne-band" is well-developed in the Raman spectrum of a material prepared by carbonization of poly(tetrafluoroethylene) with sodium amalgam (C-NaF, see Chapter 3.5 of this book for preparative details). However, the C-NaF spectrum also indicates the presence of non-carbynoid impurities (sp^2, sp^3) with a broad band of amorphous carbon between 1000-1700 cm^{-1} (cf. Eq. 3 and discussion thereof). This situation is typical for all other carbynoid materials studied so far. Actually, all IR and Raman spectra of carbyne-like materials indicate the presence of considerable amount of amorphous- or graphite-like carbon, which makes the detailed spectroscopic assessment problematic.

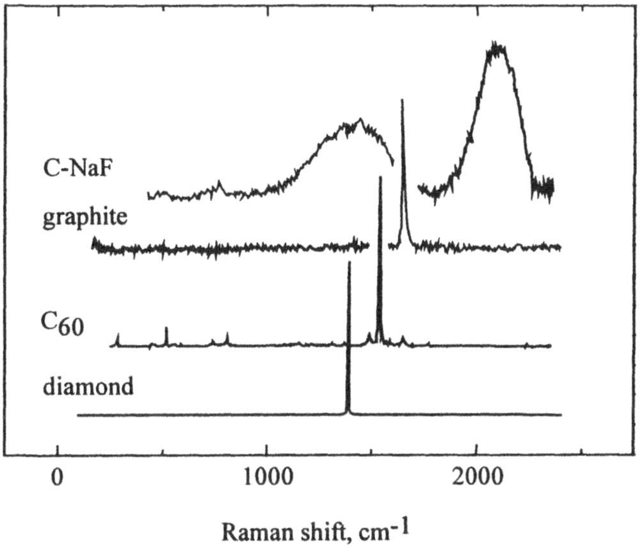

Figure 2: Raman spectra of polycrystalline C-NaF film (made by carbonization of PTFE with sodium amalgam) as compared to the Raman spectra of single crystal graphite, buckminsterfullerene C_{60} and diamond. All spectra were excited with $\lambda = 457.9$ nm. Reprinted from Ref. [28].

2. Raman spectroscopy of carbyne-like materials

Raman spectroscopy has been first used to study carbynoid materials by Walker *et al.* in 1974 [29]. This pioneering work was aimed at a "white carbon" made by laser treatment of graphite, which was believed to be identical with the Ries-crater carbyne (cf. Chapter 3.4). The Raman spectrum of "white carbon" indeed showed weak features at 2140 cm^{-1} in addition to the usual doublet of polycrystalline graphite (cf. section 6.1 - 1) [29]. A subsequent report also mentioned a Raman band at 2150 cm^{-1} in graphite after bombardment with H^+, D^+, He^+ (10-15 keV) [30]. On the other hand, the detection of this Raman band failed in thin (0.1 - 2 μm) transparent carbyne (chaoite) films [31]. Analogously, sputtered carbon films, containing reportedly carbyne, did not show any Raman signal at 2000-2200 cm^{-1} [32].

Higher-quality Raman spectra were obtained for carbyne-like materials synthesized by soft chemical and electrochemical transformations. Table 2 presents an overview of the relevant data.

Table 2: Position, relative intensity and dispersion of the carbyne Raman band for various materials.

Precursor	Preparation method	λ (excitation) [nm]	ν(carbyne) [cm^{-1}]	I_C/I_{bypr}	Dispersion [cm^{-1}/eV]	Ref.
graphite	CO_2 laser treatment	488.0	2140	≈0.03	-	[29]
	laser ablation	-	≈2100	-	-	[50]
	ion bombardment	488.0	2150	-	-	[30]
PVDC	dehydrochlorination	514.5	2159	≈0.3	-	[51]
		457.9	2179	≈0.4	54	[36]
		-	2136	≈0.4	-	[52]
PVC	laser treatment	647.1	2188	≈0.2	-	[38]
PTFE	reduction with K - vapor	514.5	2016	0.5	-	[22]
	reduction with Cs - vapor	457.9	2060	0.4	-	[21]
	reduction with liquid Cs	457.9	2045	0.4	-	[21]
	reduction with Na/K alloy	457.0	2023	1.1	-	[21]
	electrochemical reduction	488.0	2100	0.4	-	[34]
	electrochemical reduction	488.0	2177	0.2	-	[35]
	reduction with K amalgam	457.9	2002	2.2	31	[14]
	reduction with Na amalgam	457.9	2006	1.6	39	[14]
	reduction with Li amalgam	457.9	2042	0.55	73	[14]
FEP	reduction with K amalgam	457.9	2050	0.35	-	[21]
	reduction with Cs vapor	457.9	2080	0.2	-	[21]
PCTFE	reduction with Na amalgam	457.9	2074	0.23	-	[21]
	reduction with Li amalgam	457.9	2010	0.10	-	[21]
$(CHCl)_n$	dehydrochlorination	457.9	2157	≈0.1	59	[33]
C_9F_{20}	reduction with Na amalgam	457.9	2085	0.2	-	[21]
$C_{20}F_{42}$	reduction with Na amalgam	457.9	2021	0.75	-	[21]
$C_{24}F_{50}$	reduction with Na amalgam	457.9	2080	1.1	-	[21]
IC≡CI	electrochemical	-	2150	≈ 0.3	-	[20]
HCBD	electrochemical	-	1950	-	-	[39]

FEP = poly(tetrafluoroethylene-co-hexafluoropropene)
HCBD = hexachlorobuta-1,3-diene
PCTFE = poly(chlorotrifluoroethylene)
PTFE = poly(tetrafluoroethylene)
PVC = poly(vinyl chloride)
PVDC = poly(vinylidene chloride)

There are five principal issues associated with the investigation of carbynoid materials by Raman spectroscopy:

1. Raman spectra of carbynoid materials were interpreted as those of polyyne [14,21,22,33-35], polycumulene [36] or mixture of both isomers [20,37-39]. The first structure is favored energetically in terms of Peierls distortion [8,14,40] but recent quantum-chemical simulations show that both isomers might have theoretically very similar Raman spectra [10,11]. Hence, a convincing distinction between polyyne

and polycumulene is not possible solely by Raman spectroscopy. (More characteristic differences are expected in the IR spectra [10,11]).

2. The intensity of the carbyne band relative to the lines of carbonaceous by-products, I_c/I_{bypr} determines the carbyne content present in the sample. However, the quantification is questionable because of the unknown contribution of resonance enhancement (*vide infra*). The need of independent chemical analysis of the carbyne content was recently addressed by Hlavatý [41].

3. The position of the carbyne band characterizes the conjugation length, i.e. the length of *sp*-conjugated regions. Some authors [20,33,38] have calculated conjugation lengths from a semiempirical correlation using end-capped oligoynes [33]:

$Ar\text{-}(C{\equiv}C)_n\text{-}Ar$; (Ar = aryl, n = 2-6)

This method gives considerably larger values than the quantum-chemical model of pure polyyne chains mentioned above (Eq. 4). Still another model of carbyne vibrations is based on poly(butatriene) [9]. Consequently, it is safer to compare the experimental frequencies of the CC stretch at defined experimental conditions (e.g. excitation wavelength, cf. next paragraph) rather than the estimated conjugation lengths from incompatible models.

4. Another useful parameter is the so-called dispersion effect, i.e., the shift of a Raman band with the energy of the exciting photon [3,4]. This effect was studied in detail for conjugated polymers like poly(acetylene) or poly(thiophene), and can be explained by a photoselective resonance effect [3,4]. The dispersion is zero for an infinite chain and maximum for a very disordered material, whose conjugation length (the region of an unperturbed *sp*-carbon chain) is small.

5. The Resonance Raman effect can be used to investigate the transition between the highest occupied molecular orbital (HOMO) and the lowest unoccupied molecular orbital (LUMO) of carbyne and to deduce the electronic band-gap energy [14,42]. In contrast to optical absorption and electron energy loss spectroscopy, this method allows to study the carbyne without interferences from by-products always present in carbynoid samples.

The vibrations around 2000 cm^{-1} (both IR and Raman active) are also detectable in the products of PTFE reduction with radical anions [43-45] and in the surface-enhanced Raman spectrum of a carbon from photoelectrochemical decomposition of ethylenediamine [46]. Table 2 shows that the highest conjugation length (20 carbon atoms, according to Eq. 4) was found for a carbynoid material prepared by electrochemical reductive dechlorination of HCBD [39]. However, the data on this product are rather scarce (Table 2 and Ref. [39]) and, surprisingly, other authors did not find any carbynoid moieties in virtually identical specimens [47,48].

2.1 RAMAN SPECTRA OF C-MF FROM PTFE

The most consistent Raman results were collected for carbynoid materials prepared by electrochemical defluorination of PTFE and other perfluoro-n-alkanes with alkali metal amalgams (cf. Chapter 3.5 for preparative details). These materials are actually composites of carbonaceous skeleton with the corresponding alkali metal fluoride, C-MF. Table 2 shows that the optimum yields (I_c/I_{bypr} ratios) and conjugation lengths are observed for PTFE as a precursor. They further increased with the molar volume of MF (LiF < NaF < KF) [14,21]. The C-KF product shows the highest known I_c/I_{bypr} ratio of 2.2 and a conjugation length of about 14 carbon atoms (from Eq. 4). Another quality-test is the dispersion effect: The lowest frequency shift per photon energy was observed for C-KF. The dispersion effect was further used to calculate the conjugation length numerically [14]. This was done by applying the conjugation length model [3,4,49] in conjunction with a quantum chemical calculation for the transition energies and matrix elements as a function of chain length.

Raman spectra of C-MF (from PTFE) indicate not only the increase of the carbyne-band frequency (dispersion effect) but also an increase of its relative intensity with increasing energy of the exciting photon (from 2.2 to 3.1 eV) [14]. This increase parallels the increase in the optical absorbance, i.e., it is a consequence of the resonance-enhancement of Raman scattering [14,42]. The absorption band at ≈ 3 eV corresponds presumably to an intrinsic HOMO-LUMO transition of carbyne, which supports the theoretical prediction that carbyne prefers the semiconducting polyyne over the metallic polycumulene configuration [40].

2.2. RAMAN SPECTRA OF ORIENTED CARBYNOID MATERIALS

Polarized Raman spectra of C-NaF from stretch-oriented PTFE indicate the presence of partly oriented *sp*-carbon chains with an average conjugation length of sixteen carbon atoms in the direction of alignment [21]. Both the band intensity and position compare favorably with those of C-NaF from ordinary PTFE.

Even better precursors for aligned carbynoid chain are highly-oriented PTFE films made by friction-deposition, also called "hot-dragging" technique [53]. They are reducible by sodium metal, but the carbonization is not complete in that the composition varied between CF and C_2F [54]. A similar conclusion follows from a reaction of "ordinary" PTFE with K- [22] and Na vapor [55]. On the other hand, the carbonization with Li amalgam leads, as expected, to a quantitative defluorination towards elemental carbon [56]. Raman spectra of highly-oriented C-LiF indicated the CC stretching band at 1988 cm^{-1} [56] which translates into a significantly larger conjugation length [14] than that of the ordinary C-LiF (2042 cm^{-1}, Table 2).

2.3. EFFECTS OF AGING AND CROSSLINKING

A more detailed Raman study indicated a downshift of the intensity of the carbyne-like CC stretch band and upshift of its frequency (i.e. decrease of the conjugation length), if a freshly prepared C-MF sample is stored in vacuum at room temperature (Fig. 3) [14]. The aging rate decreased in the series: Li > Na > K.

The aging of C-MF can be further monitored by an increase of the electronic conductivity, decrease of an UV-Vis absorption coefficient and decrease of the photoluminescence intensity [14,28,57]. These changes have been elucidated in terms of the loss of π-electron density and enhanced electron hopping during polyyne-to-graphyne crosslinking (Fig. 4).

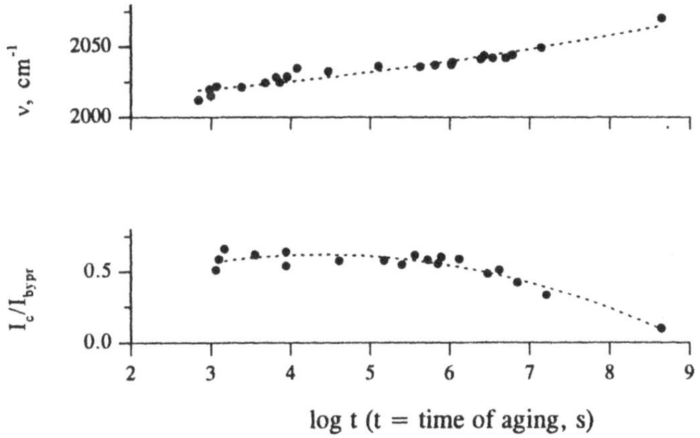

Figure 3. Aging effects in C-LiF prepared by the reaction of PTFE with Li amalgam at 25°C. The sample was stored in vacuum at room temperature. ν is the wavenumber of the Raman band of the CC stretching vibration (excitation wavelength $\lambda = 457$ nm), I_c/I_{bypr} is the Raman intensity ratio of the CC band to that of the sp^2 carbon byproducts. (The maximum time corresponds to about 14 years). Adapted from Refs. [14,28].

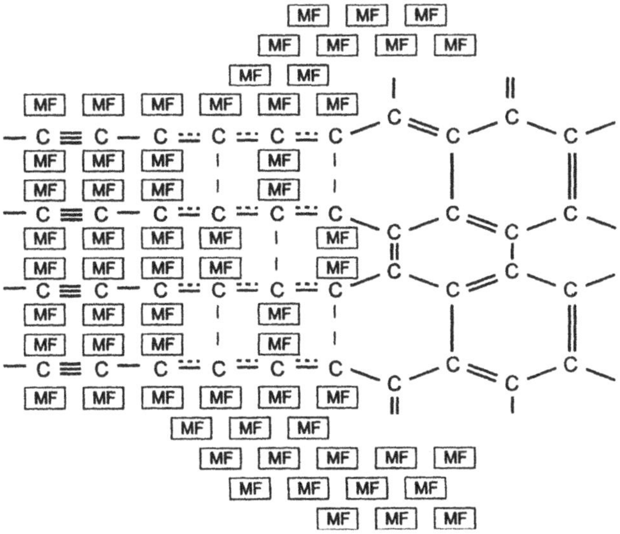

Figure 4: Scheme of polyyne crosslinking in C-MF as a result of aging. Left-hand side represents the structure of a freshly prepared C-MF; right-hand side represents the structure of an aged material.

The most interesting effect is a spontaneous increase in electronic conductivity of C-MF [59] which matches the theoretically predicted transition from semiconductor to metal as a result of crosslinking of two parallel polyyne chains [40] (cf. Fig. 4). These changes are normally slow (detectable over years, cf. Fig. 3), but they accelerate upon heating and particularly after extraction of MF from the C-MF composite [21].

All these features lead to a selfconsistent structural picture: Polyyne, formed by defluorination of PTFE, decomposes slowly to graphene; the yield of polyyne and its crosslinking rate being controlled by a spatial separation of polyyne chains with interspersed MF (Fig. 4). The stabilization of carbyne via spatial separation of chains was also addressed in numerous other studies on carbyne (cf. Chapter 4).

3. Infrared spectroscopy of carbyne-like materials

Although infrared spectroscopy is widely used to characterize carbons and carbonaceous materials [1,5,7] the data on carbynoid samples are less significant. This is caused by technical problems in recording IR spectra of these notoriously unstable and air-sensitive materials (cf. Chapter 6.1 – 1). Most reports published so far deal with IR spectra of carbynoid materials prepared by chemical or electrochemical transformation of polymers[1]. Table 3 summarizes typical examples.

IR spectra of carbynoid materials often show strong lines between ca. 900 and 1800 cm^{-1}. Theoretically, this region (800-1300 and 1500-1800 cm^{-1}) contains fundamental IR-active vibrations of polycumulene [9-11]. The most intense features occur around 1600 cm^{-1}, and many authors assign the actual IR bands between 1500-1700 cm^{-1} to polycumulene [20,23,27,33,37,51,62,63]. However, this interpretation is somewhat questionable, since various non-carbynoid structures and impurities also show strong IR absorption in the same region (e.g. C=O and C=C stretch from surface oxides and olefinic links [1,23,72]).

Interesting IR data were acquired for carbyne-like material prepared by electrochemical dehalogenation of hexachlorobuta-1,3-diene [63] (cf. Chapter 3.1.2 and Fig. 1 therein). The best sample showed only one strong polyyne-like peak in the region of CC stretch (1971 cm^{-1}). However, a sample prepared at slightly different experimental conditions (galvanostatic instead of potentiostatic reduction) gave a complicated spectrum with only a weak band at 2135 cm^{-1} and features more typical for polycumulene [63].

Still different experimental conditions of the same reaction lead to an IR spectrum of a common graphite-like material [47,48]. This example (supported also by Raman spectroscopy, see section 6.2 - 1) demonstrates the crucial role of controlling the preparation conditions in the synthesis of carbyne from polymeric precursors.

---

[1] (Note by the Editor) For further discussion of IR spectra of carbynoid materials prepared by chemical transformation of polymers see Chapter 3.2.1 of this book.

Table 3: Positions of the carbyne infrared band for various materials

Precursor	Preparation	ν (*carbyne*) [cm^{-1}]	Ref.
graphite	vapor-condensation of "carbolite"	2100	[64]
FEP	Li, Na, K amalgams	≈ 2100	[42]
$(CHCl)_n$	dehydrochlorination by DBU	2120 -2140	[33]
PVDC	dehydrochlorination by DBU	2146	[51]
	dehydrochlorination by KOH	2170	[19,27,65]
PTCB	dehydrohalogenation by KOH	2170	[23,25]
PVDF	electrochemical reduction in DMF	2116	[62]
PTFE	electrochemical reduction in THF	2059 , 2013	[35]
	electrochemical reduction in DMF	2170	[66,67]
PCTFE	electrochemical reduction in BL	1750	[24]
PA	high-pressure reaction with K	1150 - 1900?	[19]
I-C≡C-I	electrochemical reduction in DMF	2050	[20,37]
H-C≡C-H	electrochemical oxidation in DMF	2116	[68]
	catalytic oxidation (Cu/O_2)	2080 - 2150	[69]
		2065 , 2100	[70]
	catalytic oxidation (Pd/I_2)	2102 , 2120	[71]
HCBD	electrochemical reduction in DMF	1971 , 2135	[39,63]

BL = gamma-butyrolactone
DBU = 1,8-diazabicyclo[5,4,0]undec-7-ene
DMF = dimethylformamide
PA = poly(acetylene)
PTCB = poly(trichlorobutadiene)
THF = tetrahydrofuran

(For other used abbreviations see Table 2)

Acknowledgment. This work was supported by the Federal Ministry of Science, Research and the Arts of Austria, Grant Agency of the Czech Republic and Ministry of Education, Youth and Sports of the Czech Republic.

4. References

1. Zawadzki, J.: Infrared spectroscopy in surface chemistry of carbons, in Thrower, P.A. (ed.), *Chemistry and Physics of Carbon, Vol.21,* M.Dekker, New York, (1989), p. 147-380.
2. Knight, D.S. and White, W.B.: Characterization of diamond films by Raman spectroscopy, *J. Mater. Res.* **4** (1989), 385-93.
3. Kuzmany, H.: Resonance Raman spectroscopy of conjugated polymers. What can we learn? *Makromol. Chem. Macromol. Symp.* **37** (1990), 81-97.
4. Kastner, J., Winter, J. and Kuzmany, H.: Raman spectroscopy of conjugated carbon systems: polymers and fullerenes, *Conjugated Double Bond Systems: Polymers, Fullerene and Graphite,* Materials Science Forum, Warsaw, (1995), p. 161-70.

5. Zerbi, G.: New perspectives of vibrational spectra in materials science, in Mackenzie, M.W. (ed.), *Advances in Applied Fourier Transform Infrared Spectroscopy*, J. Wiley, New York, (1988), p. 247-322.
6. Kastner, J., Pichler, J., Kuzmany, H., Curran, S., Blau, W., Weldon, D.N., Delamesiere, M., Draper, S. and Zandbergen, H.: Resonance Raman and infrared spectroscopy of carbon nanotubes, *Chem. Phys. Lett.* **221** (1994), 53-8.
7. Kuzmany, H., Matus, M., Pichler, J. and Winter, J.: Vibrational spectroscopy of fullerenes and fullerides, *NATO ASI-C Series* **82** (1993), 274-95.
8. Rice, M.J., Phillpot, S.R., Bishop, A.R. and Campbell, D.K.: Solitons, polarons, and phonons in the infinite polyyne chain, *Phys. Rev. B* **34** (1986), 4139-49.
9. Korshak, V.V., Kudryavtsev, Y.P., Evsyukov, S.E., Korshak, Y.V., Guseva, M.B., Babaev, V.G. and Kostishko, B.M.: IR spectra of carbyne, *Dokl. Akad. Nauk SSSR* **298** (1988), 1421-4. (Dokl. Phys. Chem. 1988, 298, 199).
10. Kurti, J., Magyar, C., Balazs, A. and Rajczy, P.: Calculation of the infrared and Raman spectra of short linear carbon chains, in Kuzmany, H., Fink, J., Mehring, M. and Roth, S. (eds.), *Electronic Properties of Novel Materials: Progress in Fullerene Research*, World Sci. Publ. London, (1994), p. 362-5.
11. Kurti, J., Magyar, C., Balazs, A. and Rajczy, P.: Vibrational analysis for short carbon chains with alternating and cumulenic structure, *Synth. Metals* **71** (1995), 1865-6.
12. Hawthorne, M.F.: Comments on certain reports of carbonaceneous linear polyacetylenes, *Mater. Res. Counc. Summer Conf.* **1** (1973), 135-40.
13. Smith, P.P.K. and Buseck, P.R.: Carbyne forms of carbon, *Science* **229** (1985), 486-7.
14. Kastner, J., Kuzmany, H., Kavan, L., Dousek, F.P. and Kürti, J.: Reductive preparation of carbyne with high yield. An in situ Raman scattering study, *Macromolecules* **28** (1995), 344-53.
15. Weltner, W. and Van Zee, R.J.: Carbon molecules, ions, and clusters, *Chem. Rev.* **89** (1989), 1713-47.
16. Martin, J.M.L., Francois, J.P. and Gijbels, R.: Ab initio study of the infrared spectra of linear C n clusters [n=6-9], *J. Chem. Phys.* **93** (1990), 8850-61.
17. Bjarnov, E., Christensen, D.H., Nielsen, O.F., Augdahl, E., Kloster-Jensen, E. and Rogstad, A.: Vibrational spectra and force field of triacetylene, *Spectrochim. Acta* **30A** (1974), 1255-62.
18. Kloster-Jensen, E.: Reindarstellung und Elektronenspektrum des Triacetylens, Tetraacetylens und Pentaacetylens, *Angew. Chem.* **84** (1972), 483-5.
19. Udod, I.A., Shchurik, V.I., Bulychev, B.M., Sirotinkin, S.P., Guseva, M.B., Babaev, V.G., Kudryavtsev, Y.P. and Evsyukov, S.E.: Formation of carbyne in the interaction of polyacetylene with potassium under high quasi-hydrostatic pressure, *J. Mater. Chem.* **3** (1993), 413-6.
20. Kijima, M., Sakai, Y. and Shirakawa, H.: Electrochemical synthesis of carbyne catalyzed by nickel complex, *Synth. Metals* **71** (1995), 1837-40.
21. Kavan, L., Illavaty, J., Kastner, J. and Kuzmany, H.: Electrochemical carbyne: synthesis and stability studied by Raman scattering, *Carbon* **33** (1995), 1321-9.
22. Yamada, Y., Inada, Y., Shiraishi, M., Oka, T., Oya, A., Miyashita, K. and Ozawa, T.: Preparation of carbyne by chemical reduction of PTFE, *Proc. 4th IUMRS Conference, Makuhari, Japan, Sept. 16-18, 1997*, p. I3,5.
23. Evsyukov, S.E., Paasch, S., Thomas, B., Heimann, R.B. and Kudryavtsev, Y.P.: Formation of carbyne from by chemical dehydrohalogenation of PVDC. A 13-C solid-state NMR study, *Ber. Bunsenges. Phys. Chem.* **101** (1997), 837-41.
24. Stezeryanskiy, E.A., Litovcenko, K.I. and Kublanovskiy, V.S.: Effect of the electrolyte cation on electrochemical reduction of poly(chlorotrifluoroethylene), *Ukr. Khim. Zhur.* **55** (1989), 826-9. (Soviet Prog. Chem. 1989, 55, 42).
25. Vointseva, I.I., Gilman, L.M., Kudryavtsev, Y.P., Evsyukov, S.E., Pesin, L.A., Gribov, I.V., Moskvina, N.A. and Khvostov, V.V.: Chemical dehydrochlorination of polytrichlorobutadienes – a new route to carbynes, *Eur. Polym. J.* **32** (1996), 61-8.
26. Udod, I.A., Bulychev, B.M. and Sirotinkin, S.P.: An interaction of amorphous carbyne with sodium: the formation of carbyne intercalation compounds, *Synth. Metals* **60** (1993), 57-62.
27. Korshak, V.V., Kudryavtsev, Y.P., Korshak, Y.V., Evsyukov, S.E., Khvostov, V.V., Babaev, V.G. and Guseva, M.B.: Formation of carbyne by dehydrohalogenation, *Makrom. Chem. Rapid Commun.* **9** (1988), 135-40.
28. Kavan, L. and Kastner, J.: Carbyne forms of carbon: continuation of the story, *Carbon* **32** (1994), 1533-6.
29. Nakamizo, M., Kammereck, R. and Walker, P.L.: Laser Raman studies on carbons, *Carbon* **12** (1974), 259-67.

30. Wright, R.B., Varma, R. and Gruen, D.M.: Raman scattering and SEM studies of graphite and silicon carbide surfaces bombarded with energetic protons, deutrons and helium ions, *J. Nuclear Mater.* **63** (1976), 415-21.
31. Vidano, R.P., Fischbach, D.B., Willis, L.J. and Loehr, T.M.: Observation of Raman band shifting with excitation wavelengh for carbons and graphites, *Solid State Commun.* **39** (1981), 341-4.
32. Tither, D., Matthews, A., Fitzgerald, A.G., Storey, B.E., Henderson, A.E., Moir, P.A., Dines, T.J., Bower, D.I., Lewis, E.L.V., Doughty, G. and Foster W.: Characterisation of carbon films containing boron and nitrogen, *Carbon* **27** (1989), 899-907.
33. Akagi, K., Nishiguchi, M. and Shirakawa, H.: One-dimensional conjugated carbyne - synthesis and properties, *Synth. Metals* **17** (1987), 557-62.
34. Matsui, H., Yamaguchi, C., Yasuda, A. and Kawase, N.: Preparation of carbyne by electrochemical reduction of PTFE with reactive anode, *Proc. 4th IUMRS Conference, Makuhari, Japan, Sept. 16-18, 1997*, p. I 5,20.
35. Oka, T., Araki, T., Oya, A., Yamada, Y., Shiraishi, M., Miyashita, K. and Ozawa, T.: Preparation and characterization of carbyne derived from aliphatic compounds, *Proc. 4th IUMRS Conference, Makuhari, Japan, Sept. 16-18, 1997*, p. I 5,19.
36. Bedryugin, V.V., Kudryavtsev, Y.P., Evsyukov, S.E. and Korshak, Y.V.: Raman spectra of beta-carbyne, *Dokl. Akad. Nauk SSSR* **305** (1989), 362-5. (Dokl. Phys. Chem. 1989, 305, 217).
37. Kijima, M., Sakai, Y. and Shirakawa, H.: A novel approach for synthesis of carbyne by electroreductive polymerization of diiodoacetylene catalyzed by Ni complex, *Chem. Lett.* (1994), 2011-4.
38. Shimoyama, M., Niino, H. and Yabe, A.: A KrF excimer laser induced dehydrochlorination of chlorinated poly(vinylchloride): preparation of conjugated polyene and polyyne, *Makrom. Chem.* **193** (1992), 569-74.
39. Kijima, M., Toyabe, T. and Shirakawa, H.: Novel approach to synthesis of a carbyne film by electrochemical reduction of hexachlorobuta-1,3-diene, *Chem. Commun.* (1996), 2273-4.
40. Springborg, M. and Kavan, L.: On the stability of polyyne, *Chem. Phys.* **168** (1992), 249-58.
41. Hlavaty, J. and Kavan, L.: Modification of electrochemical carbon by in-situ generated carbenes, *Carbon* **35** (1997), 127-31.
42. Kavan, L. and Dousek, F.P.: Carbynoid species in electrochemical polymeric carbon, *Synth. Metals* **58** (1993), 63-72.
43. Bening, R.C. and McCarthy, T.J.: Surface treatment of poly(tetrafluoroethylene-co-hexafluoropropene), introduction of alcohol functionality, *Macromolecules* **23** (1990), 2648-55.
44. Iqbal, Z., Ivory, D.M., Szobota, J.S., Elsenbaumer, R.L. and Baughman, R.H.: Polyacetylene formed by reduction of PTFE with benzoin dianion, *Macromolecules* **19** (1986), 2992-6.
45. Costello, C.A. and McCarthy, T.J.: Surface selective introduction of specific functionalities onto PTFE, *Macromolecules* **20** (1987), 2819-28.
46. Mernagh, T.P., Cooney, R.P. and Spink, J.A.: Chemical origin of SERS by water on Ag and carbon overlayered Ag electrodes, *J. Chem. Soc. Faraday Trans. I* **80** (1984), 3469-79.
47. Nishihara, H., Harada, H., Tateishi, M., Ohashi, K. and Aramaki, K.: Graphite synthesis by electrochemical reduction of hexachlorobuta-1,3-diene, *J. Chem. Soc. Faraday Trans.* **87** (1991), 1187-92.
48. Nishihara, H., Harada, H., Kaneko, S., Tateishi, M. and Aramaki, K.: Synthesis of graphite film by electrochemical reduction of hexachlorobuta-1,3-diene, *J. Chem. Soc. Chem. Commun.* (1990), 26-7.
49. Kastner, J., Kuzmany, H., Paloheimo, J. and Dyreklev, P.: Resonance Raman scattering from spin-coated and Langmuir-Blodgett poly(3-alkyl-thiophene) films, *Synth. Metals* **55-57** (1993), 558-63.
50. Saito, K. and Koga, Y.: Carbon produced by laser ablation of graphite in He and N2, *Nucl. Instr. Meth. Phys. Res. B* **121** (1997), 400-3.
51. Komatsu, T., Nomura, M., Kakudate, Y., Fujiwara, S. and Heimann, R.B.: Characterization of dehydrochlorinated poly(vinylidene chloride) and the shock-compressed material, *Macromol. Chem. Phys.* **196** (1995), 3031-40.
52. Guseva, M.B., Babaev, V.G., Kudryavtsev, Y.P., Alexandrov, A.F. and Khvostov, V.V.: Medical material based on carbyne, *Diamond Rel. Mater.* **4** (1995), 1142-4.
53. Dietz, P., Hansma, P.K., Ihn, K.J., Motamedi, F. and Smith, P.: Molecular structure and thickness of highly oriented PTFE measured by atomic force microscopy, *J. Mater. Sci.* **28** (1993), 1372-6.
54. Schott, M.: Preparation and properties of highly oriented PTFE films, *Synth. Metals* **67** (1994), 55-61.
55. Tasker, S., Chambers, R.D. and Badyal, J.P.S.: Surface defluorination of PTFE by sodium atoms, *J. Phys. Chem.* **98** (1994), 12442-6.

56. Kavan, L.: Electrochemical preparation of hydrogen-free carbyne-like materials, *Carbon* in press.
57. Kavan, L.: Electrochemical carbonization of fluoropolymers, in Thrower, P.A. (ed.), *Chemistry and Physics of Carbon, Vol. 23*, Marcel Dekker, New York, (1991), p. 69-171.
58. Kavan, L., Dousek, F.P., Weber, J. and Micka, K.: Electronic resistivity of carbon in the products of electrochemical reduction of poly(tetrafluoroethylene), *Carbon* **26** (1988), 235-43.
59. Kavan, L., Dousek, F.P. and Micka, K.: Time-dependent electrical resistivity of carbon, *J. Phys. Chem.* **94** (1990), 5127-34.
60. Kavan, L., Dousek, F.P. and Kubelková, L.: Chemisorption of oxygen on carbon produced by electrochemical reduction of poly(tetrafluoroethylene), *Carbon* **24** (1986), 671-6.
61. Kavan, L. and Dousek, F.P.: Reaction of hexafluorobenzene and hexadecafluorohexane with lithium amalgam, *J. Fluorine Chem.* **41** (1988), 383-91.
62. Kijima, M., Toyabe, T. and Shirakawa, H.: Electrocatalytic dehydrofluorination of PVDF, *Chem. Lett.* **1995** (1995), 553-4.
63. Kijima, M., Toyabe, T., Shirakawa, H., Kawata, S., Kyotani, H., Nakamura, Y. and Endo, M.: Synthesis of carbyne by cathodic electrolysis of hexachlorobutadiene, *Synth. Metals* **86** (1997), 2279
64. Barker, D.J., Brewis, D.M., Dahm, R.H. and Hoy, L.R.J.: Anisotropic electrochemical reduction of PTFE, *J. Mater. Sci.* **14** (1979), 749-51.
65. Kudryavtsev, Y.P. and Evsyukov, S.E.: Formation of diamond from carbyne, *Diamond Rel. Mater.* (1997),
66. Pud, A.A. and Shapoval, G.S.: Cathodic transformation of PTFE in the presence of electron mediator, *Theor. Exp. Khim.* **32** (1996), 40-2. (Theor. Exp. Chem. 1996, 32, 38).
67. Pud, A.A. and Shapoval, G.S.: Electrochemistry as the way to transform polymers, *Macrom. Reports* **A32(Suppls. 5&6)** (1995), 629-38.
68. Ohmura, K., Kijima, M. and Shirakawa, H.: Synthesis of conducting polymers with conjugated carbon-carbon triple bonds by electrochemical condensation of acetylene derivatives catalyzed by coper complex, *Synth. Metals* **84** (1997), 417-8.
69. Sladkov, A.M., Kasatochkin, V.I., Kudryavtsev, Y.P. and Korshak, V.V.: Synthesis and properties of linear carbons, *Izvest. Akad. Nauk Ser. Khim.* **12** (1968), 2697-704.
70. Cataldo, F.: Synthesis of carbonaceous material rich in carbyne, a carbon allotrope by coupling reaction of copper acetylide, *Eur. J. Solid State Inorg. Chem.* **34** (1997), 53-63.
71. Hattori, T., Kijima, M. and Shirakawa, H.: Oxidative polycondensation of acetylene by iodine in the presence of a palladium-copper catalyst, *Synth. Metals* **84** (1997), 357-8.
72. Tanuma, S. and Palnichenko, A.V.: Synthesis of low density carbon crystal "carbolite" by quenching of carbon gas, *J. Mater. Res.* **10** (1995), 1120-5.

6.2. ELECTRON SPIN RESONANCE SPECTROSCOPY

GENERALIZED SPIN-PEIERLS TRANSITION IN CARBYNOID FILMS

D.P. ERTCHAK
Department of Physics,
Byelorussian State University, Minsk, Byelorussian Republic

Abstract - Some ideas of generalized Peierls as well generalized spin-Peierls (s-P) transitions in conjugated (quasi) 1D--polymers are discussed. ESR, IR and microwave photoconductivity studies on oriented carbynoid samples prepared by chemical dehydrofluorination of poly(vinylidene fluoride) (PVDF) agreed rather well with the assuption of a room temperature generalized S-P system. Concepts of s-P solitons and s-P soliton lattice formation in carbynoid films at room temperature are proposed and discussed.

1. Introduction

Low-dimensional systems possess a number of remarkable properties in comparison with corresponding three-dimensional (3D) analogues. Carbyne is an example of a very simple quasi-one-dimensional (1D) system containing atoms of only one kind. Consequently, carbyne and carbynoid structures may serve as models to study the physics of low-dimensional systems. At the same time, the 1D-structure of carbyne suggests promising potentials for their practical use, *e.g.* in electronics or medicine. ESR studies on oriented carbynoid samples prepared by (incomplete) chemical dehydrofluorination of poly(vinylidene fluoride) films showed two types of paramagnetic centers (PC), denoted as C-M1 and C-M2, and an anisotropic broad line (BL) [1]. The ESR spectrum of the PC C-M2 consists of a series of unsaturated lines, and is anisotropic both in relation to the static magnetic field and the magnetic component of the microwave field. The study of the interaction dynamics of the paramagnetic centers with the microwave field, the absorption dependence with increasing microwave field, and their temperature dependencies have shown that these PC are mobile quasi-particles. The results described in Ref. [1] are consistent with the notion that a ground state of the carbon chains in carbynoid samples is a state with a spin density wave (SDW), and that the PC C-M2 can be associated with quasi-particles related to kinks (domain walls) in the alternated spin density distribution, tentatively considered solitons in the SDW. The absorption by PC C-M1, well-saturated at room temperature, was attributed either to free radical states [1] or pinned quasi-particles similar to Su-Schrieffer-Heeger topological solitons [2]. Recently it was assumed by

R.B. Heimann et al. (eds.), Carbyne and Carbynoid Structures, 357–369.

Ertchak *et al.* [3] that these states may represent new members of the soliton family very similar to those described by the Korteweg de Vries (KdV) equation. It has also been found that spin waves associated with the PC C-M2 can be observed in samples treated thermally at 120°C for 2 hours [1]. The samples had a fluorine to carbon atomic ratio of F/C = 3/7 and contained up to 20 at.%oxygen.[1] The results were similar to quasi-1D-carbon systems formed by ion-beam modification of films and single crystals of diamond. During irradiation, presumably a superlattice of quasi-1D spatially ordered nanotubes was formed in the 3D-diamond matrix. [3,4,5].

Presently, so-called s-P compounds are intensively studied by a number of experimental methods, in particular by ESR and NMR. Among these systems are even inorganic compounds, *e. g.* $CuGeO_3$ and α'-NaV_2O_5. An s-P system is an interacting (quasi 1D-antiferromagnetically ordered) spin system coupled to phonons in such a way that below a certain temperature T_{SP} the translational spin symmetry is spontaneously broken and the spin system becomes dimerized, *i.e.* spin coupling alternation takes place.

The development and generalisation of the ideas presented in Ref. [1] shown below revealed that the samples studied can be referred to as generalized s-P systems.

2. Experimental Technique

Identical samples as investigated in Ref. [1] were used, *i.e.* both unoriented and uniaxially oriented samples were prepared by chemical dehydrofluorination of poly(vinylidene fluoride) (PVDF) films. The samples contained a rather high concentration of residual fluorine atoms, and also oxygen atoms introduced during the experimental treatment. Three sample series were studied: series 1 with a F/C ratio of 3/10, and 10% oxygen content, series 2 with a F/C ratio of 3/7 and 20% oxygen content, both thermally treated at 120°C for 2 hours. The contents of F and O of series 3 were intermediate between the former ones though not determined exactly. Statistically uniformly distributed impurities display, as will be discussed below, spin and/or charge dopant effects that will lead to spin-/charge conductivity increase with increasing impurities contents. According to the classification proposed for doped transpolyacetylene (*trans*-PA [5]), the samples studied can be tentatively considered doped carbynes with doping levels > 11 mol % as argued in Ref. [1][2].

ESR measurements were performed using VARIAN E-Line, E-4 and BRUKER spectrometers with a TE_{102} cavity. The absorption of Cr^{3+} ions of a ruby standard sample permanently located in the cavity was used for (i) fine tuning the phase of the

[1] The extremely high fluorine content points to a very incomplete dehydrofluorination of the precursor poly (vinylidene fluoride) (PVDF). Hence the results of the ESR measurements and their interpretation by the present author should be viewed with caution and still considered the subject of future revision. Nevertheless, the editors have decided to include this chapter despite of its shortcomings since it provides useful details with regards to experimental strategy and also to statements that are of general interest for the study of quasi 1D-dimensional systems (eds.).

[2] These concentrations are clearly beyond the range generally considered for doped conducting polymers as they result from incomplete dehydrofluorination of the starting polymer (eds.).

modulation field, (ii) the calibration of the relative values of the magnetic component H_1 of the microwave field, and (iii) the calibration of the dependence of the line intensity on the modulation frequency. In addition, IR studies were performed at room temperature but will not be reported here.

3. Results

Figure 1 shows general ESR absorption spectra observed in samples of the 3rd series. The enveloping lines of ESR absorption by the C-M2 paramagnetic centers have an asymmetrical shape similar to the Dysonian-type shapes observed in conductive materials. A considerable change of the intensities of the ESR absorption with time was observed. The intensity decreases within 12 days by a factor of 2 for PC C-M2, and by a factor of 7.5 for the broad line (BL).

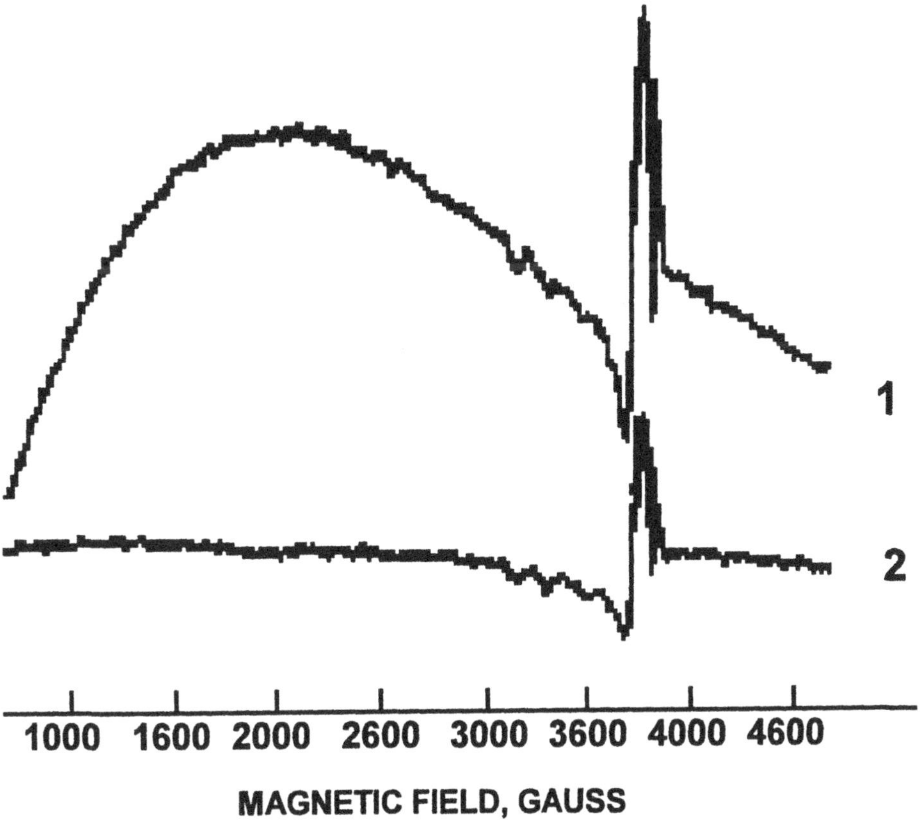

Figure 1. ESR absorption in an oriented doped carbynoid film of the 3rd series, immediately after sample preparation (1) and after 12 days storage time (2).

It is interesting that in the absorption derivative curve the sequence of the weak-amplitude peak and the strong-amplitude peak with increasing field is opposite to that predicted by the common Dyson effect [6] observable in conductive solids, *e.g.* 3D metals [7]. Note also that the shape of the ESR absorption curve in doped *trans*-PA corresponds to a Dysonian with the usual sequence of weak and strong amplitude peaks with field increase [8].

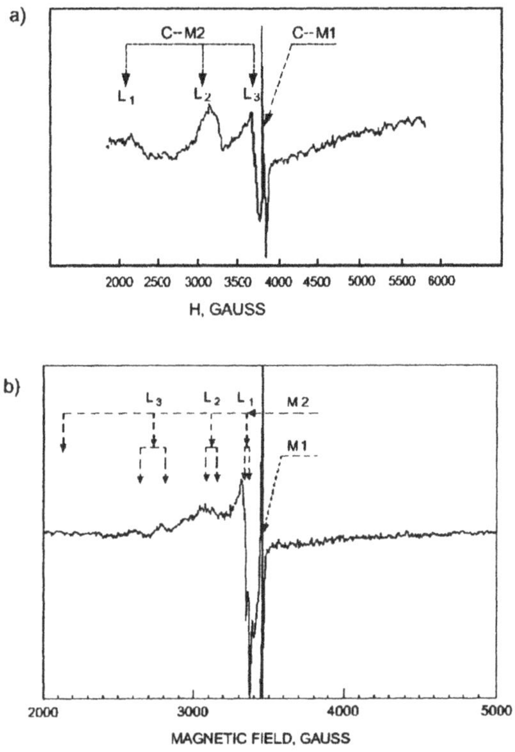

Figure 2. ESR spectra of an oriented carbynoid film of the 2nd series after sample preparation (a) and after storage for 1.8 year (b). The sample plane is perpendicular to the microwave magnetic field $\vec{H_l}$, the static magnetic field $\vec{H}$ is parallel to the chain orientation axis, ν = 10.7324 GHz (a), ν = 9.6447 GHz (b). Upper arrows in figure 2b indicate the positions of the center of gravity of the spectral distribution of line groups L_1 (n = 0, 1), L_2 (n = 3), and L_3 (n = 5) and the possible position of the 7th mode L_4. Lower arrows indicate the positions of the two dominating lines in groups L_1, L_2 and L_3.

In all oriented samples an anisotropy was observed in the PC C-M2 spectrum in relation to the magnetic component of the microwave field [1]. The particulars of this dependency are very sensitive to the level of impurities, *i.e.* fluorine and oxygen, and to the thermal treatment. In the samples of the 2nd series (Fig. 2), two additional groups of lines L_2 and L_3 appear in the C-M2 spectrum when $\vec{H_l}$ is perpendicular to the sample plane. These additional lines may be spin wave excitations. The group L_1 corresponds to the main quasi-particle resonance mode discussed in Ref. [1]. The emergence of spin waves correlates with the larger contents of fluorine and oxygen in the samples of the 2nd series compared with the first and the third series.

These results appear to be related to the doping properties of residual impurities. The thermal treatment at 120°C probably triggers an impurity redistribution and hence the formation of some dopant complexes. Mobility of impurity atoms, not included in the main carbon backbone chain, is a general property of 1D organic polymers [9]. Note also that doping of organic 1D-polymers with halogens, halogen-oxygen (cp. footnote 2) complexes as well as oxygen is well known [10].

If the $\vec{H_l}$--component of the microwave field is parallel to the sample plane, the spectrum of PC C-M2 is almost isotropic, and consists of the group L_1 only, *i.e.*, spin waves are not excited when the sample plane orientation is parallel to $\vec{H_l}$. Instead of spin wave resonance, an additional broad line ($\Delta H \sim 600$ G) is clearly visible (cp. Ref. [1]).

In samples of the 1st and the 3rd series spin waves were not observed. However, in these samples a broad line (BL) also appears simultaneously with the C-M1 and C-M2 spectra when the sample plane was parallel to $\vec{H_l}$. This line is much broader when compared with that observed in samples of the 2nd series, and is registered only partly in the field scan range accessible to the Varian E-4 spectrometer (Fig. 1). The lines of groups L_2 and L_3 (Fig. 2) show also considerable anisotropy in relation to the static $\vec{H}$-field.

Studies of the interaction dynamics of the PC C-M2 with the microwave field were performed in a manner analogous to that reported by Ertchak *et al.* [4,5,11,12], and were discussed in Ref. [1] and are represented in Figure 3.

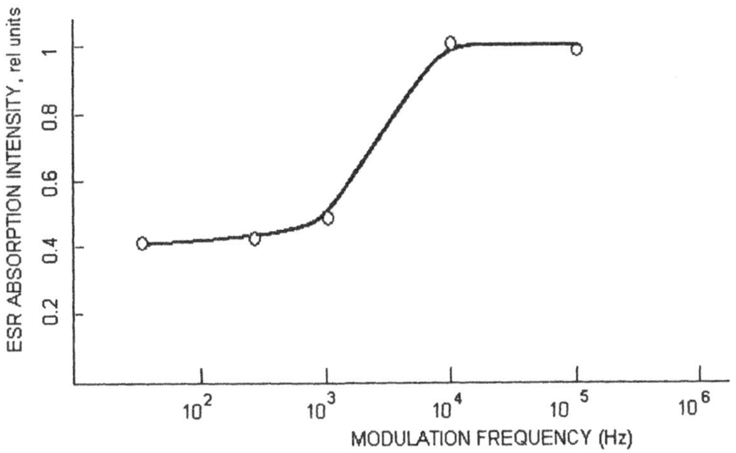

Figure 3. Dependence of the intensity of ESR lines of PC C-M2 on the modulation frequency f_m of the static magnetic field *H*.

It is characteristic that there are essential angular dependencies of the amplitudes of the broad line BL and of the PC C-M2 envelope in contrast to PC C-M1, the intensity of which behaves practically isotropic (Fig. 4).

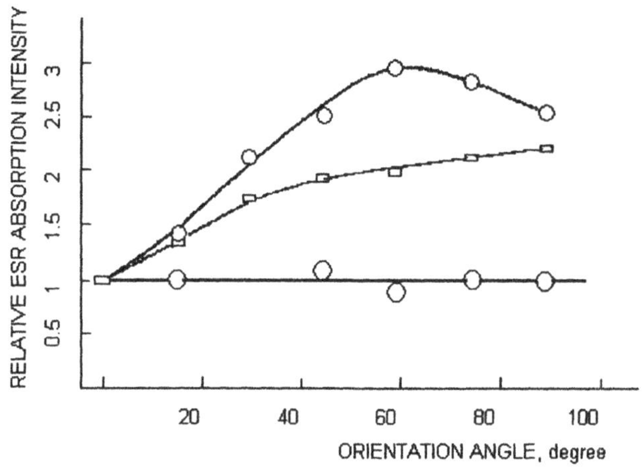

Figure 4. Relative ESR absorption intensities of the amplitudes of PC C-M1 (bottom curve), PC C-M2 (center curve) and the broad line (BL) (top curve) as a function of the angle between the static magnetic field $\vec{H}$ and the normal $\vec{n}$ to the sample plane. The microwave magnetic field $\vec{H_l}$ is parallel to the sample plane.

4. Discussion

There is a longstanding discussion concerning the principal possibility of the existence in a stable solid state of carbyne as an example of pure 1D solids. Based on the theoretical work of Peierls [14] and Landau [15], according to which there is an infrared divergence for the mean square of the lattice atomic displacements u_i^2 at $k = 0$ for 1D-solids, many researchers consider the existence of carbyne principally impossible. However, as recently found by Apostol [16], the mean square deviation of the lattice constant $(u_i - u_m)^2$ is finite thus providing a theoretical possibility for the existence of 1D-solids in a dynamical regime. Therefore, the theoretical results of Chugreev and Misurkin [17] concerning the special case of carbyne satisfy Apostol's suggestion by explaining the existence of a total energy minimum of the carbon chain as the result of the alternation in both the distribution of bond lengths, and the spin or charge densities. In other words, a number of dynamical wave states of a chain, *e.g.* a coupling-order wave (COW), spin density wave (SDW) and a charge density wave (CDW) as well as mixed states can be realized in pure as well as in doped carbon (quasi) 1D-chain system. The variability of these states seems to relate to the experimental conditions of the carbyne synthesis.

There is an earlier suggestion by Heimann *et al*, that the stabilization of pure carbyne chains can be realized by kink formation in their geometrical structure [18-21][3]. The actual driving forces of kink formation can be understood on the basis of recent results by Ung *et al.* [22] and Hu *et a.l* [23]. The main idea is as follows. The common Peierls transition leads to bond length dimerisation. When the electron-electron interaction is sufficiently strong, the Peierls transition leading to conventional dimerisation can be suppressed, but the transition corresponding to a larger number of atoms in the dimerised unit can still be realized. For instance, when both on-site Coulomb repulsion and nearest-neighbour electron-electron interactions are included, tetramerization will be enhanced [23]. It is thus suggested that the varying numbers of carbon atoms in the rectilinear chain between kinks in a number of polytypic carbyne phases is determined by the degree of electron-electron interaction along the chain, *e.g.* 6-atom-kinks can be realized in the case when additionally second-nearest neighbour interaction is included. Peierls transition in this form can be considered a generalized Peierls transition. In fact, the work of Heimann *et al.* suggests a generalized Peierls transition in quasi 1D-systems such as in carbyne [19,20]. The notion of conjugation lengths introduced for carbyne chains [19,20] then obtains an additional physical sense as the correlation lengths for 1D-electron-electron interatomic interaction in 1D-systems subjected to a generalized Peierls transition.

Furthermore, the superlattice produced as the result of the action of a generalized Peierls transition can be formally renormalized and hence reduced to a structure such as the *trans*-PA carbon backbone described in Refs. [24,25]. Accordingly, calculations analogous to those shown in Ref. [26] can be made. It can be predicted *a priori* that the dimerised bond structure of the superlattice units is energetically more favourable. Since the distance between superlattice units can be described as a vector quantity, the present case corresponds to simultaneous dimerisation for bond lengths and bond angles between superlattice units. Consequently, the formation and propagation of quasi-particles similar to *topological* solitons in *trans*-PA, *i.e.* solitons acting as domain walls in the bond length distribution is conceivable also in the superlattice of a kinked chain-structure. Essentially, kink angles and kink offsets are subject to statistical fluctuations as suggested previously [18-20].

Thus the idea implies a two step-process: 'kink-superlattice' production and subsequent bond dimerisation between superlattice unit elements. Both steps can contribute to stabilization of the pure carbyne chain depending on the synthesis conditions, for example employing laser ablation-, shock wave-, and chemical vapour deposition technologies, or technologies using chemical or electrochemical reactions with widely varying process parameters, *i.e* temperature, pressure *etc*.

Note that the 1D-superlattice formation with translationally invariant units equal to interkink distances provides a base for the observation of many physical phenomena characteristic for crystals with superlattices. For instance, temporal and spatial Bloch oscillations, defined as quantum beats of Bloch electron wave functions that are determined by interatomic and interkink translation invariances, can be predicted similarly to those observed in Ref. [27,28].

---

[3] Here the term 'kink' refers to a non-zero degree bond angle between two adjacent fragments of the same chain. See Chapter 4.2 of this book.

We will here reproduce briefly the arguments advanced in Ref. [1] concerning the SDW formation in the samples studied. Since the band gap values due to a Peierls transition in *trans*-PA and a Mott-Hubbard transitions in carbyne (carbynoid) samples appear to be comparable, the energy states of quasi-particles such as solitons in alternating spin density distributions and alternating chemical bond length distributions, in particular in COWs and in SDWs (CDWs) are expected to be *a priori* similar, *viz*., deep states in a gap. The assumption of a different origin of quasi-particles in carbyne compared to *trans*-PA is supported by the deviation of the *g*-value range of the PC C-M2 from the *g*-value of 2.00263 characteristic of solitons in polyacetylene [2,8,29]. Indeed, since the spin-orbital interaction parameter λ for carbon has a relatively low value, at least compared to silicon and germanium, the range of the *g*-changes $\Delta g = g - 2.0023$ for all known intrinsic PCs in pure carbon materials is also relatively narrow ($\Delta g < 10^{-2}$). Thus it is expected that the *g*-value of the solitons predicted to exist in the (bond-length alternating) polyyne chains [30] should be close to that in *trans*-polyacetylene (g = 2.0026), *i.e.* should possess a different value compared to the *g*-values observed for PC C-M2 for which the Δg-value, ranging from 2.1×10^{-2} to $\sim 8.3 \times10^{-2}$, is relatively large. The alternated spin and charge distributions can essentially change the orbital motion of the spin carriers and, consequently, can considerably change the *g*-values. Thus, the large Δg-values of *ca.* $(2 - 8) \times\cdot 10^{-2}$ observed for PC C-M2 [1] agree well with the assumption of an alternating spin or charge density distribution for the ground state of the chains.

The observation of the anisotropy of the ESR spectra related to the magnetic component of the microwave field, and the absence of a similar anisotropy related to the electric component of the microwave field is a manifestation of the resonance absorption effect by the quasi-particles associated with SDWs rather than CDWs. The analogous quasi-particles in simultaneously formed CDWs can be spinless and hence are not observable by ESR. In conclusion, the effect of the anisotropy of the ESR spectra related to the magnetic component of the microwave field can be explained by an anisotropy of the spin wave excitation, and indicates that the samples studied are systems of oriented 1D-elements.

Furthermore, the assumption of formation of a SDW in carbyne is believed to be confirmed directly by the observation of the broad line, BL (Fig. 1). This line may be associated with collective spin absorption in the SDW. The possibility of the collective magnetic absorption in SDWs means that SDWs themselves can move along the chains. This suggestion allows to explain the angular dependencies of the intensities of the broad line (BL) and the PC C-M2 (Fig. 4). The measured angular dependencies can be explained by an anisotropy of excitation of motion both of mobile quasi-particles, suggested to be solitons in a SDW (corresponding to the PC C-M2 spectra; Fig. 4, center curve), and of the SDW itself (corresponding to the broad line. BL; Fig. 4, top curve). The qualitative similarity of the degree of asymmetry of the angular dependency values of BL and PC C-M2 can be easily accounted for by referring to systems with similar absorption and spin energy transfer properties.

The ground state of carbynoid chains with SDWs is presumably a result of the effect of the residual impurity atoms such as fluorine and hydrogen, and oxygen. The doping effect of these impurities can be tentatively realized in relation to both spin and charge conductivities. If the number of fluorine atoms is less than that of carbon atoms per unit length they can stimulate a commensurate or incommensurate transitions in the spin or charge distributions leading correspondingly to formation of SDWs or/and CDWs. Which of these transition will occur presumably depends on the relative concentration of the doping atoms and their distribution along the carbon chains. It should be noted that the SDWs and CDWs in the samples studied can be simultaneously produced, similarly to those observed, in the structure of the perovskite phase $La_{1-x}Sr_xFeO_3$ [31]. However, as stated above, CDWs cannot be directly detected by ESR since charged quasi-particles such as charged solitons in *trans*-PA [5] are expected to be spinless. In other words, while the presence of impurity atoms (F, O, H) appears to preserve the common and generalized Peierls bond length (angle) instability, it can also activate an s-P transition and a generalized s-P transition, respectively. This takes place in a manner analogous to the generalized common Peierls transition. According to Mazumdar *et al* [32,22], the presence of elements of a two-dimensionality in the interaction is inherent in quasi-1D chains with SDWs. Thus, the effect of the interaction of the main atomic chain with impurity atoms located in side chains seems to favour these conclusions.

s-P transitions as well as the state with SDWs can be considered special cases of a generalized s-P transition.

ESR- transitions observed in s-P compounds are attributed to transitions in excited states, corresponding to „spin-conductivity'' band states [33] or interband transitions from ground to excited states [34]. However, the detailed nature of interband transitions is under intense discussion at present (cp. [33,34]). *A priori*, domain walls in alternating spin interaction system with, for example SDWs, *i.e.* in generalized s-P systems do correspond to a state in a s-P band gap with deep levels. As a result, the temperature dependence of the absorption intensity ought to be described by the Curie law or its close approximation. However, the temperature dependence of the absorption intensity observed in some s-P systems is different. While the intensity of ESR absorption in a „spin-conductivity'' band has a temperature dependence typical for excited sites [33], the intensity of the ESR absorption for interband transitions is only weakly dependent on the temperature. Moreover, interband transitions can be observed at high frequencies, *e.g* in a range of 1100-1600 GHz for $CuGeO_3$ [33]. The intensity change according to the Curie law observed for the PC C-M2-absorption corresponds to deep level states. Therefore, the observed dependence can be considered an additional argument in favour of the conclusion that the ESR absorption expressed in the PC C-M2 spectrum is determined by quasi-particles in SDWs, *i.e.* in systems that can also be referred to as generalized s-P systems.

Recently, the conjecture that s-P solitons can be considered elementary exitations in s-P quasi-1D systems was treated theoretically [35]. Spin-soliton characteristics were calculated within the framework of the generalized t-J model taking into account both fermion-phonon and spin-phonon interactions, and considering also spin-charge separation. The model advanced in Ref. [34] can be employed to a comparatively wide range of quasi-1D systems with very strong electron-electron interactions and moderate electron-phonon interactions.

Note that the dynamical studies presented in Figure 3 display in fact the character of the staggered distribution of the distances between individual spins or cooperative spins within spin units for s-P solitons or generalized s-P solitons, respectively.

The experimental confirmation that two different types of elementary excitation, *viz.*, spin waves and a system of mobile quasi-particles such as solitons, are actually connected to each other has been obtained in Ref. [1]. Hence a system of mobile quasi-particles can be the source of spin waves at appropriate conditions that are determined by the strength of the π-electron-electron correlations between the atoms in the chains. This conclusion can be understood theoretically taking into account that, for example for the nonlinear Korteweg de Vries (KdV) equation, the localized initial disturbance can be divided into two parts: a solitary wave part, and a part with an oscillating wave whose amplitude decreases with time [36,37]. On the other hand, the nonlinear equation describing the propagation of soliton-like quasi-particles in the samples studied ought to be similar to the KdV-equation as well as to the equation describing the soliton propagation in *trans*-PA as discussed above.

The formation of the well-resolved structure of the PC C-M2 (Fig. 2b) also points to a process of self-ordering of the structure. This process resembles the crystallization of amorphous substances, *e.g.*, amorphous SiO_2 or common glasses with time. A possible explanation of this self-ordering effect is as follows. According to Heimann *et al.* [19,20], the carbyne chains represent a superposition of atomic linear C-C fragments (clusters) of definite lengths (cp. Chapter 4.2). The C-C fragments are connected to each other, producing alternating kinks in chemical bonds. The kink angles and kink offset distances are proposed to be statistically fluctuating and hence satisfy the concept of a paracrystal [18,21]. The kinks in the carbon chains are terminating the crystallographical identity period and thus define discrete *c*-axis lengths („conjugation lengths'') [18,19,20]. There exist also the concept of a temperature-dependent „equilibrium'' of chain lengths [19,38]. This concept may lead to a random distribution of the chain lengths around a mean value. It should be expected that immediately after their preparation the (oriented) carbyne (carbynoid) films have some distribution in atomic cluster lengths around the length corresponding to the conditions of synthesis. Irregularities in the distribution of impurities, and hence the difference of the chain lengths as well as the presence of other chain defects [39] can lead to an initial set of spin (and charge) density distributions which, presumably, can explain the presence of the superposed paramagnetic absorption lines in groups L_1, L_2 and L_3 assigned to PC C-M2 (Figs. 1, 2). With this concept, the differences of the experimental spectra obtained after a storage time of 1.8 years (Fig. 2b), may correspond to a time-dependent chain „self-rearrangement'' at fixed conditions, *e.g.*, temperature. This is also well in accord with observations by Kavan and Kastner [40] on carbynoid samples stored up to 14 years that showed systematic changes in their physical properties with time (see Chapter 6.1). The differences of the experimental spectra obtained after a storage time of only 12 days (Fig. 1) may correspond to a time-dependent impurity redistribution, a process requiring much less time compared to the situation above.

The results obtained allow to tentatively propose a stabilization mechanism for the carbon chains in carbyne and carbynoid structures. The stabilization can be achieved dynamically through the formation of mobile SDWs (and probably CDWs) as well as through the formation of mobile quasi-particles in spin density distribution (and probably in charge density distribution, too) in full accord with theoretical predictions of the dynamical stabilization in pure 1D-solids [16]. It is suggested that commensurate-incommensurate transitions take place, *i.e.* spin density distributions along the atomic chains can have different spatial order in comparison with the chemical bond distribution. The role of impurity atoms affecting some physical properties such as charge transfer, spin transfer *etc.* can also be taken into account in correspondence with Refs. [18-20,43-47] as additional factor stabilizing the carbyne chains. However, dynamic stabilization seems to be preferred since the dopant atoms as well as other impurities not incorporated in the C-C chain directly are also mobile in (quasi) 1D-polymers [9].

5. Conclusions

ESR, and IR and microwave photoconductivity studies (discussed in Ref. [1]) on oriented films of incompletely dehydrofluorinated poly(vinylidene fluoride) appear to confirm the notion of the existence of a generalized spin-Peierls (s-P) system at ambient conditions.

It is tentatively proposed that a chain state exists with a spin density wave (SDW) as a ground state of the carbon chains, as well as a paramagnetic center (PC C-M2) associated with kinks (domain walls) in the spin density distribution. The latter has been assigned to a soliton in the SDW.

Impurities related to either incomplete dehydrofluorination (F,H) or contamination during sample preparation (O) are considered mediating agents of the formation of a s-P soliton lattice that is expressed by a 'spin-conductive' band in the generalized s-P band gap.

6. Acknowledgments

The author gratefully acknowledges a grant-in-aid by the Saxon State Ministry of Sciences and the Arts, Dresden, Germany in support of the present work. He is also indebted to Professor Dr R.B Heimann (Freiberg) for helpful discussions and editorial comments, and to Professor Yu.P Kudryavtsev and Dr Yu.A Koksharov (Moscow), and Dr.S.E.Evsyukov (Ludwigshafen) for help with the experiments and their interpretation.

7. References

1. Ertchak, D.P., Kudryavtsev, Yu.P., Guseva, M.B., Alexandrov, A.F., Evsyukov, S.E., Babaev, V.G., Krechko, L.M., Koksharov, Yu.A., Tichonov, A.N., Blumenfeld, L.A., Bardeleben, v. H.J.: Electron spin resonance in doped carbynes, submitted to *Chem.Phys.Letters*.
2. Heeger, A.J., Kivelson, S., Schrieffer, J.R. and Su, W.-P.: Solitons in conducting polymers, *Rev. Mod. Phys.* **60** (1988), 781-850.

3. Ertchak, D.P., Efimov, V.G., Stelmakh, V.F., Martinovich, V.A., Alexandrov, A.F., Guseva, M.B., Penina, N.M., Varichenko, V.S., Karpovich, I.A., Zaitsev, A.M., Fahrner, W.R., and Fink, D.: Origin of dominating ESR absorption in ion implanted diamonds, *Physica Status Solidi*, **203** (1997), 529-548.
4. Ertchak, D.P., Efimov, V.G.and Stelmakh, V.F.: ESR studies of low dimensional structures in natural diamonds and synthetic diamond films (Review), *J. Appl. Spectroscopy* **64** (1997), 433-460.
5. Erchak, D.P., Guseva, M.B., Aleksandrov, A.F., Alexander, H.and Pilar von Pilchau, A.: Spin waves in boron-implanted polycrystalline diamond, *JETP Lett.* **58** (1993), 275-278.
6. Dyson, F.J.: Electron spin resonance absorption in metals. II. Theory of electron diffusion and the skin effect, Phys. Rev. **98** (1955), 349--359.
7. Feher, G. and Kip, A.F.: Electron Spin Resonance Absorption in Metals. I. Experimental, Phys. Rev. **98** (1955), 337--348.
8. Goldberg, I.B., Crowe, H.R., Newman, P.R., Heeger, A.J., and MacDiarmid, A.G.: Electron spin resonance of polyacetylene and AsF5-doped polyacetylene, *J. Chem. Phys.* **70** (1979), 1132-1136.
9. Kirova N. and Brazovskii S.: Dopant complexes and their effect on optical and contact processes, Synth. Metals, **85** (1997).
10. Yoshino, K., Harada, S. Kyokane, J., Iwakawa, S. and Inuishi, Y.: Carrier mobilities and breakdown in halogen doped polymers, J.Appl. Phys. **51** (1980), 2714--2717.
11. Erchak, D.P., Efimov, V.G., Zaitsev, A.M., Stelmakh, V.F., Penina, N.M., Varichenko, V.S., and Tolstykh, V.P.: Peculiarities of damage in diamond irradiated by high energy ions, *Nucl. Instrum. Methods Phys. Res., Sect. B* **B69** (1992), 443-451.
12. Erchak, D.P., Efimov, V.G., Azarko, I.I., Denisenko, A.V., Penina, N.M., Stelmakh, V.F., Varichenko, V.S., Zaitsev, A.M., Melnikov, A.A., Ulyashin, A.G., Shlopak, N.V., Bouilov, L.L, Varnin, V.P., Botev, A.A., Sokolina, G.A., and Teremetskaya, I.G.: Electron paramagnetic resonance of boron-implanted natural diamonds and epitaxial diamond films, *Diamond Relat. Mater.* **2** (1993), 1164-1167.
13. Boivin, C., Jacolin, C., and Savard, J.Y.: Detection of paramagnetic resonance by the saturation effect, *Rev. Sci. Instrum.* **44** (1973), 191-196.
14. Peierls, R.: Transition temperatures, Helvetica Physica Acta, Suppl. 2, **7** (1934), 81-83.
15. Landau, L.: On the theory of phase transition , *Zh. Exp. Teor. Fiz*, 7 (1937) 627-632, *Phys. Z. Sowjetunion*, **11** (1937), 545 (in Russian).
16. Apostol M.: On the low-dimensional solids and their melting, *Synth. Metals*, **79** (1996), 253--257.
17. Chugreev, A.L. and Misurkin, I.A.: Electronic structure of carbyne in the unrestricted Hartree-Fock approximation, *Zh. Strukt. Khim.* 30, (1989), 24-28 (in Russ.)[*J. Struct. Chem.* **30** (1989), 377-380 (Engl. Transl.)]
18. Heimann, R.B.: Linear finite carbon chains (carbynes): their role during dynamic transformation of graphite to diamond, and their geometric and electronic structure, *Diamond Relat. Mater.* **3** (1994), 1151-1157.
19. Heimann, R.B., Kleiman, J., and Salansky, N.M.: A unified structural approach to linear carbon polytypes, *Nature* **306** (1983), 164-167.
20. Heimann, R.B., Kleiman, J., and Salansky, N.M.: Structural aspects and conformation of linear carbon polytypes (carbynes), *Carbon* **22** (1984), 147-156.
21. Heimann, R.B.: Should carbyne be described by paracrystal theory?, *Carbon* **35** (1997) 1669-1671.
22. Ung, K.C., Mazumdar, S. and Toussaint, D.: Metal-insulator and insulator- insulator transitions in the quarter-filled band organic conductors, *Phys. Rev. Lett.* **73** (1994), 2603--2606.
23. Hu, L., Szeto and K.Y, Sun, X.: Influence of strong electron-electron interaction on the Peierls transition, *Chin. Phys. Lett.* **14** (1997), 63-66.
24. Mendez, E.E., Agullo-Rueda, F. and Hong, J.M.: Stark localization in GaAs—GaAlAs superlattices under an electric field, *Phys. Rev. Lett.* **60** (1988), 2426-2429.
25. Lyssenko, V.G., Valusis, G., Lowser, F., Hasche, T. and Leo, K.: Direct measurement of the spatial displacement of Bloch-oscillating electrons in semiconductor superlattices, *Phys. Rev. Lett.* **79** (1997), 301-304.
26. Su, W.P., Schrieffer, J.R. and Heeger, A.J.: Soliton excitations in polyacetylene, *Phys. Rev. B* **22** (1980), 2099-2111.
27. Da Rosa, A.L.S. and de Melo, C.P.: Electronic properties of polyacene, *Phys. Rev. B* **38** (1988), 5430-5437.

28. Chen,Y., Xiong, S.-J. and Evangelou, S.N.: Transport of carriers in a semiconductor polymer random chain: a renormalization approach, *J. Phys.:Condens. Matter* **9** (1997), 4823-4830.
29. Weinberger, B.R., Ehrenfreund, E., Pron, A., Heeger, A.J. and MacDiarmid, A.G.: Electron spin resonance studies of magnetic soliton defects in polyacetylene, *J. Chem. Phys.* **72** (1980), 4749-4755.
30. Rice, M.J., Phillpot, S.R., Bishop, A.R., and Campbell, D.K.: Solitons, polarons and phonons in the infinite polyyne chain, *Phys. Rev. B* **34** (1986), 4139-4149.
31. Li, J.Q., Matsui, Y., Park, S.K. and Tokura, Y.: Charge ordered states in La_{1-x} Sr_x FeO_3, *Phys. Rev. Lett.* **79** (1997), 297-300.
32. Mazumdar S., Lin H.Q. and Campbell, D.K.: The phase diagram of 2:1 cationic charge transfer salts, *Synth.Met.*, **41-43** (1991), 4047-4050.
33. Brill, T.M., Boucher, J.P., Voiron, J., Dhalenne, G., Revcolevschi A. and Renard, J.P.: High-field electron spin resonance and magnetization in the dimerized phase of $CuGeO_3$., *Phys. Rev. Lett.*, **73** (1994), 1545-1548.
34. Uhrig, G.S.: Symmetry and dimension of the magnon dispersion of inorganic SP systems, *Phys. Rev. Lett.*, **79** (1997), 163-166.
35. Terai, A.: Solitons and phonons in the spin - Peierls background, *Synth. Metals*, **85** (1997), 1055-1058.
36. Solitons in Action, Ed. by Lonngren, K. and Scott, A., Academic Press, New York, 1978.
37. Soliton theory. Inversal Problem Method. Ed. by Novikov, S.P., Moscow, Nauka, 1980.
38. Whittaker, A.G.: Carbon: A new view of ist high temperature behaviour, *Science* **200** (1978), 763-764.
39. Evsyukov, S.E., Paasch, S., Thomas, B., and Heimann, R.B.: Formation of carbynoid structures by chemical dehydrochlorination of poly(vinylidene chloride). A C13 solid-state NMR study, *Ber. Bunsen-Ges. Phys. Chem.* **101** (1997), 837-841.
40. Kavan, L. and Kastner, J.: Carbyne forms of carbon: continuation of the story, *Carbon* **32** (1994), 1533-1536.
41. Korshak, V.V., Kudryavtsev, Yu.P., Evsyukov, S.E., Korshak, Yu.V., Guseva, M.B., Babaev, V.G., and Kostishko, B.M.: Infrared spectra of carbyne, *Dokl. Akad. Nauk SSSR* **298** (1988), 1421-1424 (in Russian).
42. Kudryavtsev, Yu.P., Evsyukov, S.E., Guseva, M.B., Babaev, V.G. and Khvostov, V.V.: Carbyne - the allotropic form of carbon, *Izv. Akad. Nauk, Ser. Khim.*, (1993), 450-463, (in Russ.).[Russ. Chem. Bull. **42** (1993), 399-413 (Engl. Transl.)]
43. Eastmond, R., Johson,T.R. and Walton D.R.M.: Silylation as a protective method for terminal alkynes in oxydative couplings. Ageneral synthesis of the parent polyynes H $(C{\equiv}C)_n$ H (n = 4-10,12),*Tetrahedron* **28** (1972), 4601-4616.
44. Whittaker, A.G. and Wolten, M.: Carbon: a suggested new hexagonal crystal form, *Science* **178** (1972), 54-56.
45. Hay, A.S.: Oxidative polymerization of diethynyl compounds, *J.Polym. Sci, Part A-1* 7 (1969), 1625-1634
46. Kudryavtsev, Yu.P., Evsyukov, S.E., Babaev, V.G., Guseva, M.B., Khvostov, V.V. and Krechko, L.M.: Oriented carbyne layers, *Carbon* **30** (1992), 213-221.
47. Udod, I.A., Shchutvik, V.I., Bulychev, B.H., Sirotinkin, S.P., Guseva M.B., Babaev, V.G., Kudryavtsev, Yu.P. and Evsyukov, S.E.: Formation of carbyne in the interaction of polyacetylene with potassium under high quasi-hydrostatic pressure, *J. Mater. Chem.* **3** (1993), 413-416.

6.3. ELECTRON SPECTROSCOPY

L.A. PESIN
Chelyabinsk State Pedagogical University
454080 Chelyabinsk, Russia

Abstract – A comparative examination and discussion of electron and photon spectroscopy data obtained at present for a variety of carbynoid materials are given. Special attention is being paid to the problem of identification of chain-like atomic arrangements by means of XPS, XES, EELS and AES techniques. The validity of the commonly used AES self-deconvolution routine to extract direct information on the valence DOS is discussed. Some prospects of the application of synchrotron radiation for studying the electronic structure of carbyne-like materials are also considered.

1. Introduction

Electron and photon emission spectral analyses have nowadays become a common and widely employed tool to study surfaces of solids including various forms of carbon. However, comparatively little work was devoted to carbyne-like modifications of condensed carbon, caused primarily by the extreme difficulty to synthesize chemically pure samples of equilibrium chain-like material. As a rule, only amorphous materials with a significant content of carbon atoms with bond hybridization distinct from an *sp* configuration were available for spectroscopic research. Also, samples with high concentrations of heteroatoms were commonly studied. The fact that these structures exist in a non-equilibrium state has been confirmed by the pronounced effect on their chemical composition and spectroscopic parameters of experimental factors such as vacuum, particle bombardment, temperature and other external conditions. As a result, information on the electronic structure of chain-like carbon has a fragmentary and often controversial character.

The aim of this chapter is to examine, compare and discuss the emission data of various carbon materials with presumably a one-dimensional mode of atomic arrangement. Since semi-empirical interpretation of spectra is usually based on the experimentally determined intensity values, represented as the product of the combined density of states (DOS) and the emission cross-section, the chapter begins with a short review of one-dimensional carbon electron structure calculations. These theoretical results are then compared with the experimental data obtained by means of x-ray photoemission (XPS), soft x-ray emission (XES), Auger emission excited by

R.B. Heimann et al. (eds.), Carbyne and Carbynoid Structures, 371–394.

electron bombardment (AES) and by x-ray *K*-level photoionization (XAES), and electron energy loss (EELS) spectroscopy techniques. The *2s* carbon states photoionization cross-section is at least 10 times greater than that of *2p*-electrons if x-ray radiation with photon energies of about 1-1.5 keV is used. Thus carbon XPS provides information about energy distribution of predominantly *s*-like valence states. In contrast, dipole selection rules restrict the XES sensibility to mostly *p*-symmetrical states. A correct combination of both techniques may not only allow to obtain total valence band DOS, but also to study the shape and energy location of the subbands formed by wave functions of different symmetry configurations as previously done successfully for graphite and diamond. The combined conduction and valence DOS and collective electron excitations are commonly studied by EELS technique which is also widely employed for various carbons investigations.

One of the most complicated problems is to gain information about valence DOS from AES line shapes. In spite of the close relation between them it will be shown below that results of an inverse self-folding routine are essentially wrong when directly compared with DOS because of effects of the multiple shape distortion during the final stage of Auger relaxation.

The problem of carbyne identification needs more careful consideration. It actually always arises in carbynoid studies and thus graphite-like and diamond-like samples are necessary for data comparison. New spectroscopic criteria applicable to elucidate the carbon atoms hybridization mode have been found recently and seem to have good prospects for further application. A brief description of this technique based on XPS and XAES principal features and intensity analysis will be given and the results obtained will also be discussed.

The chapter concludes with suggestions to apply electron and photon spectroscopy utilizing synchrotron radiation to carbyne study.

2. Theoretical Study of One-dimensional Carbon Electronic Structure

The band structure of one-dimensional crystals is the most frequently discussed topic in textbooks on solid state physics[1]. Both the near-free electron assumption as well as the tight-binding approximation [1] predict similar behaviour with respect to a near parabolic or cosine dependence of the energy on the wave vector. The least curvature of it and equally the greatest DOS are observed in the Brillouin zone (BZ) centre and on its edge. The bottom and the top of the valence band are localized exactly at these very points of inverse space.

The energy levels of cumulene-type chains with 2 to 8 double bonded atoms have been calculated by the MO LCAO technique [2] (Fig. 1). The larger the number of atoms n in a chain, the wider become the energy intervals in which electron states exist, with a saturation tendency evident. For n=8 the energy width between the lowest and the highest bonding hybrid levels (σ) reaches 7.5 eV, that between non-hybrid

[1] For additional details *cf.* Chapter 4.1 of this book.

bonding (π) and antibonding (π*) states is about 12 eV. The upper part of the latter range almost entirely includes antibonding hybrid states (σ*). There exists an energy gap of about 2 eV separating σ- and π-levels. If general considerations [1] are valid, one should expect for $n \geq 8$ DOS maxima at 22, 15 and 13 eV below the highest σ*-level. These energy values correspond to the bottom and the top of σ-states, and the bottom of the π-subband. In addition, there must appear a pronounced DOS feature in the π*- and σ*-states overlap region [3].

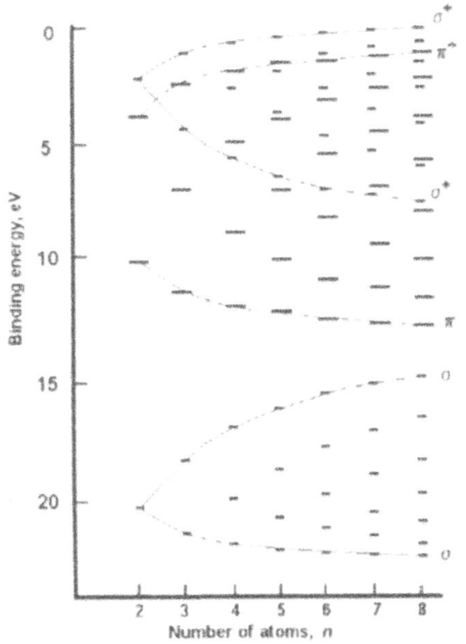

Figure 1. Energy levels of cumulene-type carbon chains containing different number of atoms [2]. The origin of the energy scale coincides with the highest antibonding states.

The energy dispersion law has been determined for an infinite linear carbon chain with interatomic distances of 128 pm using the tight-binding approximation by Baitinger *et al.* [4]. Hamiltonian matrix elements have been calculated as usual by the extended Hückel technique. To simplify the analysis of the data in comparison with those obtained for graphite and diamond, a diatomic unit cell was used. The results of these calculations (Fig. 2) show the existence of two doubly degenerated hybrid dispersion branches σ_1 and σ_2. The spherical wave functions comprise an overwhelming contribution of the first branch. The *2p*-states yield in the second branch σ_2 is at its minimum at the one-dimensional BZ edge, but it increases as the energy approaches the valence band top and reaches its maximum at the BZ centre. The σ_1- and σ_2-branches touch each other at the BZ edge; in its centre they produce the valence hybrid subband bottom and top, respectively. This subband width is approximately 15 eV, *i.e.*, twice the value that can be anticipated from the results discussed above [2]. The other two dispersion branches π_1 and π_2 are also doubly degenerated. They are formed by quasi independent *p*-like wave functions oriented in space normally to each other as well as to the atomic chain axis. Their dispersion laws are analogous: the larger the wave number, the higher the π-states energy becomes. If interchain forces are absent both branches completely coincide. A small splitting takes place if one takes into account the interaction of atoms belonging to the adjacent chains. Its magnitude depends on the interchain distance and may reach up to 2 eV [4]. Valence π-states form the top and the bottom of the subband whose maximum width is about 7 eV at the BZ centre. At the same point σ- and π-states overlap slightly (about 1 eV). At the BZ edge the valence π-subband and the empty π*-states touch or overlap slightly in the case of interchain interaction. Thus there is no gap in the band structure of the model β-carbyne chain. This conducting structure is unstable and transition to a dielectric state occurs. Chain conformations and interactions stabilize the carbyne structure when initiated by such a phase transition [4]. Moreover, in one-dimensional crystals the electron occupation of the bands may have a specific character. It is possible if the Coulomb repulsion of two electrons in the same orbit overcomes the entire chain energy advantage owing to compact Pauli-like band filling. Antiferromagnetic ordering as well as interatomic distance alternation may create an energy gap between the π- and π*-states [5].

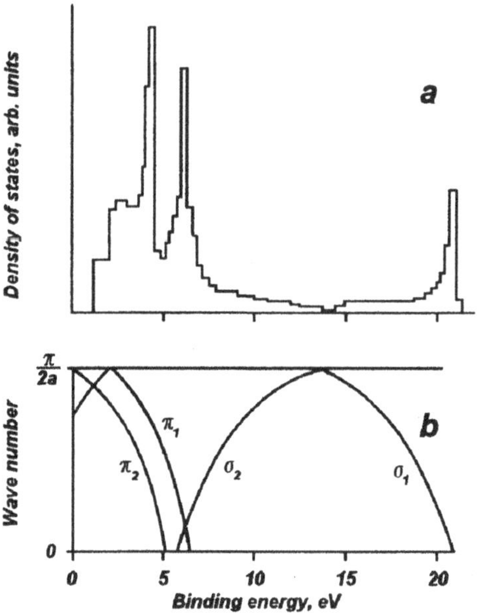

Figure 2. The infinite chain DOS (a) and energy band dispersion curves (b) [4]. The origin of the energy scale coincides with the Fermi level position.

In the contributions by Baitinger *et al.* [4-6] a thorough description of the results described above has been given with special attention being paid to the states symmetry since this is extremely important for direct comparison with experimental spectroscopic results. One-dimensional energy band spectrum calculations were accomplished using information on partial *s*- and *p*-like electronic states DOS. The data have been obtained not only for β-carbyne, but also for conformed *cis*- and *trans*-like chains [4,6]. They are qualitatively similar to each other and agree with general conclusions [1,3]: DOS is largest at the subband boundaries. The same is true for many other chain-like carbon systems [6]. Figure 3 shows the α-carbyne model which has a more complicated DOS structure with some additional features.

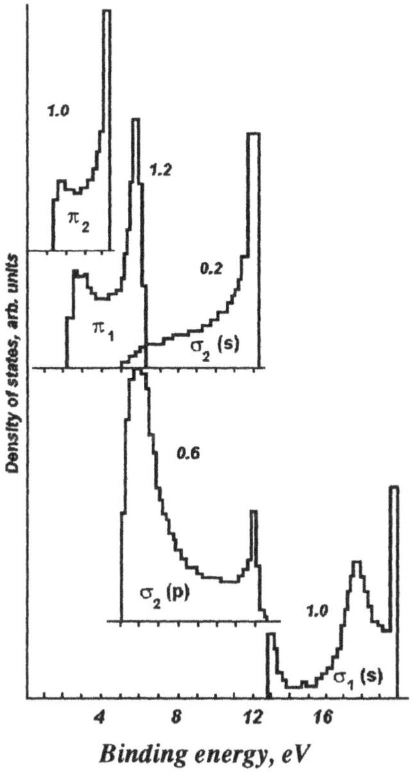

Figure 3. α-carbyne partial DOS normalized to the maximum height. The partial contribution of each subband to the total DOS is given in electron units [6]. The energy scale starts from the Fermi level position.

Finally the results of *ab initio* band spectrum calculations for polyethylene [7] will be presented (Fig. 4). The carbon atoms in this infinite one-dimensional hydrocarbon chain exist in sp^3 hybrid state. The total width of the valence band is about 20 eV. It is very close to that predicted for graphite and diamond [5]. It consists of low-energy (C-

C) and high-energy (C-H) subbands of nearly the same width (7 eV) separated from each other by a gap (5 eV). Qualitative analysis of the dispersion curves predicts DOS features at 0, 2 to 4, 6 to 7, 12 to 14, and 18 to 19 eV below the valence band top, *i.e.*, actually at the subband boundaries with the exception of the second and the last feature whose position correspond neither to a bottom nor to a top of any subband. This fact is likely to mean that a more complicated band spectrum exists in the case of an sp^3- compared to an *sp*-hybridization mode. Since a direct DOS determination unfortunately was not the scope of the work [7], the last conclusion seems to have only tentative validity.

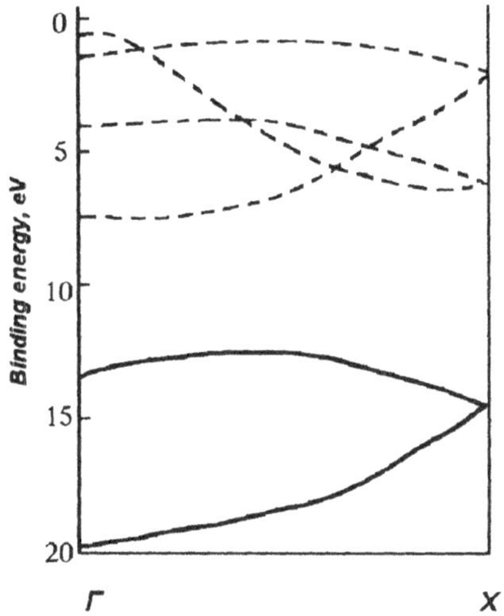

Figure 4. Energy band dispersion of polyethylene [7]. Energy values are measured from the top of the valence band.

Thus various one-dimensional electron structure calculations show in most of the cases a qualitative consistency of the results obtained. Quantitative distinctions concern the subband widths and their relative energy positions. In this respect the data reported by Baitinger *et al.* [4] considerably disagree with those obtained by other authors but are quite similar to those for graphite and diamond.

3. Experimental Study of Carbyne-like Structures by Various Spectroscopic Techniques

3.1. XPS

XPS and XES on carbyne-containing films derived from poly(vinylidene fluoride) (PVDF) and acetylene were studied by Baitinger *et al.* [4] and Sergushin *et al.* [8]. In the latter contribution [8] samples of natural diamond and graphite as well as a sample of carbon black obtained from a propane plasma were also investigated to provide data for comparison. Special attention was paid to the description of significant differences between valence and core electron XPS of the carbynoid samples, and the graphite and carbon black samples. It was confirmed that the carbynoid electronic structure differs greatly from that of graphite-like materials. However, the problem of carbyne identification could not be solved by Sergushin *et al.* [8] since the results showed essentially a close similarity of the carbynoid and diamond XPS shape. In both cases in the binding energy range of 10 to 20 eV there exists a doublet valence band feature. In the vicinity of the C*1s*-line there is no feature typical of graphite, which is caused by π-electron excitations during *K*-level photoionization [9]. The detailed XPS shape correlation of these two samples was not analysed [8]. The shift observed for the carbynoid XPS to the lower binding energies in comparison with the diamond spectrum may be due to different electrostatic charging of dielectric samples and thus does not reflect directly a distinguishing feature of the electronic structure. Since a high oxygen content has been observed at the carbynoid surface [8], a shape distortion may occur of the low energy shoulder (8-10 eV) of the valence band XPS resulting from the O2*p*-states. Considering this effect would lead to a closer resemblance of the XPS of these samples.

The data obtained by Baitinger *et al.* [4,10,11] with monochromatic Al K_α-radiation confirm the valence band XPS doublet shape of the carbynoid materials studied and its essential similarity to diamond [12,13], pristine PVDF [11] and polyethylene [9,14] spectra (Fig. 5). The XPS of pristine PVDF displays the existence of oxygen and fluorine on the film surface. Dehydrofluorination for 18 minutes reduces the heteroatom content but does not induce drastic modifications of the spectrum shape. All principal spectral features and their energy positions agree well with the results obtained for polyethylene [7] and diamond [15] band structure calculations. Some quantitative discrepancies between DOS and XPS shapes originate from the difference in ionization cross-sections for the various initial electron states, as has been demonstrated earlier for graphite and diamond [13]. Since the general shape of the valence band XPS is governed by the nearest neighbour environment of the emitting atom [13], one can conclude that the remarkable similarity of the spectra displayed in Fig. 5 is primarily due to an identical dominant sp^3-bond mode for all samples investigated. On the contrary, the type of structural dimensionality is ruled by long range order of the atomic arrangement. As shown by McFeely *et al.* [13], long range order effects have little influence on the spectral shape in general but display more clearly the fine structure features.

From this point of view a conservation of the prevalent sp^3-bond type in carbynoid materials becomes obvious. This fact may testify whether an essentially incomplete hydrogen removal during chemical carbonization of PVDF (hydrogen cannot be detected by XPS) takes place or, most probably, a pristine one-dimensional structure bifurcation due to side reactions and/or polymer chains cross-linking. These processes are indicated in Fig. 5 by features *c* and *d* showing an intensity redistribution in the direction consistent with the XPS shape modification from polyethylene to diamond. Only the appearance of feature *e* in spectrum 3 confirms the formation of a carbyne structure during dehydrofluorination. Its energy position (23 eV) closely corresponds to the calculated bottom of the valence band. Though its low intensity reflects a small content of *sp*-type structural fragments in the film, the energy resolution is sufficient as not to confuse this spectral feature with an O2*s*-state. The energy location of the latter is almost identical in the variety of oxygen bond modifications (24-26 eV) [11,16]. This fact is useful for the alignment of the spectra when charging effects take place that may produce XPS displacement [16].

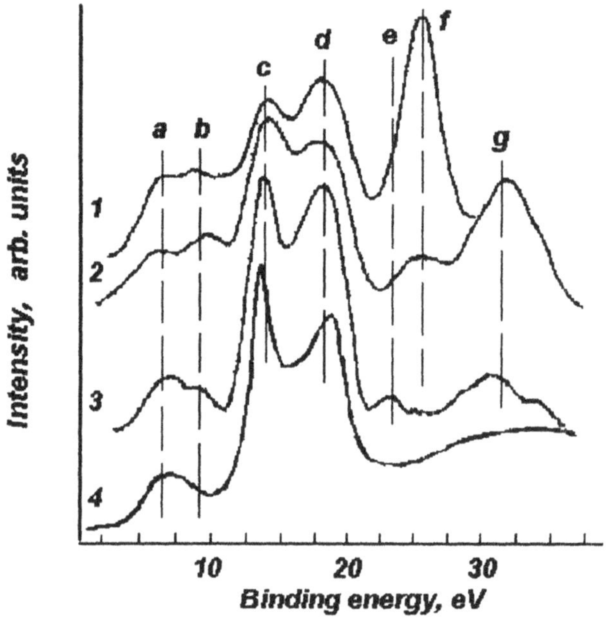

Figure 5. Valence electron XPS of diamond (curve *1*), a pristine PVDF film (*2*), a carbynoid material derived from PVDF by dehydrofluorination for 18 minutes (*3*) [4,10-12] and polyethylene (*4*) [9,14]. The broken vertical lines *a-g* mark spectral fine structure features, *f* and *g* reflect O2*s*- and F2*s*-state yields, respectively. It is difficult to separate the O2*p*- and F2*p*-state contributions in the energy range close to feature *b*.

An alternative interpretation of the data of Fig. 5 may be based on the qualitative correspondence between the DOS of spectrum 3 and α-carbyne (see Fig. 3). However, in this case feature *e* would dominate because of the great magnitude of the DOS and photoemission cross section for *s*-like state products near the valence band bottom. This is rather unlikely and thus the first interpretation seems to be more probable, though it is not absolutely free from ambiguities since the valence electron hybridization mode sometimes cannot exactly predict the typical XPS shape. This fact is evident from the data for polypropylene [9].

Observations on the valence band structure of three diacetylene polymers have been reported by Stevens *et al.* [16]. XPS data were recorded with a spectrometer using Mg K_α-radiation. The use of different crystalline materials allowed to identify the spectral features produced by the polymer chains and to compare them with theoretical models. Spectral shapes of two of these polymers with side groups CH_2-O-SO_2-∅CH_3 and $(CH_2)_3$-O-CO-NH∅ closely resemble those of carbynoid structures [8], diamond [13] and polyethylene [14] because of the doublet character of the C2*s*-peak. The valence band XPS shape of the third diacetylene with the simple side group CH_2OH is almost featureless having an intense and broad maximum around 17.5 eV. Nevertheless, weak features are noticeable in the binding energy region of 21-22 eV suggesting a contribution of the states near the bottom of the σ-subband and thus indicating the existence of some *sp*-hybrid bonds in all three polymers examined. The XPS of diacetylene polymers display a valence band width that is in good agreement with calculated results (about 20 eV). However, intercomparison of the spectra leads to the assumption that only the lowest binding energy π-band, just above the spectra oneset, and the σ-band at about 17.5 eV can be attributed to the polymer backbone. Interference by side group bands prevents a detailed comparison of the theoretical band structures of the polymer skeleton with simple side groups attached to it with the experimental data.

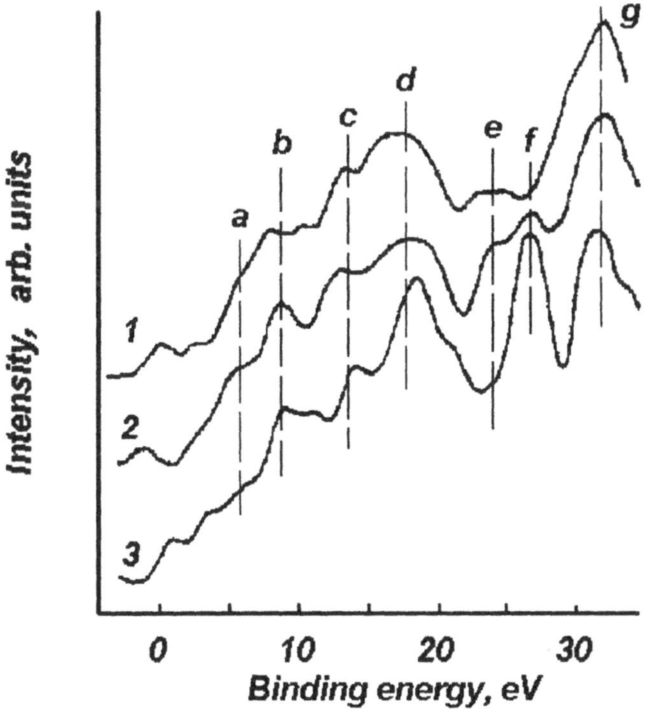

Figure 6. Valence electron XPS of carbynoid materials derived from PVDF by dehydrofluorination for 6 (curve *1*), 12 (*2*) and 16 min (*3*). The energy position features *a*-*g* are in a satisfactory accordance with the data of Figure 5.

The main features of the intensity redistribution effect are more clearly pronounced in the case of fluorine (and, possibly, hydrogen) substitution by oxygen or oxygen-containing species during chemical carbonization of PVDF [17] (Fig. 6). If the synthesis time does not exceed 12 min, the spectra display features *c* and *d* which are specific for diamond and polyethylene XPS (see Fig. 5). Longer reaction times lead to a significant decrease of the intensity of feature *c* and results in a graphite-like spectrum [13]; reduction of the intensity of the F2*s*-peak (*g*) and an increase of the O2*s*-peak (*f*) intensity are observed simultaneously. Though these spectra have been obtained with lower energy resolution than reported by Baitinger *et al.* [4,10,11], the separation of the *e* and *f* features is clearly visible indicating their different origin. Apparently, the creation of oxygen-containing "bridges" between adjacent chains promotes formation of sp^2-hybrid orbitals. Thus a modification of a short range-order mode drastically changes the valence electron XPS shape. The analysis of the data of Fig. 6 shows the complicated nature of the effect in question. At the first stage linear chain fragments of sufficient length appear that are free from fluorine and hydrogen. In fact, while the synthesis time increases from 6 to 12 min, the intensity of the *e*-feature increase is accompanied by the simultaneous depression of the F2*s*-state yields indicating *sp*-bonds formation. Longer reaction times lead to the disappearance of the *e*-feature because oxygen-containing species initiate bond transformations to a local sp^2-configuration by chain bending and cross-linking with neighbouring chains.

3.2. XES

The shape and energy position of the CK_α-emission spectrum have been considered by Sergushin *et al.* [8] as the most prominent criterion of the carbyne-like mode of atomic arrangement. Actually their data analysis (Fig. 7) demonstrates a considerable shift of carbynoid XES to the region of the lower photon energy (or of the higher binding energy) relative to diamond and graphite spectra. A shift of the short wave length-feature *A* was interpreted as a result of a strong π-electron interaction within a carbyne chain. However, the displacement of the carbynoid spectrum as a whole, and particularly that of the main maximum *B* formed by the hybrid *p*-like states $\sigma_2(p)$ (*cf.* Fig. 3) could not been explained by the authors [8]. The shift of feature *B* appears to be contradicted by the results presented later by Khvostov *et al.* [3]. The latter show equivalent energy positions of the hybrid subbands in any form of one-dimensional carbon.

Several reasons for this discrepancy may exist. First, a small number of atoms within a chain result in a narrower hybrid subband width [2,3] (*cf.* Fig. 1). Thus its top corresponding to feature *B* will be observed at higher binding energies with reference to the Fermi level than in the case of an infinite chain. Then, despite the fact that the final state of the CK_α-emission process must be quite similar in carbyne, graphite and diamond, the initial states may significantly differ due to a specific response of these carbons valence electrons to a sudden creation of *K*-vacancies and different rates of relaxation of excited states. Finally, a failure of the spectrometer calibration cannot be excluded. This latter possibility is suggested as an explanation for an abnormal short wave length-threshold position (Fig. 7) for graphite and diamond XES (about 288 eV instead of 284-285 eV that is consistent with XPS data [8]). The carbynoid and graphite thresholds matching equalizes the positions of feature *B* and diminishes the energy differences of feature *A* of both carbons, but the latter shift insists on being observed in accordance with the interpretation given in [8]. It should be particularly

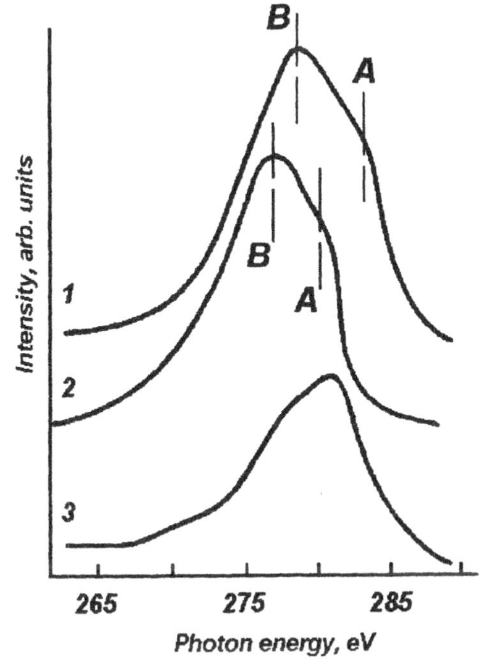

Figure 7. XES of carbon samples: graphite (*1*), carbynoid material (2) and diamond (*3*) [8].

noted that an absolutely controversial result has been obtained by Baitinger *et al.* [4]: the shift of the photon energy values to higher values relatively to the graphite spectrum was observed in XES of a carbynoid material derived from PVDF. This remarkable discrepancy of the data by Sergushin *et al.* [8] and Baitinger *et al.* [4] may have occurred because of the irreversible destruction of the surface of the pristine sample due to the heavy electron bombardment required for XES excitation [4]. This assumption is supported by the crucial observation that after XES analysis [4] the carbynoid film surface had changed its colour from shiny black to a dull grey colour [18]. The qualitative similarity of the XES shapes for carbynoid and graphite [4,10] suggests, at first sight, a transformation of the sample surface under electron bombardment from a chain-like to a graphitic arrangement. The effect can provide a rationale for the *p*-like electron energy distribution, but fails to explain both the shift of the carbynoid spectrum and the change of the colour of the surface observed.

3.3. EELS

EEL spectra of carbynoid structures were investigated by several authors [3,4-6,10-11]. Two carbynoid samples studied by Khvostov *et al.* [3] were synthesized by dehydrochlorination of poly(vinylidene chloride) (PVDC) at 258 K as well as by graphite sputtering under low energy argon ion irradiation of the growing film [19]. Experimental data were presented for a primary electron beam energy of 150 eV [3]. It turned out that at an excitation energy of less than 100 eV, electron energy losses are not observed at all. A rapid increase of the EELS intensity was observed when the excitation energy increases up to 200 eV. The EELS integral intensity yields values up to a quarter of that observed for an elastic peak. This inelastic contribution is approximately six times greater than that of π-π^* excitations in graphite. EELS on carbynoid structures display a well-defined fine structure whose features are shifted to 4.2, 7 and 8-9 eV relative to the elastically backscattered electron energy. Kramers-Krönig routine has been applied to the evaluation of EEL spectra. This allowed to separately determine a real and an imaginary part of the dielectric function. The results were then compared with band structure calculations to determine the origin of the EELS fine structure. It turned out that the feature at 4.25 eV appears to be the result of a collective π-electrons excitation; the imaginary part peculiarities at 1, 7.8 and 9.9 eV reflect one-electron interband transitions between π- and π^*-subbands, and between σ- and σ^*-subbands at the BZ edge, and between the bottom of the π-subband and the top of the σ^*-subband in the BZ centre, respectively.

Since the experimental geometry [3] allowed only to register direct transitions, *i.e.*, those without a change of the wave number, one would also expect the maxima of the imaginary part to be at 12 eV (the transition between the tops of σ- and π^*-subbands at the BZ edge) and at 20 eV (from σ-bottom to σ^*-top in the BZ centre): However, EELS data in this energy range were not presented [3]. If the carbyne band structure which was a basis for EELS interpretation by Khvostov *et al.* [3] is valid, no interband transitions would exist with energies exceeding 20 eV and total valence band widths in excess of 13 eV. This is, however, in shocking contrast with XPS results that indicate a width of the occupied part of the bands of more than 20 eV for all carbynoid materials studied independent of their synthesis conditions [4-6,8,10,11,17]. This contradiction leads to the conclusion that either the band structure determined by Khvostov *et al.* [3] is not correct or that the samples studied are not genuine carbynes, or, possibly, that the manner of the band occupation significantly differs from Pauli's law [5].

EEL spectra in the vicinity of the $C1s$-line excited by monochromatized Al K_α x-ray radiation (1486.6 eV) were likewise investigated [4-6,10,11]. The carbynoid samples used were the same as in the valence electron XPS study (*cf.* Fig. 5). The loss spectra were obtained by subtraction of the model $C1s$-line from the total core electron XPS registered in a wide energy range (50 eV). The model line had a symmetrical shape, its half-height-width and the binding energy corresponding to respectively, 2.5 and 285 eV, and can be constructed rigorously enough as the sum of Gaussian and Lorentzian contributions with different apparent weights but corresponding positions of the maxima. The subtracted spectrum is displayed in Fig. 8 (curve 1) that shows the EEL spectrum extending over a wide energy range as a broad and rather intensive satellite tail and showing a well-resolved fine structure with the main maximum at 21 eV relatively to the *C1s*-peak position, *i.e.*, apparently far from it. This is the most striking difference of the data presented by Baitinger and co-workers [4-6,10,11] and those obtained by Khvostov *et al.* [3]. On the other hand, the fine structure features of the loss spectrum found at 1 and 4.2 eV [4] are in a good agreement with [3]. Kramers-Krönig treatment has shown features at 4, 10.5, 14, 22, 24, 28 and 34 eV attributed to the imaginary part of the loss function. The zero value of the real part is not located at 4 eV as stated by Khvostov *et al.* [3], but at 22 eV. This fact confirms that the characteristic energy value for the generation of the collective electron excitations (plasma oscillations) in carbynoids is 22 eV [4].

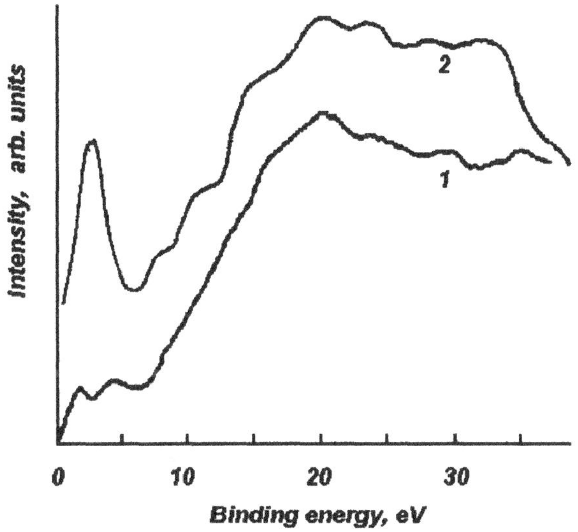

Figure 8. EELS of carbynoid material (*1*) and diamond (2) [4,20]. The energy scale begins with the *C1s*-maximum position.

Comparative analyses of EELS and valence electron XPS have shown the "mirror-like" relative character of the occupied and empty electronic states of the carbynoid studied [4], in qualitative agreement with the band scheme developed by Khvostov *et al.* [3].

The comparison of the EEL spectra of carbynoid material [4] and diamond [20] measured under identical conditions (*cf.* Fig. 8) demonstrates that though the main maximum energy positions of both spectra are very close to each other, remarkable differences exist of their shape: diamond EELS near 33 eV has a specific feature usually attributed to volume plasmon excitations [13]. The absence of this peculiarity in the carbynoid spectrum confirms the low dimensionality of the structural elements forming the film. Additional evidence supporting this conclusion comes from a close resemblance between the carbynoid and polyethylene EELS [21]. Thus the shape and energy position of the fine structure features of the C*1s*-line satellite spectrum allow to obtain information about the dimensionality of the carbon allotrope investigated.

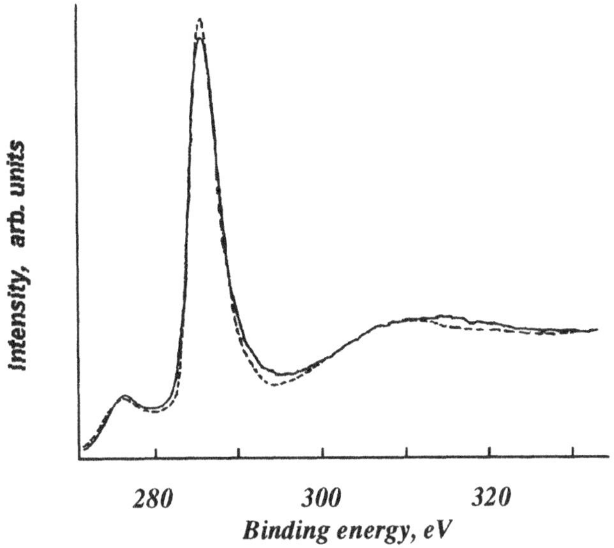

Figure 9. The effect of Ar^+ bombardment on C*1s* XPS of a carbynoid film. Dashed and solid curves correspond to doses of 10^{17} and $2{\cdot}10^{17}$ ions/cm^2, respectively.

We have observed this effect [22] when studying carbynoid film derived from PVDF. It occurs during irradiation with Ar^+ ions with a kinetic energy of 1200 eV whose cumulative dose exceeds 10^{17} ions/cm^2 (Fig. 9). The transformation of the satellite shape in the nearest vicinity of the C*1s*-line as well as in the maximum region of the loss curve is clearly visible. The effect originates from cross-linking of adjacent chains caused by the heavy ion bombardment and results in the creation of sp^2-bonded two-dimensional structural fragments. At a dose value of $2{\cdot}10^{17}$ ions/cm^2 the EEL spectra of the bombarded film and polycrystalline graphite become indistinguishable at a spectrometer resolution of about 3 eV. Below an irradiation dose of less than 10^{17} ions/cm^2 the shape of the EEL spectrum of carbynoid material is consistent with the notion of a predominantly one-dimensional character of the carbon atom arrangement. The irradiation process only promotes purification of the film surface by removing fluorine and oxygen atoms.

3.4. AES

Most publications are concerned with the acquisition of valence DOS information for carbynoids using AES technique [3,23-27]. In most cases samples were studied that have been either derived from PVDF and PVDC by chemical carbonization or synthesized by graphite sputtering combined with the low energy ion irradiation. Only one of the samples investigated has been obtained by the deposition of carbon evaporated in an electric arc on the cold substrate [25]. In some papers [3,23] heat treatment up to 900 K has been employed to modify the structure of pristine samples. All the experimental data demonstrated a measurable shift of the AES spectra of carbynoid materials to the low kinetic energy region at different values up to 10 eV in comparison with graphite and diamond spectra (Fig. 10). The shape of the low energy shoulders in the AES also differs significantly from that of graphite and diamond.

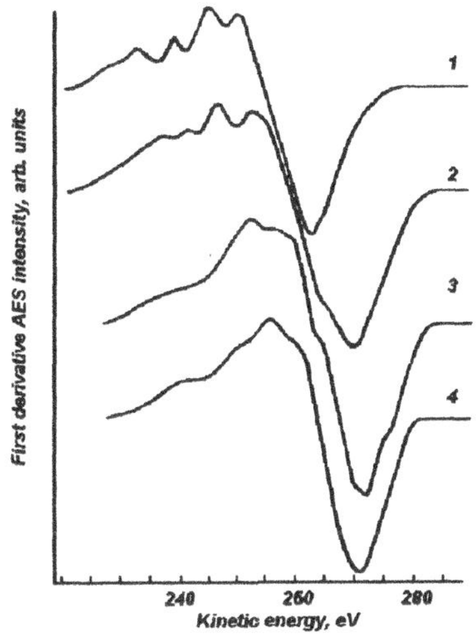

Figure 10. AES of carbynoid materials (*1*,*2*), graphite (*3*) and diamond (*4*) taken in the derivative mode [3].

To extract valence DOS information experimental spectra were subjected to an inverse self-folding routine (Fig. 11). The shape of self-deconvoluted AES of carbynoid materials also show some specific features whose energy positions agree well with the calculated data [2]. Attempts to determine the average number of atoms in a chain with the help of such a direct comparison (Fig. 11) have been made.

However, the employment of the self-unfolding routine for the extraction of DOS is essentially incorrect. Hence the coincidences obtained by other research groups [3,23-27] seem to be fortuitous. A principal reason for this unpleasant conclusion originates in the multiple nature of the Auger relaxation that leads to the breakdown of the DOS information inherent in the AES shape. The C *KVV*-shapes of various carbons differ strikely from the self-folded DOS found by superposition of experimental XPS and XES data [28]. The most pronounced effect of this discrepancy is seen as the AES shift discussed above (Fig. 12). Its magnitude increases in the order: diamond - graphite - polyethylene, and reaches its highest value for gas-phase molecules. This phenomenon is caused by Coulombic repulsion of two valence holes created at the same site in the final state of the process [28]. Therefore the lower the probability is of hole delocalization, the higher the shift value becomes, and thus is entirely governed by the number of the nearest neighbours of the emitting atom having

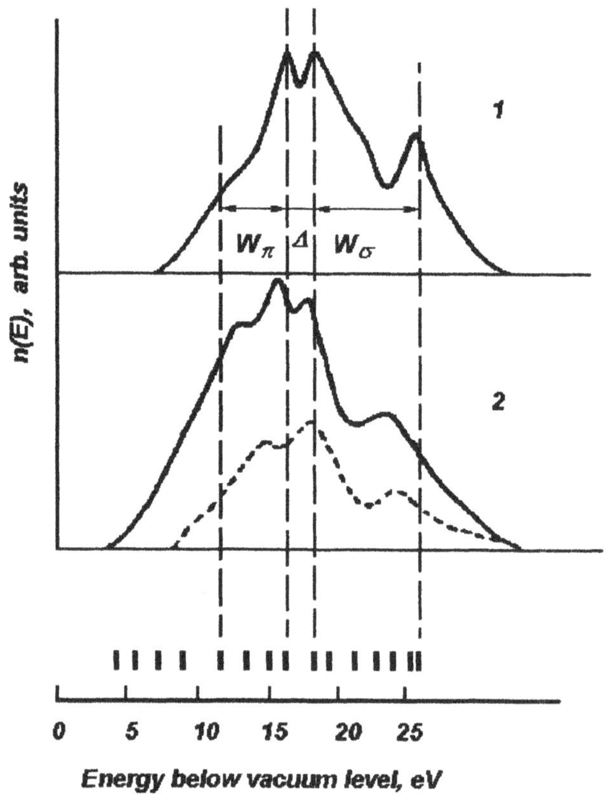

Figure 11. Results of the inverse self-folding of the data of Fig. 10 (solid curves *1*, *2*). A tentative DOS profile derived from the second spectrum after subtracting the contribution from graphite-like fragments is shown as a broken curve [3]. The features of the inverted AES are related to the theoretical data [2] of a cumulene chain comprising eight atoms. The σ- and π-subband widths as well as the energy gap Δ between them are also shown.

values of, respectively 4, 3, 2 and 0 for diamond, graphite, polyethylene and an isolated atom. The terminating hydrogen atoms in polyethylene can hardly be considered probable delocalization channels. That is why the AES shift value is only informative when determination is required of the type of carbon structure dimensionality. It depends on more or less pronounced deviations of the type of structural arrangement from a chain-like configuration in various carbynoid samples, not only on the different content of graphite-like ordered fragments as previously assumed by Khvostov *et al.* [3].

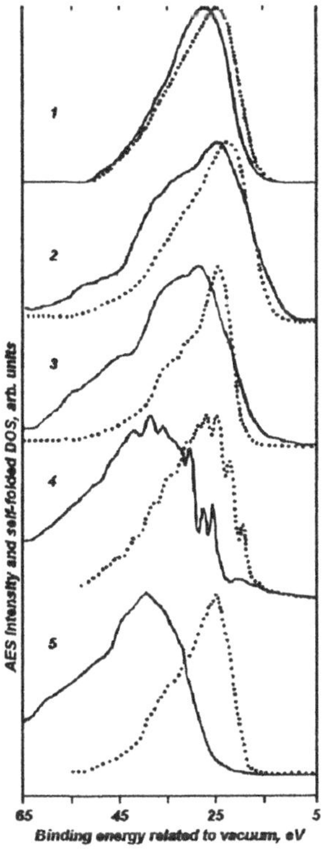

Figure 12. Comparison of the experimental C *KVV* line shapes (solid curves *1-5* for diamond, graphite, polyethylene, benzene and cyclohexane, respectively) with the self-folded DOS (dotted curves) [28].

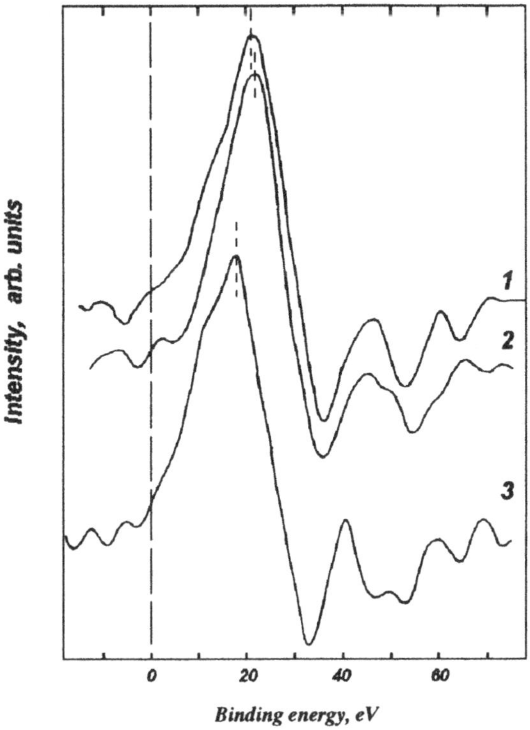

Figure 13. The effect of the Ar^+ bombardment dose on the XAES modification of carbynoid films: non-irradiated (curve *1*), irradiated with doses of 10^{17} and $2{\cdot}10^{17}$ ions per square centimetre (curves *2* and *3*, respectively). The XAES are displayed in the derivative mode. The origin of the energy scale was found independently for each sample using their C*1s*-peak positions. Short vertical broken lines mark the main XAES features.

The effect discussed above makes impossible the coincidence in energy positions of the self-unfolded AES features with their valence band "originals", *i.e.*, the corresponding DOS peculiarities. This statement is even more correct if low-dimensional structures are examined. Moreover, the assumption of a nearly equal probability of the participation of any valence electron in the Auger process [24] has been invalidated by the results of diamond and graphite studies [29,30] that clearly demonstrated that both electrons taking part in the process originated predominantly from *p*-like initial states.

Nevertheless, many fruitful results have been obtained by several workers in the field [3,23-27]. The discovery of the AES shift has become a reliable confirmation of the one-dimensional structure of the samples examined. Besides, the fact of the „polyamorphousness" of carbon was established in that amorphous carbon films may have a distinct short-range order (SRO) mode corresponding to the existence of 2, 3 or 4 adjacent atoms in the first coordination sphere [24].

The possibility of an irreversible carbon target surface modification during AES excitation by electron bombardment has been shown by Mizokawa *et al.* [31]. This is even more true for carbynoid materials. Electron diffraction studies showed amorphization of the films if the electron beam exposure time exceeds 5 seconds [25]. Thus x-ray AES excitation (XAES) seems more preferable [31]. Analyses of the XAES data obtained by Pesin *et al.* [22] and displayed here in the derivative mode (Fig. 13) indicate a spectral shift to lower binding energy caused by ion irradiation [32]. This confirms the idea of chain cross-linking at doses higher than 10^{17} ions/cm^2 previously deduced from the analysis of $C1s$-spectra.

4. New Spectroscopic Criteria for Identification of Carbon Structures

The results described above indicate that the problem of identification of the carbyne form of carbon is far from being solved. A rather large number of inconsistencies arise when attempting to unequivocally explain the results of a relatively wide set of experimental and theoretical data obtained by many workers in the field. Thus a comprehensive search is urgently needed for new criteria and parameters which can distinguish carbon structures belonging to different modes of the arrangement of carbon atoms.

In this spirit investigations into the integral intensity of electron emission spectra appear to become a new and promising method to study the character of electronic states in carbon. The extreme sensitivity of the photoionization cross-section with respect to the valence state symmetry is well known; it modulates the relative intensity of the spectral fine structure features which reflect the corresponding peculiarities of the DOS [13]. The prominent part of the latter is formed by hybrid wave functions. Their symmetry strongly depends on the $2p$-state yield and changes in a regular fashion from carbyne to graphite to diamond.

This effect is manifest in the differences of the integral intensity of the valence electron XPS of various carbon forms [33].

However, several difficulties arise when employing this technique in practice. They are mostly related to rather low XPS intensities in the valence band energy range. Besides, heteroatoms always present on carbon surfaces contribute to the total XPS. Their yield is extremely difficult to account for and hence to subtract in a proper way.

From this point of view measurements of the carbon XAES integral intensities seem to yield better prospects. The numerical values of XAES intensities are on the same order as the core electron XPS and thus large enough to be measured more accurately. Moreover, XAES and XPS can be registered almost simultaneously as required for energy and intensity calibrations [31,33,34].

Indeed, the experimental data [34] (Fig. 14, 15 and Table 1) demonstrate the sensitivity of the normalized XAES integral intensity S to the type of the carbon structure investigated. The values of this parameter are well reproduced within an error of <7 %. The parameter S regularly increases with increasing contribution of the $2p$-states yield to the hybrid orbitals or, equivalently, when the number of atoms in the first coordination sphere increases. This fact appears to be in qualitative agreement with the one-electron prediction [34]. The discovered effect is unlikely to be governed by different probabilities of the general Auger relaxation process. It is, however, supposed to be determined by variations of multiple Auger ionization probabilities in various carbon forms depending on the valence states hybridization mode. This leads to different XAES intensities in the energy region where processes with only one escaped electron are counted.

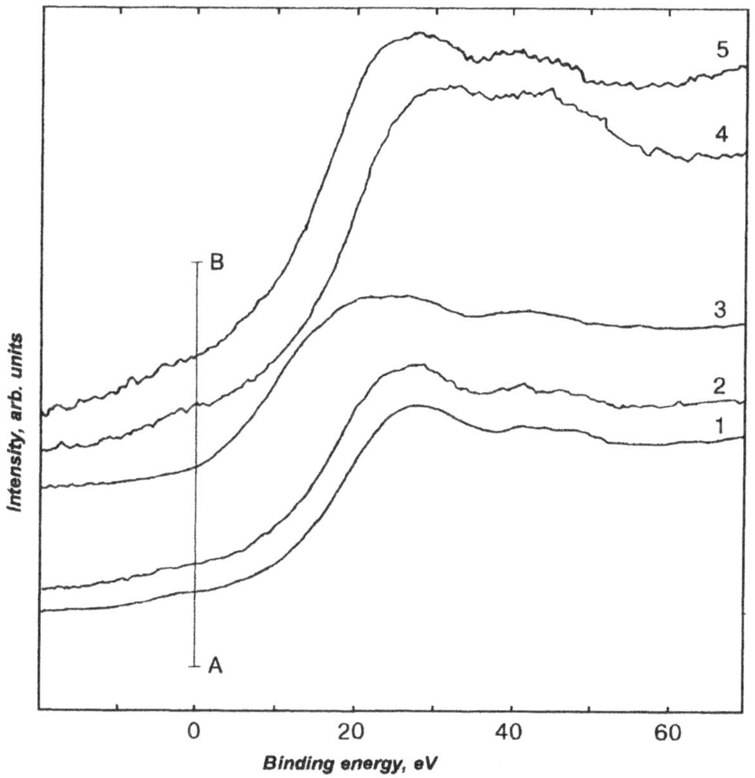

Figure 14. XAES of carbon samples: Carbynoids derived from PTCB 1,1,2 (24 hours synthesis time) and PVDC (240 min) (curves 1 and 2, respectively); HOPG (3); artificial and natural diamonds (4, 5). The length of the vertical bar *AB* corresponds to the maximum intensity of the $C1s$-spectra used for energy and intensity calibrations.

It should be particularly noted that the correct application of the described techniques demands a highly stable x-ray source used for excitation of XPS and XAES.

TABLE 1. Average values of the parameters S, F and R of various carbon samples : 1, 2 - carbynoids obtained by dehydrochlorination of PVDC (synthesis time: 260 min) [35] and PTCB 1,1,2 (synthesis time: 24 hours) [36]; 3 - HOPG; 4, 5 - natural and artificial diamonds

Structure type	Carbynoid		Graphite	Diamond	
Sample number	1	2	3	4	5
S, arb. units	0.54	0.56	0.60	1.06	0.97
F, arb. units	0.66	0.64	0.71	0.55	0.51
R, arb. units	5.16	5.10	5.40	4.66	4.38

Additional criteria for the determination of the type of the carbon structural arrangement are based on a comparative phenomenological analysis of the contributions of elastic and inelastic processes to core electron XPS and XAES (Fig. 15): average magnitudes of the parameters F and R (Table 1) are rather specific for each carbon form studied. Measuring errors collected from many experiments are within 3-4 % for XPS, and 2-3 % for XAES.

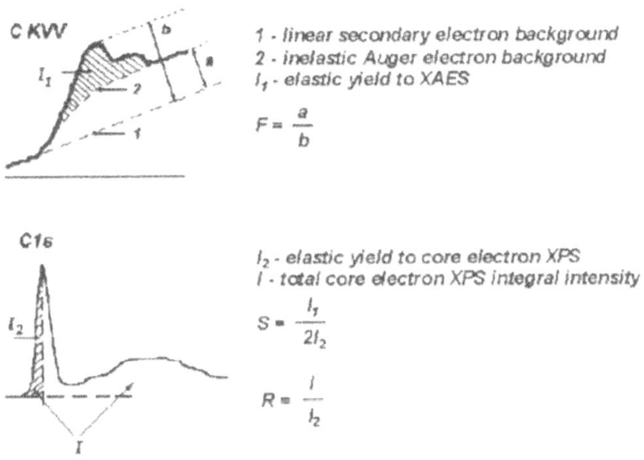

Figure 15. Schematic of determination of the parameters S, F and R based on the intensity analysis of the principal features of the spectra.

The XAES first derivative maxima of the carbynoid films 1 and 2 are shifted to the higher binding energy region relatively to the HOPG spectrum. This is a strict confirmation of relatively low dimensionality of their atomic structure. Oxygen and nitrogen atoms are able to increase the dimensionality of the carbynoid structure providing covalent "bridges" between the adjacent chains. Chlorine atoms retained in the film due to the incompleteness of the dehydrochlorination process can also disturb the carbon chain and produce local sp^2- or (and) sp^3- hybrid orbitals depending on the number of heteroatoms (one or two) attached to a carbon site. The first of these effects is clearly seen as a general tendency of the decrease of the aforementioned XAES shift

ΔE. However, both effects elevate the S parameter of the films with increasing heteroatom concentration (Table 2).

TABLE 2. The heteroatom surface concentration effect on the ΔE and S parameters of the XAES of carbynoids: 1, 2 - the same as in Table 1; 3, 4 - carbynoids obtained by dehydrochlorination of PVDC (synthesis time: 60 min) and PTCB 1,2,3 (synthesis time: 24 hours) [36]

Sample number	1	2	3	4
Oxygen, at. %	6.9	6.7	7.9	6.6
Nitrogen, at. %	1.4	0.0	0.8	1.4
Chlorine. at. %	0.3	0.4	1.5	1.6
ΔE, eV	5.3	6.1	4.4	4.9
S, arb. units	0.54	0.56	0.70	0.64

When analysing the data of Table 2 it should be kept in mind that some quantity of oxygen and nitrogen will always be adsorbed at the carbon surface already without any chemical interaction and thus does not participate in the phenomena discussed. Information about the total concentration of surface heteroatoms can be gained from XPS but the relatively poor resolution required for sufficient spectral intensity prevents the separation of purely chemical effects. Hence the validity of the treatment of the data of Table 2 is restricted by its semiquantative character.

Nevertheless, carbon core electron XPS and XAES parameters obtained as a results of spectra intensity evaluations are evidently a sensitive and informative tool for the carbon hybrid state analysis. This effect is expected to have good prospects for the new techniques of short-range order study, but presently its correct application is severely limited owing to the absence of a chemically pure chain-ordered sample of carbyne among the variety of those investigated.

5. Conclusion and Prospects

This review clearly shows that empirical data and theoretical results accumulated within a long period of carbyne studies with electron and photon spectroscopy made a considerable contribution to the understanding of its structure and physical properties. Now it has become clear that the XPS shape in general as well as the XPS and XAES intensities can provide a short-range order (SRO) structural information determined primarily by the type of valence electron hybridization. On the other hand, XPS, EELS and AES (XAES) fine structure features are sensible to the predominantly long-range order (LRO) arrangement of carbon atoms and reflect the type of the structural dimensionality. But matter-of-factly, these achievements will certainly not end research on this unique and wonderful material that promises many interesting and fruitful applications in various branches of science and technology. Indeed, it is only the beginning of a hard and long way.

The main obstacle preventing us from gaining better knowledge of the electronic structure of carbyne-like systems lies essentially in the field of synthesis. It has been shown above that up to now neither crystalline carbyne nor samples having aligned chain regions of sufficient size are available for spectroscopic research. Solving this problem would open up opportunities to not only obtain a real picture of the one-dimensional valence band as a whole, but also to examine the π- and π^*-states in detail using the extremely anisotropic character of their spatial distribution as it was previously done for graphite [37,38].

General prospects are undoubtedly connected with the employment of synchrotron radiation for band structure study. Carbon photoionization cross-sections for emission from the initial *2s*- and *2p*-states are equal for photon energies about 120 eV [5]. This fact shows the most proper way to receive correct information about the DOS distribution along the entire valence band. In this case the spectral features closely match the corresponding DOS peculiarities no matter what the type of the wave function symmetry is. To separate the states with different symmetry and to study only the *p*-like part of the DOS one can use the effect of x-ray fluorescence. It is absolutely analogous to XES, and for the purpose of this analysis any radiation with energies in excess of 285 eV is sufficient to create a hole in *K*-level. Unfortunately, a low characteristic radiation probability relative to that of the Auger process is typical for the light atoms and greatly restricts the photon emission intensity. Thus, the employment of this method demands a powerful primary x-ray beam source. To avoid surface damage of carbyne it is advisable to replace the electron gun by traditional Al or Mg anode sources with which many laboratories all over the world are well equipped. Nevertheless, utilization of synchrotron radiation promises obvious advantages due to polarization effects. For the same reasons the study of empty states should also preferably be made by means of ultrasoft x-ray absorption spectroscopy (XAS) using synchrotron radiation [37].

Acknowledgements – The author gratefully acknowledges Drs. Yu. P. Kudryavtsev, S. E. Evsyukov, O. B. Sokolov and E. M. Baitinger for supplying carbon samples and for helpful discussion.

6. References

1. Kittel, Ch.: *Introduction to solid state physics,* John Wiley & Sons, New York, 1953.
2. Leleyter, M. et Joyes, P.: Etude experimentale et theoretique de l'emission secondaire d'ions moleculaires. Cas des elements du groupe IV-B, *J. de Physique* **36** (1975), 343-355.
3. Korshak, V.V., Kudryavtsev, Yu.P., Khvostov, V.V., Guseva, M.B., Babaev, V.G., and Rylova, O.Yu.: Electronic structure of carbynes studied by Auger and electron energy loss spectroscopy, *Carbon* **25** (1987), 735-738.
4. Kudryavtsev, Yu.P., Baitinger, E.M., Kugeev, F.F., Korshak, Yu.V., and Evsyukov, S.E.: Electronic structure of carbyne studied by x-ray photoelectron spectroscopy and x-ray emission spectroscopy, *J. Electron Spectrosc. Relat. Phenom.* **50** (1990), 295-307.
5. Baitinger, E.M.: *Electronic structure of condensed carbon,* Urals State University, Sverdlovsk, 1988 (in Russian).

6. Baitinger, E.M.: *Electronic structure of low-dimensional carbon,* D. Sc. Thesis, Leningrad, 1988 (in Russian).
7. Perkins, P.G., Marwaha, A.K., and Stewart, J.P.: An improved LCAO SCF method for three-dimensional solids and its application to polyethylene, graphite, diamond and boron nitride, *Theoret. Chim. Acta (Berl.)* **57** (1980), 1-23.
8. Sergushin, I.P., Kudryavtsev, Yu.P., Elizen, V.M., Sadovsky, A.P., Sladkov, A.M., Nefedov, V.I., and Korshak, V.V.: Carbyne study by means of XPS and XES, *Zhurn. Strukt. Khim.* **18** (1977), 698-700 (in Russian).
9. Briggs, D. and Seach, M.P. (Eds): *Practical surface analysis by Auger and x-ray photoelectron spectroscopy,* John Wiley & Sons, Chichester - New York - Brisbane - Toronto - Singapore, 1983.
10. Korshak, V.V., Kudryavtsev, Yu.P., Evsyukov, S.E., Korshak, Yu.V., Baitinger, E.M., Kugeev, F.F., and Karasov, V.Yu.: Peculiarities of the carbyne electron structure, *Dokl. Akad. Nauk SSSR* **297** (1987), 130-132 (in Russian).
11. Korshak, V.V., Baitinger, E.M., Kugeev, F.F., Kudryavtsev, Yu.P., Evsyukov, S.E., Korshak, Yu.V., and Teterin, Yu.A.: Modification of the chain electronic structure during a course of the carbyne synthesis, *Dokl. Akad. Nauk SSSR* **303** (1988), 894-897 (in Russian).
12. Baitinger, E.M. and Kugeev, F.F., private communication.
13. McFeely, F.P., Kowalczyk, S.P., Ley, L., Cavell, R.G., Pollak, R.A., and Shirley, D.A.: x-Ray photoemission studies of diamond, graphite, and glassy carbon valence bands, *Phys. Rev.* **B 9** (1974), 5268-5278.
14. Pireaux, J.J. and Caudano, R.: Experimental picture of the band structure formation in a solid, *Am. J. Phys.* **52** (1984), 821-826.
15. Painter, G.S., Ellis, D.E., and Lubinsky, A.R.: *Ab initio* calculation of electronic structure and optical properties of diamond using discrete variational method, *Phys. Rev.* **B 4** (1971), 3610-3622.
16. Stevens, G.C., Bloor, D., and Williams, P.M.: Photoelectron valence band spectra of diacetylene polymers, *Chem. Phys.* **28** (1978), 399-406.
17. Pesin, L.A. and Kuznetsov, V.L.: Valence band XPS of carbynoids derived from PVDF, *CSPI Report* No. 427/92 (1992).
18. Kugeev, F.F. and Baitinger, E.M., private communication.
19. Savchenko, N.F., Guseva, M.B., Babaev, V.G., and Palatnik, L.S.: Modification of carbon films structure induced by ion irradiation, *Poverkhnost'. Fizika, khimiya, mekhanika,* No. 6 (1985), 106-111 (in Russian).
20. Baitinger, E.M., Teterin, Yu.A., and Kugeev, F.F.: On the nature of crystalline carbon *C1s* XPS fine structure, *Fizika Tverd. Tela* **31** (1989), 316-319 (in Russian).
21. Pireaux, J.J., Caudano, R., and Verbist, J.: Carbon *1s* inelastics in polymers, *J. Electron Spectrosc. Relat. Phenom.* **5** (1974), 267-272.
22. Pesin, L.A., Kuznetsov, V.L., Kudryavtsev, Yu.P., and Evsyukov, S.E.: The effect of ion bombardment on the amorphous carbyne XPS, *Khim. Vysok. Energ.* **30** (1996), 243-248 (in Russian).
23. Korshak, V.V., Kudryavtsev, Yu.P., Khvostov, V.V., Guseva, M.B., Babaev, V.G., and Rylova, O.Yu.: Experimental confirmance of a new structural hypotesis of carbyne, *Dokl. Akad. Nauk SSSR* **293** (1987), 393-396 (in Russian).
24. Khvostov, V.V., Babaev, V.G., and Guseva, M.B.: Auger spectroscopy of the amorphous carbon films, *Fizika Tverd. Tela* **27** (1985), 887-891 (in Russian).
25. Kudryavtsev, Yu.P., Evsyukov, S.E., Babaev, V.G., Guseva, M.B., Khvostov, V.V., and Krechko, L.M.: Oriented carbyne layers, *Carbon* **30** (1992), 213-221.
26. Kudryavtsev, Yu.P., Evsyukov, S.E., Guseva, M.B., Babaev, V.G., and Khvostov, V.V.: Carbyne: the third carbon allotrope, *Russ. Chem. Bull.* **42** (1993), 399-413.
27. Korshak, V.V., Kudryavtsev, Yu.P., Khvostov, V.V., Guseva, M.B., and Babaev, V.G.: The electron structure of carbyne study by means of Auger spectroscopy, *Dokl. Akad. Nauk SSSR* **280** (1985), 402-403 (in Russian).
28. Ramaker, D.E.: Chemical effects in the carbon KVV Auger line shapes, *J. Vac. Sci. Technol.* **A 7** (1989), 1614-1622.
29. Aleshin, V.G., Bugaets, O.P., Kucherenko, Yu.N., and Nemoshkalenko, V.V.: Theoretical study of the diamond Auger spectrum, *Dokl. Akad. Nauk SSSR* **287** (1986), 611-614 (in Russian).

30. Agostino, R.G., Kuttel, O.M., Fasel, R., Osterwalder, J., and Schlapbach, L.: *KVV* Auger-electron diffraction patterns from carbon solids, *Phys. Rev.* **B 49** (1994), 13820-13825.
31. Mizokawa, Y., Miyasato, T., Nakamura, Sh., Geib, K.M., and Wilmsen, C.W.: The C *KLL* first derivative x-ray photoelectron spectra as a fingerprint of the carbon state and the characterization of diamondlike carbon films, *J. Vac. Sci. Technol.* **A 5** (1987), 2809-2813.
32. Pesin, L.A.: The effect of ion bombardment on XAES parameters of carbynoids, *CSPU Report* No. 3448 (1996).
33. Kugeev, F.F., Baitinger, E.M., Teterin, Yu.A., and Gagarin, S.G.: On the carbon fibres structure from the XPS data, *Khim. Tverd. Topl.* No. 3 (1991), 120-125 (in Russian).
34. Baitinger, E.M., Pesin, L.A., and Cherednichenko, A.V.: The effect of the electron hybridization mode on the carbon Auger spectra intensity, *Izv. Vysshykh Uchebn. Zaved., Fizika* No. 8 (1996), 111-115 (in Russian).
35. Korshak, V.V., Kudryavtsev, Yu.P., Korshak, Yu.V., Evsyukov, S.E., Khvostov, V.V., Babaev, V.G., and Guseva, M.B.: Formation of β-carbyne by dehydrohalogenation, *Macromol. Chem., Rapid Commun.* **9** (1988), 135-140.
36. Vointseva, I.I., Gil'man, L.M., Kudryavtsev, Yu.P., Evsyukov, S.E., Pesin, L.A., Gribov, I.V., Moskvina, N.A., and Khvostov, V.V.: Chemical dehydrochlorination of polytrichlorobutadienes. A new route to carbynes, *Eur. Polym. J.* **32** (1996), 61-68.
37. Skytt, P., Glans, P., Mancini, D.C., Guo, J.-H., Wassdahl, N., and Nordgren, J.: Angle-resolved soft-x-ray fluorescence and absorption study of graphite, *Phys. Rev. B* **50** (1994), 10457-10461.
38. Beyreuther, Ch. and Wiech, G.: x-Ray K-emission and electronic structure of graphite, *Phys. Fenn.* **9** (1974), 176-178.

6.4 ELECTRON ENERGY LOSS SPECTROSCOPY STUDIES OF CARBYNOID STRUCTURES

JACOB I. KLEIMAN
Integrity Testing Laboratory Inc./University of Toronto Institute for Aerospace Studies, North York, Ontario, M3 5T6, Canada

Abstract - A number of carbyne precursor samples were analysed using Electron Energy Loss Spectroscopy (EELS) technique together with a few carbon-bearing compounds including graphite, diamond, and amorphous carbon in an attempt to better understand the structure of carbyne precursors. and the mechanisms of their formation. The EELS data were compared and associated with the molecular orbital structure of the compounds investigated.

1. Introduction

During last years substantial progress has been achieved in technologies associated with synthesis of carbon-bearing materials that contain atoms with various types of sp-hybridization, i.e sp^3, as in diamond and diamond films; sp^2 - sp^3, as in fullerenes; sp - sp^3 , as in diamond-like carbon (DLC), and sp or sp-sp^2 , as in carbynes. While some of the structures mentioned above are quite clear and were studied extensively, others are only beginning to be understood.

Most of the data on the structure of carbyne came from x-ray diffraction, electron diffraction and some spectroscopic techniques [1-3]. A classification scheme for all known carbyne forms was proposed [1] that was based solely on data from the above mentioned analytical techniques. The proposed scheme classified all known carbyne forms into two major groups having either conjugated triple or cumulated double bonds [1]. Information on the type of the C-C bonds can be obtained only from studies of their electronic properties. The band structure of one-dimensional solids has specific features. The electronic structure of organic molecules with sp^3, sp^2 and sp-type of hybridization of the bonds and with different number of atoms in the chain, is quite variable. It was shown that, while the σ-subband energy positions and bandwidths are approximately the same, the π-subband structure depends on the chain characteristics [4]. Based on these conclusions, a few studies were conducted using various electron spectroscopic techniques to investigated the electronic structure of carbynes [4-7].

X-ray photoelectron results were not conclusive enough to fully identify the electronic structure and the type of bonding, and appreciable differences were found to exist between carbynes and graphite, diamond, and carbon black samples [5]. It was

R.B. Heimann et al. (eds.), Carbyne and Carbynoid Structures, 395–407.

shown that a characteristic π-system exists in carbyne. In a more recent study of the electronic structure of carbynes [4,6], experimental carbon Auger lineshapes of two carbyne samples produced in two separate laboratories by different methods were compared. The electronic structure of carbynes were calculated by self-deconvolution of the experimental Auger spectra. In addition, the loss spectra of carbynes were measured at different primary electron energies, E_p. It was found that at E_p<100 eV, losses were not observed, but for E_p> 100 eV, the intensity of losses increased very rapidly and reached a maximum at $E_p = 200$ eV [4].

Experimental evidence obtained from EELS analysis [6] confirmed the linear structure of the carbyne macromolecules. The data revealed one-electron transitions and plasmons corresponding in energy to a one-dimensional chain. Hence it corroborated nicely the linear chain structure and the semiconducting properties of the investigated carbyne. Such features of the electronic structure of carbyne as the continuity of the energy spectrum within the band, which attests to a large chain length, on one hand, and a band width corresponding to short chains, on the other hand [6], suggested that the carbon chains in carbyne consist of short, periodically repeated linear fragments that agreed with the hypothesis of a carbyne structure with periodically arranged kinks in the carbon chain [1].

Electron energy loss spectroscopy (EELS) combined with transmission and scanning transmission electron microscopy analysis proved to be a very useful technique for the analysis of light elements. The first EELS transmission results on the electronic structure of noble metals were reported almost 70 years ago [8]. Since then, numerous studies on plasmons, interband transitions and core-level excitations were performed using EELS. The reader is referred to Refs. [9,10] and references therein for an excellent review of such activities and the principles of the EELS technique. In its essence, the properties of the bulk material can be determined by measuring the transmission of electron beams through thin crystal films. The EELS technique, therefore, is based on the fact that the energy loss spectrum of electrons having passed through a thin crystal foil can be described by the dielectric theory [10]. The dielectric function ξ (q, ω) is a well-known quantity to describe excitations of an electron system. The real part Re {ξ (q, ω) -1} ≡ ξ_1 (q, ω) -1, is a measure of the polarizability of the system, while the imaginary part Im {ξ (q, ω)} ≡ ξ_2 (q, ω), is a measure for absorptions occurring in the electron system. The loss function Im{$-1/\xi$}= $\xi 2/(\xi_1 + \xi_2)$ has a peak when $\xi_1 = 0$ and ξ_2 is small [9].

In a typical EELS spectrum from a thin solid foil, the low energy part of the loss spectrum contains information on the plasmons, and on the excitation or ionization of electrons from various bound states. The low energy loss part of the EELS spectrum can be a very useful tool in the analysis of various carbon structures with different atom hybridization states since the lowest energy losses (below 15 eV) come from the excitation of electrons in molecular orbitals [4, 13-15]. Above 15 eV the spectrum shows structures which are mostly due to valence shell excitations.

At higher energies, the losses are due to excitation of atomic core electrons to states above the Fermi level. In case of carbon, these transitions are between 1 s levels (K-shell) in the valence or conduction band [11]. It was shown that differences exist in the K-ionisation spectra of graphite, diamond, and amorphous carbon [11], suggesting the possibility of using this range of the EELS spectra to distinguish among different forms of carbon. It should be noted that the K-ionisation spectrum of amorphous carbon was shown to resemble to some extent that of graphite, but with the structure "smeared out" by several eV [11]. It was also demonstrated [16] that the K-shell ionisation region can provide very useful information on the presence of different functionalities in polymers. A comparison of the carbon and oxygen core edge spectra of PET polymer with those of benzene, benzaldehyde, terephthalaldehyde, ethylbensoate, terephthaloyl chloride, and phosgene, showed the dominance of benzene-like transitions, peak splitting due to ring substitution, and differences in core edge fingerprints for various functionalities [16].

As evident from this short review, methodologies that give information on the electronic structure of materials, and EELS in particular, can contribute significantly to the understanding of the electronic structure of materials. For carbynes, very scarce data on EELS exists. The low energy plasmon range is discussed in Refs. [4,6]. No data was found on the high energy (K-shell) range. For that reason, a number of carbon-bearing compounds including spectroscopically pure graphite, fine grained diamond, and a number of carbyne precursors were analysed using the EELS technique in an attempt to better understand the structure of carbyne. The EELS data was compared and associated with the molecular orbital structure of the investigated compounds. Initial results of this study will be presented together with a discussion on future work in this field.

2. Experimental

The carbyne precursors used in the present EELS study were obtained from the National Institute of Materials and Chemical Research (NIMC), Tsukuba, Japan. The description of sample preparation obtained from NIMC can be found in Refs. [17,18]. The elemental atomic composition of the products of this reaction was determined to be 88.6% C, 5.4% H, 3.7% N, and 2.3% Cl showing that a part of the original poly(vinylidene chloride) was not completely reacted to dehydrogenated carbyne, and that some nitrogen from the DBU (1,8-diazobicyclo[5,4,0]undeca-7-ene) remained in the product. From the above values an atomic ratio of carbon:hydrogen can be calculated to be approximately 1.4 which suggests a carbon chain configuration of $-(C{=}C)_m-(CH{=}CH)_n$ with m:n=0.4 [18].

In addition, samples of diamond powder, < 2 μm in size, a sample of spectroscopic grade graphite, and amorphous carbon were prepared and analyzed as well. Table I lists all samples that were analysed and discussed in this paper.

TABLE I. Summary Data on Samples Analyzed in the Present Work.

Sample ID	Description	Comments
#4, CND/0-2	Diamond Powder	natural diamond, < 2 μm
#5, SPGR	Pyrolytic Graphite	Spectroscopic grade
#8, P/C-D7	Precursor to Carbyne	obtained from NIMC
#7, P/C-A1	Precursor to Carbyne	obtained from NIMC
#11, P/C-C7	Precursor to Carbyne	obtained from NIMC

The samples for EELS analyses were prepared using conventional TEM preparation techniques. The film samples were prepared by using a standard microtome technique. The EELS and TEM analyses were performed with a Phillips CM12 transmission electron microscope. The EEL spectra were recorded in transmission using an energy dispersive lens, single electron counting by means of a silicon detector, and analysis in multichannel storage system. Auto background subtraction routine was used. The X-axis was calibrated at 0.3 eV/channel.

3. Results

3.1 EELS OF SPECTROSCOPICALLY PURE GRAPHITE POWDER

In graphite, the two-dimensional planes are composed of C atoms in sp^2 configuration. The three trigonally directed σ-bonds form the hexagonal lattice structure of the graphene planes. In addition there is weaker π-bonding between the carbon atoms in the planes. Between the planes there is only van der Waals bonding. The band structure of a carbon plane shows bonding and antibonding σ- and σ*-bands, and a bonding π- and π*-band. There is no gap between the π- and π*-band. Rather, at the corners of the hexagonal Brillouin zone they form electron pockets at the K point and hole pockets at the H point.[1] Thus a semi-metal with about 10^{-4} charge carriers per C atom is formed. At the Fermi level there is a very low density of states which almost linearly increases when changing to positive or negative binding energies.

The loss function of pristine spectroscopically pure graphite (Figs. 1a and 1b) shows a peak at 27 eV that arises from plasma resonance of all valence electrons (π + σ plasmon). A calculation of the energy loss (assuming a simple free electron gas model approximation) due to plasma oscillation puts the position of the plasmon peak at 25.1 eV [19]. The observed higher experimental value (also reported in other studies) can be explained by single-electron excitations that were not taken into account in the free electron gas approximation calculations [19]. The peak at 6 eV can be attributed to two phenomena. There is a high joint density of states between the π and π* bands at the M point causing a strong π - π* interband transition at 6 eV. In the loss function the interband transition leads to a pronounced π-interband plasmon at 7 eV [9].

From the spectrum of the K-shell excitation region (Fig. 1b) four regions can be distinguished, i.e. relatively sharp peaks at 284.9 eV, and 285.5 eV, and two much broader peaks around 292 eV and 326 eV. For pristine graphite the onset that corresponds to the Fermi level is at 284.2 eV which corresponds with the 284.9 eV value of the first peak in the loss spectra [9]. The peaks in the region from 285 eV to 290 eV result from excitation of electrons in the ground-state 1s core levels to the vacant π*-like antibonding states. Excitations to the higher lying σ*-states occur above 290 eV. The ratio of the integrated areas under these two energy windows is proportional to the relative number of π and σ* orbitals.

---

[1] *cf.* Chapter 6.3.

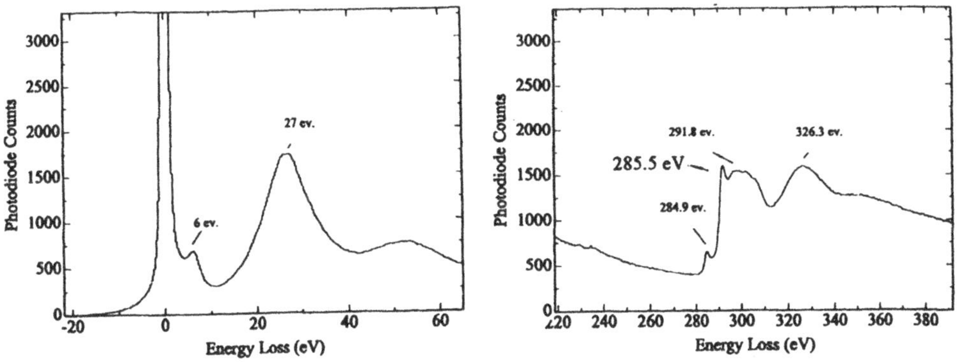

Figure 1. EELS spectra of spectroscopically pure graphite powder. a) EELS spectrum of the low energy range, b) the K-shell excitation region.

3.2 EELS OF DIAMOND POWDER

In diamond, the carbon atoms possess sp^3-type hybridization with all four valence electrons contributing to the formation of four σ-bonds. A sample prepared from a diamond powder sample (< 2 μm size distribution) was analysed. Typical EELS spectra in the low and high energy ranges are shown in Figures 2a and 2b, respectively.

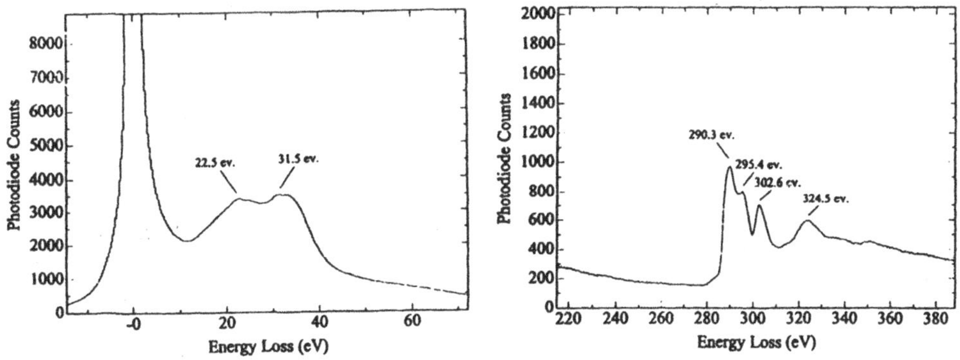

Figure 2. EELS spectra of the diamond powder (< 2 μm size distribution). a)EELS spectrum of the low energy range, b) EELS of the K-shell excitation region.

The low energy region associated with plasmons (Fig. 2a) can be characterized by two peaks around 22 eV and 32 eV. The intensity ratio between these peak was found to change during the analysis of various samples.

Figure 3 shows examples of two EELS spectra obtained from two different diamond grains in the same sample. A calculation of the energy loss for a diamond-type structure, similar to that mentioned above using the free-electron gas model [19] puts the theoretical position of the plasmon peak at 31 eV.

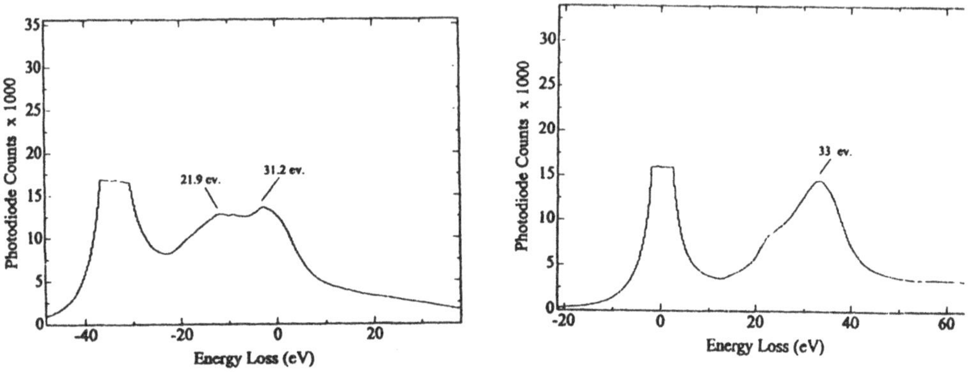

Figure 3. Low energy (plasmon) range EELS spectra of diamond powder (< 2 μm). a) EELS spectrum from an edge of a particle, b) EELS spectrum from an edge of a different particle in the sample.

TABLE II. Positions of the plasmon peaks in analysed diamond grains.

Grain	Low energy side	High energy side
1	21.9	31.2
2	22.5	31.5
3	22.8	32.1
4	23.1	32.4
5	-	32.7
6	-	33

Table II summarizes the data on observed changes in the positions of the plasmon peaks in analysed diamond grains. This phenomenon can be attributed, in part, to the sample thickness effect [19]. It is also possible that amorphous areas of the carbon support film contributed to the intensity of the peak at 22 eV due to sample shift [20]. The K-edge spectra of the analysed diamond did resemble the reported structures [11] and contained three well-resolved peaks at 290.3 eV, 295.4 eV, 302.6 eV, and one broader peak at 324.5 eV.

3.3 EELS OF AMORPHOUS CARBON FILMS

Figures 4a and 4b show typical EEL spectra recorded from the support carbon film. As can be seen, the EEL spectra of the amorphous film are different from those of graphite (Figs. 1a and 1b), and diamond (Figs. 2a and 2b), which is in agreement with published data [11,15].

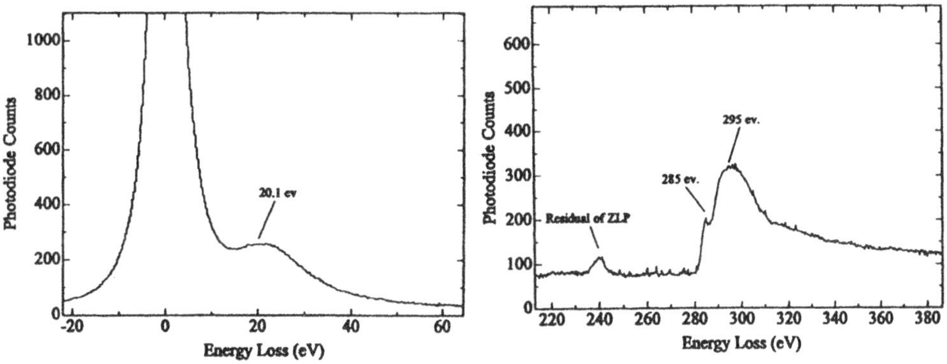

Figure 4. EELS spectrum of the amorphous support carbon film. a) EELS spectrum of the low energy range, b) EELS of the K-shell excitation region.

3.4 EELS OF CARBYNOID SAMPLES

Figures 5 through 7 show three typical EEL spectra acquired from a number of carbynoid obtained from NIMC that were analysed in an attempt to better understand their structure.[2] It should be noted that a similar EELS analysis was performed on these samples after they underwent a shock wave treatment. The results of this analysis will be discussed elsewhere.

In all analysed samples the K-shell excitation edge was found to be close to that of graphite.

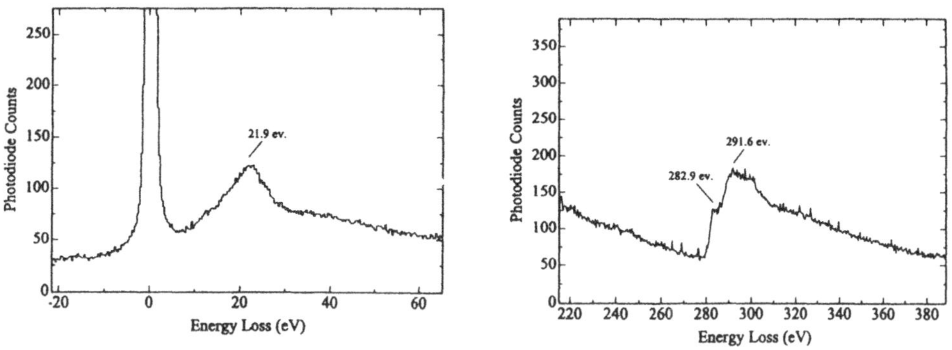

Figure 5. EELS spectra of the precursor to carbyne (sample P/C-D7 in table I). a) EELS spectrum of the low energy range, b) EELS of the K-shell excitation region.

[2] For further discussion on EELS data of carbynoid materials see section 3.3. of Chapter 6.3. and section 7 of Chapter 2.1. of this book (eds).

The plasmon peaks, however, were found to be shifted to somewhat lower energy values. Table III summarizes the data on measured EELS bands in the plasmon and the K-shell energies ranges. The maxima of the carbon low energy loss peaks in the spectra from the carbyne precursors as shown in Table III are in the range 20.4 to 24.0 eV that puts them between the values of amorphous carbon (20.1 eV, Fig. 3a) and graphite (27 eV, Fig. 1a).

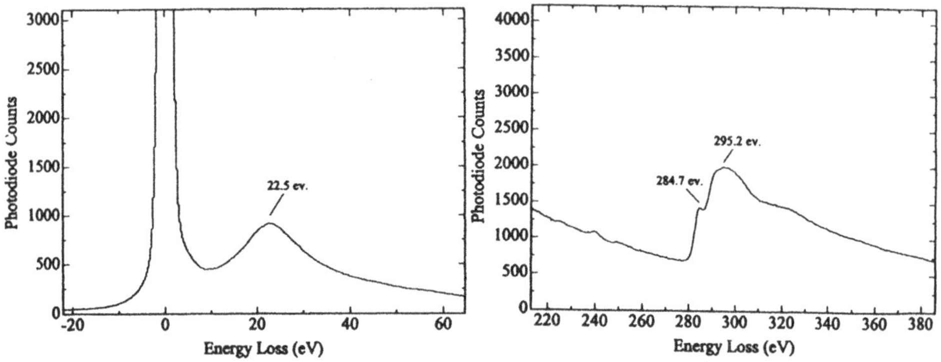

Figure 6. EELS spectra of the to carbynoid material (sample P/C-A1 in Table I). a) EELS spectrum of the low energy range, b) EELS of the K-shell excitation region.

The position of this maximum, as discussed above, is thought to be a characteristic structure parameter of this type of material [5,16] and can be attributed to the $\pi + \sigma$ plasmon of all valence electrons. It should be noted, that a peak near 23 eV has been previously observed for carbyne [5]. This result was reported as tentative only, mainly due to possible presence of oxygen. No oxygen, that has its K-edge at 530 eV, was found in the analyzed samples.

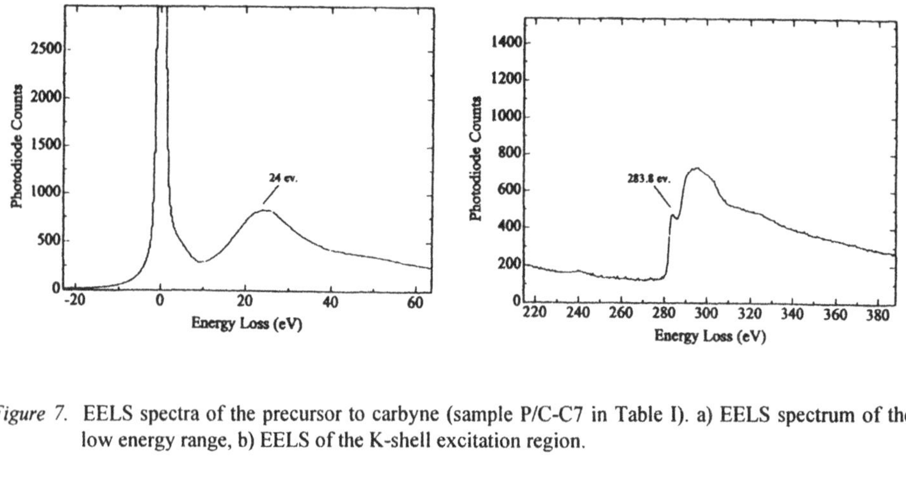

Figure 7. EELS spectra of the precursor to carbyne (sample P/C-C7 in Table I). a) EELS spectrum of the low energy range, b) EELS of the K-shell excitation region.

TABLE III. Summary of the data on measured EELS bands in the plasmon and the K-shell ranges of energies for carbyne precursor samples

Sample #	Low energy range (eV)	High energy range (eV)
P/C-A1	20.4	283.4, 294.0
	6.0, 22.5	283.8, 295.5, 320.0
	6.0, 21.9	284.1, 296.4, 321.0
	22.5	284.7, 295.2
	21.9	284.7, 294.9
	22.5	284.4, 296.0
P/C-C7	6.5, 24.0	283.8, 294.0, 324.0
	23.7	N/A

4. Discussion

To better understand the results of the EELS analysis described above, a brief overview of the structural features of carbynes and the carbynoid materials analysed in the present work is in order.

As the analysis of the data had shown, the carbynoid material obtained from NIMC exhibited in the low energy range only one broad peak centred around 20-24 eV and a peak in the K-edge range that resembled the amorphous carbon peak shown in Figure 3b. No peaks were found in the 3-10 eV range that are characteristic of the strong π - π^* interband transition and π-interband plasmon [9]. In some of the spectra a shoulder around 5-6 eV was observed.

A theoretical calculation of the position of the plasma frequency oscillation for carbyne requires the knowledge of the density of the material. The position of the bulk (π + σ) plasmon which is one of the notable features in the EELS spectra can vary depending on factors like deposition or formation conditions [23] and physical properties such as density [24]. For carbynes, densities ranging from as low as 1.97 g/cm^{-3} and 2.25 g/cm^{-3} [21] to as high as 2.9-3.4 g/cm^{-3} [22] were reported. Applying a free electron gas model of valence electrons, a simple relationship was established between the energy of plasmon modes, E_p, and the carrier charge density, n, as follows [25]:

$$\omega^2 = 4\pi \times n \times e^2/\varepsilon \times m \tag{1}$$

and

$$E_p = \hbar \times \varpi_p \tag{2}$$

where $n = n_e \bullet N_a$, n_e- being the number of valence electrons per atom, N_a - the atomic density with the rest of parameters having their regular meanings.

The variations in the theoretical position of the plasmon peak, depending on density and concentration of impurities in the samples, were estimated using the relationships above. For calculation of the bulk plasmon peak position, the value of the number of valence electrons per atom is required. For structures having a certain concentration of other atoms, a simple "law of mixtures" can be used to calculate the average value of the number of valence electrons per atom that contribute to the plasmon intensity [23]. Figure 8 shows the results of the calculations of the plasmon energy for a hypothetical carbynoid structure with 12.5% of impurities, plotted versus the density of the material. Since no data is actually available on the valency state of the impurity atoms, a number of cases were considered with valencies ranging from m=1 to m=3. As visible in Figure 8, the position of the bulk plasmon can change considerably depending on the density of the material and the nature of the impurity atoms in it. Figure 9 shows the results of a calculation of the dependence of the plasmon energy position on the structure of the carbyne unit cell and its density. Comparing the data in Table III and Figures 8 and 9, it can be suggested that the densities of all analysed carbynoid samples should lie in the range from 1.4 to 2.2 g/cm^3.

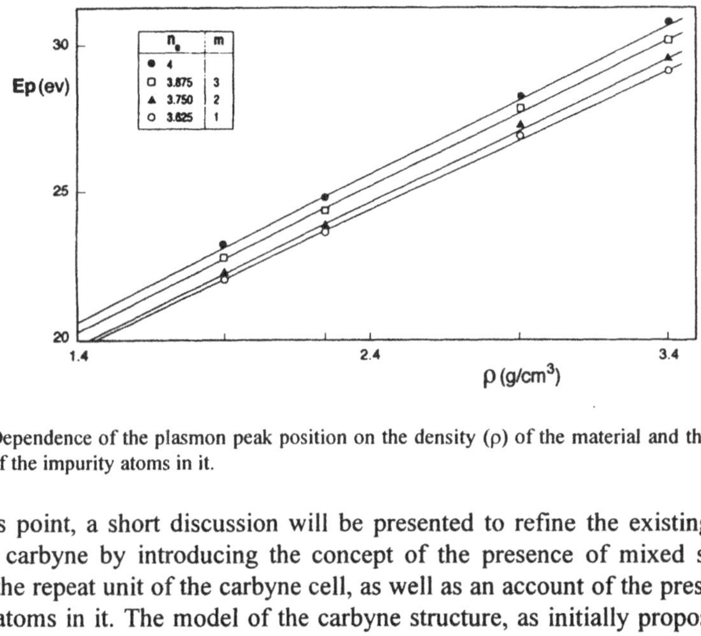

Figure 8. Dependence of the plasmon peak position on the density (ρ) of the material and the valency (m) of the impurity atoms in it.

At this point, a short discussion will be presented to refine the existing structural model of carbyne by introducing the concept of the presence of mixed sp- and sp^2 bonds in the repeat unit of the carbyne cell, as well as an account of the presence of the impurity atoms in it. The model of the carbyne structure, as initially proposed [1] and later confirmed and developed further [26], consists of repeated linear chain fragments separated by kinks.[3]

Impurity atoms that are thought to be present in many such structures have a stabilizing effect and may play the role of a catalyst in the crystallization of carbyne. It was assumed in these models that all carbon atoms in the unit cell possess sp-type of hybridization. No attempt was made to explain the formation of kinks in sp-hybridized atoms which imply a change of the angle between bonds that is impossible without a change in the hybridization type of the atom. A model is proposed here of the carbyne repeat unit, along the concepts discussed above, that introduces a mix of sp- and sp^2 atoms into the repeat unit in order to explain the formation of kinks.

[3] *cf.* Chapter 4.2.

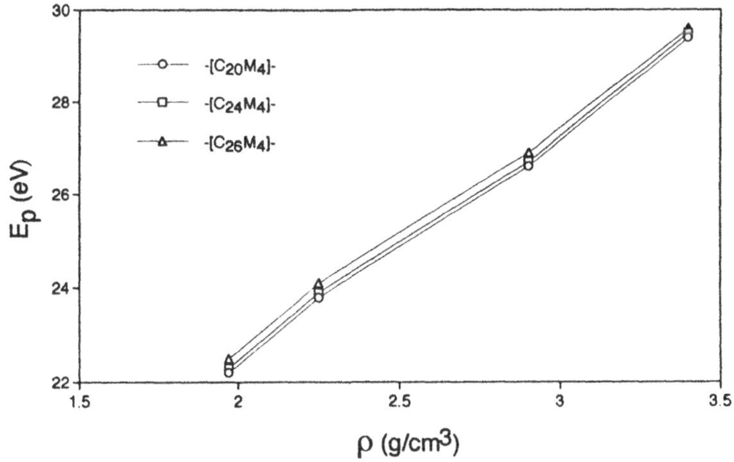

Figure 9. Dependence of the plasmon peak position on the density (ρ) of the material and the structure of the unit cell.

According to the model proposed here (Fig. 10) linear chain fragments containing three carbon atoms in sp-hybridization (cumulene or polyyne structure) are connected to two carbon atoms that possess sp^2-type of hybridization. A repeat unit is shown consisting of two such chains, each containing 20 carbon atoms and four impurity atoms corresponding to a unit cell $-[C_{20}M_4]_n-$.

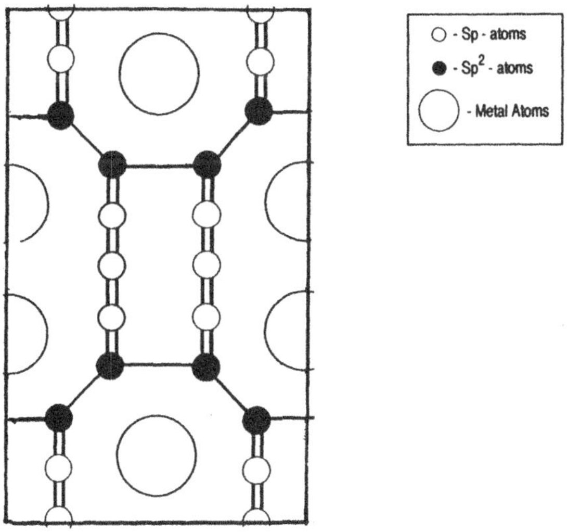

Figure 10. Schematic representation of a unit cell of a carbyne-type material.

It is assumed that the proposed repeat unit contains a total of 20 carbon atoms and four impurity atoms ($C_{20}M_4$), and that 8 of the carbon atoms are in sp^2-hybridization and the remaining 12 in sp-hybridization. The $-[C_{24}M_4]-$ and $-[C_{26}M_4]-$ structures shown in Figure 9 reflect possible carbyne structures with longer chains.

It should be noted that the position of the bulk plasmon can also shift depending on the relative concentrations of atoms with sp- and sp^2-types of hybridization in the analysed structure. Thus, for instance, the free electron model described above was used to calculate the fraction of sp^2- and sp^3-bonds in EELS spectra of carbon films [23]. It was shown that depending on the amount of sp^2 bonds in the film, the plasmon energy was shifted considerably [23]. A similar calculation for the case of materials with mixed sp and sp^2-types of bonding is not as straightforward due to the presence of π-electrons in both types of atoms.

5. Conclusions

Clear differences exist between the EELS spectra of various carbon-bearing materials with sp-, sp^2-, and sp^3-type of bonding. Even with the very low electron dose afforded by parallel- EELS detection, changes in small particles of carbon-bearing compounds can occur.

Caution should be used, therefore, in the interpretation of spectra in the case of complicated structures containing carbon atoms with different types of hybridization or impurities in the structure. The position of the plasmon peak will depend on parameters such as the density of the material, the relative concentration of impurities, and the ratio between carbon atoms with different types of hybridization The best approach is to compare the real samples with a series of standard materials with known types of carbon atoms in their structure. It was recently suggested that the carbyne forms reported hitherto might correspond to polytypes of polydiacetylene phases, where the packing sequence in the chiral phase can be interrupted by that in the trans-hinged phase [27].[4] Work is now in progress to compile more EELS data on different forms of carbon and polydiacetylene, to compare these structures.

Acknowledgements

I wish to thank Prof. R.B. Heimann for his comments on the carbyne model. Thanks are also due to Dr. S. Evsyukov, Prof. D. McCullough and Prof. G. Weatherly for their helpful suggestions and advise during the work and interpretation of the spectra, and to Prof. Z. Iskanderova for useful comments on the free-electron model and some of the calculations performed.

6. References

1. Heimann, R.B, Kleiman, J.I., and Salansky, N.: Structural aspects and confirmation of linear carbon polytypes (carbynes), *Carbon* **22(2)** (1984) 147-155.
2. Kavan, L. and Kastner J.: Carbyne forms of carbon: continuation of the story, *Carbon* **32(8)** (1994), 1533-1536.

---

[4] *cf.* Chapter 4.2.

3. Korshak, V.V., Kudryavtsev, Yu.P., Evsiukov, S.E., Korshak, Yu.V., Guseva, M.B., Babaev, V.G., and Kostishko, B.M.: Infrared spectra of carbyne, *Dokl. Akad. Nauk SSSR* **298(6)** (1988), 1421-1424.
4. Korshak, V.V., Kudryavtsev, Yu.P., Khvostov, V.V., Guseva, M.B., Babaev, V.G., and Rylova, O.Yu: Electronic structure of carbynes studied by auger and electron energy loss spectroscopy, *Carbon* **25(6)** (1987), 735-738.
5. Sergushin, I.P., Kudryavtsev, Yu.P., Elizen, V.M., Sadovskii, A.P., Sladkov, A.M., Nefedov, V.I. and Korshak, V.V: X-ray electron and x- ray spectral study of carbyne, Zhurnal Strukturnoi Khimii **18(4)** (1977), 698-700.
6. Kudryavtsev, Yu.P., Evsiukov, S.E., Guseva, M.B., Babaev, V.G.and Khvostov, V.V.: Carbyne- the third allotropic form of carbon, *Russian Chem. Bull.* **42(3)** (1993), 399-413.
7. Kleiman, J.I.: EELS study of carbon-precursors and other carbon bearing materials, Proceedings of the Symposium on Super Carbon, The 4-th IUMRS International Conference, Makuhari, Chiba, Japan, (1997), 43-47.
8. Rudberg, E.: EELS measurements in reflection on noble metals, *K. Svenska Vet. Akad. Handl.* **7** (1929), 1.
9. Fink, J.,: Transmission Electron Energy-Loss Spectroscopy", in: 'Unoccupied Electronic States", ed. J.C. Fuggle and J.E. Inglesfield, *Springer-Verlag* **v.69** (1992), 203-241.
10. Daniels, J., Festenberg, C.V., Raether, H. and Zeppenfeld, K.,: Optical Constants of Solids by Electron Spectroscopy, in: "Springer tracts in Modern Physics, ed. G. Höhler, *Springer-Verlag* **v.54** (1970), 77-135.
11. Egerton, R.F. and Whealan, M.J.: Electron Energy Loss Spectra of Diamond, Graphite and Amorphous Carbon, *J. of Elec. Spec. and Related Phenomena* **3** (1974) 232-236.
12. Koma, A. and Miki, K.: Core electron excitation spectra of diamond, graphite and glassy carbon, *Appl. Phys.* **A34** (1984), 35-39.
13. Gensterblum, G., Pireaux, J.J., Thiry, P.A. Caudano, R., Vigneron, J.P., Lambin, Ph., Lucas, A.A. and Krätschmer, W.: High-resolution electron-energy-loss spectroscopy of thin films of C60 on Si(100), *Phys, Rev. Lett.* **67(16)** (14 Oct.1991), 2171-2174.
14. Kovarik, P., Bourdon, E.B.D. and Prince, R.H.: Electron-energy-loss characterization of laser-deposited a-C, a-C:H and diamond films, *Phys. Review* **B 48**(16) (15 Oct.1993-II), 12 121-12 129.
15. Weissmantel, Chr., Bewilogua, K. and Schürer, C.: Characterization of hard carbon films by electron energy loss spectrometry, *Thin Solid Films* **61** (1979), L1-L4.
16. Rightor, E.G., Young, G.P., Urquhart, S.G. and Hitchcock, A.P.: Correlation of polymer (PET) parallel-EELS spectra with spectra from related molecular analogues, Microscopy: The Key Research Tool, March (1992), 67-71.
17. Heimann, R.B., Fujiwara, S., Kakudate, Y., Koga, Y., Komatsu, T. and Nomura, M.: A new carbon form obtained by weak shock compression of carbyne, *Carbon* **33(6)** (1995), 859-863.
18. Komatsu, T., Nomura, M., Kakudate, Y., Fujiwara, S., Heimann, R.B.: Characterization of dehydrochlorinated poly(vinylidene chloride) and the shock-compressed material, *Macromol. Chem. Phys.* **196** (1995), 3031-3040.
19. Egerton R.F. and Whealan, M.J.: The electron energy loss spectrum and band structure of diamond, *Phil. Mag.* **30** (1974), 739-749.
20. McCulough, D., Private communication:
21. Sladkov, A.M., Kasatochkin, V.I., Kudryavtsev, Yu.P. and Korshak, V.V.: Synthesis and properties of valuable polymers of carbon, Izvestiya Academii Nauk SSSR, *Seriya Khimicheskaya* **No.12** (1968), 2697-2704.
22. Wittaker, A.G.: Carbon: Occurrence of carbyne forms of carbon in natural graphite, *Carbon* **17** (1979), 21-24.
23. Cuomo, J.I., Doyle, J.P., Bruley, J. and Liu, J.C.: Sputter deposition of dense diamond-like carbon films at low temperature, *Appl. Phys. Lett.* **58(5)** (1991), 466-468.
24. Berger, S. D., McKenzie, D.P. and Martin, P.J.: EELS analysis of vacuum arc-deposited diamond-like films, *Phil. Mag. Lett.* **57(6)** (1988), 285-290.
25. Colliex, C.: Electron energy loss spectroscopy in the electron microscope, in Advances in Optical and Electron Microscopy, ed. By R. Barer and V.E. Cosslet, (Academic, New York) **v.9** (1984), 65-177.
26. Kudryavtsev, Yu.P, Evsiukov, S.E., Babaev, V.G., Guseva, M.B., Khvostov, V.V. and Krechko, L.M.: Oriented carbon layers, *Carbon* **30**, (1992), 213-221.
27. Baughman, R.H., Galvao, D.S., Cui, C. and Dantas, S.O.: Hinged and shiral polydiacetylene carbon crystals, *Chem. Phys. Lett.* **269** (1997), 356-364.

Chapter 7: Suggested technical applications of carbyne materials

7.1 DIAMOND SYNTHESIS FROM CARBYNE

R.B. HEIMANN
Department of Mineralogy, Freiberg University of Mining and Technology, D09596 Freiberg, Germany

Abstract - Linear carbon allotropes (carbynes) with 1D-sp σ-hybridization were found to transform readily into diamond at comparatively weak dynamic pressures below 5 GPa without external thermal activation. The mechanism involved may be based on shock-induced close interaction of the initially weakly van der Waals-type bonded carbon chains to permit overlap of wave functions of neighbouring chains and generation of coupled π-π* bonding states. Further increasing interaction is expressed through 3D-cross linking by σ-bond formation of periodically arranged carbon chains with either kinks or screw symmetry axes.

1. Introduction

The transformation of the 2D-planar sp^2 graphitic carbon structure into the 3D sp^3 diamond network with chair-type carbon ring arrangement is energetically difficult to achieve since it involves a change in the first coordination sphere of the carbon atoms as well as in the type of the chemical bonds. Hence high static or dynamic pressures and temperatures are required. At very high static (isotropic) pressures (>80 GPa) rhombohedral graphite may transform to diamond without any thermal or catalytic activation [1]. However, high temperatures and addition of solvents for carbon such as nickel or iron will accelerate the transformation rates and provide substantially lower equilibrium pressures [2].

On the other hand, the transformation of graphite to diamond involving shock waves has been studied for over thirty years and large-scale industrial production of such diamonds is now being undertaken in several countries. Competing hypotheses about the mechanism of shock transformation include ultrafast annealing of a glassy carbon phase created by total destruction of the crystal lattice of graphite [3-5], martensitic shear transformation of 3R-graphite [6-8], dislocation reactions [5,9], carbon self-diffusion [10] and a complex solid-vapour-liquid-solid mechanism [11]. Here we will report on other plausible mechanisms involving linear sp-hybridized carbon clusters, i.e. carbynes and carbynoid structures as precursor phases.

'Carbynes' by definition are "systems consisting only of (straight [13] or kinked [12]) carbon chains held together by van der Waals forces to form a crystal"[14].

R.B. Heimann et al. (eds.), Carbyne and Carbynoid Structures, 409–425.

Theoretically, linear carbon chains with conjugated triple (polyyne-type) and/or cumulated double (polycumulene-type) intrachain bonds are assembled in a hexagonal array without any chemical bonds between the chains ('pencil-in-box' model). In real samples, however, cross-linking with various periodicities as well as kinks will always occur resulting in a layered lattice arrangement. Hence the definition above can be extended by stating that carbynes are "partially cross-linked three-dimensional polymer(s) having a more or less regular (layered) crystalline structure and (are) predominantly composed of extensive fragments of linearly polymerised carbon"[14]. In the spirit of this, graphite and diamond, respectively are limiting conformations of network and chain structures that are regularly periodically cross-linked by three(graphite)- and four(diamond)-function branching points. Indeed, the graphite plane can be viewed as composed of 2D-cross linked polyyne [15] or polycumulene [16] chains whereas the diamond structure consists of a 3D-array of an equal number of screw axes of opposite chirality, i.e. of helically twisted carbon chains [17].

Based on the (tentative) carbon phase diagram developed by Whittaker (cp. Fig. 4) carbynes should transform to diamond at comparatively weak (shock) pressures without external thermal activation. If the assumption is correct that diamond formation can proceed at such weak pressures through a low energy rearrangement of 1D sp-hybridized carbon chains than the deliberate addition of carbynes to graphite or even using more or less pure carbyne as precursor material should lower substantially the energy barrier required to attain 3D sp^3-hybridization.

2. Experimental Methods

2.1 MATERIALS

A variety of methods have been developed in the past to synthesize carbynes including oxidative dehydropolycondensation of acetylene [18], condensation of carbon vapour obtained by sublimation of graphite by a laser beam [19], argon ion sputtering [20] or an electrical current [21], chemical dehydrohalogenation of stereospecifically chlorinated polyacetylene [22] or poly(vinylidene fluoride) [23], and eximer laser-induced dehydrochlorination of chlorinated poly(vinylchloride) [24]. Other potential routes to carbynes were proposed recently involving pyrolysis or photolysis of 3,4-dichloro-3-cyclobutene-1,2-dione adducts [25]. A comprehensive review of carbyne syntheses was given recently by Kudryavtsev *et al* [26]. For additional details the reader is referred to Chapter 3 of this treatise.

To obtain the carbyne samples used in the present investigation a procedure similar to the Shirakawa method [22] was adopted [27] since the *trans*-elimination of HCl should result in only polyyne- but not in polycumulene-type chain segments. The latter are thought to be less useful for low energy transformation routes. Poly(vinylidene chloride) (PVDC) was dehydrochlorinated with DBU (1,8-diazabicyclo[5,4,0]undeca-7-ene). The solvent for PVDC was N,N-dimethylacetoamide (DMAC). A 10% (2.1 molar) solution of DBU was slowly added to a 1% (1 molar) solution of PVDC (Asahi

Chemical Industry Co.Ltd., Saran polymer) under nitrogen and stirred for 5 hours at 70°C. After treatment the mixture was cooled and poured into an excess amount of methanol. The black reaction product was filtered off and repeatedly washed with methanol, acetone, and THF in a dry box. Any remaining contaminants were removed by Soxhlet extraction for 48 h under nitrogen using methanol as a solvent. The product was dried in vacuum for 48 h at ambient temperature.

2.2 MATERIALS CHARACTERIZATION

The carbyne was investigated by high resolution transmission electronmicroscopy (HRTEM) using a JEOL JEM 4000-FX. The samples were prepared on a carbon microgrid coated with a carbon-collodion membrane in the customary manner. The acceleration voltage used was 400 kV. Diffuse Reflectance Infrared spectra of carbyne samples embedded in a KBr matrix were acquired using a Fourier-transform infrared spectrometer JEOL JIR 100. X-ray diffraction spectra were obtained with a Philips PW 1800 instrument using CuK_a radiation at 40 kV, 40 mA. The specimens were prepared on a Si wafer coated with vaseline.

The elemental atomic composition of the product was determined to be 88.6% C, 5.4% H, 3.7% N and 2.3% Cl showing that a part of the original poly(vinylidene chloride) was not completely reacted to (dehydrogenated) carbyne, and that some nitrogen from the DBU remained in the product. From the values above an atomic ratio can be calculated to be C:H:N:Cl = 7.4:5.4:0.26:0.07 or approximately C:H ≈ 1.4. This ratio suggests a carbon chain configuration of $-(C{\equiv}C)_m-(CH{=}CH)_n-$ with m:n = 0.4, i.e. n > m. The synthesized material might therefore conceivably be called a carbynoid structure with a high proportion of polyene-type segments [24].

Figure 1 shows the FTIR spectrum of the starting material. The absorption peak at 2146 cm^{-1} (peak 02) can be assigned to the $\nu_{C\equiv C}$ stretching vibration of the polyyne chain [28]. The peak at 1670 cm^{-1} (peak 03) is presumably related to -(C=C-C=C)- resonance vibrations with a possible contribution of the $\nu_{C=O}$ stretching vibration of residual oxygenated material in the sample [29]. The existence of carboxyl groups has been also confirmed by a broad asymmetric NMR peak centered around 182 ppm, and the polyyne-cumulene resonance structure by a symmetric peak at 134.8 ppm. The broad band ranging from 1630 to 1200 cm^{-1} can likely be assigned to an overlap of infrared active $\nu_{C=C}$ stretching vibrations of various segments of different chain lengths [23,30]. The intensive band centered around 2933 cm^{-1} (peak 01) presumably belongs to the CH_3 asymmetric stretching vibration with contributions of the symmetric stretching vibration of the $=CH_2$ end group of polyene chains [30]. The peak at 1000 cm^{-1} (peak 06) may belong to the $\nu_{C\text{-}C}$ stretching vibration of *trans*-polyene. Finally, the peak at 1440 cm^{-1} (peak 04a) is presumed to be a deformation vibration confirming the existence of kinks in the carbon chain. These results also confirm the inhomogeneous nature of the carbyne reaction product consisting of polyyne-, poly(cumulene)- and polyene-type chain segments and their derivatives [31].

The x-ray diffraction spectrum of the carbynoid material (Fig. 2a) shows a very broad peak centred around 470 pm and an even broader shoulder around 236 pm buried in the scattered background radiation. These spacings resemble the (10.0) and (20.0) lattice planes of highly disordered Carbon IX [32]. From these results it can be concluded that the microstructure of the carbyne resembles a more or less disordered, i.e. glasslike short range ordered (SRO) material.

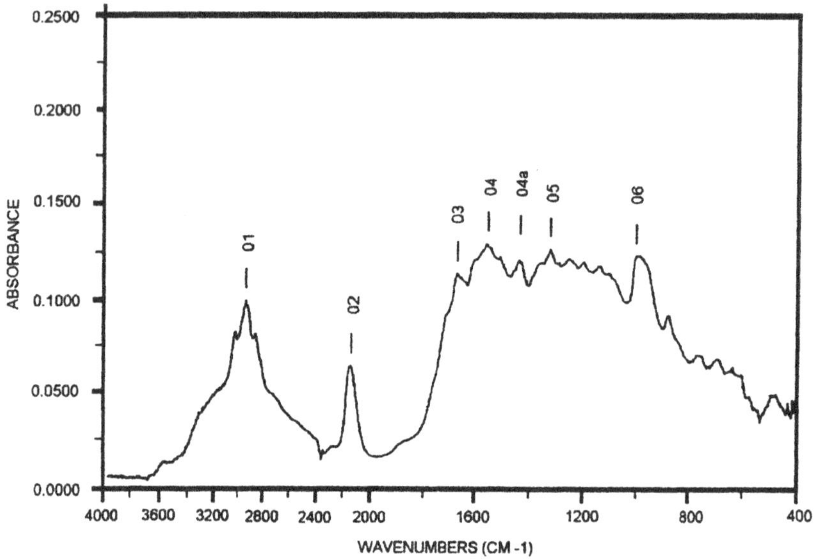

Figure 1 Fourier-transform Infrared spectrum of precarbyne, i.e. a carbynoid structure with mixed triple and double bonds ($-(C{\equiv}C)_m-(CH{=}CH)_n-$, m : n = 0.4).

3. Shock Compression of Carbyne

Carbyne samples were mixed with copper powder (carbyne/copper = 2/98 by weight) and pressed into steel capsules to form discs 5 mm in thickness and 20 mm in diameter. A planar shock wave was created using a copper flyer plate accelerated by a plastic explosive containing 85% HMX with an explosive lens of 65 mm diameter, and a momentum trap recovery system [33,34]. The recovered samples were machined and immersed in nitric acid to remove the copper matrix.

Since the Hugoniot-Rankine EOS is unknown for carbynes that of copper powder was used in the pressure and temperature calculations. The incident shock pressures to the samples were estimated to be below 5 and $10<P<15$ GPa, respectively.

X-ray diffraction spectra of the shock-compressed carbynoid material are shown in Figures 2b and c. Figure 2b shows that at pressures of about 5 GPa or below a rather barren spectrum occurs with broad peaks at 343, 206 and 107 pm. The first peak can be attributed to the (00.2) spacing of turbostratically disordered graphite [35,36]. The second and third peaks are the (111) and (311) spacings of cubic diamond. Other minute peaks occuring in the spectrum belong to SnO_2 that originated from the container material used in the shock compression experiment. On increasing the shock pressure to about 10-15 GPa the x-ray interplanar spacings become more prominent (Fig. 2c). The spacing of the (00.2) lattice plane of graphite has decreased to 334 pm corresponding to a well-ordered 2H-polytype. Table 1 shows the observed and calculated interplanar spacings of the diamond formed.

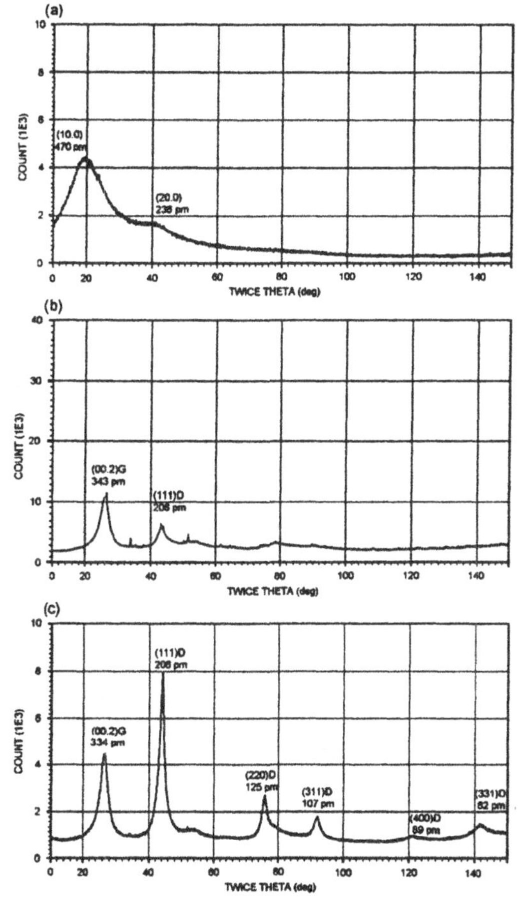

Figure 2. XRD diffraction pattern of (a) carbynoid structure (cp. Fig. 1), (b) material of (a) shocked to < 5 GPa, and (c) material of (a) shocked to 10 < P < 15 GPa.

TABLE 1. Observed (Fig. 2c) and calculated interplanar spacings of diamond. The spacings were calculated using a lattice parameter $a_0 = 356.6$ pm.

No.	(hkl)	d_{obs} (pm)	d_{calc} (pm)	I_{obs}	I_{calc}
1	111	206	205.9	100	100
2	220	125	126.1	25	25
3	311	107	107.5	14	16
4	400	89	89.2	3	8
5	331	82	81.8	7	16

Fourier-transform laser Raman spectroscopy (diameter of the Ar laser beam: 1 mm) of the shocked samples failed to detect any diamond presumably because of its small crystal size. Instead, the characteristic Raman bands of graphitic carbon at 1362 and 1600 cm^{-1} were observed. On the other hand, EELS pattern shown in Figure 3 revealed (a) for the unshocked carbynoid material a carbon low energy loss peak centred around 24 eV attributable to carbyne [37], (b) a broad signal around 37 eV attributable to disordered diamond as well as (c) evidence for newly formed graphite as confirmed by a strong extended peak at 28 eV. Also, a new hexagonal carbon phase with a_0= 338 pm was discovered presumably structurally related to carbyne [38].

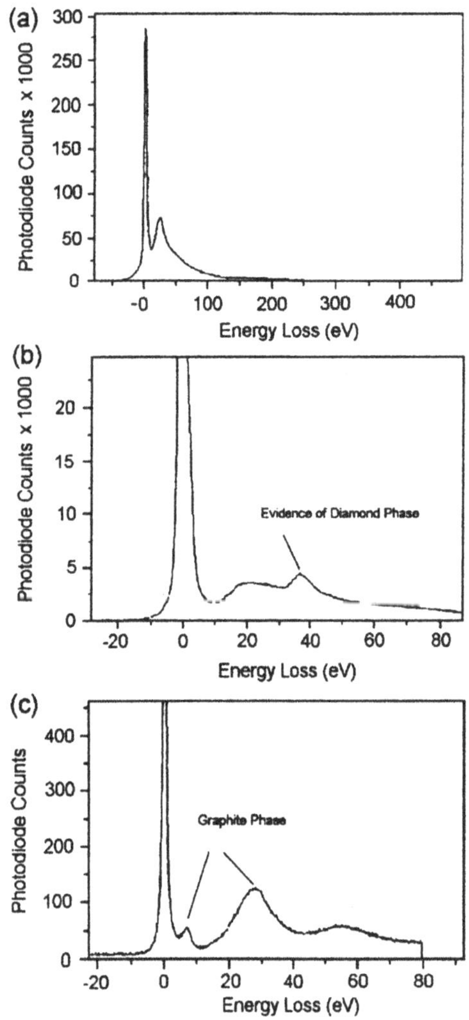

Figure 3. EELS spectra of (a) unshocked carbynoid material, and (b, c) the same material shocked to < 5 GPa (Courtesy: J.I. Kleiman, U. of Toronto), see also Figure 11, Chapter 3.4.3

The presently accepted phase diagram of carbon suggests that stable diamond should form from graphite at ambient temperature already at pressures below 2 GPa [2,39]. There is, however, a huge kinetic barrier to overcome at low temperatures that historically has had hampered successful diamond synthesis for quite a long time. The transformation of the planar hexagonal rings of the graphite structure into the chair-form hexagons of diamond requires high compressional forces since the stiff planar hexagonal rings of graphite can store a large amount of energy without changing into buckled (lonsdaleite) or puckered (diamond) conformations [40]. Thus out-of-plane shifting of carbon atoms of (rhombohedral) graphite to produce the puckered chair-like hexagons of diamond without thermal or catalytic activation can be achieved only at very high static pressures exceeding 80 GPa [1]. However, in the presence of temperatures exceeding 3,000 K this release of excess energy stored in bond compression can be achieved without catalytic activation already at a static pressure of 12.5 GPa resulting in a spontaneous collapse of the graphite lattice [41].

Transformation by shock compression of a large variety of carbons and carbonaceous precursor materials has been investigated at pressures between 20 and 100 GPa, and temperatures between 500 and 4,000 K [42]. The high temperatures required are either generated thermodynamically by adiabatic and/or entropic processes in porous precursor materials or by an external high temperature pulse. Strong 1D-shear stresses typical for shock compression also aid the transformation. High diamond yields up to 80% of the recovered material have been reported [33,43].

The mechanism of shock-assisted transformation has been controversially discussed ever since the first successful diamond synthesis by DeCarli and Jamieson [6]. The martensitic shear transformation of rhombohedral (3R) graphite suggested by these authors was later abandoned in favour of processes involving ultrafast annealing of a collapsed SRO-carbon phase [3-5]. Competing theories assumed dislocation reactions [5,9], selfdiffusion of carbon atoms into interstitial positions between planar graphite hexagons [10], and a complex solid-vapour-liquid-solid sequence [11]. Recently, however, the idea of a martensitic phase transformation mechanism has been revived and confirmed by time-resolved experiments [7,8]. At the heart of this process is a two-wave structure obtained by the anomalous behaviour of the Hugoniot adiabat in the region of the phase transition. At shock-generated temperatures below 2,000 K first a hexagonal (lonsdaleite) phase is formed by martensitic, i.e. diffusionless displacive shear transformation of adjacent planar hexagonal graphite planes. In a second step another mechanism, possibly also martensitic, transforms the hexagonal lonsdaleite into the cubic diamond. This process actually may be activated by the shock wave-driven passage of a train of Shockley partial dislocations through the crystal lattice as suggested earlier by Wheeler and Lewis [44]. These ideas should be contrasted with recent findings by Borodina *et al* [87] who subjected thin (10 μm) and thick (150 μm) amorphous carbyne films, obtained by dehydrofluorination of poly(vinylidene fluoride), to shock-waves at peak pressures of 18, 26, and 36 GPa. As in the case of the shock-pressure-induced transformation of graphite, a two-stage process was invoked with the first stage leading within the shock front to a martensite-type transformation of the amorphous carbyne to crystalline α-carbyne. During the second stage, transformation to diamond took place, presumably by diffusive kinetics, i.e. a reconstructive-type of phase transformation. The diamond yield was rather low, likely influenced by graphitization of diamond owing to postshock annealing. It was also concluded from the experimental results that the temperature conditions throughout the amorphous carbyne films may play a role in the transformation kinetics. In thin films the amorphous carbyne transforms to crystalline carbyne, in thick films a mixture of graphite and nanocrystalline diamond forms. This suggests that rapid quenching of the thin films creates conditions conducive to retain crystalline carbyne whereas in thicker films high temperatures prevail that eventually lead to transformation of diamond into graphite.

During previous extensive studies of the shock-induced conversion of graphite to diamond it was found that a separation in time of a high temperature (up to 3,200 K) and a high pressure (up to 30 GPa) pulse created favourable conditions to form diamond along the metastable melting curve of graphite [11,41,45]. Moreover, an optimum delay time of the pressure pulse was determined to be around 100 μs at 18 GPa. Supported by these experimental facts an alternative mechanism was developed involving the idea of thermally induced bond splitting of the graphite atomic arrangement on a (00.2) plane parallel <10.0> to form linear polyyne-type carbon chains [15] or parallel <11.0> to form polycumulene-type chains [12]. This idea has been exhaustively described in Chapter 4.2. It has been proposed [45] that it is energetically much more favourable to unravel the 2D-planar graphite structure into 1D-linear carbon chains and rearrange these chains than to compress the intact graphite lattice into a 3D-diamond configuration. The former mechanism constitutes a low energy path involving only the shifting of electrons and small displacement of carbon atoms [12,15]. The suggested bond splitting can be accomplished within the 100 μs interaction time of the high temperature pulse, for example by application of a discharge pulse of a laser [46,47] or a capacitor [11] with very high energy densities. Subsequent dynamic compression lasting some 20 ns generates the driving force for close interaction of the sp-linear carbon chains to permit overlap of molecular bonding π-orbitals of one chain with empty antibonding π*-orbitals on the neighbouring chain. Recent chain-interaction potential calculations using an exponential-six (exp-6) formalism [48] showed indeed that with increasing interaction, i.e. compression the sp-chain structure becomes unstable because of the proximity of π-orbitals on neighbouring chains, resulting in cross-linking of the adjacent chains and eventually in a collapse into sp^3-diamond by formation of interchain σ-bonds.

How this cross-linking occurs in detail is still largely a matter of conjecture. It has been proposed [45] that during shock conversion of graphite to diamond (hypothetical) crystalline allotropes of carbon [49,50] could be intermediate structures in the transformation process. For example, "polyyne-diamond" [51], i.e. layers of *trans*-decalin rings with sp^3-hybridization connected by sp-polyyne chains perpendicular to the layers could form trigonal arrays of interstitial carbon atoms on shock-triggered collapse of the polyyne chains. These interstitials can rearrange and generate chair-form hexagons of a distorted diamond lattice by shear-induced puckering. This mechanism is entirely equivalent to the collapse of strain-free carbenoid triradicals, i.e. diamond-like cyclohexane rings connected by triangular nets of linear sp-diradicals (carbenes) [52].

Another feasible mechanism may involve an array of periodically arranged carbon chains with kinks and two-fold screw symmetry axes, respectively [12,16,48]. Details are available in Chapter 4.2. Many of such hypothetical structures have been investigated by Stankevich *et al* [50] with respect to their ability to form strain-free buckled or puckered 3D-carbon networks. Analogous structures consisting of infinite 2D-cross linked polyene chains were proposed by Hoffmann *et al* [53]. Since it was found that the Mulliken bond order of carbon atoms located in adjacent polyene chains is positive a strong interaction is expected, likely through a change from a (repulsive) interchain π-orbital interaction to an (attractive) regime of σ-bonds between adjacent chains.

Lastly, the introduction by excitation of strings of double bonds into polyyne chains may lead to chain segments of the type $-C{\equiv}C-[C^{\bullet}{=}C{=}C^{\bullet}]-C{\equiv}C-$ [45,54] whose unpaired electrons (positive-sign solitons[1]) provide coupling sites for the cross-linking reaction. In this case a chair-type 6-ring configuration could be locally formed without the need for large shear forces. It will be noted that the polyyne-polycumulene hybrid structure shown above is equivalent to a dehydrogenated polyyne-polyene configuration of the type used in the present study. In this connection it should also be emphasized that the actual structure of the carbyne, in particular their chain lengths [12], spatial arrangement, nature of chain-terminating end groups [57], and kink angles and kink offset distances [16] will play a decisive role in the ease of transformation to diamond. According to Kudryavtsev *et al* [48] polyyne-type chains become more stable against collapse with increasing chain lengths since their equilibrium interchain distance increases. This makes it more difficult to establish σ-links among neighbouring chains. Since the structural properties of the carbynes are highly dependent on the starting material and the method of preparation care must be taken to obtain the precursor material most suited for a low energy path transformation. For example, short chains will transform more easily than longer chains. Because there is an indication that the chain lengths decrease with increasing temperature by a process thought to be analogous to cracking of hydrocarbon chains [15,16,21,58], transformation reactions should be favoured at high temperature and low pressure.

With this in mind we therefore suggest, based on our preliminary experimental findings, that weak shock pressures <5 GPa suffice to achieve an sp^3-hybridization by cross-linking of periodically arranged sp-hybridized short carbon chains. This does explain the ample experimental evidence of several polytypic forms of carbyne occuring during the synthesis of diamond in dynamic pressure experiments [38,59-62] as well as under reduced pressure [63]. It further appears that the carbon phase diagram proposed by Whittaker [15] (see also Derjagin *et al* [64]) is essentially correct in that

[1] The concept of a dynamic stabilization of carbon chains by spin density and charge density waves [55] has been extended by postulating a strong intrinsic electron-phonon coupling in polyyne chains that is being expressed through the occurence of solitons [54]. Mobile paramagnetic solitons have been identified by ESR spectroscopy in carbynes recently [56] (see Chapter 6.2).

the region of thermodynamic stability of graphite is separated from that of liquid carbon by a rather broad stability field of carbyne polytypes (Fig. 4). Beyond temperatures of approximately 2,600 K the Berman-Simon line turns into the carbyne/diamond equilibrium line. The invariant point graphite+carbyne+diamond located at P = 6 GPa, T = 2,600 K defines the minimum *equilibrium* condition at which carbyne under pressure can be transformed to diamond. This phase diagram also gives a rationale for the low pressure/high temperature decay of the graphite structure into that of carbyne [15,16][2] since it was concluded from estimated heat of formation and energy transition calculations that at high temperatures exceeding 2,500 K carbynes ought to be thermodynamically more stable than graphite [66].

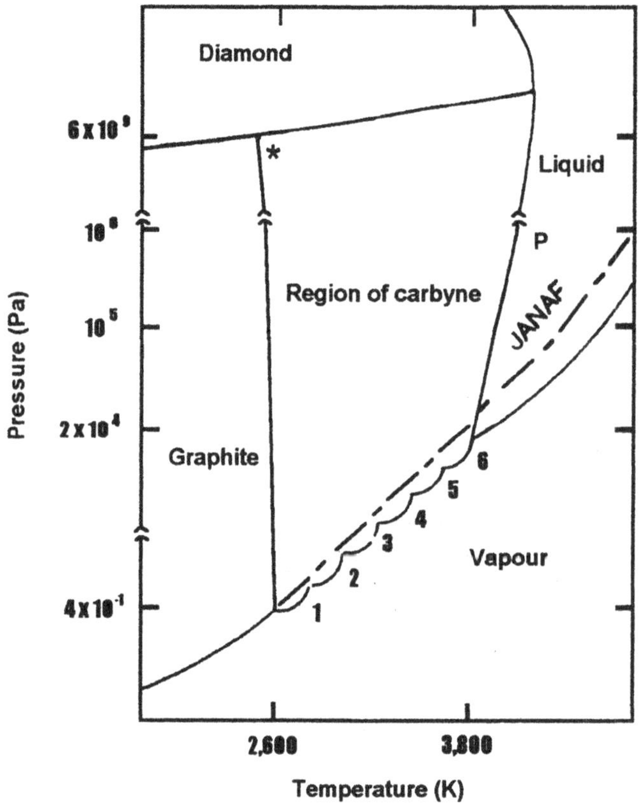

Figure 4. High temperature - low pressure region of the phase diagram of carbon as proposed by Whittaker [15]. The numbers 1 to 6 refer to carbyne polytypes of different chain lengths. The asterisk is the P, T-region to which a natural diamond has been treated (see Text)

[2] This view should be contrasted with the hypothesis by Yamada *et al* [65] that carbynes form during condensation of a high pressure-high temperature carbon vapour between 2,600 K and the triple point.

The conjecture expressed above that diamond, at low pressure and high temperature, can transform into graphite by the way of carbyne has been corroborated by a unique experiment already fifteen years ago [67]. A small natural diamond was heated at a temperature and pressure denoted by the asterisk in Figure 4 depicting Whittaker's proposed carbon phase diagram at high temperature and low pressure [15]. Subsequently the sample was quenched. Painstaking XRD and ED studies revealed the coexistence of diamond, graphite and carbyne suggesting that the P,T-conditions were at, or close to, the triple point diamond/graphite/carbyne at about 2600 K and 6 GPa. The sample was then progressively chemically oxidized, and after each step XRD and ED pattern recorded. It was found that the graphite lines disappeared but the diamond and carbyne lines persisted, attesting to their higher stability against oxidation. The assumption of carbyne as an intermediary phase between diamond and graphite is also supported by thermodynamic arguments. The activation volume of the diamond-graphite transition of about 170 cm^3/mole at 2400 K and 2 GPa reported by Bundy *et al* [68] appears to be exceptionally large considering that the molar volumes of diamond and graphite are, respectively 3.42 and 5.33 cm^3/mole. This means that the system must expand by a large factor for the transition to take place. Bundy's explanation of this inconsistency relied on a domain or 'zipper'-like reaction in which the high activation energy and volume are fed from the diamond to the graphite phase in an organized manner across the reaction wavefront. Considering a succession of different carbyne polytypes forming the 'domains', it can be shown that the activation volume of the diamond-graphite transition is the sum of all activation volumina of the intermediate steps, thus providing a rationale for the large value found by Bundy *et al* [68].

According to van Thiel and Ree [69], the explanation of the measured detonation velocity vs. initial density curve in shock wave-assisted transformation of graphite, by the explosion of TNT, requires the stabilization of a low-density phase over a large pressure range (4 GPa) to account for the non-equilibrium behaviour of the graphite. The density of this intermediate phase is close to 2.48 g/cm^3 at 1 bar, suggesting mixed carbyne polytypes as intermediate carbon phases [70]. Using this density, the $T = 0$ K isotherm can be calculated from which in turn the thermal part of the EOS can be inferred. Then the theoretical curve of the detonation velocity vs. initial TNT density can be determined which predicts two possible detonation velocities for the same initial density above 1.55 g/cm^3 where the EOS starts to deviated from equilibrium. These two detonation velocities are the minima of the Hugoniot shock velocity loci. Further analyses showed that neither a diamond-like nor a carbyne-like EOS could account for the slope of the Chapman-Jouguet detonation pressure vs. density curve. This slope, however, can be reduced when the transition is made second order, i.e. when it spreads out over a larger pressure range. This implies that the system must be described by *a range of standard heats of formation* that would correspond to a series of carbyne polytypes with decreasing chain length consistent with the concept of chain 'cracking'.

An indication of the ease with which carbyne can transform to diamond is found in a set of experiments performed by Kudryavtsev and coworkers [71]. When crystalline silicon, or NaCl or KBr single crystal surfaces are coated by „amorphous" carbyne, i.e.

carbyne consisting of polyyne chains mixed with polyene segments obtained by dehydrofluorination of PVDF, the chain direction is perpendicular to the substrate surface. At temperatures above 150°C the multiple bonds in the chains will be broken and extensive cross-linking occurs that results in the formation of „quasi-1D diamond". The bond breaking mechanism releases heat that provides additional activation energy for the crystallization process [86]. Here is a mechanism that offers an explanation to the curious fact that Whittaker [72] obtained an extremely hard layer on selected substrates, referred to a chaoite, when coated with carbon obtained by condensation of carbon gas. It would be difficult to understand why an essentially van der Waals-bonded carbyne material should exhibit a hardness exceeding that of corundum. When we invoke a mechanism as suggested above, the extensive cross-linking of conjugated carbon chains aligned perpendicular to the substrate surface could produce an at least partially puckered array of carbon hexagons akin to diamond-like carbon (DLC).

According to Kudryavtsev *et al* [71], carbyne prepared by chemical dehydrochlorination of poly(vinylidene chloride) transforms to diamond in the presence of minor amounts of ultradisperse diamond (UDD) already at ambient pressure on heating to 350-400°C in a stream of argon. Figure 5 shows the x-ray diffraction pattern of the diamond formed in this process (Courtesy: Yu.P.Kudryavtsev). Although it is difficult at present to reconcile this finding with the discussions above the UDD may have provided nucleation sites for diamond by drastically reducing the activation energy barrier for the carbyne-diamond transformation reaction in very short ($n < 5$) chains. Also, the extended Berman-Simon line may have shifted towards lower (ambient?) pressure for very small carbon clusters [73].

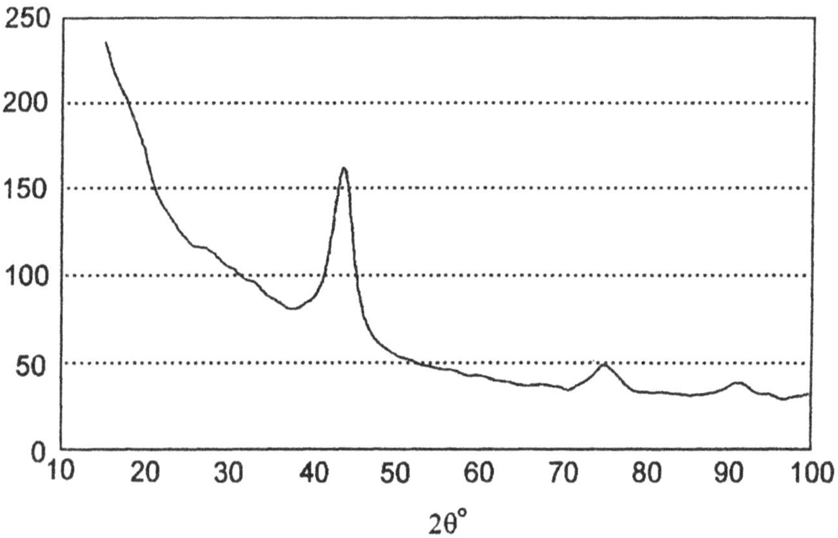

Figure 5. XRD pattern of diamond formed by pressureless thermally activated transformation of carbyne (Courtesy: Yu.P. Kudryavtesev [71]).

Pressureless formation of diamond was also obtained from carbyne produced by dehydrochlorination of a PVDC film by eximer laser irradiation (KrF; $\lambda = 248$ nm) at a laser fluence of 40 mJ/cm^2 per pulse (total number of pulses: 1,000) at a pressure of 10^{-3} Pa. The Raman spectrum showed broad peaks of disordered diamond as well as graphite [74] likewise suggesting a low energy transformation route. The Raman spectrum of DLC deposited by a biased rf-plasma [75] revealed a broad band centred at 1550 cm^{-1} that was attributed to a C=C carbon bond stretching vibration [76] suggesting the involvement of carbyne-like phases in the DLC deposition process.

Finally, it is intriguing to speculate whether interstellar diamond nanoclusters recovered from meteorites [77] as well as lonsdaleite found in chondritic porous (CP) aggregates and carbonaceous chondrites [78] have been formed, at low pressure and temperatures, by radiation-induced cross-linking of linear short-chained cyanopolyyne clusters found ubiquitously in cold interstellar molecular-cloud complexes [79-81]. This assumption would be an alternative explanation to the proposed low pressure CVD-like process of metastable diamond formation from a hot gas [77] as well as a process relying on catalytically activated hydrous pyrolysis of hydrocarbons [78], respectively.

This would also be generically analogous to the recently reported formation of graphite-shell carbon "onions" by electron beam irradiation [82] as well as thermal annealing of carbon soot [83,84] whose absorption feature at 217.5 nm bears a striking resemblance to that obtained from interstellar matter (see Chapter 2.1 for details).

4. Conclusions

It has been shown that weak shock pressures (< 5 GPa) are sufficient to transform linear sp-hybridized carbon chains (carbynes) into diamond without external thermal activation. This is consistent with a modified phase diagram of carbon suggested by previous authors [15,64] that includes carbynes as high temperature, low pressure allotropes of carbon.

The mechanism of the transformation of an array of periodically arranged 1D sp-hybridized carbon chains of polyyne- or mixed polyyne/polyene-type ('pencil-in-box' model) with van der Waals bonds into the puckered chair-type 3D sp^3-hybridized diamond configuration occurs presumably by extensive cross-linking. This may involve the introduction of strings of double bonds into kinked polyyne chains that provide unpaired electrons as coupling sites for the σ-links between adjacent chains.

Hence carbynes constitute a means for a low energy path towards diamond. The very high activation energy required to convert the stiff planar carbon hexagons of graphite into a puckered mode by out-of-plane movement of carbon atoms [85] is not required when carbyne is used as precursor phases whose soft shear deformation mode towards sp^3-hybridization, assisted by the strong shear component of a shock wave, requires much lower activation energy. In fact, there is recent experimental evidence that carbyne can convert to diamond already at ambient pressure on heating to intermediate temperature. The technological implications of such a low energy path to diamond are obvious and substantial. There are, however, more data required to substantiate the conditions for a pressureless transition of carbynes to diamond.

5. References

1. Fahy, S., Louie, S.G. and Cohen, M.L. (1986) Pseudopotential total-energy study of the transition from rhombohedral graphite to diamond, *Phys.Rev.* **B 34(2)**, 1191- 1199.
2. Bundy, F.P., Bassett, W.A., Weathers, M.S., Hemley, R.J., Mao, H.K. and Goncharov, A.F. (1996) The pressure-temperature phase and transformation diagram for carbon: updated through 1994, *Carbon* **34**(2), 141-153.
3. DeCarli, P.S. (1967) in:Science and Technology of Industrial Diamonds, Vol.1 (IDIB, London, 1967) p.49.
4. Horie, Y. (1980) Thermodynamics of dislocations and shock compression of solids, *Phys.Rev.* **B 21**, 5549-5557.
5. Pujols, H. and Boisard, F. (1970) Effects of an intense shock wave on graphite, *Carbon* **8**, 781-782.
6. DeCarli, P.S. and Jamieson, J.C. (1961) Formation of diamond by explosive shock, *Science* **133**, 1821.
7. Erskine, D.J. and Nellis, W.J. (1991) Shock-induced martensitic phase transformation of oriented graphite to diamond, *Nature* **349**, 317-319.
8 Hirai, H. and Kondo, K. (1991) Modified phases of diamond formed under shock compression and rapid quenching, *Science* **253**, 772-774.
9. Solov'ev, V.A. (1976) Stacking faults: plasticity and transformation kinetics under high pressure, *High Temp.High Press.* **8**, 706-707.
10. Chomenko, A.A., Gankevich, T.L., Kotonosov, A.S. and Kuteinikov, A.F. (1975) *Sint.Almazy* **3**, 3-.
11. Kleiman, J., Heimann, R.B., Hawken, D. and Salansky, N.M. (1984) Shock compression and flash heating of graphite/metal mixtures at temperatures up to 3200 K and pressures up to 25 GPa, *J.Appl.Phys.* **56**(5), 1440-1454.
12. Heimann, R.B., Kleiman, J. and Salansky, N.M. (1983) A unified structural approach to linear carbon polytypes, *Nature* **306**, 164-167.
13. Sladkov, A.M., Kasatochkin, V.I., Korshak, V.V. and Kudryavtsev, Yu.P. (1971) Inventor's Certificate, No.107 (7 December 1971), (Priority date 4 November 1960).
14. Mel'nichenko, V.M., Sladkov, A.M. and Nikulin, Yu.N., (1982) Structure of polymeric carbon,. *Russ.Chem.Rev.* **51**(5), 421-438.
15. Whittaker, A.G. (1978) Carbon: a new view of ist high-temperature behavior, *Science* **200**, 763-764.
16. Heimann, R.B., Kleiman, J. and Salansky, N.M. (1984) Structural aspects and conformation of linear carbon polytypes, *Carbon* **22**(2), 147-156.
17. Zdanov, G.S., Ilyushin, A.S. and Nikitina, S.V.(1980) Symposium "Difraktionnyi i Rezonansnyi Strukturnyi Analiz" (Diffraction and Resonance Methods of Structural Analysis), (Nauka, Moscow, 1980) p.45. Quoted after [14].
18. Sladkov, A.M., Kasatochkin, V.I., Kudryavtsev, Yu.P. and Korshak, V.V. 1968) Synthesis and properties of linear carbons, *Izv. Akad.Nauk SSSR, Ser.Khim.* **12**, 2697-2704.
19. Kasatochkin, V.I., Kazakov, M.E., Savranskii, V.V., Nabatnikov, A.P. and Radomir ,N.P. (1971) Synthesis of new allotropic forms of carbon from graphite, *Dokl.Akad.Nauk SSSR* **2011** 1104-1105 (in Russian).
20. Guseva, M.B., Savchenko, N.F. and Babaev, V.G. (1985) New data of carbyne structure, *Dokl.Akad.Nauk SSSR* **283**(6), 1336-1339.
21. Whittaker, A.G. and Kintner, P.L. (1969) Carbon: observations on the new allotropic form, *Science* **165**, 589-591.
22. Akagi, K., Nishigushi, M., Shirakawa, H., Furukawa, Y. and Harada, I. (1987) One-dimensional conjugated carbyne-synthesis and properties, *Synth.Metals* **17**, 557-562.
23. Korshak, V.V., Kudryavtsev, Yu.P., Korshak, Yu.V., Evsyukov, S.E., Khvostov, V.V., Babaev, V.G. and Guseva, M.B. (1988) Formation of ß-carbyne by dehydrohalogenation, *Macromol.Chem. Rapid Commun.* **9**, 135-140.

24. Shimoyama, M., Niino, H. and Yabe, A. (1992) A KrF excimer laser induced dehydrochlorination of a chlorinated poly(vinyl chloride): preparation of conjugated polyene and polyyne, *Macromol.Chem.* **193**, 569-574.
25. Dietrich, F. and Rubin, Y. (1992) Synthetic approaches towards molecular and polymeric carbon allotropes, *Angew.Chem.Intl.Edn.engl.* **31**(9), 1101-1123.
26. Kudryavtsev, Yu.P., Heimann, R.B. and Evsyukov, S.E. (1996) Carbynes: advances in the field of linear carbon-chain compounds, *J.Mater.Sci.* **31**(21), 5557-5571.
27. Komatsu, T., Nomura, M., Kakudate, Y., Fujiwara, S. and Heimann, R.B. (1995) Characterization of dehydrochlorinated poly(vinylidene chloride) and the shock-compressed material, *Macromol. Chem Phys.* **196**, 3031-3040.
28. Stang, P.J. and Ladika, M. (1981) Unsaturated carbenes. 16. Extended unsaturated carbenes. Generation, nature, and chemistry of alkatetraenylidene carbenes, *J.Am.Chem.Soc.* **103**, 6437-6443
29. Shirakawa, H. and Ikeda, S. (1973) Raman scattering of electronic spectra of polyacetylene, *Polym.J.* **4**, 460-462.
30. Furukawa, Y., Arakawa, T., Takeuchi, T. and Harada, I. (1984) Vibrational spectroscopic study on partly hydrogenated trans-polyacetylenes, *J.Chem.Phys.* **81**(7), 2907-2914.
31. Evsyukov, S.E., Paasch, S., Thomas, B. and Heimann, R.B. (1997) Formation of carbynoid structures by chemical dehydrohalogenation of poly(vinylidene chloride). A ^{13}C solid-state NMR study, *Ber.Bunsenges.Phys.Chem.* **101**(5), 837-841.
32. Whittaker, A.G., Neudorffer, M.E. and Watts, E.J. (1983) Carbon: a rhombo-hedral carbyne form, *Carbon* **21**(6), 597-599.
33. Fujihara, S., Narita, K., Saito, S., Tatsumoto, K., Fujiwara, S., Yoshida, M., Aoki, K., Kakudate, Y., Usuba, S. and Yamawaki, H. (1991) Explosove shock synthesis of polycrystalline diamond powders, in: Proc.18th Symp. on Shock Waves, Sendai, Japan, July 21-26, 1991.
34. Fujiwara, S. (1992) Explosive Technique for Generation of High Dynamic Pressure, in: Shock Compression Technology and Materials Science, edited by A.B.Sawaoka, KTK Scientific Publishers/Terra Scientific Publishing Comp., Tokyo, 1992, p.7.
35. Kleiman, J., Salansky, N.M. and Heimann, R.B. (1984) On phase changes of carbon under dynamic conditions as observed by STEM, *J.Mater.Sci. Letters* **3**, 117-120.
36. Kelly, B.T. (1981) *Physics of Graphite* , Applied Science, London, New York.
37. Sergushin, I.P., Kudryavtsev, Yu.P., Elizen, V.M., Sadovskii, A.P., Sladkov, A.M., Nefedov, V.I. and Korshak, V.V: (1977) X-ray electron and x-ray spectroscopic investigation of carbyne, *Zhurn.Strukt.Khim.* **18**(4), 698-700.
38. Heimann, R.B., Fujiwara, S., Kakudate, Y., Koga, Y., Komatsu, T. and Nomura, M. (1995) A new carbon form obtained by weak shock compression of carbyne, *Carbon* **33**, 859-863.
39. Bundy, F.P. (1980) The P,T phase and reaction diagram for elemental carbon 1979, *J.Geophys.Res.* **85** (B 12), 6930-6936.
40. Riter, J.R. (1970) Interpretation of diamond and graphite compressibility data using molecular force constants, *J.Chem.Phys.* **52**, 5008-5010.
41. Bundy, F.P. (1963) Direct conversion of graphite to diamond in static pressure apparatus, *J.Chem.Phys.* **38**, 631-643.
42. Heimann, R.B. and Kleiman, J. (1988) Shock-Induced Growth of Superhard Materials, in: *Crystals. Growth, Properties and Applications*, Vol.**11** ,Springer, Berlin, Heidelberg, New York, p.1-73.
43. Balchan, A.S. and Cowan, G.R. (1972), *U.S.Patent 3,667,911* (June 1972).
44. Wheeler, E.J. and Lewis, D. (1975) Structure of a shock-quenched diamond, *Mater.Res.Bull.* **10**, 687693.
45. Heimann, R.B. (1994) Linear finite carbon chains (carbynes): their role during dynamic transformation of graphite to diamond, and their geometric and electronic structure, *Diamond Relat.Mater.* **3**(9), 1151-1157.
46. Whittaker, A.G. and Tooper, B. (1974) Single-crystal diffraction patterns from vitreous carbon, *J.Am.Ceram.Soc.* **57**, 443-446
47. Heath, J.R., Zhang, Q., O'Brian, S.C., Curl, R.F., Kroto, H.W. and Smalley, R.E. (1987) The formation of long carbon chain molecules during laser vaporization of graphite, *J.Am.Chem.Soc.* **109**, 359-363.
48. Kudryavtsev, Yu.P., Evsyukov, S.E., Babaev, V.G., Guseva, M.B., Khvostov, V.V. and Krechko, L.M. (1992) Oriented carbyne layers, *Carbon* **30**(2), 213221.

49. Bochvar, D.A., Nikerov, M.V. and Stankevich, I.V. (1987) Problems in the structural chemistry of crystalline carbon, *Sov.Sci.Rev.B. Chem.* **9**, 483-527.
50. Stankevich, I.V., Nikerov, M.V. and Bochvar, D.A. (1984) The structural chemistry of crystalline carbon: geometry, stability, and electronic spectrum, *Russ.Chem.Rev.* **53**(7), 640-655.
51. Mel'nichenko, V.M., Nikulin, Yu.N. and Sladkov, A.M. (1985) Layer-chain carbons, *Carbon* **23**(1), 3-7.
52. Hoffmann, R., Eisenstein, O. and Balaban, A.T. (1980) Hypothetical strain-free oligoradicals, *Proc.Natl.Acad. Sci. USA* **77**(10), 5588-5592.
53. Hoffmann, R., Hughbanks, T., Kertesz, M. and Bird, P.H. (1983) A hypothetical metallic allotrope of carbon, *J.Am.Chem.Soc.* **105** , 4831-4832.
54. Rice, M.J., Philpot, S.R., Bishop, A.R. and Campbell, D.K. (1986) Solitons, polarons, and phonons in the infinite polyyne chain, *Phys.Rev.* **B 34**(6), 4139-4149.
55. Kertesz, M., Koller, J. and Azman, A. (1978) *Ab initio* Hartree-Fock crystal orbital studies: II. Energy bands of an infinite carbon chain, *J.Chem.Phys.* **68**(6), 2779-2782.
56. Kudryavtsev, Yu.P. (1994) INEOS, RAS, Moscow. Personal Communication, 31 May1994;
57. Hoffmann, R. (1966) Extended Hückel theory. V. Cumulenes, polyenes, polyacetylenes and C_n, *Tetrahedron* **22**, 521-538.
58. Sladkov, A.M. (1981) Carbyne-a new allotropic form of carbon, *Soviet Sci.Rev.* **B3**, 75-110.
59. Seal, N M. (1960) Graphitization of diamond, *Nature* **185**, 522-523.
60. Whittaker, A.G (1975) Behavior of diamond under positive ion bombardment and CO_2 laser irradiation at very high power density,. *Abstr.12th Biennial Conf. Carbon*, Pittsburgh, PA, July 28 to August 1, 1975, p.43.
61. Whittaker, A.G. (1977) A kinetic study of the transformation graphite = linear carbon forms by x-ray diffraction, *Abstr.13th Biennial Conf.Carbon*, Irvine, CA, July 18 to 22, 1977, p.413.
62. Setaka, N. and Sekikawa, Y. (1980) Chaoite, a new allotropic form of carbon, produced by shock compression, *J.Am.Ceram.Soc.* **63**(3/4), 238-239.
63. Sokolowski, M., Sokolowska, A., Gokieli, B., Michalski, A., Ruse, A. and Romanowski, Z. (1979) Reactive pulse plasma crystallization of diamond and diamond-like carbon, *J.Cryst.Growth* **47**, 421-426.
64. Derjagin, B.V., Fedoseev, D.V., Varnin, V.P. and Vnukov, S.P. (1977) The nature of metastable phases of carbon, *Nature* **269**, 398-399.
65. Yamada, K., Kunishige, H. and Sawaoka, A.B. (1991) Formation process of carbyne produced by shock compression, *Naturwiss.* **78**, 450-452.
66. Kasatochkin, V.I., Kazakov, M.E., Kudryavtsev, Yu.P., Sladkov, A.M. and Korshak, V.V.(1968) Enthalpy of chain-like polymers of carbon-carbyne, *Dokl.Akad.Nauk SSSR* **183**(1), 109- 111 (in Russian).
67. Whittaker, A.G. (1982), written communication, June 20, 1982.
68. Bundy, F.P., Bovenkerk, H.P., Strong, H.M. and Wentorf, R.H. (1961) Diamond-graphite equilibrium line from growth and graphitization of diamond, *J.Chem.Phys.* **35**(2), 383-391.
69. van Thiel. M. and Ree, F.H. (1987) Properties of carbon clusters in TNT detonation products: graphite-diamond transition, *J.Appl.Phys.* **62**(5), 1761-1767.
70. Kasatochkin, V.I., Korshak, V.V., Kudryavtsev, Yu.P., Sladkov, A.M. and Sterenberg, I.E. (1973) On crystalline structure of carbyne, *Carbon* **11**, 70-71.
71. Kudryavtsev, Yu.P., Bystrova, N.A., Zhirova, L.V. and Rusanov, A.L. (1996) Formation of diamond from carbyne, *Russ.Chem.Bull.* **45** (1), 233-235.
72. Whittaker, A.G. (1981) Chaoite coating process, U.S.Patent 4,248,909 (Feb 3, 1981).
73. Trefilov, V.I. and Savvakin, G.I. (1979) Direct conversion of noncrystalline carbon to diamond and increased rate of diffusion transformation at high pulsed pressures and temperatures, *Dokl Akad. Nauk SSSR*,. **246**(5), 1115-1119. (in Russian)
74. Niino, H. (1994), NIMC ,Tsukuba, Japan. Personal Communication.
75. Setaka, N. (1989) Diamond synthesis from vapor phase and its growth process, *J.Mater.Res.* **4**(3), 664-670.
76. Kuzmany, H. (1981) Resonance Raman scattering from neutral and doped polyacetylene, *Phys.Stat.Sol. B* **97**, 521-531.
77. Lewis, R.S., Ming, T., Wacker, J.F., Anders, E. and Steel, E. (1987) Interstellar diamonds in meteorites, *Nature* **326**, 160-162.

78. Rietmeijer, F.J.M. and Mackinnon, I.D.R. (1987) Metastable carbon in two chondritic porous interplanetary dust particles, *Nature* **326**, 162-165
79. Turner, B.E. (1971) Detection of interstellar cyanoacetylene, *Astrophys.J.* **163** (1971) L35-L39.
80. Kroto, H.W., Kirby, C., Walton, D.R.M., Avery, L.W., Broten, N.W., MacLeod, J.M. and Oka, T. (1978) The detection of cyanohexatriyne, $H(C{:}C)_3CN$ in Heiler's Cloud 2, *Astrophys.J.* **219,** L133-L137.
81. Webster, A.S. (1979) in: Meeting of the Royal Astronomical Society, London, *The Observatory* **99,** 29-30.
82. Ugarte, D. (1992) Curling and closure of graphitic networks under electron-beam irradiation, *Nature* **359,** 707-709.
83. de Heer, W.A. and Ugarte, D. (1993) Carbon onions produced by heat treatment of carbon soot and their relation to the 217.5 nm interstellar absorption feature, *Chem.Phys.Letters* **207,** 480-486.
84. Greiner, N.R., Phillips, D.S., Johnson, J.D. and Volk, F. (1988) Diamonds in detonation soot, *Nature* **333,** 440-442.
85. Coulson, C.A. and Haigh, C.W. (1963) Theoretical calculations of steric effects in conjugated systems, I. Phenanthrene, triphenylene, and chrysene, *Tetrahedron* **19,** 527-544.
86. Smirnjagina, N.N., Semenov, A.P., Batoev, V.B. and Korsun, V.P. (1995) On the crystallization of carbyne by an electron beam in thin carbon films grown by ion-assisted sputtering, Proc.6th Int.Symp.,'Thin Films in Electronics', Vol.2: Diamond, Diamond Films, Moscow, Kiev, Cherson, pp.50-53.
87. Borodina, T.I., Fortov, V.E., Lash, A.A., Zhuk, A.Z., Babaev, V.G. and Guseva, M.B. (1996) Shock-induced transformations of carbyne, *J.Appl.Phys.* **80**(7), 3757-3759.

7.2. MEDICAL APPLICATIONS OF CARBYNOID MATERIALS

V.I. KIRPATOVSKY
Scientific Research Institute of Urology,
Ministry of Health of Russia,
105425 Moscow, Russia

Abstract – The experience accumulated to date in the field of biomedical applications of carbyne-containing polymeric materials has been summarized. The properties of two types of materials, viz., Vitlan and Carbylan, have been surveyed. These materials have been shown to possess high biocompatibility, low litho- and trombogenic activity, and complete resistance to urine salt incrustation. The prospects of practical use of carbyne-containing materials as sutures and prosthetic implants in reconstructive surgery have been discussed. The materials are particularly promising for utilization in cases of long-term contacts with blood and chemically aggressive biological liquids such as bile and urine.

1. Introduction

The utilization of synthetic materials for replacing human or animal tissues or as suture threads has become more and more widespread as medical science develops.

On one hand, the broadening of the spectrum of surgical operations requires the development of new synthetic biocompatible materials that can be implanted into human organism without negative consequences. On the other hand, the availability of such materials increases the possibilities of a surgeon to perform more aggressive surgical interventions while using artificial organs and tissues.

At present, it is possible to substitute some parts of the organism to implant, for example, metallic joints; synthetic (Capron, Lavsan)[1] ligaments; aponeurotic, pericardial and other kinds of connective tissue; blood vessel prostheses fabricated from Dacron™, PTFE and other materials; artificial hearts (total or left ventricle replacements); artificial lenses; and other organs. In addition, there are different kinds of surgical thread that are used for suturing different tissues during surgical operations.

All these materials posses some biocompatibility, *i.e.*, they do not induce severe inflammation after implantation and long-term contact with biological tissues. But nevertheless, one observes inflammatory reactions to several degrees around

[1] Poly-ε-capronamide and poly(ethylene terephthalate), respectively.

R.B. Heimann et al. (eds.), Carbyne and Carbynoid Structures, 427–435.

implanted artificial materials the magnitude of which depends on the kind of synthetic material. It is for this reason that investigations are being continued into new materials with higher degrees of biocompatibility or novel ways explored to diminish inflammatory reactions of known materials.

Carbyne is a promising material to be used for medical applications. It is considered to be a polymeric compound of carbon with linear structure consisting of sp-hybridized carbon atoms [1].

The investigation of the chemical properties of carbyne reveals a chemical and biological inertness comparable to that of diamond [1, 2]. This property is considered to be useful for utilization in medicine. Investigations performed over many years lead to the development of a technology of manufacturing hydrocarbon suture threads with carbyne coating, named "Vitlan" [2]. Subsequently there were some modifications and simplifications of the carbyne coating method and the development of a technology for low-temperature carbonization of poly(vinylidene fluoride)-based[2] threads or films by chemical dehydrohalogenation [3, 4]. The synthetic material obtained by this technology was named "Carbylan".

In the following sections results will be presented of experimental and clinical investigations related to a possible application of Vitlan or Carbylan in medical practice.

2. Investigation of Vitlan

2.1. MECHANICAL STRENGTH AND STERILIZATION

The investigation of of Vitlan threads showed their tensile strength to be in the same range as for other synthetic nonabsorbable surgical threads. Vitlan's mechanical properties did not change after sterilisation by autoclaving at 122 °C, but its hardness and hence its brittleness increase after sterilisation at 136 °C [2].

2.2. ANTIADHESIVE PROPERTIES

The investigation of blood platelet adhesion onto the surface of Vitlan films with different PVDC contents showed after treatment with sodium amide in liquid ammonia in an autoclave (the technology of Vitlan manufacture) [2, 5] that there is a negative correlation between the PVDC content in the film and the magnitude of platelet adhesion, i.e. specimens with higher PVDC content show more pronounced antiadhesive activity (Table 1).

There were no changes of pH nor haemolysis after long-term contact of blood with Vitlan films thus confirming the biological inertness of these films [5, 6].

These studies became the basis for using Vitlan as a material for producing blood vessel prostheses that were tested in acute and chronic experiments on dogs [2, 7–9].

[2] PVDF-based filamentous material is also known in the Russian literature as 'Ftorin'.

TABLE 1. Platelet adhesion on the surface of polyethylene films with different PVDC contents [2]

Content of PVDC in polyethylene film (%)	0	7.3	14.9	23.0	31.5	43.8	64.9	87.7
Mean amount of platelets per unit of area (8000 μm^2)	37.2	24.6	21.5	8.5	8.1	6.7	1.0	1.1

2.3. VASCULAR PROSTHESES FROM VITLAN

In six acute experiments fragment of the intrathoracic part of the *vena cava inferior* were replaced by wicker Vitlan prostheses with inner diameters of 8-12 mm and lengths of 30-40 mm. During 2-6 hours of observation after prostheses implantation it was detected that Vitlan prostheses possess higher thromboresistance properties, compared to control prostheses from Ftorolon[3]-Lavsan blend. These experiments also showed that prostheses with a high content of Vitlan filaments, although being highly thromboresistant, were very hard and often developed deformation by anastomoses after suturing with thin venous walls which leads to a decrease of their functionality after prostheses implantation. Using more compliant prostheses with lower content of Vitlan fibers lead to better functional results [7].

As shown by the results of 14 analogous chronic animal experiments with a postoperative observation time between 2 month and 5 years there was absolute patency of all prostheses. Histological investigation, performed after one year, confirmed that the Vitlan prostheses were completely incorporated into the surrounding tissues. At the inner surface of the prostheses a new functional layer developed that consisted of elastic fibres and smooth muscle cells. The *neointima* was covered with endothelial cells on all inner surfaces and showed good vascularisation. Inflammatory reaction of surrounding tissue due to contact with Vitlan was very weak [8].

After replacement of the canine abdominal aorta and femoral arteries by Vitlan prostheses it was noted that the optimum content of Vitlan threads in vascular prostheses is less than 10%. In experiments with such prostheses no cases of thrombosis were observed during 5 months of observation. Histologically, with prostheses with 3-10% of Vitlan fibers a rapid regeneration of the *intima* near anastomosis of the arterial wall was noted as well as the absence of periprosthetic inflammation that happens usually after implantation of other kinds of vascular prosthesis [2].

Clinical experience with Vitlan prostheses implantation of patients with obliterative diseases of the arteries of the lower extremities also confirmed their positive properties. Several authors have had positive results of replacing large arteries of patients with aorta bifurcation and femoral artery obliteration that were better than the results obtained with prostheses fabricated from other synthetic materials [9].

[3] Polyfluoroolefin. The Ftorolon-Lavsan fibers are commonly used in vascular prostheses.

3. Investigation of Carbylan

3.1. TISSUE REACTION TO CARBYLAN

After implantation of Carbylan films into soft tissues of experimental animals (rats, rabbits) there was a low-graded inflammatory reaction with weak leukocyte infiltration during the first 3 days. By the fifth day the intensity of exsudation and hyperaemia diminished with a prevalence of the infiltrative-proliferative component of the inflammatory reaction. This dynamic was more rapid compared to other synthetic materials conventionally used in medicine. In later stages fibroblast-like cells appeared around the implants forming a thin collagen capsule after two months implantation. Supraperiostal implantation of Carbylan films did not induce severe pathological disorders in underlying osseous tissue [10].

After implantation of Carbylan fibers in hepatic or pancreatic tissues only a minimal inflammatory response occurred, lasting for three days, compatible with the duration of the reaction to surgical trauma. After seven days there were intensive encapsulative processes around the fibers that abated after implantation for 30 days with the formation of a mature thin capsule. No inflammatory signs could be discerned up to 120 days [11].

Hence, Carbylan fibres induce minimal and fast inflammatory responses that are even less pronounced than those of Carbylan films.

The comparison of the magnitude of the inflammatory reaction of gastro-intestinal organ tissues to Carbylan, PVDF (original material from which Carbylan is produced), Capron or polyester threads has shown a weaker tissue response to Carbylan fibres early after implantation. Inflammatory reactions to Carbylan disappeared after 30 days in contrast to those induced by the other investigated materials that were pronounced even after 180-360 days [12]. After long-term exposure there were delicate coatings by connective tissue around Carbylan filaments, however in experiments with PVDF, Capron or polyester there was excessive collagen formation leading to the development of massive fibrous adhesions and severe deformation of organs of the gastro-intestinal system [12, 13].

These data let the authors to confirm the high biocompatibility of Carbylan that consequently was proposed for utilization as suture material to connect biological tissues during surgical operations, and a method was developed for manufacturing such sutures to be used in surgery [14, 15].

3.2. SURGICAL THREADS FROM CARBYLAN

Carbylan threads with thicknesses of 5/0[4] possess absolute tensile strengths of 3.0±0.2 N and show a relative lengthening of 24.2±4.2%. Threads with thicknesses of 2/0[4] possess absolute tensile strengths and relative lengthening of, respectively 26.3±1.9 N and 17.7±1.5% [13]. These parameters comply with international standards for

[4] Conditional counts 5/0 and 2/0 correspond to diameters of 0.13±0.01 and 0.33±0.01 mm, respectively.

synthetic nonabsorbable surgical threads, as indicated in European Pharmacopoeia (1984).

According to their tensile strength and elasticity, Carbylan threads occupy an intermediate position between silk and Capron threads of comparable calibre [16].

Investigations were performed on the influence of different methods of sterilisation on the mechanical properties of Carbylan sutures. It was determined that the properties did not change after boiling in distilled water for 30 minutes, incubation in 70% alcohol for 3 hours or after incubation in 6% hydrogen peroxide for 6 hours. However, after sterilisation at 160 °C the poly(vinylidene fluoride) base of the threads was destroyed. Consequently the structure of the material changed and the threads became deformed, fragile and nonsuitable for surgery application [16].

Sterile Carbylan threads were tested in operations on experimental animals. It was noted that Carbylan threads used for suturing of stomach, intestine or colon wounds lead to a good healing response and to repaired tissues without significant adhesions at the operation region. Histological changes during wound healing were more favourable compared to experiments using Capron for this purpose. When using Carbylan threads no fistula, ulcer or microabscess formation were observed while these complications frequently developed when using Capron threads [11, 13].

After suturing of abdominal wounds by Carbylan ligatures after laparotomy, performed on 96 rats and 8 dogs, complications such as wound suppuration, wound dehiscence, seroma or ligature fistulas developed in only 3.8% of experiments, but when using Capron threads (76 experiments on rats and 12 on dogs) such complications developed in 10.2% [13].

Suture materials for operations on the liver and extrahepatic bile ducts must posses not only mechanical rigidity and biocompatibility but also low cholelithogenic properties to prevent the formation of gall stones at the surface of ligatures. In this aspect Carbylan also has significant advantages over some other suture materials. It was found that there were much smaller amounts of bile acids adsorbed at the surface of Carbylan filaments, immersed in bile from a patient with cholelithiasis, as compared with PVDF and polyester threads. The difference was insignificant when compared with Capron threads (Table 2).

TABLE 2. Cholelithogenic activity of some surgical threads after different time of incubation in bile [13]

Suture material	Amount of adsorption sites of bile acids on 1 cm of length		
	1 day	10 days	75 days
Carbylan	1,2±1,0	1,7±0,7	7,9±1,8
Capron	1,5±1,3	1,7±0,8	10,9±2,3
PVDF	5,1±1,0*	3,9±0,8	16,5±1,5*
Polyester	6,5±1,8*	7,4±1,3*	15,9±2,9*

* the difference is significant when compared to Carbylan.

Low reactivity of Carbylan with minimal cholelithogenic activity was the background to obtain good functional results after using of Carbylan sutures in operations on the liver, the bile ducts, pancreas and duodenum of dogs. In these experiments no gall stone formation, bile leakage, pancreatitis or any other specific complications were observed [15, 17]. It can be concluded that the usage of Carbylan for hepatic, bile duct and pancreatic surgery is preferable over other conventional materials.

The problem of stone formation at the surface of suture materials is also pertinent in urological operations, where surgical threads are in long-lasting contact with urine. Urinal salts are impregnating the threads and their crystals act as nuclei for stone growth. As revealed in experiments on dogs, Carbylan has the capability to resist urinal salt impregnation and to prevent ligature stone formation. In conjunction with its high biocompatibility this material appears to be very useful for urological operations [18, 19].

Carbylan threads were used in the following experimental operations on the organs of the upper and low urinary tract of dogs: (i) ureteral resection in the middle or lower portion with reanastomosis by end-to-end fashioning (4 operations), (ii) resection of the prevesical portion of the ureter and ureterovesical anastomosis by the Politano-Leadbetter method (2 operations), (iii) suturing of the urinary bladder after cystotomy (8 operations) and (iv) suturing of the urethra after its transsection (4 operations) [19]. During observation from 1 week to 1 year after the operations not any complications induced by this suture material were obvious. All anastomoses were patent without signs of stenosis. There were no cases with urinal salt incrustation of sutures or ligature stone formation even after long-term contact of the threads with urine. Meanwhile usage of chromic catgut, traditionally used in urological operations, lead to some specific complications. In one out of 8 urethral resections with reanastomosis by end-to-end fashioning a urinary leakage developed with formation of an urinoma in the retroperitoneum due to anastomosis dehiscence, and in two of such cases an anastomosis stricture formed. In one out of 10 cases of cystotomy and suturing of the urinal bladder with chromic catgut, a dehiscence of the vesical wound occurred with formation of urinal peritonitis. In one case the formation of a ligature stone was noted.

Histological investigations revealed that Carbylan in urinary tract tissue, as in other tissues, induced a minimal inflammatory reaction with small focal lymphomacrophagal infiltrates around sutures. Inflammatory responses were short-lived and disappeared after 1 month. After long-term implantation nearly normal structures of ureteral, urethral or bladder tissues were observed by histological techniques. All anastomoses were completely covered with epithelium [19].

3.3. URETERAL AND URETHRAL PROSTHESES FROM CARBYLAN

Until recently, no synthetic materials existed that could substitute the urinal tract because of the high reactivity of the urinal epithelium against foreign substances (synthetic prostheses), and owing to incrustation of the inner prostheses surfaces by urinary salts and prosthesis occlusion. Because of the particularly low Carbylan reactivity and its tolerance to urine salt incrustation an attempt was made to use Carbylan as a material for manufacturing of ureteral and urethral prostheses that were implantated in dogs [18].

In this work, 3 to 5 cm long segments in the middle part of the ureter or 3 to 4 cm long parts of the urethra were replaced with wicker prostheses from Carbylan fibres (inner diameter of 6 mm) in 6 experiments on dogs. For 1 month these prostheses provided adequate transport for urine in all cases. After 2 months, two of the 6 dogs revealed radiographic signs of ureteral prosthesis passability disorder, and after 3 month a radiographic picture of ureteral prosthesis obstruction was obtained in all experimental animals. At this time, in two dogs the kidneys were completely blocked. These changes were provoked by ureteral bending at the site of proximal or distal anastomosis, or formation of massive adhesions around the prosthesis that induced prosthesis deformation and constriction. There were no cases of urine salt incrustation of the inner prosthetic surface. Histological results showed that inflammatory reaction of tissue was almost absent at the contact sites with the Carbylan prosthesis. It was concluded that Carbylan prosthesis disfunction was due to prosthesis inhermeticity and urine leakage through the prosthetic walls. This leakage induced the inflammatory response and an intensification of the adhesive process that lead to formation of massive constrictive adhesions. Hence the inhermeticity is the main cause of insufficient functional results, however, Carbylan by itself is a suitable material for urological purpose with indispensable condition for formation of prosthesis hermeticity.

This assumption was later confirmed in experiments on dogs (Kirpatovsky *et al.*, unpublished data). Utilizing the thromboresistant properties of Carbylan and its capability to function in the blood vascular system, prostheses fashioned from this material were initially implanted into the blood vessel system instead of using fragment of the femoral artery. After 3 to 7 days the hermetic prostheses, covered with a fibrin layer, were removed and implanted as a segment of the ureter. This lead to excellent results: the prosthesis did function normally for one year. After 13 months the urine passage through the prosthesis was impared owing to an adhesion process near the prosthesis and thus its constriction. After a long-term period the development of a tube of connective tissue outside the prosthesis was observed that was covered on the inside with epithelium and contained intramural smooth muscle cells.

4. Conclusion

The analysis of the literature covering experimental investigations into the application of carbyne-containing material in medicine revealed a very high biological inertness of these materials. This was manifest after carbyne implantation in different forms into some tissues of different experimental animals including subcutaneous fat, muscle, periosseal layers, liver, pancreas, stomach, intestines, colon, urinal tract, and blood vessel system. Carbylan is characterized by a unique combination of high biocompatibility, thromboresistance, low capability of surface adsorption of bile acid salts, and complete resistance to urine salt incrustation. Hence this novel carbyne-based material opens up a wide perspective of utilization in cases of contact with blood and chemically aggressive biological liquids such as bile and urine.

5. References

1. Kudryavtsev, Yu.P., Heimann, R.B., and Evsyukov, S.E.: Carbynes: Advances in the field of linear carbon chain compounds, *J. Mater. Sci.* **31** (1996), 5557-5571.
2. Kudryavtsev, Yu.P.: Synthesis and properties of carbyne, *D.Sci. Thesis*, The A.N.Nesmeyanov Institute of Organoelement Compounds, USSR Academy of Sciences, Moscow, 1987 (in Russian).
3. Evsyukov, S.E.: Polymers with a system of multiple bonds on the basis of dehydrohalogenation reactions, *Ph.D. Thesis*, The A.N.Nesmeyanov Institute of Organoelement Compounds, USSR Academy of Science, Moscow, 1989 (in Russian).
4. Evsyukov, S.E., Kudryavtsev, Yu.P., and Korshak, Yu.V.: Chemical dehydrohalogenation of halogen-containing polymers, *Usp. Khim.* **60** (1991), 764-798 (in Russian) [*Russ. Chem. Rev.* **60** (1991), 373-390 (Engl. Transl.)].
5. Dobrova, N.B. and Navrotskaya V.V.: Present status of studying of thromboresistant properties of artificial materials for endoprostheses in cardio-vascular surgery, *Proc. 5th Symp. "Synthetic Polymers for Medical Applications"*, Riga, USSR, 1980, 90-95 (in Russian).
6. Dobrova, N.B., Il'ina, M.B., and Shekhter, A.B.: Behaviour of vascular prostheses in experimental studies, *Abstr. 1st All-Union Symp. " Synthetic Polymers for Medical Applications"*, Alma-Ata; USSR, 1983, 128-129 (in Russian).
7. Dobrova, N.B., Il'ina, M.B., Shekhter, A.B., Navrotskaya V.V., Kudryavtsev, Yu.P., Snimshchikova, A.A., Plotkin, L.L., and Sil'kis, E.M.: Investigation of vascular prostheses from carbon-containing fibre Vitlan, *Abstr. All-Union Conf. "Topical Questions of Organisation, Prophylaxis, and Surgical Treatement of Magistral Vascular Deseases", Part 1*, Moscow, USSR, 1985, 199-200 (in Russian).
8. Dobrova, N.B., Il'ina, M.B., Tsetlin, B.L., and Sil'kis, E.M.: New domestic vascular prostheses from Vitlan, *Abstr. All-Union Workshop "Application of Polymeric Materials and Manufactured Articles in the Health Care"*, Moscow, USSR, 1985, 22-23 (in Russian).
9. Krikovtsov, A.S., Ivanov, S.V., Kapustin, A.A., Gorodkov, A.Yu., and Sidorenko, E.S.: Clinical characterization of woven bifurcational prostheses Vitlan in operations for Lerich syndrome, *Abstr. 3rd All-Union Conf. of Young Scientists "Diagnosis and Surgical Treatment of Cardio-Vascular Deseases"*, Moscow, USSR, 1990, 116 (in Russian).
10. Grigorian, A.S., Panikarovsky, V.V., Khamraev, T., Sarkisyan, I.E., Isakova, A.I., Kudryavtsev, Yu.P., Evsyukov, S.E., and Aleksandrov, A.F.: Reaction of tissue structures of a recipient bed after Carbylan implantation, *Proc. 1st Int. Conf. "Modern Approaches to the Development of Effective Dressing Materials and Polymeric Implants"*, Moscow, Russia, 1992, 209-210 (in Russian).
11. Bondarev, A.A., Biryukov V.I., and Grydasov Yu. A.: Reactions of tissue structures of the recipient bed after implantation of synthetic threads with carbyne coating in experiment model, *Proc. Conf. "Topical Questions of Medicine and Pharmacy"*, Kursk, Russia, 1996, 86 (in Russian).
12. Myasnikov, A.D., Kudryavtsev, Yu. P., Bezhin, A.I., Edemsky, A.I., and Bel'sky, V.V.: Aprobation of a new polymeric suture material, Carbylan, *Proc. Conf. „Actual Problems of Emergency Surgery"*, Sklifosovsky Institute of Emergency Medicine, Moscow, Russia, 1994, 97-99 (in Russian).
13. Bondarev, A.A.: Experimental and morphological backgrounds of application of the synthetic thread with carbyne coating in abdominal surgery, *Ph.D. Thesis*, Kursk Medical University, Voronezh, Russia, 1995 (in Russian).
14. Kudryavtsev, Yu. P., Babaev, V.G., Bezhin, A.I., Bondarev, A.A., Guseva, M.B., Evsyukov, S.E., Myasnikov, A.D., and Zhirova, L.V.: Surgical suture material, *Russ. Pat. RU 2.076.739*, 1997.
15. Myasnikov, A.D., Kudryavtsev, Yu. P., Bezhin, A.I., and Bondarev, A.A.: Perspectives of usage of synthetic threads with carbyne coating in operations on organs of biliopancreatoduodenal zone, *Proc. Int. Conf. "New Technologies in Diagnosis and Surgery of Organs of Biliopancreatoduodenal Zone"*, Moscow, Russia, 1995, 66-67 (in Russian).
16. Bezhin, A.I.: Pathogenetic backgrouds of operations on pancreas after its trauma and pancreatitis, the prophylaxis and treatement of postoperational complications, *D.Sci. Thesis*, Kursk Medical University, Voronezh, 1995 (in Russian).

17. Myasnikov, A., Kudryavtsev, Yu., Bezhin, A., and Bondarev, A: Possibility of the synthetic thread "Carbylan" employment by the liver, biliary ducts and pancreas surgery, *IWA Int. Ass. the 4-th Int. Congr.*, Tel-Aviv, Israel, 1996, 101-102.
18. Kirpatovsky, V.I., Kudryavtsev, Yu.P., Mudraya, I.S., Kudryavtsev, Yu.V., Evsyukov, S.E., Babaev, V.G., and Aleksandrov, A.F.: The first experience of application of synthetic Carbylan prostheses for reconstruction of urinary tract in experimental model, *Proc. 1st Int. Conf. "Modern Approaches to the Development of Effective Dressing Materials and Polymeric Implants"*, Moscow, Russia, 1992, 213-215 (in Russian).
19. Kirpatovsky, V.I., Kudryavtsev, Yu.P., Mudraya, I.S., and Evsyukov, S.E.: The usage of suturing material from Carbylan in urological experimental operations, *Experimental urology and nephrology. General problems of pathology*, Scientific Research Institute of Urology, Moscow, 1996, 185-188 (in Russian).

INDEX

Physics and Chemistry of Materials with Low-Dimension Structures

Previously published under the Series Title:

PHYSICS AND CHEMISTRY OF MATERIALS WITH LAYERED STRUCTURES

1. R.M.A. Lieth (ed.): *Preparation and Crystal Growth of Materials with Layered Structures.* 1977 ISBN 90-277-0638-7
2. F. Lévy (ed.): *Crystallography and Crystal Chemistry of Materials with Layered Structures.* 1976 ISBN 90-277-0586-0
3. T. J. Wieting and M. Schlüter (eds.): *Electrons and Phonons in Layered Crystal Structures.* 1979 ISBN 90-277-0897-5
4. P.A. Lee (ed.): *Optical and Electrical Properties.* 1976 ISBN 90-277-0676-X
5. F. Hulliger: *Structural Chemistry of Layer-Type Phases.* Ed. by F. Lévy. 1976 ISBN 90-277-0714-6
6. F. Lévy (ed.): *Intercalated Layered Materials.* 1979 ISBN 90-277-0967-X

Published under:

PHYSICS AND CHEMISTRY OF MATERIALS WITH LOW-DIMENSIONAL STRUCTURES
SERIES A: LAYERED STRUCTURES

7. V. Grasso (ed.): *Electronic Structure and Electronic Transitions in Layered Materials.* 1986 ISBN 90-277-2102-5
8. K. Motizuki (ed.): *Structural Phase Transitions in Layered Transition Metal Compounds.* 1986 ISBN 90-277-2171-8

PHYSICS AND CHEMISTRY OF MATERIALS WITH LOW-DIMENSIONAL STRUCTURES
SERIES B: QUASI-ONE-DIMENSIONAL STRUCTURES

B1. P. Monceau (ed.): *Electronic Properties of Inorganic Quasi-One-Dimensional Compounds.* Part I: Theoretical. 1985 ISBN 90-277-1789-3
B2. P. Monceau (ed.): *Electronic Properties of Inorganic Quasi-One-Dimensional Compounds.* Part II: Experimental. 1985 ISBN 90-277-1800-8
B3. H. Kamimura (ed.): *Theoretical Aspects of Band Structures and Electronic Properties of Pseudo-One-Dimensional Solids.* 1985 ISBN 90-277-1927-6
B4. J. Rouxel (ed.): *Crystal Chemistry and Properties of Materials with Quasi-One-Dimensional Structures.* A Chemical and Physical Synthetic Approach. 1986 ISBN 90-277-2057-6

Discontinued.

PHYSICS AND CHEMISTRY OF MATERIALS WITH LOW-DIMENSIONAL STRUCTURES
SERIES C: MOLECULAR STRUCTURES

C1. I. Zschokke (ed.): *Optical Spectroscopy of Glasses.* 1986 ISBN 90-277-2231-5
C2. J. Fünfschilling (ed.): *Relaxation Processes in Molecular Excited States.* 1989 ISBN 07923-0001-7

Discontinued.

Physics and Chemistry of Materials with Low-Dimension Structures

9. L.J. de Jongh (ed.): *Magnetic Properties of Layered Transition Metal Compounds*. 1990 ISBN 0-7923-0238-9
10. E. Doni, R. Girlanda, G. Pastori Parravicini and A. Quattropani (eds.): *Progress in Electron Properties of Solids*. Festschrift in Honour of Franco Bassani. 1989 ISBN 0-7923-0337-7
11. C. Schlenker (ed.): *Low-Dimensional Electronic Properties of Molybdenum Bronzes and Oxides*. 1989 ISBN 0-7923-0085-8
12. Not published.
13. H. Aoki, M. Tsukada, M. Schlüter and F. Lévy (eds.): *New Horizons in Low-Dimensional Electron Systems*. A Festschrift in Honour of Professor H. Kamimura. 1992 ISBN 0-7923-1302-X
14. A. Aruchamy (ed.): *Photoelectrochemistry and Photovoltaics of Layered Semiconductors*. 1992 ISBN 0-7923-1556-1
15. T. Butz (ed.): *Nuclear Spectroscopy on Charge Density Wave Systems*. 1992 ISBN 0-7923-1779-3
16. G. Benedek (ed.): *Surface Properties of Layered Structures*. 1992 ISBN 0-7923-1961-3
17. W. Müller-Warmuth and R. Schöllhorn (eds.): *Progress in Intercalation Research*. 1994 ISBN 0-7923-2357-2
18. L.J. de Jongh (ed.): *Physics and Chemistry of Metal Cluster Compounds*. Model Systems for Small Metal Particles. 1994 ISBN 0-7923-2715-2
19. E.Y. Andrei (ed.): *Two-Dimensional Electron Systems*. On Helium and other Cryogenic Substrates. 1997 ISBN 0-7923-4738-2
20. A. Furrer: *Neutron Scattering in Layered Copper-Oxide Superconductors*. 1998 ISBN 0-7923-5226-2
21. R.B. Heimann, S.E. Evsyukov and L. Kavan (eds.): *Carbyne and Carbynoid Structures*. 1999 ISBN 0-7923-5323-4

KLUWER ACADEMIC PUBLISHERS – DORDRECHT / BOSTON / LONDON